Bio-Based Plastics

Wiley Series in Renewable Resources

Series Editor

Christian V. Stevens – Faculty of Bioscience Engineering, Ghent University, Ghent, Belgium

Titles in the Series

Wood Modification – Chemical, Thermal and Other Processes
Callum A. S. Hill

Renewables – Based Technology – Sustainability Assessment
Jo Dewulf & Herman Van Langenhove

Introduction to Chemicals from Biomass
James H. Clark & Fabien E.I. Deswarte

Biofuels
Wim Soetaert & Erick Vandamme

Handbook of Natural Colorants
Thomas Bechtold & Rita Mussak

Surfactants from Renewable Resources
Mikael Kjellin & Ingegärd Johansson

Industrial Application of Natural Fibres – Structure, Properties and Technical Applications
Jörg Müssig

Thermochemical Processing of Biomass – Conversion into Fuels, Chemicals and Power
Robert C. Brown

Biorefinery Co-Products: Phytochemicals, Primary Metabolites and Value-Added Biomass Processing
Chantal Bergeron, Danielle Julie Carrier and Shri Ramaswamy

Aqueous Pretreatment of Plant Biomass for Biological and Chemical Conversion to Fuels and Chemicals
Charles E. Wyman

Forthcoming Titles

Introduction to Wood and Natural Fiber Composites
Douglas Stokke, Qinglin Wu & Guangping Han

Cellulosic Energy Cropping Systems
Doug Karlen

Cellulose Nanocrystals: Properties, Production and Applications
Wadood Hamad

Introduction to Chemicals from Biomass, 2nd edition
James Clark & Fabien Deswarte

Lignin and Lignans as Renewable Raw Materials: Chemistry, Technology and Applications
Francisco García Calvo-Flores, José A. Dobado, Joaquín Isac García, Francisco J. Martin-Martinez

Bio-Based Plastics

Materials and Applications

Editor

STEPHAN KABASCI

Fraunhofer-Institute for Environmental, Safety, and Energy Technology UMSICHT, Germany

WILEY

This edition first published 2014

Registered office
John Wiley & Sons Ltd, The Atrium, Southern Gate, Chichester, West Sussex, PO19 8SQ, United Kingdom

For details of our global editorial offices, for customer services and for information about how to apply for permission to reuse the copyright material in this book please see our website at www.wiley.com.

Library of Congress Cataloging-in-Publication Data

Bio-based plastics : materials and applications / editor Stephan Kabasci.
pages cm
Includes index.
ISBN 978-1-119-99400-8 (cloth)
1. Biopolymers. 2. Plastics. I. Kabasci, Stephan.
TP248.65.P62B5184 2014
668.4–dc23

2013026528

A catalogue record for this book is available from the British Library.

ISBN: 978-1-119-99400-8

Cover images © Fraunhofer UMSICHT

Set in 10/12pt Times by Aptara Inc., New Delhi, India
Printed and bound in Malaysia by Vivar Printing Sdn Bhd

1 2014

Contents

Series Preface

Renewable resources and their modification are involved in a multitude of important processes with a major influence on our everyday lives. Applications can be found in the energy sector, chemistry, pharmacy, the textile industry, paints and coatings, to name but a few fields.

The broad area of renewable resources connects several scientific disciplines (agriculture, biochemistry, chemistry, technology, environmental sciences, forestry . . .), but it is very difficult to take an expert view on their complicated interactions. The idea of creating a series of scientific books focusing on specific topics concerning renewable resources is therefore very opportune and can help to clarify some of the underlying connections in this field.

In a very fast-changing world, trends do not only occur in fashion and politics; hype and buzzwords occur in science too. The use of renewable resources is more important nowadays; however, it is not hype. Lively discussions among scientists continue about how long we will be able to use fossil fuels, opinions ranging from 50 years to 500 years, but they do agree that the reserve is limited and that it is essential to search not only for new energy carriers but also for new material sources.

In this respect, renewable resources are a crucial area in the search for alternatives to fossil-based raw materials and energy. In the field of energy supply, biomass and renewable-based resources will be part of the solution alongside other alternatives such as solar energy, wind energy, hydraulic power, hydrogen technology and nuclear energy.

In the material sciences, the impact of renewable resources will probably be even bigger. Integral crop use and the use of waste streams in certain industries will grow in importance, leading to a more sustainable way of producing materials.

Although our society was much more based on renewable resources centuries ago (almost exclusively so), this disappeared in the Western world in the nineteenth century. Now it is time to focus again on this field of research. This should not mean a *retour à la nature*, but it does require a multidisciplinary effort at a highly technological level to perform research on new opportunities, to develop new crops and products from renewable resources. This will be essential to guarantee a level of comfort for a growing number of people living on our planet. The challenge for coming generations of scientists is to develop more sustainable ways to create prosperity and to fight poverty and hunger in the world. A global approach is certainly favoured.

This challenge can only be met if scientists are attracted to this area and are recognized for their efforts in this interdisciplinary field. It is therefore also essential that consumers recognize the fate of renewable resources in a number of products.

Furthermore, scientists do need to communicate and discuss the relevance of their work so that the use and modification of renewable resources does not follow the path of the genetic engineering concept in terms of consumer acceptance in Europe. In this respect, the series will certainly help to increase the visibility of the importance of renewable resources.

Being convinced of the value of the renewables approach for the industrial world, as well as for developing countries, I was delighted to collaborate on this series of books focusing on different aspects of renewable resources. I hope that readers will become aware of the complexity, interactions, interconnections, and challenges of this field and that they will help communicate the importance of renewable resources.

I would like to thank the staff from Wiley's Chichester office, especially David Hughes, Jenny Cossham and Lyn Roberts, in seeing the need for such a series of books on renewable resources, for initiating and supporting it and for helping to carry the project through to the end.

Last but not least I want to thank my family, especially my wife Hilde and children, Paulien and Pieter-Jan, for their patience and for giving me the time to work on the series when other activities seemed to be more inviting.

Christian V. Stevens
Faculty of Bioscience Engineering
Ghent University, Belgium
Series Editor 'Renewable Resources'
June 2005

Preface

The world is becoming increasingly aware of the fact that fossil raw materials are a finite resource. Their use needs to be reduced considerably in order to achieve sustainable development, defined by the UN Brundtland Commission in 1987 as: 'development that meets the needs of the present without compromising the ability of future generations to meet their own needs.'

In the chemical products sector, bio-based raw materials are the only renewable alternative to replace fossil carbon sources. In some product categories, such as detergents, renewable resources already hold a large share of the used raw materials due to their superior suitability and functionality. In the major chemical product category (with respect to the annually produced amount) of plastics, however, renewable resources still play a very small role. Nonetheless, steadily increasing numbers of bio-based polymers and products thereof have been developed. Moreover, the number of scientific papers for this topic is growing rapidly.

This book, as a part of the 'Wiley Series on Renewable Resources' presents a wide range of bio-based plastics and highlights some of their applications. Emphasis is placed on materials that are presently in use or show a significant potential for future applications. The book contains an up-to-date, broad, but concise overview of basic and applied aspects of bioplastics. The main focus is on thermoplastic polymers for material use. Elastomers, thermosets and coating applications, like natural rubber or alkyd resins, will be covered in other volumes in the series.

The book is organized in several chapters and deals with the most important biopolymer classes like the different polysaccharides (starch, cellulose, chitin), lignin, proteins and (polyhydroxy alkanoates) as raw materials for bio-based plastics, as well as with materials derived from bio-based monomers like lipids, poly(lactic acid), polyesters, polyamides and polyolefines. Additional chapters on general topics – the market and availability of renewable raw materials, the importance of bio-based content and the aspect of biodegradability – provide important information related to all bio-based polymer classes.

On behalf of all the authors, I would like to invite you to enter the world of bio-based plastics. Enjoy reading!

Stephan Kabasci
Fraunhofer-Institute for Environmental, Safety,
and Energy Technology UMSICHT, Germany
July 2013

Preface

List of Contributors

Catia Bastioli Chief Executive Officer, Novamont S.p.A., Italy

Gerhart Braunegg ARENA Arbeitsgemeinschaft für Ressourcenschonende und Nachhaltige Technologien, Austria

Benjamin Brehmer Evonik Industries AG, Germany

Michael Carus nova-Institut GmbH, Germany

Margarida Casal CBMA (Centre of Molecular and Environmental Biology), Department of Biology, University of Minho, Portugal

Stuart Coles International Digital Laboratory, WMG, University of Warwick, United Kingdom

António M. Cunha IPC (Institute of Polymers and Composites), Department of Polymer Engineering, University of Minho, Portugal

Wilhelm Eckl Fraunhofer Institute for Chemical Technology ICT, Germany

Norbert Eisenreich Fraunhofer Institute for Chemical Technology ICT, Germany

Daan S. van Es Wageningen University and Research Centre – Food and Biobased Research, Netherlands

Hans-Peter Fink Fraunhofer Institute for Applied Polymer Research IAP, Germany

Johannes Ganster Fraunhofer Institute for Applied Polymer Research IAP, Germany

Sebastià Gestí Garcia R&D-Physical Chemistry Laboratory, Novamont S.p.A., Italy

Emilia Regina Inone-Kauffmann Fraunhofer Institute for Chemical Technology ICT, Germany

Stephan Kabasci Fraunhofer Institute for Environmental, Safety, and Energy Technology UMSICHT, Germany

Frits van der Klis Wageningen University and Research Centre – Food and Biobased Research, Netherlands

Rutger J. I. Knoop Wageningen University and Research Centre – Food and Biobased Research, Netherlands

Martin Koller Graz University of Technology, Institute of Biotechnology and Biochemical Engineering, Austria and ARENA Arbeitsgemeinschaft für Ressourcenschonende und Nachhaltige Technologien, Austria

R.J. Koopmans Dow Europe GmbH, Switzerland

Raul Machado CBMA (Centre of Molecular and Environmental Biology), Department of Biology, University of Minho, Portugal

Paolo Magistrali R&D-Physical Chemistry Laboratory, Novamont S.p.A., Italy

Karin Molenveld Wageningen University and Research Centre – Food and Biobased Research, Netherlands

Helmut Nägele Tecnaro GmbH, Germany

Ramani Narayan Department of Chemical Engineering and Materials Science, Michigan State University, United States

Jürgen Pfitzer Tecnaro GmbH, Germany

Stephan Piotrowski nova-Institut GmbH, Germany

Achim Raschka nova-Institut GmbH, Germany

Marguerite Rinaudo Biomaterials Applications, 6 rue Lesdiguires, France

Anna Salerno Graz University of Technology, Institute of Biotechnology and Biochemical Engineering, Austria

Lolke Sijtsma Wageningen University and Research Centre – Food and Biobased Research, Netherlands

Hideto Tsuji Department of Environmental and Life Sciences, Graduate School of Engineering, Toyohashi University of Technology, Japan

Jacco van Haveren Wageningen University and Research Centre – Food and Biobased Research, Netherlands

Lars Ziegler Tecnaro GmbH, Germany

1

Bio-Based Plastics – Introduction

Stephan Kabasci
Fraunhofer Institute for Environmental, Safety, and Energy Technology UMSICHT, Germany

The world is becoming increasingly aware of the fact that fossil raw materials are a finite resource. Around the year 2010, circa 7×10^9 t of fossil carbon were being extracted from oil, coal and natural gas reservoirs annually. This demand has led to a considerable increase in fossil raw material prices, threatening the world's economy, and has been responsible for the rise in atmospheric carbon dioxide concentration over the past two centuries, affecting the world's climate. The massive use of fossil materials also presents an ethical problem. It can be foreseen that within a few generations these resources will be depleted. Their use needs to be reduced considerably in order to reach a sustainable level of development, defined by the UN Brundtland Commission in 1987 as: 'development that meets the needs of the present without compromising the ability of future generations to meet their own needs'.

More than 90% of raw fossil material utilization is for the purpose of satisfying the world's energy demand. A small fraction is converted to chemical products. Regarding the energy sector, several alternative technologies have already been developed. Wind, water, solar and geothermal sources can be used to set up a sustainable energy supply. Worldwide they already constitute, for example, 20% of electricity generation. Increasing the proportion of energy that is produced from renewable sources is a social and political goal in a lot of countries.

In the chemical products sector bio-based raw materials are the only renewable alternative to replace fossil carbon sources. In some chemical product categories like, for example, detergents, renewable resources already make up a large share of the raw materials used due to their superior suitability and functionality. In the major chemical product category (with respect to the annually produced amount) of plastics, however, renewable resources still play a very small role. Nonetheless, steadily increasing numbers of bio-based polymers and products have been developed in recent years. The number of scientific papers on this topic is still growing rapidly while it remains at a constant level for traditional fossil-based polymeric materials.

Bio-Based Plastics: Materials and Applications, First Edition. Edited by Stephan Kabasci.

This book covers a wide range of different bio-based plastics and highlights some of their applications.

1.1 Definition of Bio-Based Plastics

According to the Technical Report 15392, drawn up by the Technical Committee CEN/TC 249 of the European Committee for Standardization (CEN) in August 2009, 'bio-based plastics' are plastics derived from biomass. 'Plastics', as laid down in EN ISO 472, are materials that contain as an essential ingredient a high polymer and which at some stage in their processing into finished products can be shaped by flow. 'Biomass' means nonfossilized and biodegradable organic material originating from plants, animals and micro-organisms. Biomass is considered as a renewable resource as long as its exploitation rate does not exceed its replenishment by natural processes.

Although the above definition describes bio-based plastics rather unambiguously, some confusion still can be noticed, mainly due to the use of the inaccurate term 'bioplastics'. The prefix 'bio-' in bioplastics sometimes is used not to indicate the origin of the material ('bio-based') but to express a 'bio'-functionality of the material, in general either biodegradability or biocompatibility.

Biodegradable plastics can undergo decomposition processes induced by micro-organisms in composting or anaerobic digestion processes. Decomposition must proceed down to the ultimate stage of small molecules like methane (CH_4) and/or carbon dioxide (CO_2), water (H_2O) and mineral salts. Different national and international standards (e.g. ASTM D 6400, EN 13 432, ISO 17 088) have been developed, in which the process criteria (e.g. temperature and time) of test procedures and methods to analyse ultimate decomposition are laid down. Only if materials tested according to one of the standards yield more than the required minimum decomposition rate may they be designated as 'biodegradable' with reference to the testing method. The process of biodegradation is closely linked to the molecular structure of the polymer, it does not depend on the origin of the material. Some fossil-based polymers, like polycaprolactone (PCL), or poly(butylene adipate terephthalate) (PBAT), are biodegradable according to these standards. On the other hand, there are bio-based plastics, like polyethylene (PE), from sugar cane, which are resistant to biodegradation.

Biocompatible plastics are used in medicinal applications, and the prefix 'bio' indicates that the polymer, when being immersed in a living organism (human or animal), does not harm the body or its metabolism in any way. These biopolymers can also be based on fossil raw materials or on renewable resources. They may be durable in the body, as in the case of artificial blood vessels, or they may disintegrate and be resorbed in the body, as in the case of resorbable suture threads.

Another form of ambiguity arises from the definition of 'biopolymers' in biochemistry. These are polymers synthesized by living organisms (animals, plants, algae, micro-organisms) like polysaccharides, proteins, lignin or nucleic acids. They exhibit different functions in the organisms like energy storage (starch, proteins, polyhydroxyalkanoates), structural materials (lignin, cellulose, chitin, proteins) metabolism (proteins – enzymes, nucleic acids) or information storage (nucleic acids). Direct industrial exploitation of native biopolymers is possible after extraction and purification, that is, by physical processes. Further industrial exploitation is possible by applying chemical functionalization processes to the natural polymers. Results of these physical or chemical processes can be bio-based plastics, like polyhydroxyalkanoates (PHA), or cellulose acetate (CA). On the other hand, bio-based plastics do not need to be

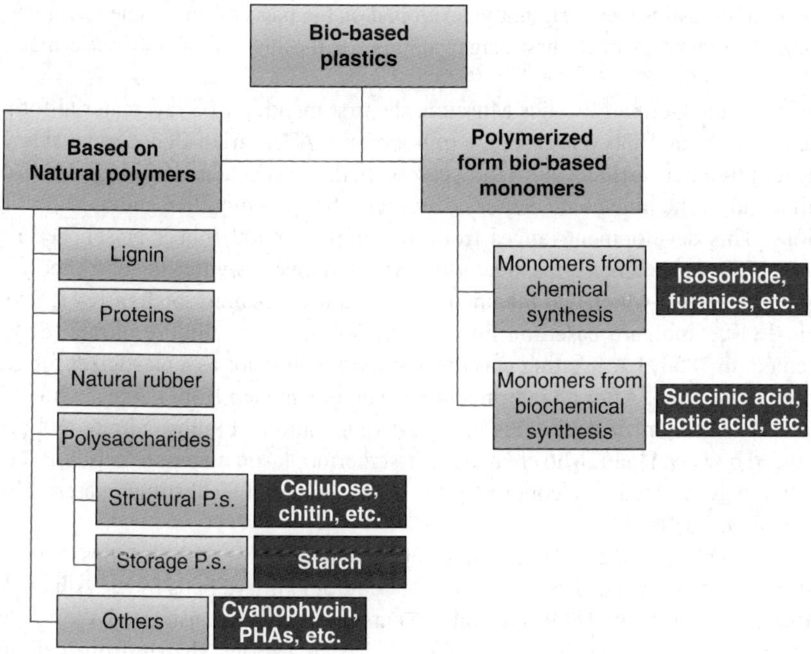

Figure 1.1 *Overview of bio-based plastics.*

derived from natural polymers. Poly(lactic acid), one of the most important bio-based plastics, is being produced by chemical polymerization of the bio-based monomer, lactic acid.

Figure 1.1 gives an overview of bio-based plastics. The distinction between materials based on natural polymers and those polymerized from bio-derived monomers can be seen from this.

Returning to the CEN definition of 'bio-based plastics' as plastics derived from biomass, while there is no difficulty in attesting a physically extracted natural biopolymer like polyhydroxybutyrate (PHB) to be 100% bio-based, applying chemical modifications to natural polymers or using bio-based monomers together with petrochemical monomers in a polycondensation reaction for example yields partially bio-based products. For example, 1,3-propanediol is being produced in the United States from corn starch using a biotechnological process. This monomer is 100% bio-based. By combining it with fossil-based terephthalic acid in a polycondensation reaction, a polyester, namely poly(propylene terephthalate) also known as poly(trimethylene terephthalate) (PTT), is being produced. This polyester is partially bio-based. International standardization on defining and measuring the bio-based fraction in such a material still is underway. Looking, for example, at the bio-based carbon atom content of PTT yields a bio-based fraction of $3/(3 + 8) = 27\%$. Nevertheless, different calculations, for example taking all chemical elements into account, are possible in principle.

1.2 A Brief History of Bio-Based Plastics

Looking at the historic development of plastics production, we can see that in the beginning it was not driven by using fossil raw materials. Quite the contrary – a lot of thermosets, elastomers and some thermoplastics were originally developed on the basis of renewable resources. Thus, the history of bioplastics in its first stages stands for the history of polymeric materials in general.

According to the German Plastics Museum, the first mention of a raw material for plastics production in the year 1530 was casein, a milk protein. A Bavarian Benedictine abbey keeps the recipe for producing artificial horn from casein. In the last decades of the eighteenth century and the first half of the nineteenth century, natural rubber was modified and used for different applications. This development ranges from the simple natural rubber eraser, described by Priestley in 1770, to Hancock's masticator in 1819, Goodyear's vulcanization process and T. Hancock's hard rubber, which was intended as a substitute for ebony, both in 1841. Soon after, in the mid-1840s, linoleum based on linseed oil (Walton) and cellulose nitrate (Schönbein) were invented. In 1854, J.A. Cutting was the first to use camphor as a plasticizer for cellulose nitrate to produce films. After an intermediate development step from Parkes, who presented the compound 'Parkesin' in 1856, this material combination, cellulose nitrate and camphor, was optimized by J. W. Hyatt, who created the first thermoplastic material, 'celluloid', in 1868. His invention was initiated by a contest for the development of a substitute material for ivory to produce billiard balls.

This bio-based plastic celluloid and its developer Hyatt gave rise to the plastics industry in the United States and worldwide. The production of celluloid billiard balls by the Albany Billiard Ball Corporation started in 1869, and three years later Hyatt constructed the predecessor of an injection moulding machine to produce parts in various shapes from celluloid. At the end of the nineteenth century the protein casein once again came under the focus of bioplastics development. In 1897 Krische and Spiteller invented Galalith, also known as Erinoid, a thermoset material from formaldehyde-hardened casein that was mainly used for the production of buttons and jewellery. In 1908, Eichengrün developed cellulose acetate, a transparent material with similar characteristics to cellulose nitrate, but with the huge advantage of being less flammable. Ten years later he also laid the foundation for the further rapid development of the plastics industry by inventing a manual piston injection moulding machine to process plasticized cellulose acetate. However, with crude oil becoming available at low prices and based on the theoretic works of Staudinger, in the 1920s and 1930s, the majority of fossil-based plastics types that are presently used (e.g. PE, PVC, PS, PA, PMMA) were developed. In these same decades, two important bio-based plastics were investigated in detail. Polyhydroxyalkanoates, which are synthesized as energy storage materials by several micro-organisms, were isolated and described by Lemoigne in 1925. Poly(lactic acid) (PLA) had been synthesized in 1913 and W.H. Carothers, one of the outstanding polymer chemists of that age, investigated the synthesis and the material in detail in 1932. Because of its biocompatibility and the ability to be resorbed in the human body, PLA and co-polyesters of lactic acid and glycolic acid have been produced for medical applications since the 1950s. Another bio-based raw material, castor oil, was exploited from the 1940s, when undecenoic acid, one of the pyrolytic degradation products of ricinoleic acid, was firstly used in the production of polyamide 11. After this, some decades of massive growth in production of fossil-based plastics followed and materials like PE, PVC, PS, PMMA and later on PP have been dominating the plastics world.

In parallel to the upcoming environmental protection movements of the 1980s, the awareness of the need for replacing fossil-based raw materials increased. The use of starch for the production of bioplastic materials was investigated and the first materials based on this research entered the market in the 1990s. In that same decade, high-volume production of PLA for nonmedicinal use started and the first tests of biodegradable PHA packaging materials were performed. Ten years later, considerable production capacity for several types of bio-based polymers had built up. With the advent, in particular, of fully bio-based drop-in materials, like bio-polyethylene (Bio-PE), and partially bio-based drop-in materials, like bio-poly(ethylene terephthalate) (Bio-PET with bio-derived ethylene glycol), production capacities of bioplastics surpassed 1 million t in 2011.

1.3 Market for Bio-Based Plastics

Looking at the different types of plastics and their applications large differences in the share of bio-based materials can be found. In 2010, the German Federal Agency for Renewable Resources presented data for the German market in the year 2007, which was analysed in three different sectors: thermoplastic and thermoset resins, elastomers, and man-made fibres (Figure 1.2).

The proportion of bio-based materials in each of the sectors of elastomers and fibres accounted for almost 40% due to the use of 290 000 t of natural rubber and 300 000 t of cellulosic fibres. The market size for thermoplastics and thermosets amounted to circa 15.8 million t, of which circa 12.5 million t accounted for rigid materials, mainly in packaging, building and construction, automotive and electronics industries as well as for furniture and consumer goods. A volume of 3.3 million t is attributed to adhesives, paints and lacquers, binders and other polymeric additives. In these areas it is estimated that bio-based materials

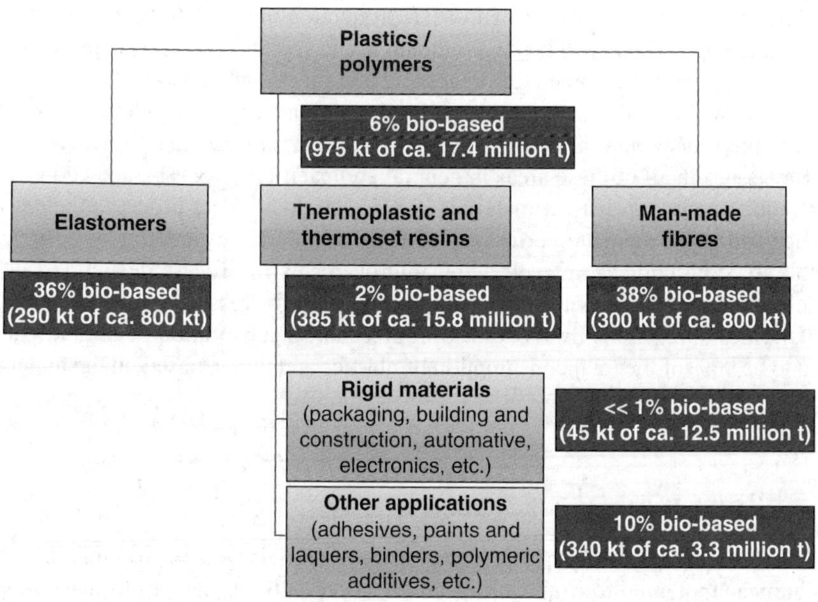

Figure 1.2 Share of bio-based plastics in the German market (as of 2007). (Data from the German Federal Agency for Renewable Resources.)

hold a share of circa 10%, while in the sector of rigid materials the bio-based plastics market size of 45 000 t merely amounts to circa 0.4% of the total market size.

A lot of market studies focus on thermoplastic bio-based plastics as rigid materials, describing the present status of these materials and predicting future growth rates. In this plastics application segment, global annual production capacity of bio-based materials surpassed 1 million t in 2011. Despite the low absolute value, bio-based plastics saw a rapid increase in production capacity, for example from 2003 (100 000 t) to 2007 (360 000 t), with a continuous average growth rate of 38% p.a.

For the near future this trend will continue. According to a study from the European Bioplastics association and the University of Applied Sciences and Arts of Hannover, this value will increase fivefold to an estimated 5.8 million t in 2016. The main driver of such an enormous expected rise is attributed to the 2011 decision of a leading worldwide soft drink company to substitute all of their PET bottles by (at least) partly bio-based PET, in which the ethylene glycol unit is derived from bioethanol – accounting for 4.6 million t. Thus, production capacity for bio-based drop-in commodity plastics (Bio-PET and Bio-PE) will largely overtake that of biodegradable materials. Poly(lactic acid), for example, was the predominant bio-based plastics material from 2000 to 2010, with annual production capacities of 100 000–150 000 t in this decade. This bio-based and biodegradable resin is predicted to undergo a twofold production capacity increase up to circa 300 000 t in 2016, too. Despite this rise, its share of the overall bio-based plastics market will drop from circa 35% to merely 5%.

Looking generally at the broad range of bio-based plastic types, most studies agree upon a growth of production capacities due to more and more companies entering and investing in the market. In 2009, circa 20 companies held 90% of the bioplastics market. By 2015 more than 250 and by 2020 over 2000 companies are expected to be at that market. Asia and South America will most likely have the highest growth rates and investments in the next decade.

Caused by the foreseen increase in Bio-PET production until 2016, bottles together with other packaging applications will be the dominant usage sectors of bio-based plastics. Nevertheless, progress in the development of processing and functional additives like, for example, flame retardants will also enhance the use of bio-based plastics in semi-durable and durable applications like transportation, construction, electronics, furniture and consumer goods in general. Key issues in all of these areas in general are material-, process-, and eco-efficiency. Common requirements for distinct materials in mass production are process efficiency, high performance and price adequate performance. They can be met by intelligent design, cheap and reliable material choices, as well as by improvements in process design and development. These optimization steps together with rising prices for fossil raw materials can allow bio-based plastics to reach the limits of maximum technical substitution potential, which was calculated to be 90% of the total consumption of plastics and fibres, based on the material mix of 2007.

1.4 Scope of the Book

This book focuses on bio-based plastics. It emphasizes materials that are presently in use or that show a significant potential for future applications. It presents a broad, up-to-date but concise overview of basic and applied aspects of bioplastics. The main focus is on thermoplastic polymers for material use. Elastomers, thermosets and coating applications, like, for example, natural rubber or alkyd resins, will be covered in other volumes of the series.

The book addresses the most important biopolymer classes like polysaccharides, lignin, proteins and polyhydroxyalkanoates as raw materials for bio-based plastics, as well as materials derived from bio-based monomers like lipids, poly(lactic acid), polyesters, polyamides and polyolefines. Additional chapters on general topics – the market and availability of renewable raw materials, the importance of bio-based content and the issue of biodegradability – will provide important information related to all bio-based polymer classes.

2
Starch

Catia Bastioli, Paolo Magistrali, and Sebastià Gestí Garcia
Novamont S.p.A., Italy

2.1 Introduction

Starch is a natural product from renewable resources, produced during photosynthesis as food reserve for plants and vegetables. It is the second most abundant biomass material in nature. It is found in plant roots, stalks, and crop seeds. The most important industrial starch sources are crops such as corn, wheat, potato, tapioca and rice. By refining these crops several byproducts can be obtained such as oil, bran, gluten, dextrin, sugar (glucose, fructose, HFCS), ethanol (for beverages and bio-fuels) and starch.

Starch is, in general, a low cost and readily available product but in recent years it has been subjected to financial speculation as several natural and fossil raw materials, so its price can fluctuate substantially from one year to the next. This fact arises from the change in nutrition habits in emerging countries and from the use of crops and byproducts as fuel sources. In this regard, the percentage of starch and ethanol for fuels production on the overall corn production in the United States, was 16 and 25% respectively in 1990, and has moved to 4 and 77% in 2009 [1].

Worldwide, the main sources of starch are corn (82%), wheat (8%), potatoes (5%) and cassava (5%), from which tapioca starch is derived. Worldwide production of corn in 2010 was approximately 800×10^6 ton. The main corn producer in 2010 was the United States (331×10^6 ton), China (158×10^6 ton), Brazil and EU-27 (57×10^6 ton each) [1].

In the United States, 39.4% of the corn production in 2010 was used as livestock feed, 10.5% was processed into food, seed and industrial products (excluding ethanol) and the 34.9% was converted to ethanol. The remaining 15.2% was exported [2].

Besides the food, pharmaceutical and paper industries, the availability of starch associated to its renewability aroused, since the late 1980s, an increasing interest in the sector of polymers. Starch can either be used as an alternative to polymers based on petrochemicals due to its

Bio Based Plastics: Materials and Applications, First Edition. Edited by Stephan Kabasci.

intrinsic polymeric structure, or as a source of glucose syrup for the production of renewable monomers through fermentation processes.

The first alternative, because of the use of the natural polymeric structure of starch as such, permits the environmental impact of resulting renewable products to be minimized, preserving at the same time the property of starch as easily biodegradable in almost all the different environments: soil, composting, water. Starch-based products are therefore particularly suitable for those applications where the risk of dispersion in the environment is high or the risk of polluting biodegradable streams – such as food and yard waste – is significant.

In starch-based bioplastics, starch is fully utilized with a yield very close to 100%, whereas in starch-derived bioplastics, synthesized from monomers resulting from fermentation of glucose syrup, the yield is generally less than 45%, and more complex processes and a less efficient use of resources are involved.

Starch-based polymers include a wide range of final properties, they can be as flexible as polyethylene or as rigid as polystyrene. They can also be soluble or insoluble in water and sensitive or mostly insensitive to humidity. Such properties explain the interest in this kind of product.

Polymeric materials are performance products that cover an impressively wide range of applications. This is the reason why many different polymers have been developed since the 1940s and combined in a very large number of alloys. Moreover, the amount of petrol used for the production of polymers is about 5% of total consumption, whereas 90% is going in energy and fuels.

This simple consideration and the fact that the annual fuel consumption worldwide is of 1.5×10^9 ton and that the worldwide production of corn is only of $0.8 \cdot 10^9$ ton illustrates the risks connected with the extensive production of ethanol from starch. The use of renewable resources becomes beneficial as much as it is oriented to performance products, with the aim of maximizing efficiency of resources, instead of mass products in a replacement logic.

The peculiar properties of starch limit the number of applications where the use of starch-based bioplastics is advisable in terms of in-use performances and end-of-life behaviour. It means that with starch-based bioplastics it is not possible to just think in terms of replacement of traditional plastics and they represent a perfect opportunity to redesign systems with attention to the efficient use of resources.

This chapter reviews the main topics related to starch in polymer technology taking into account all the different forms in which starch can appear (native, gelatinized, retrograded, destructurized, complex) giving more details for those which are related to starch-based polymers.

2.2 Starch

Starch is obtained from crops by refining in several steps, depending on the crops source. For instance, corn starch is extracted from kernel by wet milling in order to split the kernel and remove the oil-containing germ. Finer milling separates the fibre from the endosperm, which is then centrifuged to separate the less dense protein from the densest starch.

The starch slurry is then washed in a centrifuge, dewatered and dried. Either prior or subsequent to the drying step, the starch may be processed in a number of ways to improve its properties.

Starch is comprised of two major components: amylose, a mostly linear α-d-(1,4)-glucan (Figure 2.1a); and amylopectin, an α-d-(1,4) glucan (Figure 2.1b) that has α-d-(1,6) linkages

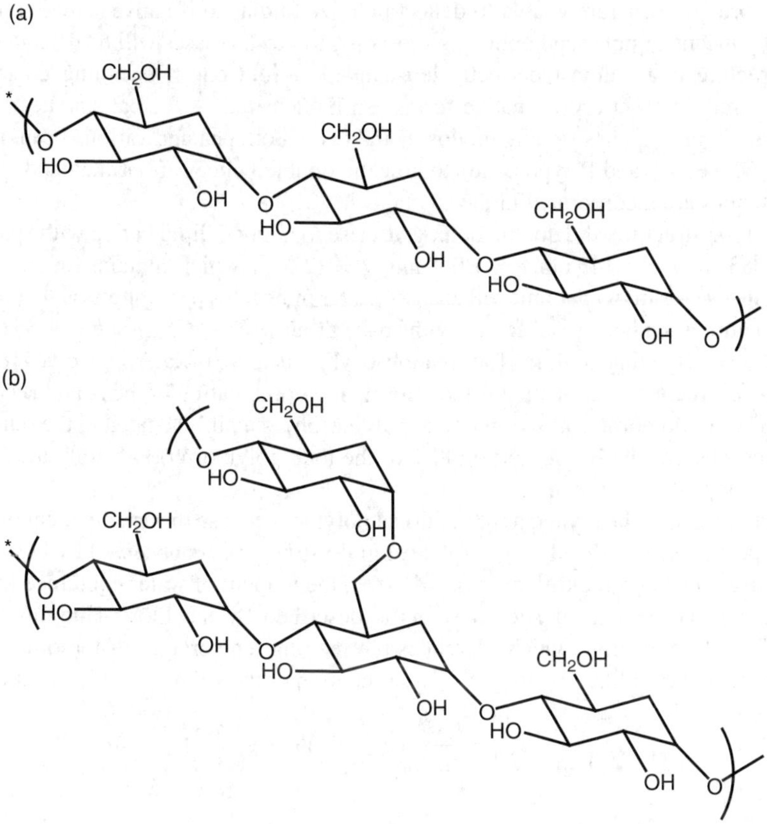

Figure 2.1 *Molecular structure of amylose (a) and amylopectin (b).*

at the branch point. The linear amylose molecules of starch have a molecular weight of 0.2–2 million, while the branched amylopectin molecules have molecular weights as high as 100–400 million [3,4].

These two macromolecules are arranged in granules having a size ranging from 2 to 100 μm depending on the source (corn, potato, etc.) and genotype. These grains have a kernel (hilum) from which an alternate lamellar structure of crystalline and amorphous shells emanates. The shells are constituted of blocklets having a size from 20 to 500 nm formed of amylopectin clusters and amylose. Blocklets are alternated lamellar structures of crystalline and amorphous amylopectin arrangements.

The ratio of the two polysaccharides varies according to the botanical origin of the starch. The so-called 'waxy' starches contain less than 15% amylose, 'normal' starches 20–35% and 'high amylose' starches more than about 40%. The moisture content of air-equilibrated starches ranges from about 10–12% (cereal) to about 14–18% (some roots and tubers).

Native granules yield X-ray diffraction patterns, which, although they are generally of low quality, can be used to identify the several allomorphs [5]. In native form, both amylopectin and amylose have a double helix structure with hydrogen bonds between these helices, either direct or through the water molecules in the unit cell.

Such an ordered structure is able to deflect polarized light, so if native starch granules are observed by means of polarized optical microscopy, Maltese crosses will be detected.

This structure is an almost perfectly left-handed sixfold one. Depending on the starch genotype, three different crystal native forms can be detected: A (cereal starches), B (tuber starches and cereal starches rich in amylose), and C (smooth pea and various beans) [6]. The difference between A and B type is due to how the double helices are packed and how many water molecules are incorporated in the crystal cell.

In the A type structure, the double helix gives rise to a monoclinic lattice with space group B_2 ($a = 2.083$ nm, $b = 1.145$ nm, $c = 1.058$ nm, $\gamma = 122.0°$), which contains four maltotriosyl units (12 glucose residues) per unit cell and two water molecules per maltotriosyl unit [7]. The B type structure has a hexagonal lattice with space group $P6_1 - C_6^2$ ($a = b = 1.852$ nm, $c =$ 1.057 nm, $\gamma = 120°$) that contains four maltotriosyl units and 36 water molecules [8].

The C type structure, or so-called C polymorph, is a combination of the A and B crystalline form [9]. The A allomorph (60% of the total polymorph) is mainly located in the outer part of the granules whereas the B allomorph (40% of the total polymorph) is mainly located in the inner part of it [10].

Native starch cannot be treated as a traditional polymer because the arrangement of amylose and amylopectin leads to the already explained grain structure. As observed by Donovan [11] this grain structure has a melting process related to the amount of water contained in it.

The melting mechanism of such systems is described by the Flory–Huggins theory of polymer-diluent interaction, which correlates the melting temperature of the dilute polymer T_m to that one of the undiluted polymer T_m^0 as a function of the volume fraction of the diluent:

$$\frac{1}{T_m} - \frac{1}{T_m^0} = \frac{R}{\Delta H_u} \cdot \frac{V_u}{V_1} \cdot (V_1 - \chi \cdot V_1^2) \tag{2.1}$$

Where:

T_m = melting point of the crystalline polymer plus diluent (water) (K)
T_m^0 = melting point of undiluted polymer crystallites (K)
R = gas constant
ΔH_u = fusion enthalpy per anhydroglucose repeating unit
V_u = volume fraction of anhydroglucose repeating unit
V_1 = volume fraction of diluent (water)
χ = Flory–Huggins interaction parameter

T_m^0 for starch crystallites is approximately 257 °C, which is above the degradation temperature of amylose and amylopectin macromolecules.

For this reason starch, unless used as a polymer filler (see section 2.3), has to be transformed in order to change its original structure allowing its macromolecules to be treated as those of a thermoplastic polymer (see section 2.4).

Besides structural modification, starch can undergo chemical modification, which can give it different hydrophilic, swelling, rheological, physical and chemical properties. Examples of chemical modification of starch are esterification, etherification or crosslinking of hydroxyl groups, or oxidation of anhydroglucose repeating units.

Both native and chemically modified starches can be structurally modified in order to be used like polymers either alone or in combination with specific synthetic polymers.

In conclusion, a starchy material is converted into thermoplastic starch (often called TPS) by melting in closed devices such as heated extruders or other closed devices capable of securing temperature, pressure and shear conditions. Starch that has undergone a thermoplastic transformation does not show the typical melting peaks of native starch at specific water content under DSC analysis in sealed vials and can be defined as thermoplastic starch.

In case of starch blends with synthetic polymers the presence of starch in form of submicron particles dispersed into the matrix is a clear demonstration that a thermoplastic transformation occurred on starch itself.

2.3 Starch-Filled Plastics

The use of native starch as filler in traditional plastics such as polyolefins nowadays is [12–15] less popular than in the past. Nevertheless, in this paragraph a short review of such kinds of products will be disclosed.

When blended with starch beads, PE films [16] deteriorate on exposure to a soil environment. The microbial consumption of the starch component, in fact, leads to increased porosity, void formation, and the loss of integrity of the plastic matrix.

Generally [17] starch is added at fairly low concentrations (6–15%); the overall disintegration of these materials is achieved by the use of transition metal compounds, soluble in the thermoplastic matrix, as pro-oxidant additives that catalyse the photo and thermo-oxidative process [18–21].

Starch-filled polyethylenes containing pro-oxidants have been used in the past in agricultural mulch film, in bags, and in sixpack yoke package. Commercial products based on this technology were first sold by Ecostar and Archer Daniels Midland Companies [22]. According to St Lawrence Starch technology, regular cornstarch was treated with a silane coupling agent to make it compatible with hydrophobic polymers, and dried to less than 1% of water content. It was then mixed with the other additives such as an unsaturated fat or fatty-acid autoxidant to form a masterbatch that is added to a commodity polymer. The polymer could then be processed by convenient methods, including film blowing, injection moulding, and blow moulding.

When processing this sort of materials, temperature has to be kept below 230 °C to prevent decomposition of the starch, and exposure of the masterbatch to air had to be minimized to avoid water absorption.

Direct addition of starch and pro-oxidant without the masterbatch step can also be used. Since this requires some specific equipment, it is only practical for large volumes [20].

It has been claimed that under appropriate conditions the disintegration time of a buried carrier bag, made of Ecostar with up to 6% starch, will be reduced from hundreds of years to three to six years [17].

The noncompliance of these materials with international standards of biodegradability in different environments, the high concern for the increasing presence of plastic debris in the marine environment, even favoured by their tendency to fragment, and their potential negative impact on recyclability of traditional plastics, should prevent this technology from being considered as a preferred industrial option.

Other attempts at fully or partially biodegradable starch-filled plastics were studied such as starch/poly(ε-caprolactone) (PCL) [23], others which are partially biodegradable, starch/poly(vinyl chloride)/poly(ε-caprolactone) (PCL) [24] or starch/modified polyesters [25]. In all these cases, starch granules are used to increase the surface area available for attack by micro-organisms.

Today, PSM (HK) Co. Ltd. (a subsidiary of Engrowth Investments Limited incorporated in BVI) is one of the few companies which still use starch-filled plastics producing raw materials for injection moulded products usually for catering purposes [26].

2.4 Structural Starch Modifications

Modifications of the native starch form can be made by means of thermal and mechanical treatments where the amount of water has an essential role. Referring to Equation 2.1, when starch is heated at a water volume fraction above 0.9 a pure gelatinization phenomenon (and afterwards gelation and retrogradation) or jet-cooking occurs. On the contrary, when the water volume fraction is low (e.g. below 0.45), a real melting of starch crystallites occurs [11, 27] and thermoplastic starch can be obtained [27].

2.4.1 Starch Gelatinization and Retrogradation

Starch gelatinization occurs when aqueous suspensions of native starch granules are heated above a characteristic temperature (the gelatinization temperature) [28]. It is an irreversible process in which granules swell, leading to loss of molecular order and crystallinity [9, 29] along with a loss of birefringence. Moreover, swelling of starch granules is also accompanied by leaching of starch molecules (mainly amylose).

The gelatinization process can be observed by the differential scanning calorimetry (DSC) technique. Depending on the amount of water present in the system, several endotherm peaks can be detected (Figure 2.2). If the water volume fraction is significantly above 0.45 (conditions that can be referred to gelatinization) a single endotherm peak, pointed as G is observed [11].

On the other hand, if the water volume fraction is below 0.45 more than one endotherm peak is observed. The first two peaks (M_1 and M_2) are the most relevant ones. The first peak, assigned as M_1 (which becomes less intense when water volume fraction decreases) occurs at the same temperature of the single endotherm G and is due to crystallites destabilization occurring when amorphous zones swell. The second peak, assigned as M_2, is related to the crystallites melting, and (as shown by Flory–Huggins theory – Equation 2.1) the higher the water volume fraction (V_1) is, the lower the melting temperature of the polymer is. For instance (Figure 2.2): for potato starch if V_1 is approximately 0.5 M_2 occurs at circa 85 °C, if V_1 is approximately 0.3, M_2 occurs at circa 110 °C [11].

When a gelatinized starch system is cooled, an increase in viscosity is observed due to re-association of starch molecules. This phenomenon, called retrogradation, is characterized by gradual increase in rigidity and a phase separation between solvent and polymer chains. The reassociation of starch molecules (both amylose and amylopectin) gives rise to short-range molecular order and crystallization of double helical aggregates in the B lattice form [9]. Starch retrogradation depends on a large variety of factors such as the source of starch, its concentration, cooking and cooling temperature speeds, pH and the presence of solutes (lipids, electrolytes, sugars).

Both amylose and amylopectin take part in the retrogradation process and network structure formation and up to four different steps of ordering have been postulated [30]. The first one, which takes place in a few hours, has been ascribed to gelation of the solubilized amylose in the intergranular space forming double helices structures. The second and the third one, for which days are needed, have been ascribed to chain reordering and crystallization of amylopectin.

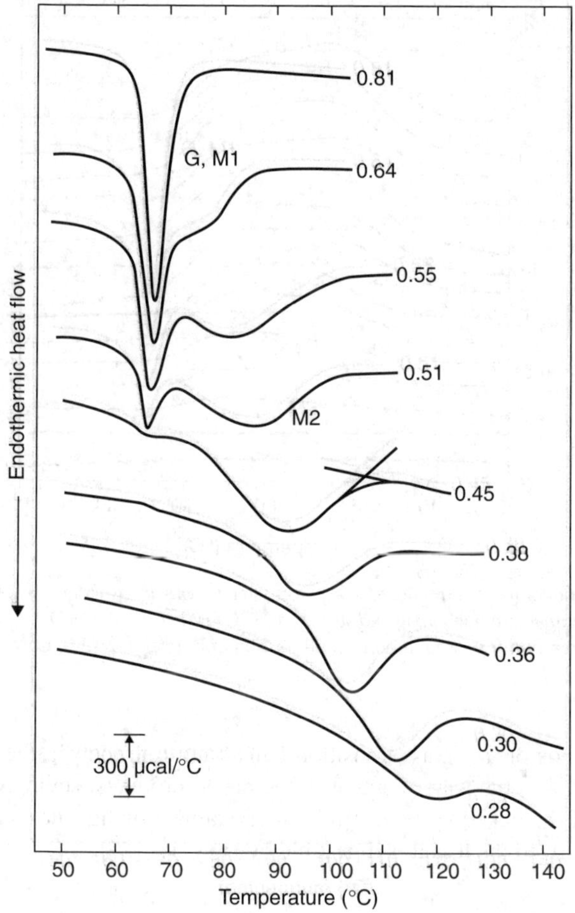

Figure 2.2 *DSC thermograms of potato starch obtained at a heating rate of 10 °C/min, labelled with volume fraction of water present. (Reprinted with permission from reference [11], Copyright 2013, Wiley.)*

The fourth step involves phase separation of water (syneresis) due to excessive retrogradation of starch chains [30–33].

Like gelatinization, retrogradation phenomenon can be studied by DSC. For instance, if gelatinized corn starch is stored in conditions suitable for retrogradation (i.e. 5 °C for several hours or days) and a DSC scan is performed, two endotherms peaks will be detected (Figure 2.3).

These two peaks are pointed as Gr and M_1r and are correlated with the G and M_1 peak in the gelatinization phenomenon. Gr and M_1r are mainly due to amylopectin, which, after the gelatinization process, can rebuild its structure under appropriate time-temperature conditions giving rise to new thermal transitions close to the original ones [34]. Gr and M_1r are detectable both in normal starches and in waxy (high amylopectin content) starches (in this case Gr and M_1r are more pronounced) [34, 35].

One practical effect of starch retrogradation is on bakery products. recrystallization of amylopectin has a great role in staling of baked products during storage. As amylopectin

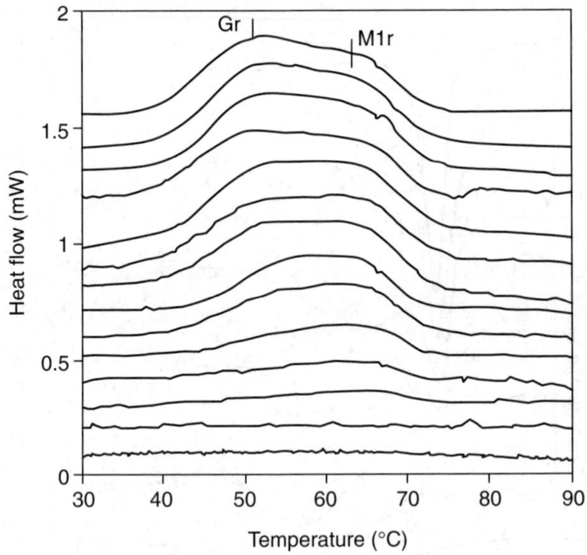

Figure 2.3 *DSC thermograms of retrograded waxy corn starch with intermediate water content after initial heating to 120 °C and immediate quenching and storage at 5 °C for 0, 1, 2, 3, 4, 5, 10, 18 h, 1, 3, 5, 10, 15 and 20 days, respectively (from bottom). (Reprinted with permission from reference [34], Copyright 2013, John Wiley & Sons, Ltd.)*

crystallization depends on T_g (glass transition temperature; the lower, the easier crystallization), which is related to the water content in the baked products, staling can be reduced by increasing the T_g of the baked system. Amylose crystallization also has an important role as the more it is crystallized the less it is digestible by enzymes [29].

2.4.2 Starch Jet-Cooking

Jet-cooking of starch is another method used to destroy the crystallinity of starch in excess of water [9]. By injecting steam into a slurry of starch while it passes through a mixing device, such as a Venturi nozzle, against a back pressure, the stream of high-velocity steam intimately and smoothly mixes the slurry and quickly raises its temperature. This system provides shear rates and temperatures needed to swell and destroy starch granules at a certain extent. Usually temperatures of 120 °C or higher are reached and starch concentrations up to 30% are used for the slurry dispersion. The final concentration is in the range of 2–14% and is obtained by adding preheated water in the desired amount in order to achieve a proper starch dilution and cooling of the slurry to 60–70 °C.

One of the main industrial uses of starch jet-cooking is in the paper industry for paper sizing [36].

2.4.3 Starch Extrusion Cooking

A different form of starch gelatinization is extrusion cooking technology [37]. As described by Conway in 1971 [38], extrusion cooking and forming are characterized by sufficient work and heat being applied to a cereal-based product to cook all the ingredients to obtain a crunchy foam though a die. The effects of processing conditions on the starch and on the texture of the extruded products have been studied by several researchers [39–45]. Cooked foamed materials with different starch viscosity, water solubility and water absorption have been prepared by altering the moisture content of the raw product and the temperature or the pressure in short extruders specifically designed for this purpose.

It has been demonstrated that an extrusion-cooked starch can be solubilized without any formation of maltodextrins and that the extent of solubilization depends on extrusion temperature, moisture content of the starch before extrusion and amylose/amylopectin ratio. For instance, Mercier *et al.* [39] determined the properties of different types of starch and considered the influence of the following parameters in a short twin-screw extruder: moisture content between 10.5 and 28%, barrel temperature between 65 and 250 °C, residence time between 20 seconds and 2 minutes. It was demonstrated that corn starch, after extrusion cooking, gave a solubility lower than 35%, while potato starch solubility was up to 80%.

Hanna *et al.* [42] observed that the degree of gelatinization of ordinary corn (30% amylose) and waxy corn (1% amylose) decreased with increasing moisture content but increased with increasing temperature. Corn starches were extruded between 17.8 and 42.2% moisture contents and extruder barrel temperatures of 116, 120, 140, 160 and 164 °C. Screw speeds of 93.5, 100, 130, 160 and 166.5 rpm were used to generate different residence times.

2.4.4 Starch Destructurization in Absence of Synthetic Polymers

In the patent literature, the terminology 'destructurized starch' [46–65] refers to a form of thermoplastic starch described as molecularly dispersed [56]. Destructurization of starch is defined as melting and disordering of the molecular structure of the starch granules as a molecular dispersion [55,56]. It means that, at the same time, the native crystallinity of starch and its granular structure disappear.

This is achieved by heating the starch above the glass transition temperature (T_g) and the melting temperatures (T_m) of its components, until they undergo endothermic transitions under shear, temperature and pressure for a time sufficient to destroy also the granular structure. As already stated, the water volume fraction required has to be below 0.45 and preferably below 0.28 [27].

Starch can be destructurized using extrusion technologies in specific conditions. Sufficient work, heat and time have to be applied to a cereal-based starch product in the presence of plasticizers to destructurize it. The best plasticizer for starch is water in quantities lower than 45%. Other plasticizers are glycols such as glycerol and sorbitol. Whereas thermoplastic starch can contain a certain amount of granular residue and a few Maltese crosses can be detected in polarized light microscopy, destructurized starch is substantially free from those features.

After having undergone destructurization, starch does not show its native crystallinity any more and loses its structure characterized by left-handed double helixes, as reported above.

Nevertheless, other forms of crystallinity, different from the native ones, induced by the interaction of the amylose component with specific molecules, can be observed. These types of crystallites are characterized by single helical structures and are known as V complexes [66].

The disappearance of the crystalline structure of the starch granule may be determined using conventional light microscopy techniques [67].

When native starch is observed using phase contrast microscopy, starch granules ranging from 2 to 100 μm (depending on starch source) in size are detected (Figure 2.4a). If the

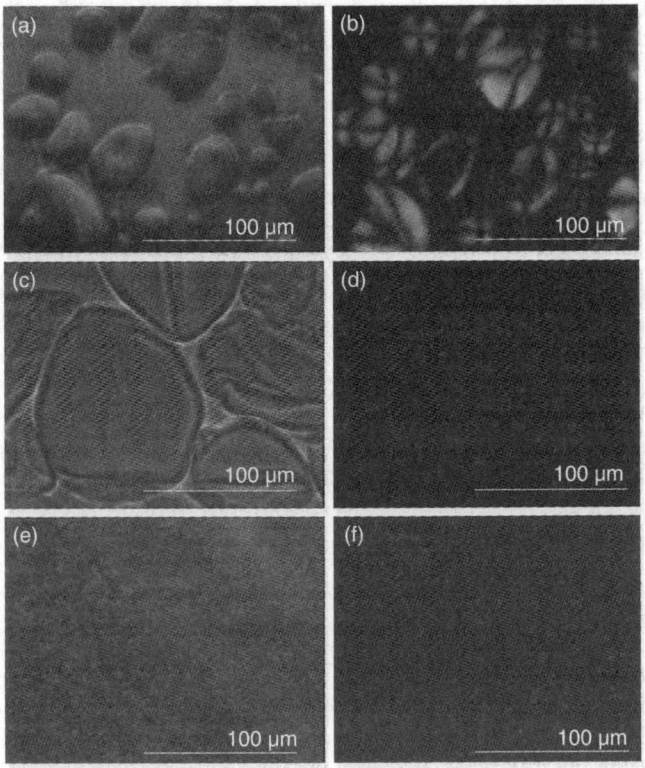

Figure 2.4 Phase contrast microscopy of native (a), gelatinized (c) and destructurized potato starch (e). Polarized optical microscopy of native (b), gelatinized (d) and destructurized (f) potato starch.

same sample is observed using polarized light microscopy, Maltese crosses due to starch birefringence will be detected (Figure 2.4b).

When gelatinized starch is observed using phase-contrast microscopy, swollen starch granules are detected (Figure 2.4c). The swelling factor (SF), which is the ratio between the volume of the swollen starch granule and the one that is not swollen, can vary from 10 to 50 and depends on the source of the starch, on the size of the granule that is not swollen and on the final temperature reached during the gelatinization process [9]. As mentioned before, gelatinization process gives rise to disordered structures but swollen granules or ghosts of swollen granules are still visible. Nevertheless, swollen granules are not able to deflect polarized light any more. So, if the same sample is observed using polarized light microscopy, a dark area will be seen (Figure 2.4d).

When destructurized starch is observed using phase-contrast microscopy, there are no more starch granules but a continuous surface is detected (Figure 2.4e). Amylopectin and amylose macromolecules, which behave like polymer macromolecules in the bulk, are arranged in a disordered way (just local points of V lattice structure can eventually be formed). This sort of amorphous structure is not able to deflect polarized light, so, if the same sample is observed using polarized light microscopy, a dark area will be seen (Figure 2.4f).

Another term that can be found in the literature is 'thermoplastically processable starch' (TPPS), defined as a thermoplastic starch that is substantially water free. TPPS is a form of destructurized starch that is obtained without water, because, instead of it, other high boiling point plasticizers or additives are used [68–73].

If plasticizers are not removed from TPS (or TPPS) in order to avoid that the glass transition temperature of these systems becomes too high, they can be processed as a traditional plastic [27, 56, 74] as their melt viscosity is comparable with those of traditional polymers [75].

This aspect makes possible the transformation of destructurized starch in finished products through traditional manufacturing technologies for plastics in absence of water. Their sensitivity to humidity, however, makes TPS or TPPS alone unsuitable for most of the applications [76].

2.4.5 Starch Destructurization in Presence of Synthetic Polymers

Starch can undergo a thermoplastic transformation up to destructurization in the presence of different synthetic polymers to satisfy a broad spectrum of market needs. Destructurized starch composites can reach starch contents higher than 50%.

With some of these polymers, a real complexation of single helical amylose with the polymer backbone can happen, which gives rise to supramolecular structures characterized by V-type crystallinity. This sort of modified amylose crystallinity is the same mentioned in section 2.4.4 and has been widely studied with small molecules such as alcohols, glycerol, dimethylsulphoxide, fatty acids, or iodine [66].

Complexation of starch by macromolecules can give rise to even more stable complexes playing an important role on the final properties of the starch-based polymer. Unlike amylose, the majority of amylopectin does not interact with the complexing agent and remains in its amorphous state.

In the following paragraphs, a resume of the state of the art related to starch destructurization with synthetic polymers will be made.

2.4.5.1 Ethylene-Acrylic Acid Copolymer (EAA)

Otey has studied the interactions between ethylene-acrylic acid copolymer (EAA) and gelatinized starch since 1977 [77–83].

In these formulations, a further compatibilization of EAA with starch can be made by using ammonium hydroxide. In fact, it has been demonstrated that a fraction of starch interacts [84, 85] with EAA after neutralization with ammonium hydroxide or other bases during gelatinization and extrusion, providing not only partial miscibility between the two polymers, but also enhancing the formation of molecular complexes.

Formulations based on thermoplastic starch, EAA and polyethylene (PE) in different ratio, are suitable for film production. In Table 2.1 a few examples of such compositions (some of them containing also polyvinyl alcohol, PVOH) with their related tensile strength properties are reported. As can be seen, the film obtained with a gelatinized starch level of about 50% shows good tensile properties.

In 1989, studies on EAA-thermoplastic starch films containing 40% by weight of EAA (acrylic acid content 20% by weight), processed at water content lower than 2%, led to destructurization of starch, improved processability and film properties with elongation at break up to 200% without the need for ammonia or sodium hydroxide [86]. Observing these materials

Table 2.1 *Influence of starch/EAA ratio and of partial replacement of EAA with PE or PVOH on the tensile strength and elongation of starch/EAA films [83, 87]. (Reproduced with permission from C. Bastioli, Handbook of Biodegradable Polymers, 2005. © 2005, Smithers Rapra.)*

Starch (%)	EAA (%)	PE (%)	PVOH (%)	Elongation (%)	UTS (MPa)
10	90	—	—	260	23.9
30	70	—	—	150	22.2
40	60	—	—	92	26.7
40	40	20	—	66	23.9
40	25	25	—	85	21.7
40	20	40	—	34	20.1
40	55	—	5	97	32.0
40	40	—	20	59	39.7

Notes: UTS: ultimate tensile strength; EAA: ethylene-acrylic acid copolymer; PE: polyethylene; PVOH: polyvinyl alcohol.

by means of microscopic analysis, at least three different phases were found: one consisting of destructurized starch, one consisting of synthetic polymer alone and a third described as 'interpenetrated', characterized by a strong interaction between the two components

It has also been seen that the use of urea in EAA-starch films enhances the tear propagation resistance and reduces ageing phenomena due to segmental motion in amorphous starch [87, 88].

Unfortunately, due to the lack of biodegradability of EAA, films based on such formulations do not comply with the international biodegradation standards, precluding their use in most packaging applications.

2.4.5.2 Ethylene-Vinyl Alcohol Copolymers (EVOH)

Thermoplastic starch and ethylene-vinyl alcohol copolymer (EVOH) copolymer systems, depending on the processing conditions, starch type and copolymer composition, can generate a wide variety of morphologies and properties [89–98]. Different microstructures have been observed as a function of different hydrophylicity of the synthetic copolymer: from a droplet-like to a layered one [92]. An in-depth description of this aspect will be presented in section 2.4.6.

The products based on thermoplastic starch and EVOH show mechanical properties suitable to meet the needs of specific industrial applications [99]. Their mouldability in traditional processing technologies is comparable with that of traditional plastics such as PS, ABS and LDPE [95]. The main limitation of these materials is the high sensitivity to low humidity, with consequent embrittlement.

To improve their characteristics, such as migration phenomena and physical ageing, specific types of plasticizers were selected [99], whereas boric acid, borax or other saline compounds were used to improve transparency [98]. With this sort of materials it is possible to obtain final products by film blowing, injection moulding, blow moulding or thermoforming and it is also possible to foam these compositions giving rise to products having cushioning characteristics close to expanded polystyrene (EPS-55).

Biodegradation of these products has been tested in different environments [90]. Ten months of aerobic biological treatment performed by a high-sensitivity respirometric test, brought to the degradation of more than 90% w/w of a product consisting of 60% of maize starch and natural additives and by 40% of EVOH copolymer (ethylene content 40% by mole) [90, 91]. If, in the same composition, an EVOH with ethylene content of 29% instead of 40% by mole was used, the reduced ability to generate interpenetrated structures, showed, in the Sturm test, a significantly higher initial biodegradation rate [90]. Furthermore, the semi continuous activated sludge test (SCAS) and biodegradation in lake water of a product consisting of 70% maize starch plus natural additives and 30% EVOH, supported the hypothesis of a substantially different biodegradation mechanism for the two components [91]:

- Starch, even though significantly shielded by the interpenetrated structure, seems to undergo hydrolysis by extracellular enzymes.
- Ethylene-vinyl alcohol copolymer appears to be biodegraded through a superficial adsorption of microorganisms, eased by the increase of available surface that occurred during the hydrolysis of starch.

The presence of starch improves the biodegradation rates of these synthetic polymers, where a fundamental role is also played by size and distribution of ethylene blocks. Attempts to speed up the biodegradation rate by modification of EVOH copolymers with carboxyl groups have been pursued but the degradation rate of thermoplastic starch and EVOH composites is too slow to consider them compostable according to international test standards.

2.4.5.3 *Polyvinyl Alcohol (PVOH)*

The combination of starch with a water soluble polymer such as PVOH (or polyalkylene glycols) has been widely considered since 1970 [100]. Since the early 1990s these compositions have been mainly studied for starch-based loose-fill production as a substitute for expanded PS [101–107], using compositional water as expanding gas. In this kind of blends, not only natural starch but also modified ones such as hydroxy propylated high amylose starch can be used, especially to improve foam resilience and density [101–105].

Optimization of composition and the foaming process have been pursued during the years in order to improve finals foams properties, leading, for example to foam densities of 8 6 kg/m^3 [108–110].

2.4.5.4 *Aliphatic Polyesters (APE)*

Starch can also be destructurized in presence of more hydrophobic polymers totally incompatible with starch, such as aliphatic polyesters [111].

It is difficult to process aliphatic polyesters having low melting points by conventional techniques for thermoplastic materials, such as film blowing and blow moulding. It has been found that the blending of starch with aliphatic polyesters allows their processability and biodegradability to be improved.

Particularly suitable polyesters considered in the past have been poly-ε-caprolactone (PCL) and its copolymers. Nevertheless, films made of thermoplastic starch and PCL are tacky as extruded, rigid, and have low melt strength at temperatures over 130 °C. Moreover, due to the slow crystallization rate of such polymers, the complete cooling process needs a long time after production of the finished articles, giving an undesirable change of properties with time.

Other APE having a higher melting point that can be used in presence of destructurized starch are those formed by the reaction of glycols such as 1,4-butandiol with succinic acid, sebacic acid, adipic acid, azelaic acid, dodecanedioic acid, or brassylic acid.

The presence of compatibilizers between starch and aliphatic polyesters is preferred. Some examples are amylose/EVOH V-type complexes [92], and starch-grafted polyesters. The use of chain extenders such as diisocyanates and epoxides is preferred. Such materials are characterized by excellent compostability, excellent mechanical properties, and reduced sensitivity to water.

2.4.5.5 *Aliphatic-Aromatic Polyesters (AAPE)*

The most important achievement of recent years in the sector of starch technology is related to the creation of micro and nanostructured composites of thermoplastic starch with polyesters of different type and particularly with aliphatic-aromatic polyesters. This technology has been developed and patented by Novamont [112, 113]. In these families of products, starch gives a technical contribution to the mechanical performances of the finished products in terms of high toughness and excellent performances stability at different humidities and temperatures. Such products can cover a wide range of demanding applications in the film sector and meet the different needs in terms of end-of-life conditions as home compostability and soil biodegradation.

The development of new aliphatic and aliphatic-aromatic copolyesters containing monomers from vegetable oils, covered by a range of Novamont's patents, have further improved and widened the performances of these products from an environmental and technical point of view.

One of the main applications of this sort of product is related to shopping bags: the substitution of traditional plastics with biodegradable ones in this field can greatly reduce the environmental impact of this sort of product. Moreover their reuse for the separate collection of organic waste can help to enhance organic waste value, converting it into high-quality compost. Again referring to 'cradle-to-cradle' applications, these products can be useful substitutes for traditional plastic in agricultural mulch films.

Such development has justified the significant industrial investment made by Novamont to build the first local biorefinery of this type in Europe, which comprises plants for the production of nanostructured starch and polyesters from vegetable oils. Moreover, new investments on monomers from vegetable oils from local crops will permit a further up-stream integration of the biorefinery.

This family of tailor-made products has permitted Novamont to work on many case studies to demonstrate the opportunity given by biodegradable and bio-based plastics to rethink entire application sectors, looking at them with a 'cradle to cradle' approach. In this way entire agro-industrial nonfood chains (or chains that are synergic with food) in the production of raw materials, the way in which the finished products are used and disposed of, and finally the effect on the local areas involved in the project are considered.

2.4.5.6 *Other Polymers*

Thermoplastic starch can also be blended with other polymers such as polyolefins [115]. In this sort of blend, a compatibilizer such as ethylene-maleic anhydride copolymer can be used in order to make hydroxyl starch groups and anhydride copolymer groups to react and obtain ester bonds. This sort of esterification helps to compatibilize the starch (hydrophilic) with polyolefins (lipophilic).

Other studies have been performed on polyamide/high amylose starch and acrylic copolymers/high amylose starch systems [55, 115–117].

The noncompliance of these products with the international norms of biodegradability and compostability, however, prevented significant market growth in Western countries.

Thermoplastic starch and cellulose derivative systems have been also reported [98, 111, 118, 119], particularly with cellulose acetate and butyrate in presence of glycerine and epoxidized soybean oil [118].

These materials are especially suitable for injection moulding technology and have been industrially used for compostable cutleries.

2.4.6 Additional Information on Starch Complexation

Novamont's Mater-Bi starch-based technology implies processing conditions able to destroy the crystallinity of amylose and amylopectin, in the presence of macromolecules, such as specific polyesters, which are able to form a complex with amylose. They can be of natural or synthetic origin. The specification of the starch, that is, the ratio between amylose and amylopectin, the nature of the additives, the processing conditions and the nature of the complexing agents allow engineering of various supramolecular structures with very different properties.

For example, destructurized starch and vinyl alcohol copolymer composites, materials containing starch with an amylose/amylopectin ratio above 20/80 w/w do not dissolve even under stirring in boiling water. Under these conditions a microdispersion constituted by microsphere aggregates is produced, whose individual particle diameter is lower than 1 μm (Figure 2.5).

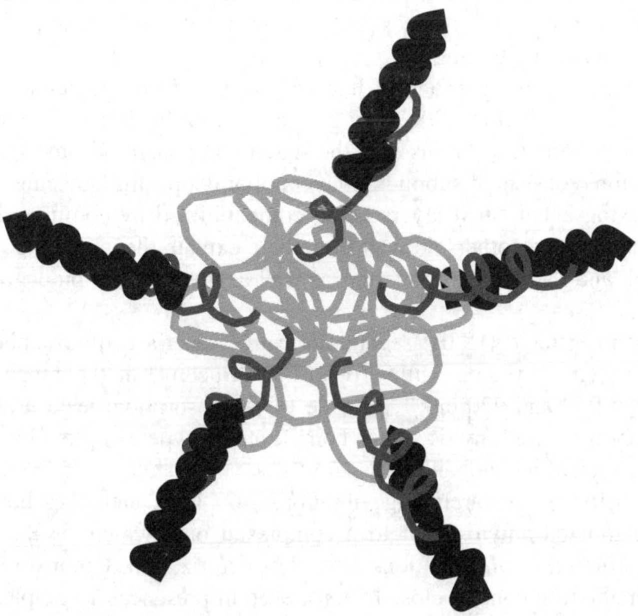

Figure 2.5 *Mater-Bi technology: Droplet-like structure (dark grey: amylopectin; light grey: amylose; black: complexing agent). (Reproduced with permission from C. Bastioli, Handbook of Biodegradable Polymers, 2005. © 2005, Smithers Rapra.)*

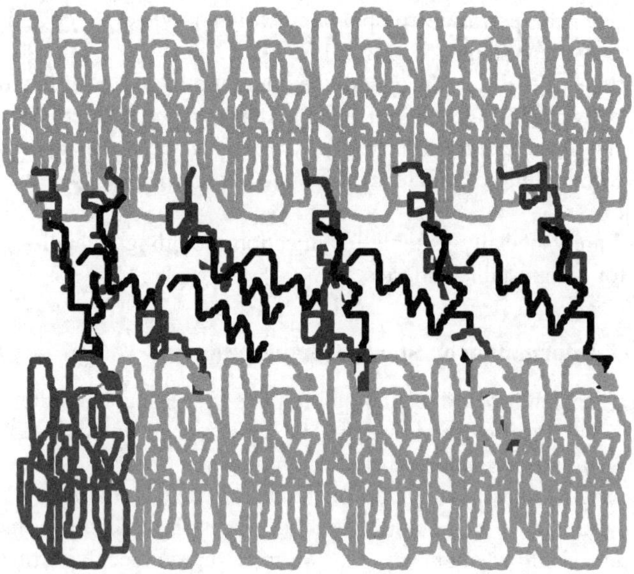

Figure 2.6 Mater-Bi technology: Layered structure (dark grey: amylopectin; light grey: amylose; black: complexing agent). (Reproduced with permission from C. Bastioli, Handbook of Biodegradable Polymers, 2005. © 2005, Smithers Rapra.)

A droplet-like structure, where a core of an almost amorphous amylopectin molecule is surrounded by complexed amylose molecules, has been confirmed by transmission electron microscopy (TEM) analysis of film slices [90]. The droplet size is comparable with that of the microdispersion obtained by boiling.

On the other hand, the morphology of materials in film form, containing starch with an amylose/amylopectin ratio below 20/80 w/w, gradually loses the droplet-like form, generating layered structures (Figure 2.6). In this case the starch component becomes partially soluble. The layered structure consists of submicron layers of amylopectin molecules intercalated by layers of complexing agent, such layers being compatibilzed by complexed amylose. The two structures and the many others derived from them explain the wide range of mechanical, physico-chemical, and rheological properties as well as the different biodegradation rates of Mater-Bi products.

Fourier transform infrared (FTIR) second-derivative spectra of thermoplastic starch and vinyl alcohol copolymer systems with droplet-like structure, in the range of starch ring vibrations between 960 and 920 cm^{-1}, provide for an absorption peak at about 947 cm^{-1} (Figure 2.7), as observed for amylose when complexed (V-type complex) by low-molecular-weight molecules such as butanol and fatty acids.

This peak is attributed to starch ring vibrations [92, 120], and does not correspond to crystalline or gelatinized amylose but to a complexed one, which results when amylose assumes V-type structures conformations after destructurization. Comparable starch-based materials with an amylose content close to zero, even in presence of complexing agents, do not show any peak at 947 cm^{-1}.

V-type structure formation can be also detected, by X-ray diffraction analysis and ^{13}CP-MAS solid state NMR [121–124].

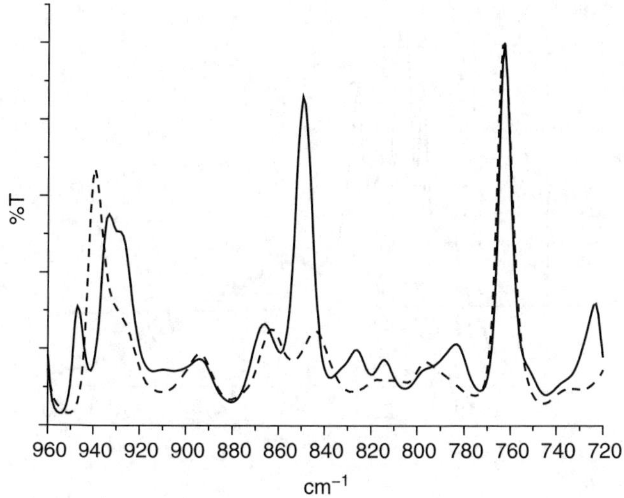

Figure 2.7 FTIR second derivative spectrum of corn starch (dashed line), and destructurized with EVOH (solid line).

As far as X-ray diffraction analysis is concerned (WAXD), characteristic peaks of different V type structures can be detected on destructurized starch in presence of complexing agents either molecular or macromolecular ones.

As stated above, V lattices forms of complexed amylose are mainly characterized by single left-handed helical forms, which surround the complexing agent. With certain complexing agents, it is possible that the left-handed structure is stabilized by the formation of supramolecular structures where the complexing agent is on the outer side of the amylose single left-handed helical [125].

V-type amylose structures nomenclature is still not univocal. Moreover, these forms have different stabilities and transitions from a less stable form to a more stable one can be achieved by changing the environmental storage conditions (such as temperature and humidity) [126–128]. Depending on the numbers of anhydroglucose repeating units present in the helical pitch, which are correlated with the complexing agent used, slightly different V-type amylose structure can be obtained.

Usually, starch destructurization in presence of synthetic polymers, gives rise to V_H and E_H structures (always a V-type one). The latter is especially obtained with high mechanical stress and low moisture content [129].

Referring to starch destructurized in the presence of EVOH copolymer, this sort of composition leads to the V_{6I} structure (WAXD patterns are reported in Figure 2.8).

Vinyl alcohol copolymers, as well as butanol, leave the amylopectin conformation unchanged. A model has been proposed considering large individual amylopectin molecules interconnected at several points per molecule as a result of hydrogen bonds and entanglements by chains of amylose/vinyl alcohol copolymer V complexes. This structure has been defined in the literature as 'interpenetrated' [92].

The amount of V-type structure present in the final formulation affects the biodegradation rate of starch. The more amylose is complexed the less the biodegradation rate is. This aspect is confirmed by second derivative FTIR spectra which show the 947 cm^{-1} peak

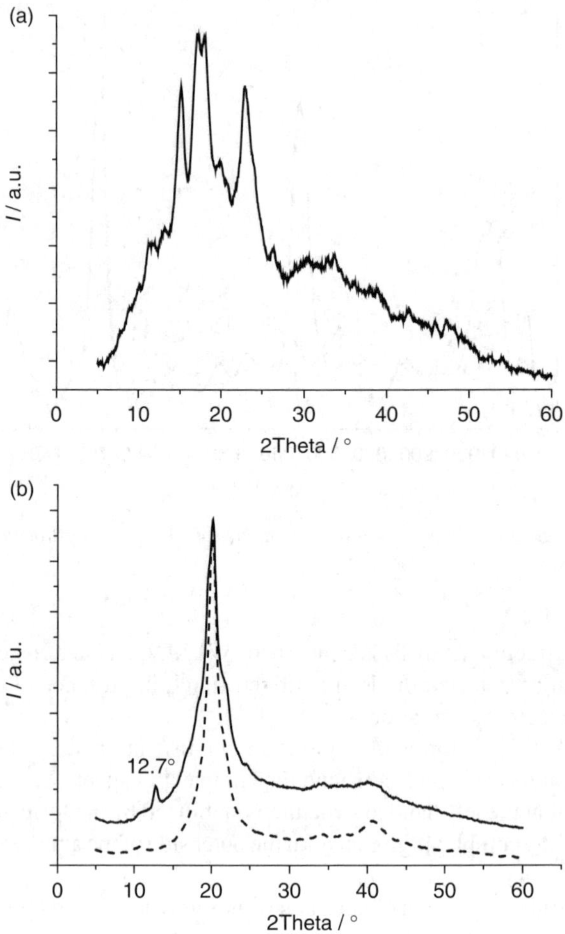

Figure 2.8 *WAXD patterns of corn starch: Native (a), destructurized with EVOH (b). The solid line pattern corresponds to the starch destructurized with EVOH, while the dashed line corresponds to the EVOH pattern.*

increasing with biodegradation, due to a delayed microbial attack of complexed amylose relative to amylopectin [92]. Furthermore the complexation degree is responsible for water vapour permeability (WVTR) of thermoplastic starch and EVOH blends ranging from 820 to 334 g m^{-2} d^{-1} (30 μm thickness, according to Lissy method) [94].

The droplet-like structure of destructurized starch, having an amylose/amylopectin ratio above 20/80 w/w in presence of EVOH copolymer, has been identified as responsible for the peculiar rheological behaviour of this sort of composition, which is characterized by strong pseudoplasticity at high shear stresses as well as yield stress at lower ones and by high levels of melt elasticity [130]. Notwithstanding the peculiar rheological behaviour shown by these systems, traditional processing techniques such as film blowing can be easily applied.

The sort of droplet-like structure developed by Novamont (see Figure 2.5) is easily dispersible in rubbers and allows a decrease of hysteresis in the final rubber compositions, which is useful for low rolling resistance treads in tyres. The first industrial application in this field was made by Goodyear in 2001, when the tyre Biotread GT3 was launched in Europe [131–133].

Novamont's recent developments in this sector have been achieved in the EU Biotyre project (LIFE06 ENV/L/000118), which have led Goodyear to produce for BMW a new tyre with a 'BioTRED' compound and a tyre structure optimized to achieve a reduction in rolling resistance of 34% (higher than the original target of 30%), which corresponds to a 5–6% decrease in fuel consumption. This tyre is used in the BMW 1-Series model and partner BMW foresees using BioTyres in other car models (e.g. for the BMW 3-Series).

2.5 Starch-Based Materials on the Market

Europe is the largest biodegradable polymers-consuming region, with about half of the global total. In 2009, despite the economic crisis, growth was in the range of 5–10%, depending on products and applications. The following factors have contributed to and will continue to contribute to growth:

- improvement of the properties and processing of biodegradable polymers;
- growing consumer interest in sustainable plastic solutions (e.g. eco-friendly packaging) and the reduction of greenhouse gas emissions;
- increasing support at the local and state levels for these kinds of products and for addressing needs about solid waste disposal; and
- existence of large-scale composting facilities.

The largest share in terms of volume corresponds to the starch-based plastic market. The main applications are compost bags, shopping bags, loose-fill packaging and mulch films.

Novamont, following the recent start-up of its seventh line dedicated to the production of Mater-Bi film grades in Terni, claims an internal production capacity of 120 000 ton/year. Established in 1990, its expertise grew out of the Fertec research facility set up by Montedison, the chemical unit of Ferruzzi agro-industrial group in 1989, expanded through the acquisition of the bioplastics IP of Warner-Lambert in 1997 (Warner-Lambert suspended the manufacture of starch-based materials in 1993) and the Eastar Bio patents of Eastman in 2004. Nowadays, the company holds a patent portfolio that includes more than 90 patent families and 800 internationally registered patents.

German company Biotec has been jointly owned by Sphere and Stanelco, from the United Kingdom, since 2005, creating Biosphere. The Sphere Group compounds starch, copolyesters and other materials into a variety of grades of bioplastic resin under the Bioplast trademark. Its factories, based in France, the United Kingdom and Spain, manufacture biodegradable carrier bags, bin liners, catering objects and refuse bags. It claims that its production capacity is 20 000 t/year.

Plantic Technologies Ltd is an Australian research company that offers products for thermoforming based on hydroxypropylated nongenetically modified high amylose corn starch. Its production capacity is of 7500 ton/year [134].

BIOP Biopolymer Technologies Ag entered the market with a starch-based material containing an additive consisting of a vinyl-alcohol/vinyl acetate copolymer [135]. The material is sold under the Biopar trademark.

Cardia Bioplastics Ltd (Australia) in 2008 acquired the resins business established in 2002 as Biograde Limited. Their Product Development Centre and manufacturing plant is in Nanjing, China. Cardia Bioplastics Ltd commercializes a biomaterial obtained mixing starch and biodegradable polyesters in the presence of a transesterification catalyst [136].

In the Netherlands, Rodenburg built up a plant for the transformation of potato wastes generated by the fried potatoes industry. The waste is fermented and it can be used in combination with natural fibres as well as being blended with other biopolymers. The resulting granulate was mainly used for the injection moulding of slow-release devices. Notwithstanding the claimed capacity of 40 000 ton/year, the products sold under the Solanyl trademark did not get significant market share [137].

2.6 Conclusions

After more than 20 years of research and development, starch-based materials have achieved specific in-use performances and can replace traditional plastics in different application sectors. They are able to offer original solutions both from the technical and the environmental point of view.

Starch-based bioplastics constitute a class of materials able to reduce significantly the environmental impact in terms of energy consumption and greenhouse effect in specific applications. If they are engineered to be compostable through the action of living organisms, these types of materials can also permit 'cradle-to-cradle' solutions through agro-industrial nonfood chains. They even offer a possible alternative to traditional materials when recycling is unpractical or uneconomical or when environmental impact has to be minimized.

Today, some of the bioplastics available in the market are used in specific applications where biodegradability is required, such as composting (bags and sacks), fast-food tableware (cutlery, cups, plates, straws, etc.), packaging (soluble foams for industrial packaging, film wrapping, laminated paper, food containers), agriculture (mulch film, nursery pots, plant labels), hygiene (nappies back sheet, cotton swabs), slow release of active molecules in the agricultural and pharmaceutical sectors.

The world market for biodegradable plastics, which started as a niche one, has grown significantly in the last few years, reaching a total capacity of 664 000 ton/year production in 2010 where a significant amount is represented by products coming partially or totally from renewable resources. The global biodegradable market in terms of volume is expected to grow to 2 330 000 metric ton in 2016 at an estimated compound average annual growth rate (CAGR) of 20.24% from 2011 to 2016 [138].

Without intending to produce a comprehensive treatise on starch, authors' desire was to illustrate the large variety of starch forms and their use in polymer technology, in order to provide basic information about this topic and a resource for further in-depth study.

References

1. US Department of Agriculture (USDA) (2010) Grains: World Markets and Trade, Corn Refiners Association 2010 Annual Report, Washington, DC.
2. US Department of Agriculture (USDA) (2010) WASDE, Corn Refiners Association 2010 Annual Report, Washington, DC.
3. Whistler, R.L. and Daniel, J.R. (1984) Molecular structure of starch, in *Starch Chemistry and Technology* (eds R.L. Whistler, J.N. BeMiller and E.F. Paschall), Academic Press, Orlando, FL, p. 154 and Snyder,

E.M. (1984) Industrial microscopy of starches, in *Starch Chemistry and Technology* (eds R.L. Whistler, J.N. BeMiller and E.F. Paschall), Academic Press, Orlando, FL, p. 667.

4. Otey, F.H. and Doane, W.M. (1984) Chemicals from starch, in *Starch Chemistry and Technology* (eds R.L. Whistler, J.N. BeMiller and E.F. Paschall), Academic Press, Orlando, FL, p. 397.
5. Kavesh, S. and Schultz, J.M. (1969) Meaning and measurement of crystallinity in polymers: A review. *Polymer Engineering and Science*, **5**, 331–338.
6. Zobel, H.F. (1988) Starch crystal transformations and their industrial importance. *Starch/Stärke*, **40**, 1–7.
7. Popov, D., Buléon, A., Burghammer, M. *et al.* (2009) Crystal structure of A-amylose: A revisit from synchrotron microdiffraction analysis of single crystals. *Macromolecules*, **42**, 1167–1174.
8. Takahashi, Y., Kumano, T. and Nishikawa, S. (2004) Crystal structure of B-amylose. *Macromolecules*, **37**, 6827–6832.
9. BeMiller, J. and Whistler, R. (2009) *Starch Chemistry and Technology*, 3rd edn, Academic Press, USA.
10. Buléon, A., Gerard, C., Riekel, C. *et al.* (1998) Details of the crystalline ultrastructure of C-starch granules revealed by synchrotron micro-focus mapping. *Macromolecules*, **31**, 6605–6610.
11. Donovan, J.W. (1979) Phase transitions of the starch–water system. *Biopolymers*, **18**, 263–275.
12. Griffin, G.J.L. (inventor) (1977) Coloroll Limited, assignee; US 4,016,117.
13. Griffin, G.J.L. (inventor) (1982) Coloroll Limited, assignee; US 4,324,709.
14. Griffin, G.J.L. (inventor) (1995) Epron Industries Limited, assignee; EP 363,383B1.
15. Griffin, G.J.L. (inventor) (1992) Epron Industries Limited, assignee; WO 9201741A2.
16. Griffin, G.J.L. (1974) Biodegradable fillers, in thermoplastics, in *Fillers and Reinforcements for Plastics* (eds R.D. Deanin, N.R. Schott) ACS Advances in Chemistry Series 134, ACS, Washington, DC, USA, p. 159.
17. Maddever, W.J. and Chapman, G.M. (1987) Making plastics biodegradable using modified starch additions, in *Proceedings of the SPI Symposium on Degradable Plastics*, Washington, DC, p. 41.
18. Gilead, D. and Scott, G. (inventors) (1985) No assignee; US 4,519,161.
19. Sipinen, A.J., Jaeger, J.T., Rutherford, D.R. and Edblom, E.C. (inventors) (1993) Minnesota Mining and Manufacturing Company, assignee; US 5,216,043.
20. Hocking, P.J. (1992) The classification, preparation, and utility of degradable polymers. *Journal of Macromolecular Science C*, **32**, 35–54.
21. Andrady, A.L., Pegram, J.E. and Nakatsuka, S.J. (1993) Studies on enhanced degradable plastics. 1. The geographic variability in outdoor lifetimes on enhanced photodegradable polyethelenes. *Journal of Environmental Polymer Degradation*, **1**, 31–43.
22. Wilder, R. (1989) Degradability I: disappearing package: pipe dream or savior? *Modern Plastics International*, **19**, 74.
23. Yutaka, T. and Masatoshi, K. (1992) Chuo Chemical Co. (assignee) Method for controlling biodegradability of plastics; JP 146953.
24. Aime, J-M., Mention, G. and Thouzeau, A. (inventors) (1989) Charbonnages de France, assignee; US 4,873,270.
25. Gallagher, F.G., Shin, H. and Tietz, R.F. (inventors) (1993) EI DuPont de Nemours, assignee; US 5,219,646.
26. Wang, S. and Zhang, X. (inventors) (2002) Wu Han Hua Li environment protection technology Co. Ltd., assignee; AU2002366251A1.
27. Colonna, P. and Mercier, C. (1985) *Pisum sativum* and *Vicia faba* carbohydrates. Part 6. Gelatinization and melting of maize and pea starches with normal and high-amylose genotypes. *Phytochemistry*, **24**, 1667–1674.
28. Blanshard, J.M.V. (1987) *Starch, Properties and Potential*, John Wiley & Sons, Inc., New York.
29. Morris, V.J. (1990) Starch gelation and retrogradation. *Trends in Food Science and Technology*, **1**, 2–6.
30. Roulet, P., MacInnes, W.M., Gumy, D. and Wuersch, P. (1990) Retrogradation kinetics of eight starches. *Starch/Staerke*, **42**, 99–101.
31. Biliaderis, C.G. and Juliano, B.O. (1993) Thermal and mechanical properties of concentrated rice starch gels of varying composition. *Food Chemistry*, **48**, 243–250.

32. Leloup, V.M., Colonna, P. and Ring, S.G. (1992) Physico-chemical aspects of resistant starch. *Journal of Cereal Science*, **16**, 253–266.
33. Van Soest, J.J.G., De Wit, D., Tournois, H. and Vliegenthart, J.F.G. (1994) Retrogradaton of potato starch as studied by Fourier transform infrared spectroscopy. *Starch/Staerke*, **46**, 453–457.
34. Liu, H., Yu, L., Tong, Z. and Chen, L. (2010) Retrogradation of cornstarch studied by DSC. *Starch/Staerke*, **62**, 524–529.
35. Liu, H., Yu, L., Chen, L. and Li, L. (2007) Retrogradation of waxy corn starch after thermal treatment at different temperatures. *Carbohydrate Polymers*, **69**, 756–762.
36. Klem, R.E. and Brogly, D.A. (1981) Methods for selecting the optimum starch binder preparation system. *Pulp and Paper*, **55**, 98–103.
37. Bastioli, C. (2005) *Handbook of Biodegradable Polymers*, Rapra Technology Ltd, Rockville, MD.
38. Conway, H.F. (1971) Extrusion cooking of cereals and soybeans. *Food Product Development*, **5** (3), 14–17.
39. Mercier, C. and Feillet, P. (1975) Modification of carbohydrate components by extrusion-cooking of cereal products. *Cereal Chemistry*, **52**(3), 283–297.
40. Chiang, B.Y. and Johnson, A.J. (1977) Gelatinization of starch in extruded products. *Cereal Chemistry*, **54**, 436–443.
41. Huber, D. and Escher, F. (1986) Structural changes of starch and starch-like products during extrusion cooking. *Mitteilungen aus dem Gebiete der Lebensmitteluntersuchung und Hygiene*, **77**, 73–84.
42. Bhattacharya, M. and Hanna, M.A. (1987) Kinetics of starch gelatinization during extrusion cooking. *Journal of Food Science*, **52**, 764–766.
43. Chinnaswamy, R., Hanna, M.A. and Zobel, H.F. (1989) Microstructural, physiochemical, and macromolecular changes in extrusion-cooked and retrograded corn starch. *Cereal Foods World*, **34**, 417–422.
44. Lai, L.S. and Kokini, J.L. (1991) Physicochemical changes and rheological properties of starch during extrusion (A Review). *Biotechnology Progress*, **7**, 251–266.
45. Orford, P.D., Parker, R. and Ring, S.G. (1993) The functional properties of extrusion-cooked waxy-maize starch. *Journal of Cereal Science*, **18**, 277–286.
46. Sacchetto, J.-P., Stepto, R.F.T. and Zeller, H. (inventors) (1990) Warner-Lambert Company, assignee; US 4,900,361.
47. Stepto, R.F.T. and Dobler, B. (inventors) (1994) Warner-Lambert Company, assignee; EP 0,326,517B1.
48. Sacchetto, J.-P., Lentz, D.J. and Silbiger, J. (inventors) (1994) Warner-Lambert Company, assignee; EP 0,404,723B1.
49. Silbiger, J., Lentz, D.J. and Sacchetto, J.-P. (inventors) (1994) Warner-Lambert Company, assignee; EP 0,408,503B1.
50. Sacchetto, J.-P., Silbiger, J. and Lentz, D.J. (inventors) (1994) Warner-Lambert Company, assignee; EP 0,409,781B1.
51. Lentz, D.J., Sacchetto, J-P. and Silbiger, J. (inventors) (1994) Warner-Lambert Company, assignee; EP 0,409,782B1.
52. Sacchetto, J.-P., Silbiger, J. and Lentz, D.J. (inventors) (1994) Warner-Lambert Company, assignee; EP 0,409,783B1.
53. Lentz, D.J., Sacchetto, J-P. and Silbiger, J. (inventors) (1994) Warner-Lambert Company, assignee; EP 0,409,788B1.
54. Egli, M. and Cole, E.T. (inventors) (1997) Warner-Lambert Company, assignee; EP 0,500,885B1.
55. Lay, G., Rehm, J., Stepto, R.F.T. and Thoma, M. (inventors) (1997) Warner-Lambert Company, assignee; EP 0,327,505B1.
56. Wittwer, F. and Tomka, I. (inventors) (1984) Warner-Lambert Company, assignee; EP 0,118,240A2.
57. Silbiger, J., Sacchetto, J.-P. and Lentz, D.J. (inventors) (1990) Warner-Lambert Company, assignee; EP 0,404,728A3.
58. Sacchetto, J.-P. and Rehm, J. (inventors) (1991) Warner-Lambert Company, assignee; EP 0,407,350A3.
59. Silbiger, J., Sacchetto, J.-P. and Lentz, D.J. (inventors) (1991) Warner-Lambert Company, assignee; EP 0,408,501A3.

60. Sacchetto, J.-P., Silbiger, J. and Lentz, D.J. (inventors) (1991) Warner-Lambert Company, assignee; EP 0,408,502A3.
61. Sacchetto, J.-P., Egil, M., Stepto, R.F.T. and Zeller, H. (inventors) (1992) Warner-Lambert Company, assignee; EP 0,391,853A3.
62. Loomis, G.L., Izbicki, M.J. and Flammino, A. (inventors) (1995) Novon International Inc., assignee; EP 0,679,172A3.
63. Fischer, H.R. and Fischer, S. (inventors) (2001) TNO, assignee; EP 1,134,258A1.
64. Bastioli, C., Bellotti, V., Del Tredici, G. *et al.* (inventors) (2002) Novamont SpA, assignee; EP 0,965,615B1.
65. Bastioli, C., Lombi, R., Nicolini, M. *et al.* (inventors) (2003) Novamont SpA, assignee; EP 1,109,858B1.
66. Le Bail, P., Rondeau, C. and Buléon, A. (2005) Structural investigation of amylose complexes with small ligands: helical conformation, crystalline structure and thermostability. *Internationl Journal of Biological Macromolecules*, **35**, 1–7.
67. Wigman, H.B., Leathen, W.W. and Brackenbeyer, M.J. (1956) Phase contrast microscopy in the examination of starch granules. *Food Technology*, **10**, 179–184.
68. Tomka, I. (inventor) (1995) Bio-tec Biologische Naturverpackungen GmbH, assignee; EP 0,397,819B1.
69. Tomka, I. (inventor) (1994) Bio-tec Biologische Naturverpackungen GmbH, assignee; EP 0,539,544B1.
70. Tomka, I., Meissner, J. and Menard, R. (inventors) (1997) Bio-tec Biologische Naturverpackungen GmbH, assignee; EP 0,537,657B1.
71. Tomka, I. (inventor) (1998) Bio-tec Biologische Naturverpackungen GmbH, assignee; EP 0,542,155B1.
72. Tomka, I. (inventor) (2001) Bio-tec Biologische Naturverpackungen GmbH, assignee; US 6,242102.
73. Tomka, I. (inventor) (1997) Bio-tec Biologische Naturverpackungen GmbH, assignee; EP 0,596,437B1.
74. Mellies, R.L. and Wolff, I.A. (inventors) (1957) USDA, assignee; US 2,788,546.
75. Willett, J.L., Jasberg, B.K. and Swanson, C.L. (1995) Rheology of thermoplastic starch: effects of temperature, moisture content, and additives on melt viscosity. *Polymer Engineering and Science*, **35**, 202–210.
76. Van den Berg, C. (1981) Vapour Sorption Equilibria and other Water-Starch Interactions; a Physico-Chemical Approach. PhD thesis. Agricultural University Wageningen, The Netherlands, p. 49.
77. Otey, F.H. and Westhoff, R.P. (inventors) (1979) USDA, assignee; US 4,133,784.
78. Otey, F.H. and Westhoff, R.P. (inventors) (1982) USDA, assignee; US 4,337,181.
79. Otey, F.H. and Westhoff, R.P. (inventors) (1984) USDA, assignee; US 4,454,268.
80. Otey, F.H. and Westhoff, R.P. (inventors) (1989) USDA, assignee; EP 0,132,299B1.
81. Fanta, G.F. and Otey, F.H. (inventors) (1989) USDA, assignee; US 4,839,450.
82. Otey, F.H., Westhoff, R.P. and Doane, W.M. (1987) Starch-based blown films 2. *Industrial and Engineering Chemistry Research*, **26**, 1659–1663.
83. Otey, F.H., Westhoff, R.P. and Doane, W.M. (1980) Starch-based blown films. *Industrial and Engineering Chemistry Product Research and Development*, **19**, 592–595.
84. Fanta, G.F., Swanson, C.L. and Doane, W.M. (1992) Complexing between starch and poly (ethylene-co-acrylic acid) a comparison of starch varieties and complexing conditions. *Carbohydrate Polymers*, **17**, 51–58.
85. Fanta, G.F., Swanson, C.L. and Shogren, R.L. (1992) Starch–poly(ethylene-co-acrylic acid) composite films. Effect of processing conditions on morphology and properties. *Journal of Applied Polymer Science*, **44**, 2037–2042.
86. Bastioli, C., Bellotti, V., Del Giudice, L. *et al.* (inventors) (1993) Novamont SpA, assignee; US 5,262,458.
87. Swanson, C.L., Shogren, R.L., Fanta, G.F. and Imam, S.H. (1993) Starch-plastic materials: Preparation, physical properties, and biodegradability (a review of recent USDA research). *Journal of Environmental Polymer Degradation*, **1**, 155–166.
88. Shogren, R.L. (1992) Effect of moisture content on the melting and subsequent physical aging of cornstarch. *Carbohydrate Polymers*, **19**, 83–90.
89 Bastioli, C., Bellotti, V., Del Giudice, L. and Lombi, R. (inventors) (1994) Novamont SpA, assignee; EP 0,400,532B1.

90. Bastioli, C., Bellotti, V., Del Giudice, L. and Gilli, G. (1993) Mater-Bi: properties and biodegradability. *Journal of Environmental Polymer Degradation*, **1**, 181–191.
91. Bastioli, C., Bellotti, V., Del Giudice, L. and Gilli, G. (1992) Microstructure and biodegradability of mater-bi products, in *Biodegradable Polymers and Plastics* (eds M. Vert, J. Feijem, A. Albertsson *et al.*), The Royal Society of Chemistry, Cambridge, UK.
92. Bastioli, C., Bellotti, V., Camia, M. *et al.* (1994) Starch / vinyl-alcohol copolymer interactions, in *Biodegradable Plastics and Polymers* (eds Y. Doi and K. Fukuda), Elsevier, Amsterdam.
93. Halley, P., McGlashan, S. and Gralton, J. (inventors) (2002) Food and Packaging Centre Management, assignee; WO 02083784.
94. Bastioli, C., Bellotti, V., Del Tredici, G.F. *et al.* (inventors) (1992) assignee; IT T092 A000199.
95. Bastioli, C., Bellotti, V. and Rallis, A. (1994) Microstructure and melt flow behavior of a starch-based polymer. *Rheologica Acta*, **33**, 307–316.
96. Bastioli, C., Bellotti, V., Del Giudice, L. and Lombi, R. (inventors) (1994) Novamont SpA, assignee; EP 0,437,561B1.
97. Bastioli, C., Bellotti, V. and Del Tredici, G.F. (inventors) (1996) Novamont SpA, assignee; EP 0,437,589B1.
98. Bastioli, C., Bellotti, V., Montino, A. *et al.* (inventors) (1995) Novamont SpA, assignee; EP 0,494,287B1.
99. Bastioli, C., Bellotti, V. and Montino, A. (inventors) (1992) Novamont SpA, assignee; WO 9214782.
100. Maxwell, C.S. (1970) Effect of ethylene copolymer-starch blends on water resistance of paper. *Tappi Journal*, **53**, 1464.
101. Lacourse, N.L. and Altieri, R.A. (inventors) (1989) National Starch and Chemical Corporation, assignee; US 4,863,655.
102. Lacourse, N.L. and Altieri, R.A. (inventors) (1990) National Starch and Chemical Corporation, assignee; EP 0,375,831.
103. Lacourse, N.L. and Altieri, R.A. (inventors) (1990) National Starch and Chemical Corporation, assignee; EP 0,376,201A1.
104. Lacourse, N.L. and Altieri, R.A. (inventors) (1991) National Starch and Chemical Investment Holding Company, assignee; US 5,035,930.
105. Lacourse, N.L. and Altieri, R.A. (inventors) (1991) National Starch and Chemical Investment Holding Company, assignee; US 5,043,196.
106. Neumann, P.E. and Seib, P.A. (inventors) (1993) Kansas State University Research Foundation, assignee; US 5,185,382.
107. Anfinsen, J.R., Garrison, R.R. (inventors) (1992) EI DuPont De Nemours & Company, assignee; WO 9208759.
108. Bastioli, C., Bellotti, V., Del Tredici, G.F. and Rallis, A. (inventors) (1995) Novamont SpA, assignee; EP 0,667,369A1.
109. Bastioli, C., Bellotti, V., Del Tredici, G.F. *et al.* (inventors) (1996) Novamont SpA, assignee; EP 0,696,611A3.
110. Bastioli, C., Bellotti, V., Del Tredici, G. and Rallis, A. (inventors) (1996) Novamont SpA, assignee; EP 0,696,612A2.
111. Bastioli, C., Bellotti, V., Del Tredici, G.F. *et al.* (inventors) (1992) Novamont SpA, assignee; WO 9219680.
112. Bastioli, C., Floridi, G. and Del Tredici, G. (inventors) (2008) Novamont SpA, assignee; WO 2008037749A2.
113. Bastioli, C., Floridi, G. and Del Tredici, G. (inventors) (2008) Novamont SpA, assignee; WO 2008037744A2.
114. Tomka, I. (inventor) (1994) Bio-tec Biologische Naturverpackungen GmbH, assignee; EP 0,539,544B1.
115. Buehler, F.S., Baron, V., Schmid, E. *et al.* (inventors) (1993) EMS-INVENTA AG, assignee; EP 0,541,050A3.
116. Schmid, E., Buehler, F.S. and Schultze, H.J. (inventors) (1993) EMS-INVENTA AG, assignee; EP 0,522,358A3.

117. Bastioli, C., Bellotti, V. and Del Tredici, G. (inventors) (1991) Butterfly Srl, assignee; EP 0,437,589A1.
118. Tomka, I. (inventors) (1994) Bio-tec Biologische Naturverpackungen GmbH, assignee; EP 0,542,155.
119. Schröter, J. (inventors) (1993) Buck Werke GmbH, assignee; EP 0,551,125A1.
120. Cael, J.J., Koenig, J.L. and Blackwell, J. (1975) Infrared and raman spectroscopy of carbohydrates. Part VI: Normal coordinate analysis of V-amylose. *Biopolymers*, **14**, 1885–1903.
121. Shogren, R.L., Thompson, A.R., Greene, R.V. *et al.* (1991) Complexes of starch polysaccarides and poly(ethylene co-acrylic acid): Structural characterization in the solid state. *Journal of Applied Polymer Science*, **47**, 2279–2286.
122. Fanta, G.F., Shogren, R.L. and Salch, J.H. (1999) Steam jet-cooking of high-amylose starch-fatty acid mixtures, an investigation of complex formation. *Carbohydrate Polymers*, **38**, 1–6.
123. Veregin, R.P., Fyfe, C.A. and Marchessault, R.H. (1987) Investigation of the crystalline 'V' amylose complexes by high-resolution 13C CP/MAS NMR Spectroscopy. *Macromolecules*, **20**, 3007–3012.
124. Gidley, M.J. and Bociek, S.M. (1988) 13C CP/MAS NMR Studies of amylose inclusion complexes, cyclodextrins, and the amorphous phase of starch granules: Relationships between glycosidic linkage conformation and solid-state 13C chemical shifts. *Journal of American Chemical Society*, **110**, 3820–3829.
125. Bias, B., Le Bail, P., Robert, P. *et al.* (2006) Structural and stoichiometric studies of complexes between aroma compound and amylose. Polymorphic transitions and quantification in amorphous and crystalline areas. *Carbohydrate Polymers*, **66**, 306–315.
126. Shogren, R.L., Fanta, G.F. and Felker, F.C. (2006) X-Ray Diffraction study of crystal transformations in spherulitic amylose/lipid complexes from jet-cooked starch. *Carbohydrate Polymers*, **64**, 444–451.
127. van Soest, J.J.G., Hulleman, S.H.D., de Wit, D. and Vliegenthart, J.F.G. (1999) Changes in the mechanical properties of thermoplastic potato starch in relation with changes in B type crystallinity. *Carbohydrate Polymers*, **29**, 225–232.
128. van Soest, J.J.G. and Knooren, N. (1997) Influence of glycerol and water content on the structure and properties of extruded starch plastic sheets during aging. *Journal of Applied Polymer Science*, **64**, 1411–1422.
129. van Soest, J.J.G., Hulleman, S.H.D., de Wit, D. and Vliegenthart, J.F.G. (1996) Crystallinity in starch bioplastics. *Industrial Crops and Products*, **5**, 11–22.
130. Herschel, W.H. and Bulkley, R. (1926) Measurement of consistency as applied to rubber-benzene solutions. *Proceedings of the ASTM*, **26**, 621.
131. Corvasce, F.G. and Linster, T.D. (inventors) (1996) The Goodyear Tire & Rubber Company, assignee; US 5,545,680.
132. Corvasce, F.G. and Linster, T.D. (inventors) (1994) The Goodyear Tire & Rubber Company, assignee; US 5,374,671.
133. Bastioli, C., Bellotti, V. and Montino, A. (inventors) (2001) Novamont SpA, assignee; EP 1,127,089B1.
134. Yu, L., Christie, G.B.Y. and Coombs, S. (inventors) (2000) Food and Packaging Centre Management Limited, assignee; WO 0036006.
135. Berger, W., Jeromin, L., Mierau, U. and Opitz, G. (inventors) (1999) BIOP Biopolymer GmbH, assignee; WO 9925756.
136. Changping, C. (inventor) (2007) Biograde (Hong Kong) PTY LTD, assignee; WO 2007012142.
137. Jongboom, R.O.J., Stuut, P.I. and Rodenburg, J.A. (inventors) (2002) Rodenburg Biopolymers B.V., assignee; US 6,482,341B1.
138. MarketsandMarkets. (2011) *Biodegradable Plastics Markets: By Type, Applications, Regulations, Prices, Trends and Forecast (2011–2016)*, Research and Markets, Dallas, USA.

3

Cellulose and Cellulose Acetate

Johannes Ganster and Hans-Peter Fink
Fraunhofer Institute for Applied Polymer Research IAP, Germany

3.1 Introduction

Cellulose is the most abundant natural polymer on earth with annual growth rates in the 10^{12} ton range [1] thus constituting a virtually inexhaustible source of raw material for creating environmentally friendly and biocompatible products that are not in competition with the food chain. Mainly produced by plants for the use as a skeletal component, cellulose is a semirigid polysaccharide of well defined chemical configuration capable of forming an extended, stable system of hydrogen bonds. Because of these intramolecular and intermolecular bonds, the hypothetical melting point of cellulose is above its degradation temperature and thus cellulose cannot be processed like a thermoplastic. The use of plasticizers, as it is common practice, for example in the case of starch (Chapter 2), does not work here either. However, via either direct or derivative-based methods, cellulose can be shaped into films, fibres, and nonwovens covering a great many of the application areas of thermoplastic materials.

On the other hand, the huge number of hydroxyl groups along the polymer chain offers the unique possibility to derivatize the carbohydrate backbone in a multitude of ways thus changing the properties of this molecule. Novel materials have been and are being developed as derivatives of this plentiful naturally occurring polymer. Main groups are ethers and esters. Cellulose ethers, often water soluble, are used, for example, as wallpaper adhesives, viscosity regulators, or additives in the building industry. Main applications for cellulose esters, on the other hand, are found in explosives, coatings, cigarette filter tow, LCD displays, membranes, and, of particular interest here, plastics. The latter use is facilitated by the inhibition of hydrogen bond formation because the hydroxyl functionalities are blocked to a large extent by the substituents. In combination with appropriate plasticizers, cellulose esters such as cellulose acetate (CA), cellulose butyrate (CB), or cellulose acetate-butyrate (CAB) can be processed

Bio-Based Plastics: Materials and Applications, First Edition. Edited by Stephan Kabasci.

like common thermoplastics. Well known examples of the use of this kind of material are frames for spectacles, combs, tool handles and jewellery.

This chapter first gives an overview of cellulose raw materials and their molecular and supermolecular structures. The principles of shaping cellulose into fibres, films, and nonwovens by means of solution techniques are then outlined followed by a section on properties and market applications of these materials. Derivatives of cellulose are presented with special emphasis on thermoplastic cellulose esters, typical plasticizers, and promising reinforcing materials. Finally, recent developments and future prospects of cellulose materials are reviewed as far as the above applications are concerned. This book does not cover the important applications of cellulose and ligno cellulose fibres for reinforcing thermoplastics, like wood plastic composites (WPC) and natural fibre reinforced plastics (NFRP), since in these cases cellulose does not substitute a thermoplastic.

3.2 Raw Materials

The main, and most important by far, source of cellulose is wood, which is a complex composite structure of three main components – cellulose, hemicelluloses, and lignin – as shown in Figure 3.1.

The naturally occurring cellulose fibres in the primary and, more importantly, the secondary cell wall provide strength and stiffness to the wood composite structure. They are 'glued' together by lignin, which is in turn linked to cellulose by hemicelluloses. Lignin is built up of the three phenylpropane derivatives p-coumaryl, coniferyl, and sinapyl alcohol, constituting an irregularly linked extended network. The third major wood component, hemicellulose, is a

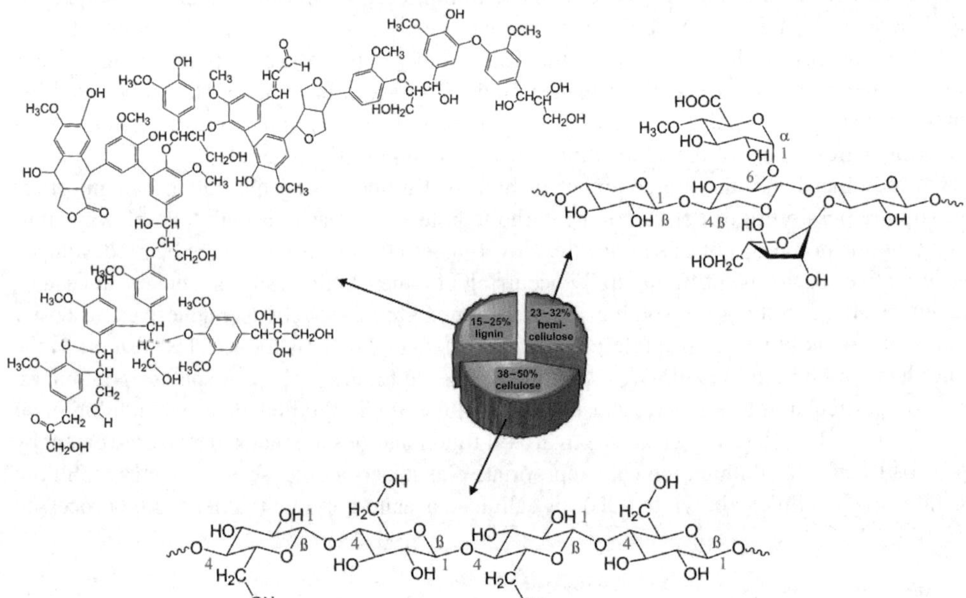

Figure 3.1 Main constituents of wood.

mixture of various branched polysaccharides with relatively low molecular weight including pentose and hexose monomers such as xylose, arabinose, mannose, and galactose.

In order to isolate the cellulose fibres, various pulping procedures are applied, of which sulfate cooking (the Kraft process) is the most common. The final product is called wood pulp and worldwide production capacities are in the order of 150×10^6 tons per year. The majority goes into paper making and related industries but roughly 2% is processed into high quality so-called dissolving pulp, having cellulose fractions higher than 93%. This dissolving pulp is used for the production of fibres, films, and cellulose derivatives, such as ethers and esters, the latter being the source of cellulose-based thermoplastics.

Another traditional source of cellulose is represented by vegetable fibres found in plants like cotton (seed fibres), hemp, flax, jute (bast fibres), and sisal (leaf fibres), to name but a few. Most applications for these fibres are found in the textile sector as fabrics but ropes, tapes, and isolating materials are also manufactured. Cotton linters, the small fibres adhering to the cotton boll after removal of the long textile fibres (lint), play a role as a raw material for producing cellulose esters. No lignin and no hemi celluloses are present and only some minor amounts of wax, fat, cuticle and ash [2] have to be removed in order to obtain a high-purity (>98% cellulose) starting material for the esterification. This kind of pulp is produced in small quantities, however, with typical worldwide amounts of 200 000 tons per year [3].

In particularly pure form and high crystallinity, cellulose is synthesized by certain bacteria (e.g. *Acetobacter xylinus*) and occurs in certain algae (e.g. *Valonia ventricosa*). These sources are important for basic research and are not used to produce thermoplastic cellulose material.

3.3 Structure

3.3.1 Cellulose

3.3.1.1 Molecular Structure

The cellulose molecule is a polysaccharide built up of six-membered sugar rings of anhydroglucose units (AGUs) which take the ${}^4C_1$ chair configuration. The AGUs are linked through $\beta(1 \rightarrow 4)$ glycosidic bonds such that, in contrast to amylose in starch which is built up of the same AGUs but is $\alpha(1 \rightarrow 4)$ linked, a rather straight chain configuration is easily generated by a rotation of the successive AGU by 180° (Figure 3.2). The chemical repeating unit is $C_6O_5H_{10}$ with a molar mass of 162 g/mol while the structural unit is cellobiose (one AGU plus another rotated by 180°) as indicated by the square brackets in Figure 3.2. A sense of direction along the chain can be defined, say, from the reducing end at C1 to the non-reducing end at C4, that is, in Figure 3.2, the chain would point from right to left. Two parallel chains can then be either parallel in the strict sense or antiparallel (see below).

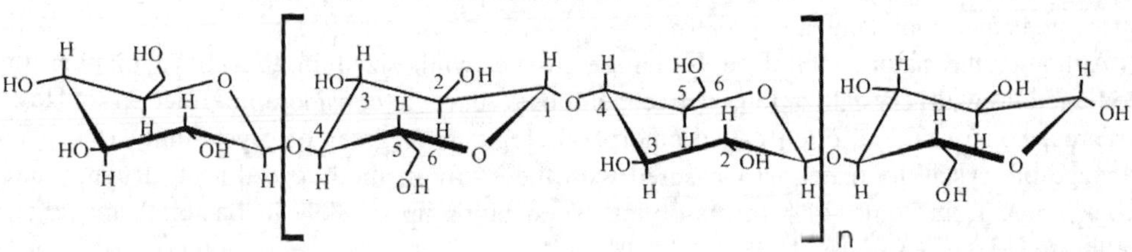

Figure 3.2 *Chemical structure of cellulose molecule with atom numbering.*

3.3.1.2 *Degree of Polymerization* (DP)

The number of AGUs constituting a chain is called the degree of polymerization (DP). The weight average DPs of native celluloses can be as high as 10 000 (intrinsic viscosity measurement of nitrated cellulose) for cotton fibres in the secondary cell wall [4]. Purified cotton linters are reported to have DPs in the range of 1000 to 3000 [5] or even 4000 [6], whereas wood pulps cover a range from 600 to 1000 according to early work in reference [5]. A comparison of wood pulp DPs for 30 representative samples of paper and dissolving grade pulps manufactured from different wood species and by means of different pulping processes is given by Siholta *et al.* [7]. As compared to ref. [5], the DP range reported here has a higher upper limit: DPs as high as 1910 for a spruce sulfite paper pulp and 1830 for a spruce sulfite dissolving pulp were found (values from intrinsic viscosity in cadoxen). More recent studies of dissolving pulps in LiCl/DMAc (lithium chloride/N,N-dimethylacetamide) solution with gel permeation chromatography using MALLS/RI (multi angle light scattering/refractive index) detection measured weight average DPs of 1500 to 1600 for acid sulfite cooking and 1000 to 1100 for the prehydrolysis Kraft process [8].

During processing and often for the sake of processing cellulose into fibres or films, the DP is reduced considerably depending on the type of processing (see below). Typical values are in the order of 300 (Viscose) to 600 (Lyocell) for this type of material, which is called regenerated cellulose [9].

3.3.1.3 *Supermolecular Structure – Natural Cellulose*

In nature, cellulose is synthesized in or at the cell membrane by large membrane-associated enzyme complexes often called terminal complexes (TC) [10, 11]. They produce crystalline arrays of many parallel (in the above strict sense) chains by adding glucose molecules to the nonreducing end of the precursor chain with rates between 300 and 1000 glucoses per minute and chain [12]. These arrays are called microfibrils. The lateral size of these microfibrils can vary from 3 nm (approximately 36 chains) in 'elementary fibrils' up to 20 nm (more than 1000 chains) in certain algae [10, 11]. The arrays have the well defined crystal structure of cellulose I with two sub-allomorphs occurring as mixtures [13]: the triclinic cellulose I_α dominant in bacterial and algal celluloses and the monoclinic (space group $P2_1$) cellulose I_β found predominantly in higher plants.

In Figure 3.3 the crystal structures of cellulose I_α (left) and I_β (right) are compared [14]. All chains are parallel, pointing from top to bottom in the lower two parts of the figure. The top part shows a view along the chain direction. Hydrogen bonded sheets are formed designated I, II, and III in the middle part of the figure, where the main difference between I_α and I_β is seen best: an upward slippage of one AGU-length (c/2 in crystallographic terms) of the hydrogen bonded sheet III in I_α produces the structure of I_β with slight conformational changes. This β structure is thermodynamically preferred.

Although the natural cellulose is thought to be synthesized in the above manner in conjunction with crystallization, native celluloses have a considerable fraction of less ordered, noncrystalline, often called amorphous regions. Degrees of crystallinity (the percentage of crystalline material) measured with the X-ray method according to Ruland and Vonk range from some 40% in dissolving wood pulps up to 80% in bacterial and algal cellulose [15].

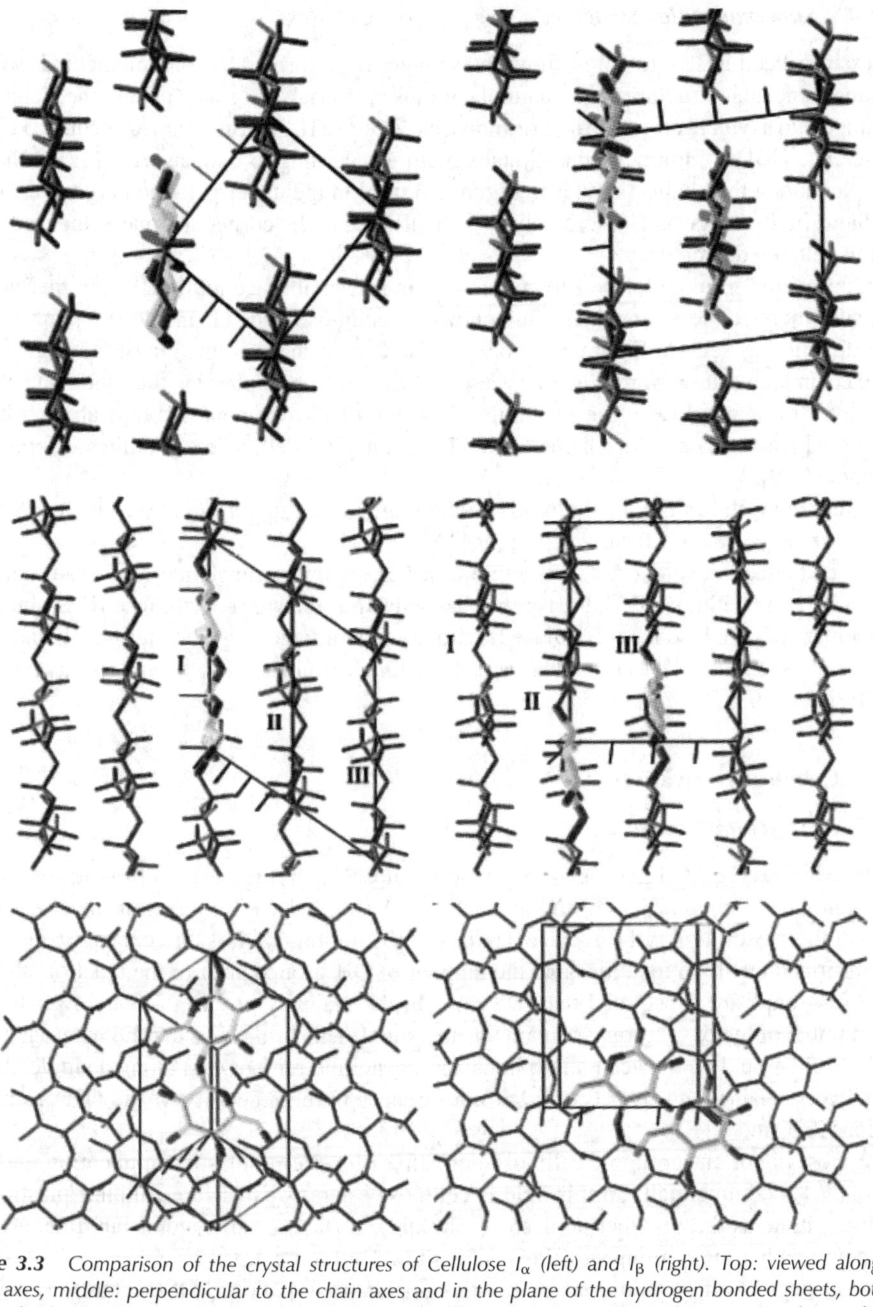

Figure 3.3 *Comparison of the crystal structures of Cellulose I_{α} (left) and I_{β} (right). Top: viewed along the chain axes, middle: perpendicular to the chain axes and in the plane of the hydrogen bonded sheets, bottom: perpendicular to the chain axis and the hydrogen bonded sheets. (Reprinted with permission from ref. [14], Copyright © 2003 American Chemical Society.)*

3.3.1.4 Supermolecular Structure – Regenerated Cellulose

Processing of cellulose into fibres, films and sponges is performed by solution methods where the supermolecular structure of the natural cellulose materials is broken up and the chains do rearrange into a typical crystal structure known as cellulose II. A more detailed account is given in reference [16]. In Figure 3.4 the crystal structure according to Langan *et al.* [17] is shown as viewed along the chains (similar to Figure 3.3, top) in the upper part and perpendicular to the chains in the lower part, where only two chains, one at the corner and one at the centre of the unit cell, are depicted.

The transition from cellulose I to cellulose II is irreversible; cellulose II is the thermodynamically more stable arrangement. In contrast to cellulose I the chains do not point in the same direction as seen in Figure 3.4, bottom. The corner chain points upwards whereas the centre chain points downwards. This poses some difficulties in explaining the solid-state phase transition from natural cellulose to cellulose II after alkali treatment and thus an interdiffusion model was proposed for chains from adjacent natural cellulose crystallites of opposite orientation [19].

Degrees of crystallinity for cellulose II products are in the range from 25%, which is typical for viscose fibres, to 45% for Lyocell fibres [16].

Besides the cellulose II crystal structure other crystalline arrangements do exist, namely cellulose III and cellulose IV [19]. Neither has industrial relevance. Cellulose III is obtained by treatment of cellulose I and cellulose II with liquid ammonia to give cellulose III_I and III_{II}, respectively. Cellulose IV is obtained by heat treating cellulose III in glycerol to give either cellulose IV_I or IV_{II}.

3.3.2 Cellulose Derivatives

3.3.2.1 Molecular Structure

The three hydroxyl groups per sugar ring in the cellulose chain (Figure 3.2) offer the possibility for a multitude of chemical substitution reactions. A simple basic concept for the characterization of the substitution is the average degree of substitution DS. It is thrice the ratio between the substituted OH functionalities and the number of OH-groups prior to the reaction, that is, if all OH-groups are substituted then DS = 3. In the majority of cases DS is lower than 3 and a number of hydroxyl groups remain unsubstituted. This brings in a certain heterogeneity, which can be found (i) between amorphous and crystalline regions, (ii) along a single chain, (iii) between chains, and (iv) according to differences in the reactivity of the OH-groups at positions 2, 3, and 6 [12].

The most important group of cellulose derivatives for thermoplastics application and the only one used commercially in this field is cellulose esters. Apart from camphor plasticized cellulose nitrate as the first thermoplastic (celluloid), acetate, propionate or butyrate, are the most common hydroxyl group substituents, as shown in Figure 3.5.

Typical degrees of substitution for plastics applications [21] are DS(acetate) = 2.5 for cellulose acetate, DS(propionate) = 2.6 and DS(acetate) = 0.1 for cellulose acetate propionate (CAP), and DS(butyrate) = 1.8 and DS(acetate) = 1.1 for cellulose acetate butyrate (CAB). In the industrial process of catalysed esterification, chain scission is a competitive reaction that can, however, be fairly well controlled under appropriate conditions [22]. Degrees of polymerization of commercial products can be estimated from literature data [22–25] to be in the range of 200 to 300.

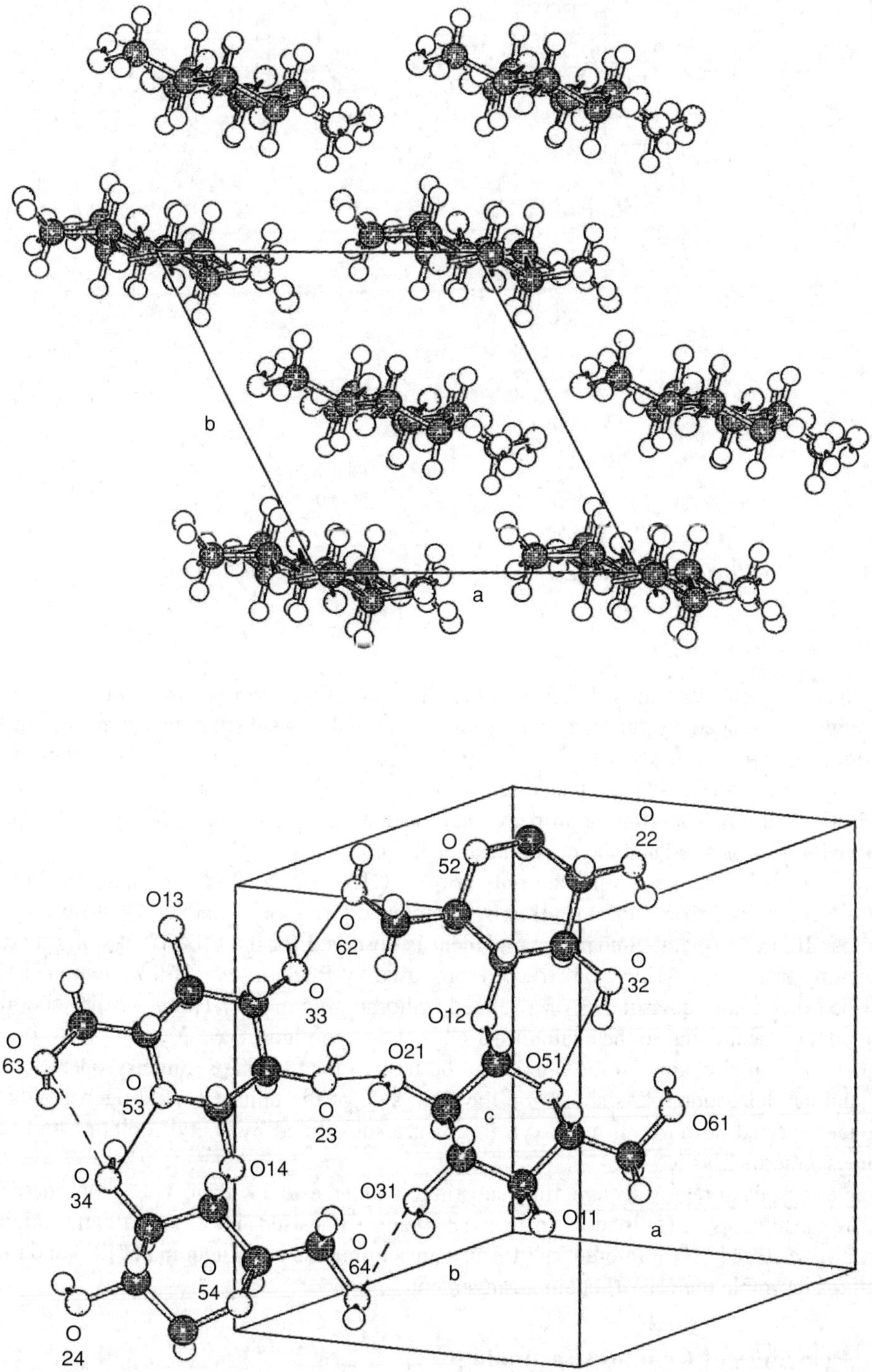

Figure 3.1 *Crystal structure of cellulose II according to ref. [17]. Top: viewed along the chain axes, bottom: perspective view perpendicular to the chain axes, one corner and one centre chain selected. (Reprinted from ref. [18] with permission from Elsevier.)*

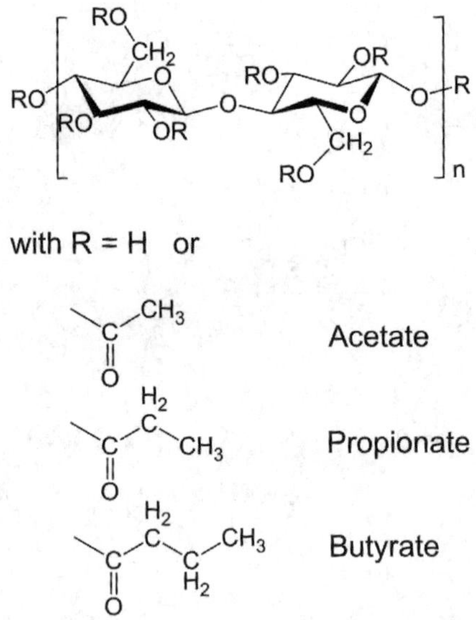

Figure 3.5 Chemical composition of most common cellulose esters.

3.3.2.2 *Crystal Structure*

The supermolecular structure of cellulose derivatives – in the present case cellulose esters – is strongly influenced by the degree of substitution and the substitution pattern. With the exception of cellulose triacetate (CTA, fully substituted, DS = 3), most commercially used cellulose esters have a DS of less than three or, in the case of mixed esters, do not have a regular substitution pattern. Therefore, a proper crystallization with well defined X-ray interferences is normally not observed and the materials appear amorphous.

For CTA, there are two crystalline polymorphs, CTA I originating from native cellulose, and CTA II obtained from Cellulose II or by complete dissolution of native Cellulose I. As for Cellulose II, an anti-parallel chain arrangement is observed for the CTA II crystal structure. Proceeding from a CTA II crystal structure proposed by Roche *et al.* [26], Zugenmaier [27] published the atomic coordinates of a refined orthorhombic model. The $a - b$ projection of this model perpendicular to the chain direction is shown in Figure 3.6.

The two parallel chains on the a-axis can be transformed by the symmetry operations to their antiparallel counterparts shown on the right side of the unit cell. Hydrogen bonds are no longer formed because all hydroxyl groups are substituted by acetyl moieties and CTA becomes a thermoplastic.

Single crystals of regioselective, fully substituted mixed esters with acetate, propionate und butyrate substituents and DPs between 46 and 84 were synthesised and structurally investigated by Iwata *et al.* (see [28] for references). Results are summarized by Zugenmaier [27] and many structures resemble the CTA II chain arrangement.

3.4 Principles of Cellulose Technology

In order to produce shaped bodies from cellulose pulp, other methods than common melt processing have to be employed (see section 3.1). The paper-making process is the most

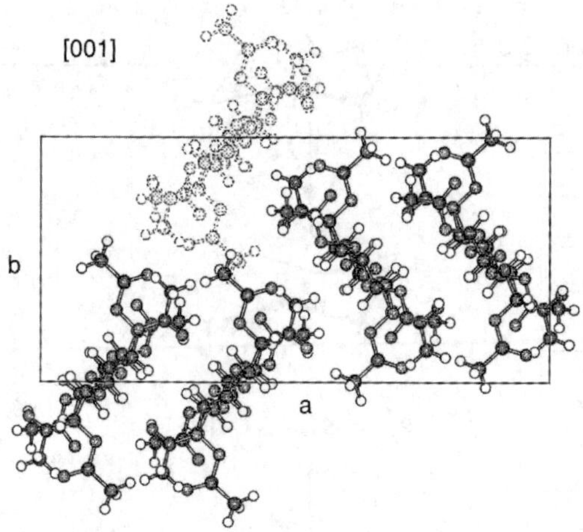

Figure 3.6 Crystal structure of cellulose triacetate according to Zugenmaier [27], projection onto the a–b plane. (Reprinted from [27] with permission from Elsevier.)

well known technology, existing, in principle, for over three thousand years. Paper is formed by aggregating intact cellulose I fibrils by aqueous methods employing basically interfibrillar hydrogen bonds to preserve the shape after drying. Recent developments have used this principle with finer fibrils after opening the natural fibrillar structure up to the microfibrillar level [29] by appropriate methods [30]. In this way, the so-called hemp stone or, more recently, Zelfo material is obtained and marketed as sheets of various densities, thicknesses, and colours (www.zelfo-technology.com, accessed 21 June 2013). While the above methods use native cellulose entities without changing their crystalline cellulose I structure and without mobilizing individual chains, solution methods often with prior derivatization are employed to produce cellulose fibres, films, nonwovens, sponges, and cellulose derivatives. This is not straightforward because the strong hydrogen intermolecular bonds (see above) render cellulose insoluble in most common organic solvents. Finding appropriate solvents for cellulose is a whole branch of cellulose chemistry and ionic liquids represent a promising recent example [31].

3.4.1 Regenerated Cellulose

If the cellulose has gone through a solubilized state, the material obtained after precipitation is called regenerated cellulose. To attain this solubilized state, two major processing principles are industrially implemented nowadays (Figure 3.7): derivative methods and direct methods. In the former, a cellulose derivative is produced after activation of cellulose by, as a rule, swelling in NaOH (alkalization). Cellulose becomes accessible to the substituents since Na^+ ions are built in to the cellulose I crystal structure disrupting the intermolecular hydrogen bonds giving the Na-cellulose I crystal structure [32]. The cellulose derivative is then dissolved in alkali and spun into a salt-containing acidic precipitation bath such that the pure cellulose is regenerated.

For the direct method special solvent systems are employed without chemical modification of the cellulose chains. Some examples are LiCl/DMAc (lithium chloride/N,N-dimethylacetamide), DMSO/TBAF (dimethyl sulfoxide/tetra-n-butylammonium fluoride),

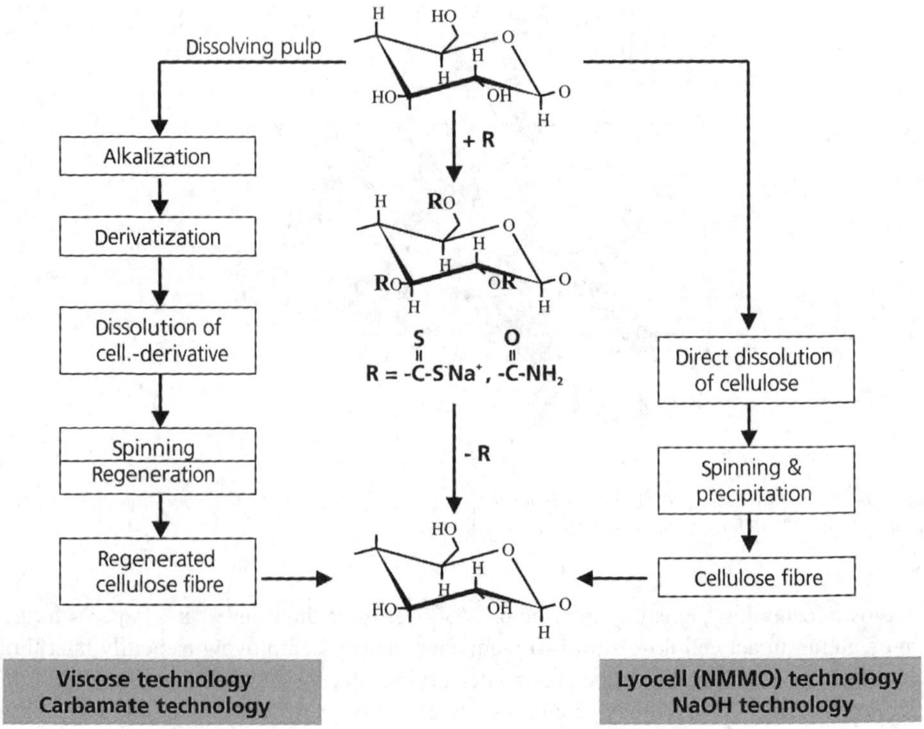

Figure 3.7 *Process principles in regenerated cellulose technology, example fibre spinning: derivative methods (left), direct methods (right). (From ref. [1].)*

sodium hydroxide/urea and superphosphoric acid. Industrial processes (Lenzing Lyocell) use N-methylmorpholine-N-oxide monohydrate (NMMO) as the solvent.

Not only fibres can be produced with such technologies. Once the cellulose or its derivative is dissolved, this dope can be shaped into other bodies like hollow fibres for membrane applications (artificial kidney), membrane films, packaging films (e.g. Cellophane), self-adhesive tapes (Sellotape), or sponges.

In particular the NMMO technology has inspired new developments, recognizing that the cellulose/NMMO/water solution can be considered in many aspects as a melt. As an example, cellulose has been shaped by a blown-film process similar to conventional thermoplastics [33,34]. A schematic of the blow-extrusion process is shown in Figure 3.8.

Using this process the longitudinal and transverse properties of the film can be adjusted by changing the orientation of the cellulose chains according to the drawdown and blowup ratios. A balanced property profile can easily be obtained without resorting to elaborate transverse film drawing equipment. Moreover, the morphology and pore structure of the films can be tailored by the precipitation regime (hard or soft) which can be different in baths 1 and 2 thus allowing gradient morphologies to be produced.

Another example of melt-like processing of cellulose in NMMO/water [35] is the so-called melt-blown nonwovens process, which is widely used for conventional thermoplastics. The basic principle is shown in Figure 3.9. As in the thermoplastic case, filaments are produced

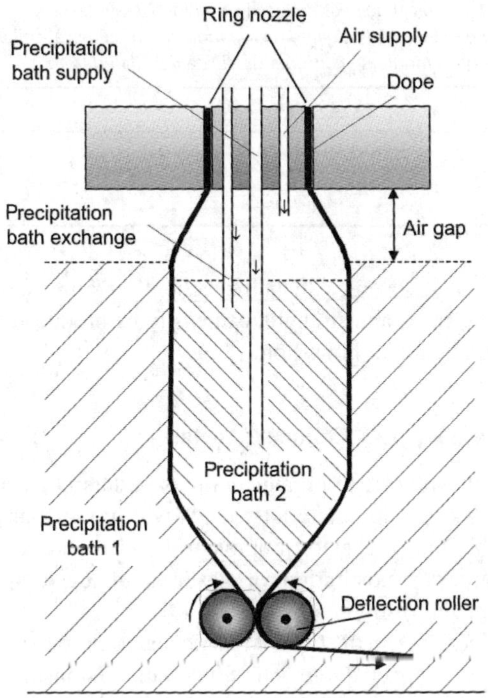

Figure 3.8 *Schematic of the blow-extrusion of cellulose-NMMO dope into blown film. (From ref. [1].)*

by extrusion through a melt-blow nozzle and aerodynamic drawing by a strong stream of hot air to increase orientation and decrease the fibre diameter. Before deposition of the final nonwoven, solidification has to be accomplished between nozzle and conveyor belt to prevent the formation of a more or less continuous film. In contrast to thermoplastics, which are basically cooled down, cellulose has to be coagulated by appropriate means. In the NMMO case, water is sprayed onto the filament curtain to accomplish coagulation and preserve the filament integrity. Further down the curtain, ambient air is used for temperature reduction.

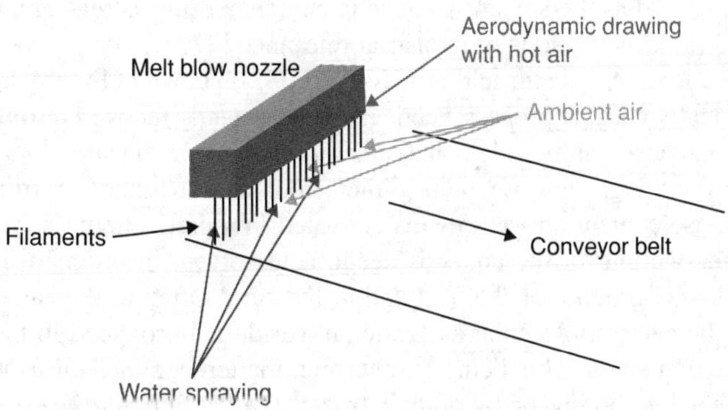

Figure 3.9 *Principle of producing melt-blown cellulose nonwovens.*

Table 3.1 Typical degrees of substitution for acetate (DS_{Ac}), propionate (DS_{Pr}), and butyrate (DS_{Bu}) moieties in technical cellulose esters used for moulding compounds. (Data calculated from ref. [22].)

	CA	CAP	CAB
DSAc	2.18–2.51	0.14–0.43	0.90–1.18
DSPr		2.11–2.71	
DSBu			1.39–1.65

In such a way, cellulose nonwovens with their excellent performance characteristics in terms of moisture absorption, softness and strength, can easily be produced for various application areas such as hygiene, medicine, or household.

3.4.2 Organic Cellulose Esters – Cellulose Acetate

The large-scale industrial synthesis of organic cellulose esters is practically restricted to the production of esters from a few aliphatic fatty acids with up to four carbon atoms, namely cellulose acetate (CA), cellulose acetate propionate (CAP), and cellulose acetate butyrate (CAB) [22]. Typical degrees of substitution for plastic moulding compounds are presented in Table 3.1 with data calculated from ref. [22].

All three have their importance in the field of plastic materials, whereas CA is by far the most widely produced derivative (expected to reach 980 000 metric tons by 2015 – see http://www.prweb.com/releases/cellulose_acetate/flake_plastics_filter_tow/prweb8067438.htm, accessed 21 June 2013) mainly used for cigarette filter tow, which is, however, not melt spun but produced by dry spinning in acetone. Since synthesis methods for CA and the mixed esters mentioned basically differ only in the starting materials (acid and acid anhydrides), only the CA process is described briefly in the following section.

3.4.2.1 Industrial Synthesis of Cellulose Acetate

Although intensive scientific work has been devoted to strategies for the synthesis of cellulose organic esters, especially with regard to homogeneous reaction conditions and in view of the advent of ionic liquids as reaction media [36], the principles employed for cellulose acetate production have remained the same for decades. Nowadays, only the 'acetic acid process', that is, the use of acetic acid as the solvent and acetic acid anhydride as reactant, in combination with sulfuric acid as the catalyst, is of industrial relevance [37].

A rough schematic of the 'acetic acid process' [37,38] is shown in Figure 3.10. In addition, to run an economically feasible process, acetic acid is constantly recovered from the respective processing steps and acetic anhydride is manufactured therefrom on site.

High-quality dissolving cellulose pulp, either from cotton linters or from high quality wood pulp (cellulose content above 95%) is activated in a pretreatment step by mechanical disintegration and swelling in the aqueous acetic acid solvent in order to make accessible the cellulose hydroxyl groups for the acetylation reaction. Often at the same time, sulfuric acid is added as the catalyst. As soon as acetic anhydride is introduced in the next step, the derivatization reaction starts. This being exothermic, the temperature must be controlled by some means, either by cooling or by adding frozen acetic acid. As the reaction proceeds,

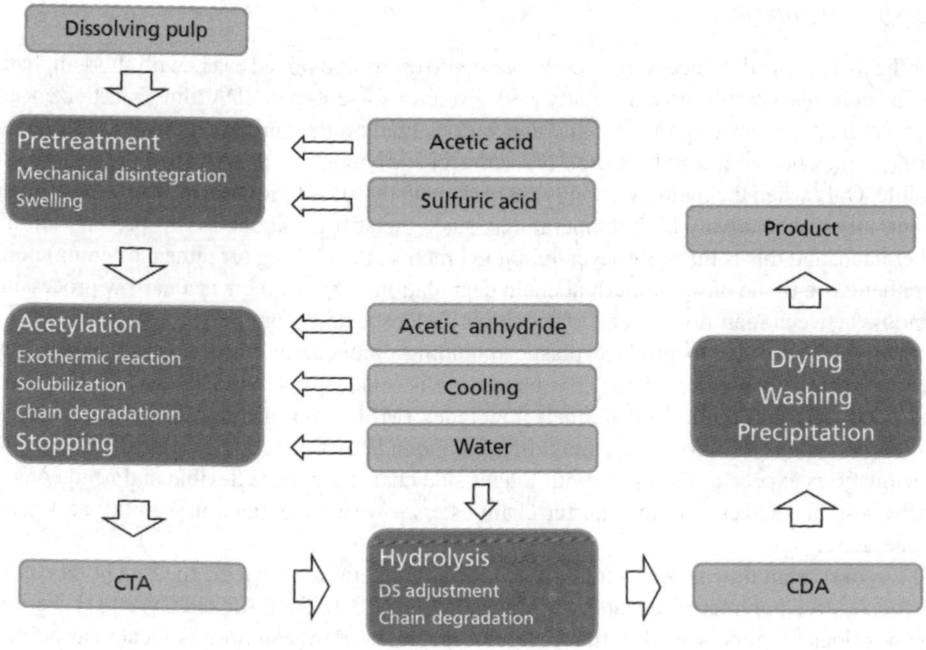

Figure 3.10 *Schematic of the industrially used 'Acetic Acid Process' to produce cellulose diacetate.*

cellulose acetate is produced and dissolved in the reaction mixture ending in a homogeneous cellulose triacetate (CTA) solution. Conditions, in particular temperature, have to be controlled carefully, because chain degradation is a competing reaction and too low a DP would result in inferior product properties. The reaction is stopped by adding water or diluted acetic acid, destroying the excess anhydride. Sulfuric acid is usually not neutralized because it is needed as the catalyst for the subsequent hydrolysis.

Both for the plastics applications in view and for the manufacture of cigarette filter tow by dry spinning, a fully substituted cellulose acetate (degree of substitution DS = 3) is not suitable. Therefore, a hydrolysis step follows the acetylation in which the DS is adjusted to 2.4 to 2.7 to give so-called cellulose diacetate (CDA). In the standard process, hydrolysis is controlled by the concentration of water (5 to 15%), usually added in the stopping step after acetylation, the reaction time (3 to 10 hours) and the temperature (between 60 and 80 °C). Again, chain degradation has to be taken into account. The faster, high-temperature hydrolysis at 140 to 150 °C with neutralized catalyst is another option.

Finally, often after a dope concentration process in which a certain amount of acetic acid is removed, the resulting CDA is precipitated in the form of flakes or powder, washed and dried. For plastics applications where the CDA is subject to high temperatures both in the plasticizing and the shaping processes, CDA is stabilized and bleached. Especially bound sulfuric acid remnants have to be removed either by boiling under pressure or by heating in 1% mineral acids during stabilization [22]. These residual, neutralized sulfate esters exacerbate chain degradation and thus further reduce the already narrow window between decomposition and melt flow temperatures [21].

3.4.2.2 *Plastification*

For the thermoplastic processing of cellulose homo esters and mixed esters with short aliphatic chain acids, plastification is a necessity [39]. Even for solvent-cast CDA films, plasticizers are incorporated to improve processability and various final product properties such as brittleness [40]. On the one hand, cellulose itself is a semirigid polymer and, as described previously, not fusible. Only when the hydrogen bonds are broken up by substitution can melt flow be brought about, albeit at relatively high temperatures, since the stiff backbone is still present. On the other hand, and this is the more severe point in practice, the melting (or rather softening) point is rather close to the onset of thermal chain degradation – that is, there is a narrow processing window between melt flow and decomposition [21], especially for CA.

Therefore, in order to produce plastic moulding compounds with reasonable melt flow characteristics, as a rule, 15 to 35% plasticizer is added for CA and 3 to 25% for CAP and CAB [22]. In Figure 3.11 the melt flow index (MFI), as a measure of melt viscosity, is shown for typical cellulose ester moulding compounds as a function of plasticizer content. Obviously, as expected, the esters with longer side chains are more flexible and need considerably less plasticizer than the shorter chain esters to reach the same flowability at a given temperature.

Moreover, melt flow is achieved for much lower temperatures than the melting points of the plasticizer-free polymers. The latter are shown in Figure 3.12 for CAP and CAB [41]. Various compositions are represented in the triangular diagrams. Pure cellulose is located at point C whereas pure CA can be found on the AC axis, with CTA at 44.8% acetyl content, pure CP on the PC axis, CB on the BC axis, and the mixed esters on the triangular plane. Melting point data are given for cellulose esters having DS from 2 to 3. Obviously, CA does not melt below 240 °C irrespective of the DS in the range between DS = 2 (34.9% acetyl content) and DS = 3. Using plasticizers (Figure 3.11) reasonable MFIs for CDA can be obtained at temperatures as low as 190 °C or, at least 210 °C with the positive effects mentioned on polymer molecular mass preservation.

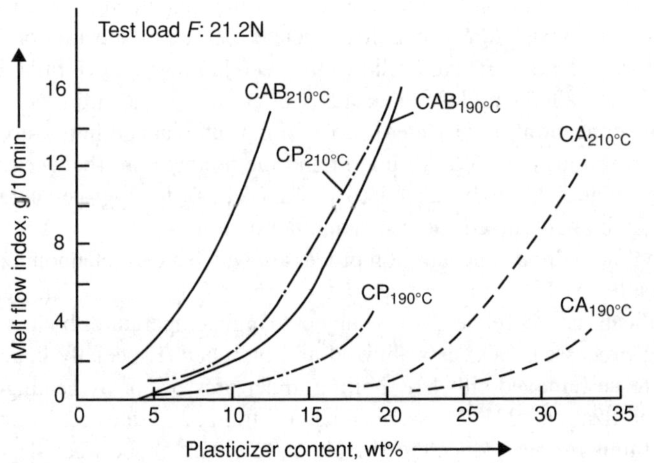

Figure 3.11 Melt flow index of cellulose ester moulding compounds as a function of plasticizer content. (Data calculated from ref. [22].)

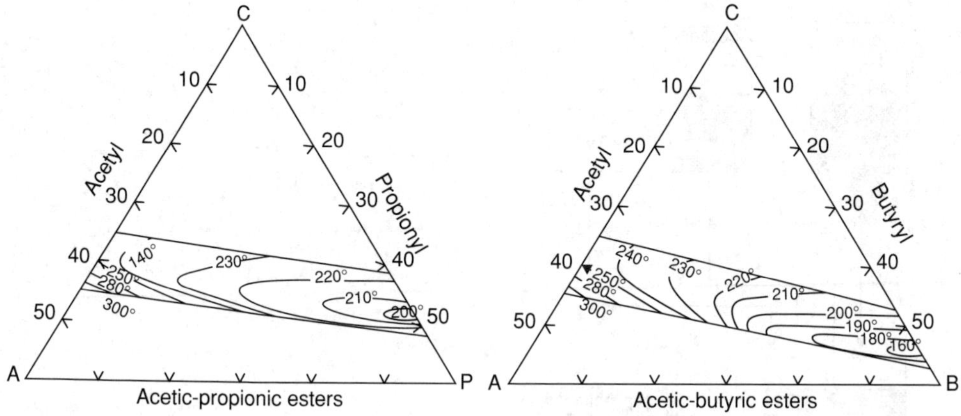

Figure 3.12 Melting points of CAP and CAB as a function of composition. (Adapted with permission from ref. [41]. Copyright © 1942 American Chemical Society.)

A lot of plasticizers typically with two ester groups have been described in the literature [39] including aliphatic diesters of phthalic acid, like diethyl phthalate, and diesters of aliphatic dicarboxylic acids like dibutyl adipate or azelate [22] as well as glycerol triacetate (triacetin), citrate esters, and phosphates [39]. Although these kind of plasticizers work well for decades in various applications, the accelerated retention test shows [39] that exudation and volatilization must be taken into account and can cause changes in the material performance. Thus, attempts have been made to synthesize long-chain esters of cellulose (LCCEs) with acid chain lengths of up to 20 carbon atoms (see [21] and references cited therein), which could be processed without external plasticizers. Positive results are reported for cellulose acetate hexanoate and cellulose acetate nonanoate [21] but these esters did not go into production.

3.4.2.3 *Processing of Cellulose Acetate*

Cellulose acetate is the most important cellulose ester and various processing techniques for forming shaped bodies exist for this polymer. The mixed esters discussed above, CAP and CAB, are used predominantly for injection moulding applications in suitably plasticized form and are marketed as pellets. This is done for CA as well, but additionally, alternative processing routes are available where solution methods play an important role. This is in particular the case for fibre spinning and (thinner) film manufacture where cellulose diacetate (CDA) is dissolved in acetone and dry spun or solvent cast. Moreover, the traditional solvent-using block process is employed for obtaining sophisticated, multicoloured thick sheets for applications like spectacle frames.

3.4.2.3.1 Fibres Both textile filament yarn and cigarette filter tow are produced from CDA by dry spinning out of acetone solution. A schematic of the dry spinning process [42] is presented in Figure 3.13. The CDA flakes are dissolved in acetone and TiO_2 is added to change the lustre from bright to dull, which is preferred in cigarette filter tow. Filtration is essential to remove the slightest impurities like gel particles, and so on. In a series of spinning cabinets, more than 10 000 filaments are spun for a typical filter tow. The solvent acetone rapidly

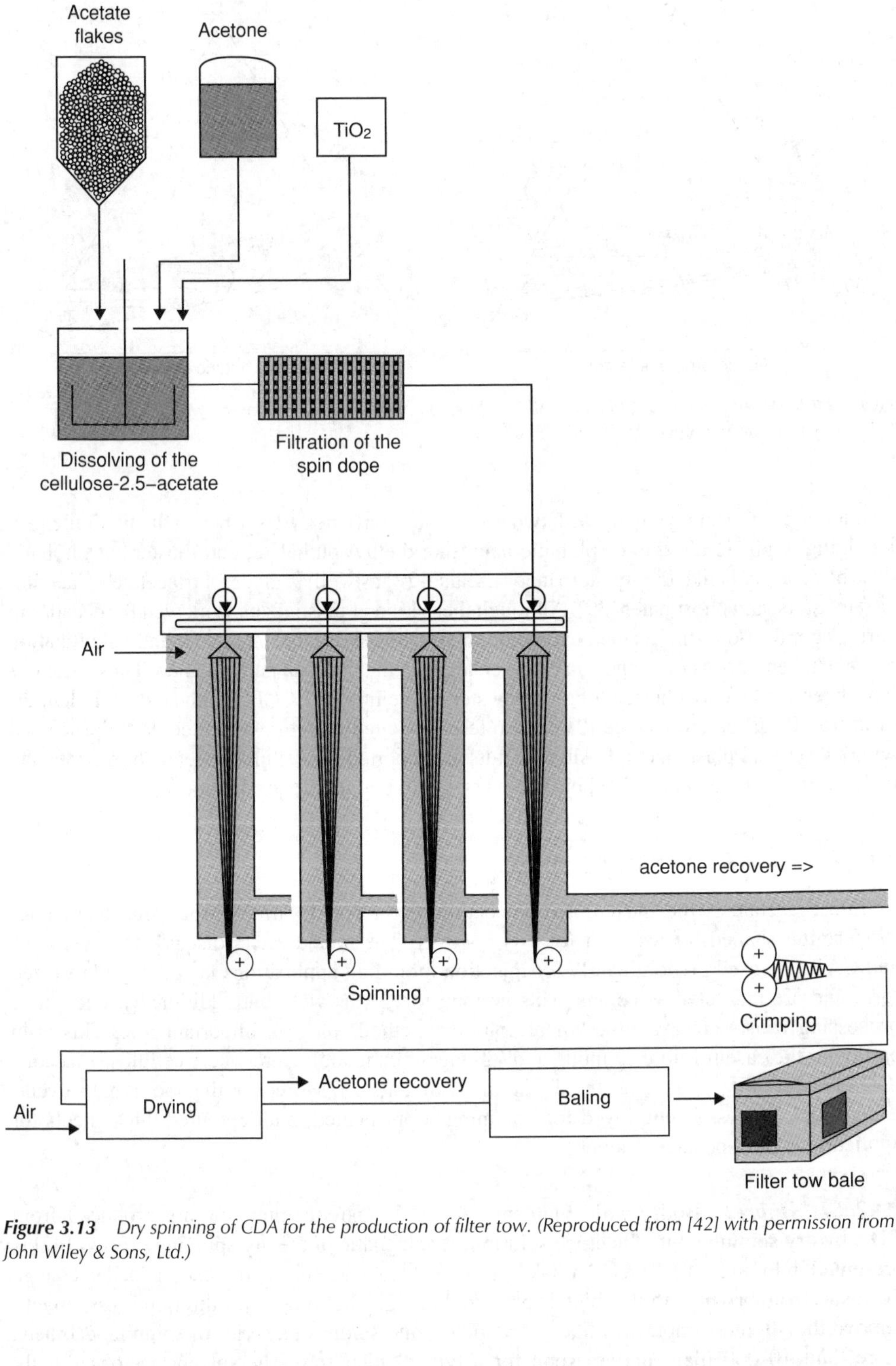

Figure 3.13 *Dry spinning of CDA for the production of filter tow. (Reproduced from [42] with permission from John Wiley & Sons, Ltd.)*

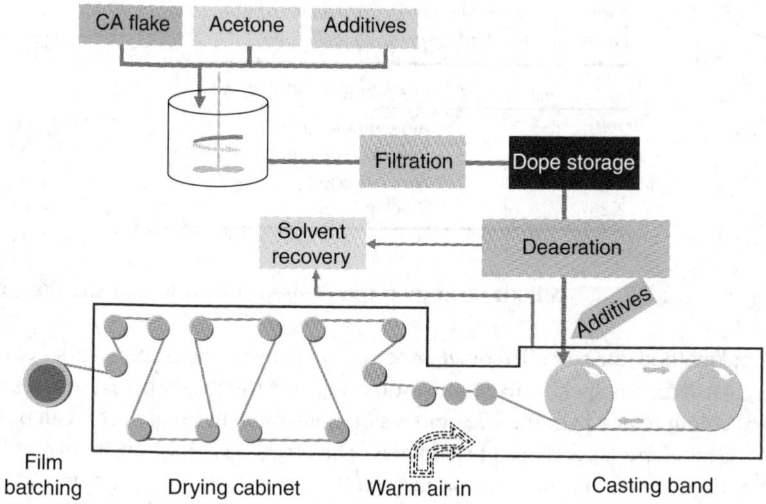

Figure 3.14 *Schematic of the Clarifoil® process for casting CDA film. (Reproduced with kind permission of Clarifoil, UK, 2012.)*

evaporates, the fibre solidifies, is stretched and acetone is recovered. Typical titers range from 1.7 to 5 dtex and a star-like cross section with three rays resembling a 'Mercedes'-star is preferred by spinning through triangular orifices. A crimp is added before drying and baling for the cigarette filter application.

Textile yarn (acetate fibres) is produced in a similar fashion but with much fewer filaments per spinning site (20 to 160) and overall yarn titers between 44 and 670 dtex [43]. Melt spinning is not, to the authors' knowledge, found in the literature.

3.4.2.3.2 Films Acetone is likewise used as the solvent for commercial cast film production from CDA [43]. A schematic of the Clarifoil® process to produce CDA cast film is shown in Figure 3.14. Dope preparation is similar to the dry spinning process but with the addition of adding plasticizer, normally diethyl phthalate in concentrations in the range between 10 and 25%.

The dope is cast on a continuously moving steel band and the solvent is removed by a countercurrent of hot air. For drying, a horizontal festoon dryer has been proven most efficient. A typical film thickness would be 50 μm. Applications are found in stationery, tapes, labels, seals and so on.

In a quite similar fashion, CTA films are produced but with an appropriate solvent mainly consisting of dichloromethane [44]. Applications are found as special photographic films and protective films for polarizing plates in LCD displays.

3.4.2.3.3 Plastic Materials For the various methods of plastic processing, special DS values are preferred for each processing technique as shown in Table 3.2 [45].

For blow and injection moulding, as well as extrusion, pellets are produced prior to the final shaping of the desired body. The CDA powder is mixed in a dry blender with plasticizer (essentially phthalic esters) and other additives such as colorants and the mixture is aged or 'matured' at 60 to 80 °C for a few hours to allow for diffusion of the plasticizer. Subsequently,

Table 3.2 *Preferred degrees of acetate substitution DS for various plastic processing techniques.*

	Processing technique
2.49	Blow moulding
2.41	Injection moulding and casting
2.34	Film extrusion
2.28	Block process

the material is extruded and pelletized to give the desired injection or extrusion moulding compound.

A traditional method, the so-called block process, to manufacture sheet material of thickness between, say, 4 to 8 mm, is specific to cellulose diacetate and found nowhere else in the industry. Sophisticated coloured and patterned designs, as demonstrated in Figure 3.15, can be obtained and used for fashion articles and eye glass frames. The process is time consuming and can take up to 6 weeks to complete. From a soft paste, which is produced by slow mixing CDA flakes, plasticizers, and solvent, subsequently filtered and coloured, a sheet is formed and placed in a press to obtain a single colour block of dimension $700 \times 1400 \times 300\ mm^3$. The block, rich in solvent, is cut into sheets which can either be kiln-dried for 10 to 20 days, or combined with other sheets of different colours to obtain another block. This can be repeated 3 to 4 times and finally, sheets are cut again and dried. The high solvent concentration, the long times involved, and the control of the temperature allow the migration of the colorants to give particularly subtle effects of shading.

3.5 Properties and Applications of Cellulose-Based Plastics

The property profile of cellulose-based plastics and cellulose products used in typical plastics applications covers a broad range where the intrinsic cellulose properties can be recognized to different degrees. So, hydrophilicity is a prominent feature for pure cellulose products, which is reduced as the number of ester substituents and their chain length is increased. Equilibrium moisture content of cellulose fibres, for example, is around 12% while the moisture regain of diacetate fibres is 6.5% and of triacetate fibres 3.2% [43]. In this respect, propionic and butyric esters are superior when fully substituted, but even mixed esters of the CAP and CAB types

Figure 3.15 *Examples of designs possible using the block process in sheet manufacture. (Reprinted with permission from Mazzucchelli 1849, Italy.)*

can have advantages over CA [41]. However, plastification has to be taken into account and the low moisture regain values of polymers like polypropylene or polyethylene terephthalate cannot be reached.

In the following, only a rough account of selected, primarily mechanical properties can be given and for more information the reader is referred to the literature cited. Applications resulting from the typical property profiles will be mentioned and some numbers will be given reflecting the respective production volumes.

3.5.1 Fibres

With the processes described in sections 3.4.1 and 3.4.2.3, fibres with excellent textile properties can be produced. The high moisture regain of the hydrophilic cellulose structure predestine these fibres for moisture regulation in textile structures either alone or in combination with synthetic fibres.

Typical tensile properties of cellulose-based fibres are shown in Table 3.3. The numbers given represent characteristic values that can be varied in a certain range. Here the viscose process is particularly versatile [16] and both textile (viscose) and technical (rayon) fibres can be spun by appropriate changes in the spinning set-up. Lyocell fibres tend to be stronger and stiffer than textile viscose but suffered from pronounced wet fibrillation in the first development stage of the textile Lyocell [48]. The data given is from the Lyocell filament yarn NewCell manufactured by Akzo Nobel NV, the Netherlands, which is out of production at present. Finally, acetate fibres are inferior in mechanical performance due to lack of chain orientation [16], which is not a critical point in the cigarette filter application. Textile applications cope with that as well and the silk-like hand of 'acetate rayon' is appreciated.

As for viscose fibres, these are mostly produced as staple fibres for textile and nonwoven applications. In 2011, world production was 3.246 million tons [49] while filament yarn for textile and technical applications reached 332 000 tons in 2011 [49] with a share of technical yarns of 56 000 tons. Technical viscose fibres, also called rayon or viscose rayon, are used mainly as carcass reinforcing fibres in fast-running and run-flat tyres. Lyocell fibres are produced only as staple and virtually exclusively by Lenzing AG, Austria, with a production capacity of 140 000 tons in 2011 [50].

Cellulose man-made fibres are typically:

- highly absorbent;
- soft and comfortable;
- easy to dye;
- they drape well.

Table 3.3 *Typical tensile properties of cellulose-based fibres.*

	Strength		Modulus			
Fibre type	[cN/tex]	[MPa]	[cN/tex]	[GPa]	Elongation [%]	Ref.
Viscose	21	308	704	11	24	[46]
Rayon	55	825	1300	19.5	13	[47]
Lyocell (NNMO)	40	603	2060	31	9	[46]
Acetate	13	195	340	5.1	20	[16]

These characteristics determine their major uses (http://www.fibersource.com/f-tutor/rayon.htm, accessed 21 June 2013):

- **Apparel:** accessories, blouses, dresses, jackets, lingerie, linings, millinery, slacks, sport shirts, sportswear, suits, ties, work clothes.
- **Home furnishings:** bedspreads, blankets, curtains, draperies, sheets, slipcovers, tablecloths, upholstery.
- **Industrial uses:** industrial products, medical surgical products, nonwoven products, tire cord.
- **Other uses:** feminine hygiene products.

Acetate fibres are used mainly for the production of cigarette filters with a global production volume of filter tow of 740 000 tonnes in 2009, which is an increase of 170 000 tonnes since the year 2000 [52]. Textile acetate yarns experienced a 30% decline to 42 000 tonnes in 2009 [52]. Main applications for the later are women's linings, woven fashion, woven satins, circular knits, men's linings, tricot knits, woven velvets, home furnishings, coffin linings, medical and ribbons [43].

3.5.2 Films

Cellulose films like Cellophane™ or NatureFlex™ (trade names of Innovia Films Ltd.) are classically manufactured via the viscose process. A wide variety of film types is commercially available, for example, from said Innovia Films Ltd with a thickness range from 19 to 45 μm, uncoated or coated with sealable layers and with typical mechanical properties of the basis films as shown in Table 3.4.

Stronger films are available, too, with machine direction tensile strengths of 180 MPa and key attributes for the uncoated (NatureFlex™) films are described (www.innoviafilms.com, accessed 22 June 2013) as:

- excellent transparency and gloss;
- highly permeable to moisture;
- excellent gas and aroma barrier;
- anti-static, easy tearing and anti-mist;
- both sides of the film are receptive to inks and adhesives;
- excellent dead fold characteristics;
- certified compostable in both industrial and home composting environments;
- suitable for anaerobic digestion;
- coloured versions are available.

Table 3.4 *Typical tensile properties of cellulose-based films in machine (MD) and transverse (TD) directions.*

	Strength (MPa)		Modulus (GPa)		Elongation (%)	
Film type	MD	TD	MD	TD	MD	TD
Cellophane[a]	125	75	4.7–5.4	3.7	25	75
NMMOa[a]	163	76	4.1	2.9	5–40	2–30
NMMOb[a]	114	82	4.7	4.4	5–40	2–30
Cast CDA[b]	90	sim.[c]	2.8	sim.[c]	20–40	sim.[c]

Notes: [a]From ref. [48]. [b]From ref. [40]. [c]Similar to MD value.

That makes the films suitable for markets and applications such as:

- Twistwrap;
- lamination;
- high-speed glued bags;
- tapes;
- premade bags.

The cellulose films can be coated to give additional property advantages in terms of sealability and as barriers. Even aluminium metallized types are available with extremely high barriers to moisture vapour, aromas, and gases still being compostable according to the industrial composting standards DIN EN 13 432, ASTM D 6400 and AS 4736, and Vinçotte OK Compost Home protocol. Innovia cellulose film production capacity is reported to be around 30 000 tonnes per year (http://www.sustainablebizness.com/PkgForum/InnoviaAnaPkgForum.pdf, accessed 22 June 2013). Markets addressed are:

- bakery;
- confectionery;
- soft cheese;
- snack food;
- cereals;
- condiments;
- medical;
- petri dishes;
- home and personal care;
- cosmetics.

New processing routes for underivatized cellulose have been addressed by developing a blown-film process using NMMO as the solvent (Figure 3.8) allowing for a broad variability of film properties in the draw-down (longitudinal, MD) and blow-up (transverse, TD) directions [48]. Tensile characteristics have been included in Table 3.4 for a more anisotropic film (NMMOa) and a film with balanced properties in MD and TD (NMMOb) demonstrating the capabilities of the blown-film process.

Cellulose diacetate cast films, also included in Table 3.4, are much less hydrophilic than uncoated cellulose but retain, due to their only partial substitution of hydroxyl with acetate groups, a certain moisture sensitivity, providing the films in many end uses with a 'breathability' to moisture vapour. Films have long been recognized as providing excellent transparency and very high gloss values [40]. But also matt films can be produced by casting on bands with roughened surfaces. The Clarifoil Company Ltd., United Kingdom, produces films in a thickness range between 14 and 500 μm addressing the following applications (www.clarifoil.com, accessed 22 June 2013):

- Print lamination in high gloss, matt and Satiné and Semitone,
- Invisible mending tape with easy tear characteristics,
- Window Cartons, especially for food packaging,
- Label stock for temper-evident and security labels,
- Optical films for polarized sunglasses, display screens,
- Scuff-resistant Semitone,
- UV-absorber film.

Finally, CTA films play an important role in optical applications due to their anti-curl property and their low in-plane birefringence even when the polymer is oriented in combination with high transmittancy (as PMMA), low haze (better than PMMA), and a refractive index of 1.48 [44]. Anti-curl is basically a bad shape memory when rolled off from the storage roll, a requirement for the colour negative film, the only left application of CTA film in the category of photography.

More important nowadays is the use of CTA as a protective film for polarizing plates particularly in LCD displays for TVs, personal computers, and mobile phones. The unique combination of properties mentioned above plus the sufficient permeability for vapourized water from the water-based adhesive used to join the polarizing and the protective film, still make CTA a choice material despite competition from other polymers [44].

Apart from the solution methods, CDA, CAP, and CAB films can be manufactured by conventional extrusion to furnish the markets mentioned above.

3.5.3 Moulded Articles

Classical thermoplastic processing methods such as injection moulding, blow moulding and extrusion, can only be performed with compounds having a distinct softening point followed by a region with flow properties required for each technique and sufficiently below the decomposition point of the polymer. This is the case for a series of plasticized cellulose esters, such as CA, CAP, and CAB, with appropriate degrees of ester substitution and plasticizer content. A set of typical properties of commercially available materials of this kind is given in Table 3.5 [52]. Obviously, the more plasticizer is used, the lower are stiffness, strength, heat distortion temperature (HDT), and Vicat softening temperature, while impact strength at room temperature is increased. This latter tendency is seen also at lower temperature (−40 °C) but is less pronounced. This and the appreciable loss in ductility might be due to the fact that the plasticizers are solidified at this temperature. Moreover, what is seen from the table is that the shortest substituent, acetate, produces the stiffest materials for a given plasticizer

Table 3.5 *Typical properties of moulding compounds from cellulose acetate (CA), acetate propionate (CAP) and acetate butyrate (CAB) [52].*

Property	CA 105–21	CA 105–27	CA 105–33	CAP 360–07	CAP 360–12	CAP 360–16	CAB 285–05	CAB 285–10	CAB 285–23
Plasticizer [%]	21	27	33	7	12	16	5	19	23
Density [g/cm^3]	1.29	1.28	1.27	1.21	1.2	1.19	1.2	1.19	1.16
Yield strength [MPa]	47	33	26	41	32	27	41	33	17
Breaking strength [MPa]	50	36	30	41	33	30	51	43	20
Elongation at break [5]	20	25	30	50	45	45	55	50	50
Flexural modulus [GPa]	2.7	2.1	1.7	1.8	1.4	1.2	1.7	1.4	0.8
Rockwell Hardness R	106	82	59	95	78	68	94	78	-
Notched Izod [J/m] at 20 °C	140	190	220	200	420	>530	160	240	450
Notched Izod [J/m] at −40 °C	43	48	53	85	107	123	85	96	123
HDT-A [°C]	84	73	63	82	75	72	83	74	52
HDT-B [°C]	92	83	74	92	83	80	93	85	68
Vicat temperature [°C]	122	109	100	107	96	92	115	104	88

concentration. For example, 21% plasticizer content in CA results in a stiffness of 2.7 GPa, while only 16% are sufficient for a stiffness level of 1.2 GPa in CAP.

Additional valuable properties of this family of cellulose moulding compounds include clarity, high gloss, colourability, a warm and pleasant feel, and suitability for direct and prolonged contact with human skin. Detailed information on chemical resistance is available [53]. The resistance against non-polar solvents such as hydrocarbons is good, but poor for more polar ones like ketones, esters, acids and alkalis in higher concentration. Applications include:

- tool handles;
- eyewear (frames and safety glasses);
- safety shields;
- appliance parts;
- pen barrels;
- displays and signage;
- toothbrush handles;
- costume jewellery;
- hair ornaments;
- combs;
- toys and sporting goods;
- casino chips;
- buttons and sequins;
- cosmetics and personal care.

3.6 Some Recent Developments

Marketwise, cellulose fibres including acetate have seen an increase of 8% from 2008 to 2009 [51]. This is due to a considerable rise in CDA filter tow production (+30% [51]) owing to the change in smoking habits towards filter cigarettes, especially in Asia. In other aspects, cellulose-based, in particular textile acetate, fibres have to cope with the competition from synthetic alternatives, such as polyester and polyamide fibres. The situation for films is likewise and mostly higher priced packaging, textile fashion, and special applications, such as protective films for polarizer plates, are served. Frames for spectacles continue to be an important application for moulding compounds and sheets. The increased awareness of environmental issues such as the limitation of fossil resources and the CO_2 footprint as well as the absence of competition with the food sector make cellulosiscs an attractive alternative for the future. Research is constantly continued both in the field of cellulose itself and with respect to cellulose esters. Some recent developments are discussed below without any claim to completeness and with consideration of own work.

3.6.1 Cellulose

Higher strength and modulus technical cellulose fibres have been spun from phosphoric acid by the Acordis research group of Boerstoel *et al.* [54] giving fibres with strengths of 1.7 GPa, moduli of 44 GPa and elongations at break of 5% [54]. This 'fibre B' never went into production though, probably due to the impractical solvent. Concentrated acids are, however, not completely uncommon in high-performance fibre spinning, as, for example, aromatic polyamides (Kevlar) are spun from anhydrous sulfuric acid.

A liquid crystalline system for fibre spinning of cellulose carbamate from NMMO solution is described in reference [55]. Cellulose is regenerated analogous to the viscose method (derivative process) and fibres are obtained with strengths of almost 1 GPa and moduli of 49 GPa. Compared to 'fibre B', a much easier process is used, which is much closer to industrially implemented dope preparation (similar to the viscose process) and spinning procedures (from the NMMO process).

The manufacture of cellulose blown films from NNMO solution [48] was mentioned in Section 3.5.2 and mechanical values of a more anisotropic film as well as a more balanced type are included in Table 3.4 demonstrating the versatility of the process.

Finally, recent developments in dissolving cellulose in ionic liquids should be mentioned along with the utilization of this kind of solvent for the derivatization of cellulose [56,57]. Even bacterial cellulose (high DP and high crystallinity) can be dissolved in suitable ionic liquids. Under mild conditions and within short reaction times at low temperatures (65 to 80 °C) and low excess of reagent, various cellulose esters and carbanilates, dendronized cellulose and trimethylsilyl cellulose can be obtained. Still, ionic liquids are expensive, industrial processes for cellulose shaping and derivatizing are not in place and NMMO is preferred as a solvent. However, with economy-of-scale effects on the part of ionic liquids and with new possibilities as to cellulose functionalization, new processes using this kind of solvent seem to have a bright future.

3.6.2 Cellulose Acetate and Mixed Esters

In the course of the recent revival of bio-based polymers [58], cellulose esters have gained attention as matrix materials both for macroscopic and nano composites. In a series of papers, Seavey, Glasser, *et al.* have investigated continuous cellulose fibre reinforced cellulose ester composites of which the last is dealing with commercial matrix and fibre options [59]. A sort of hand lay-up with acetone solutions was used as the manufacturing method. For various commercial CABs and Lyocell fibres moduli between 15 GPa and 21 GPa were obtained in unidirectional (UD) composites while for cross-ply (CP) architectures, the values were between 10 and 15 GPa. Strengths go up to 310 MPa for UD composites and to 210 MPa for CP materials.

Another topic attracting much attention in recent years is the use of nanoscaled particles for modifying polymer properties, including mechanical properties. Due to the extremely large surface and thus interaction area with the matrix polymer, small amounts of the nano fillers in the range of a few per cent are sufficient to cause appreciable changes in polymer properties. Especially nanoscaled layered silicates, such as organo-modified montmorillonite, were tested in conjunction with cellulose acetate injection moulding compounds. In a key paper, Park *et al.* [60] use CDA with a DS of 2.45 together with triethyl citrate as the plasticizer and an organo-clay (Cloisite 30B, Southern Clay Co.) modified with methyl tallow bis-2-hydrohyethyl ammonium. Adding 5% clay to a composition with 20% plasticizer, tensile strength could be improved from 70 MPa to 105 MPa, modulus from 4.1 to 4.8 GPa, while elongation at break did not change much (8.5% *vs.* 7.2%). A certain embrittlement is noticed in notched Izod impact strength the average value decreasing from 39 J/m to 26 J/m. However, the standard deviations are so high that the significance of this finding is doubtful. Nevertheless, nano-clay reinforcement has been proven to be a viable method for modifying CA properties.

In another paper [61], CDA was compounded with organo-modified and unmodified nanoclay both with and without glycerol triacetate (triacetin) as a plasticizer. The most striking results were obtained without plasticizer and with unmodified clay, Na^+ being the interlayer cations. Tensile strengths as high as 178 MPa and moduli of 8.4 GPa were found in injection moulded test bars for 5% clay loading. The bars showed a core-shell structure with highly oriented molecules in the shell part, obviously generated by the shear forces in the injection moulding process. Structure investigation revealed that the clay was not exfoliated and another reason had to be found to explain the behaviour. More detailed investigations [62] showed that the presence of the clay reduced the thermal degradation of CA enabling the longer chains to be oriented in the injection process. Moreover, it was shown that the sodium ions play the decisive role probably in neutralizing residual sulfate groups from the sulfuric acid catalyst. This novel type of bio-nano-composite might be suitable for replacing oil-based high-performance plastics in applications where high strength and stiffness is essential.

3.7 Conclusion

Cellulose is one of the essential bio-based polymers available in large quantities as a renewable feedstock for a variety of traditional applications and, as research and development progresses in this field, future applications. The synthesis work done by nature to produce this polymer chain with its characteristic features can be used, after isolation from wood, plants, and so on, either directly or taken as a basis for chemical modification. In doing so, CO_2 captured from the atmosphere during plant growth is stored in the final product. Moreover, as to cellulose esters, cellulose acetate as the most prominent example can in principle be synthesized completely from biogenic resources using acetic acid from fermentation processes. This is not done at present for economic reasons, but the financial environment may change in the future. The actual CO_2 footprint of cellulose products is strongly determined by isolation and processing conditions which are improved continuously in this respect. In comparison, for poly(lactic acid) made from maize or sugar cane, or starch itself, there is an apparent competition with the food chain. This is not the case for cellulose which is used as a structural material in nature and by men.

The extended intra- and intermolecular hydrogen bond system of cellulose rendering it non-meltable can be broken up by esterification up to a point where thermoplastic processing is possible, albeit with the use of plasticizers. So far, mainly acetate, propionate, butyrate, and mixed esters therefrom, have gained industrial importance. The use of plasticizers is not a necessity *per se* and attempts have been made to use internal plastification with longer ester chains. Certainly, the semirigid nature of the ether-linked anhydroglucose chain of cellulose will impose a certain rigidity on any derivative intended for thermoplastic processing. However, a suitable choice of the substitution pattern and the appropriate processing conditions will allow for a variety of thermoplastic shaping processes. Currently, many possible applications are precluded for economic reasons and high priced products prevail. This might change in the future when the vision of a bio-based economy starts to assume shape.

References

1. Klemm, D., Heublein, B., Fink, H.-P. and Bohn, A. (2005) Cellulose: fascinating biopolymer and sustainable raw material. *Angewandte Chemie-International Edition*, **44**, 2–37.

2. Saka, S. (2004) Cotton fibers as natural raw materials for cellulose acetate production. *Macromolecular Symposia*, **208**, 29–35.
3. Patt, R., Kordsachia, O. and Sixta, H. (1997) Eignung von Rohstoffen und Verfahren zur Herstellung von Chemiezellstoffen. *Das Papier*, **51**, 637–643.
4. Hessler, L.E., Merola, G.V. and Berkley, E.E. (1948) Degree of polymerization of cellulose in cotton fibers. *Textile Research Journal*, **18**, 628–634.
5. Kraemer, E.O. (1938) Molecular weights of celluloses and cellulose derivatives. *Industrial and Engineering Chemistry*, **30**, 1200–1203.
6. Evans, R., Wearne, R.H. and Wallis, A.F.A. (1989) Molecular weight distribution of cellulose as its tricarbanilate by high performance size exclusion chromatography. *Journal of Applied Polymer Science*, **37**, 3291–3303.
7. Siholta, H., Kyrklund, B., Laamanen, L. and Palenius, I. (1963) Comparison and conversion of viscosity and DP-values determined by different methods. *Paperi ja Puu*, **4a**, 225–232.
8. Sixta, H. (2000) Comparative evaluation of TCF bleached Harwood dissolving pulps. *Lenzinger Berichte*, **79**, 119–128.
9. Breier, R. (1997) Die Strangveredelung von Lyocell – Aktueller Stand der Praxis. *Lenzinger Berichte*, **76**, 108–111.
10. Brown, R.M. Jr., Saxena, I.M. and Kudlicka, K. (1996) Cellulose biosynthesis in higher plants. *Trends in Plant Science*, **1**, 149–156.
11. Saxena, I.M. and Brown, M. Jr. (2005) Cellulose biosynthesis: current views and evolving concepts. *Annals of Botany*, **96**, 9–21.
12. Perez, S. and Samain, D. (2010) Structure and engineering of cellulose. *Advances in Carbohydrate Chemistry and Biochemistry*, **64**, 25–115.
13. Atalla, R.H. and VanderHart, D.L. (1984) Native cellulose: a composite of two distinct crystalline forms. *Science*, **233**, 283–285.
14. Nishiyama, Y., Sugiyama, J., Chanzy, H. and Langan, P. (2003) Crystal structure and hydrogen bonding system in cellulose I_α from synchrotron X-ray and neutron fiber diffraction. *Journal of the American Chemical Society*, **125**, 14300–14306.
15. Ganster, J. and Fink, H.-P. (1999) Physical constants of cellulose, in *Polymer Handbook*, 4th edn (eds E.H. Immergut and E.A. Grulke), John Wiley & Sons, Inc, New York, pp. 135–157.
16. Ganster, J. and Fink, H.-P. (2009) The structure of man-made cellulosisc fibres, in *Handbook of Textile Fibre Structure, Volume 2: Natural, Regenerated, Inorganic, and Specialist Fibres* (eds S.J. Eichhorn, J.W.S. Hearle, M. Jaffe and T. Kikutani), CRC Press, Cambridge and Woodhead Publishers, Boca Raton, FL, pp. 379–411, ISBN 9781439820728.
17. Langan, P., Nishiyama, Y. and Chanzy, H. (2001) X-ray structure of mercerized cellulose II at 1 Å resolution. *Biomacromolecules*, **2**, 410–416.
18. Zugenmaier, P. (2001) Conformation and packing of various crystalline cellulose fibers. *Progress in Polymer Science*, **26**, 1341–1417.
19. Okano, T. and Sarko, A. (1985) Mercerization of cellulose. II. Alkali-cellulose intermediates and a possible mercerization mechanism. *Journal of Applied Polymer Science*, **30**, 325–332.
20. Hayashi, J., Sufoka, A., Ohkita, J. and Watanabe, S. (1975) The confirmation of existences of cellulose III_I, III_{II}, IV_I, and IV_{II} by the X-ray method. *Journal of Polymer Science: Polymer Letters Edition*, **13**, 23–27.
21. Edgar, K.J., Buchanan, Ch.M., Debenham, J.S. *et al.* (2001) Advances in cellulose ester performance and application. *Progress in Polymer Science*, **26**, 1605–1688.
22. Basler, K., Hoppe, L., Eicher, Th. *et al.* (2005) Cellulose esters, in *Ullmann's Encyclopedia of Industrial Chemistry*, Wiley-VCH. doi 10.1002/14356007.a05_419.pub2
23. Wibowo, A.C., Misra, M., Park, H.-M. *et al.* (2006) Biodegradable nanocomposites from cellulose acetate: mechanical, morphological, and thermal properties. *Composites A*, **37**, 1428–1433.
24. Sand, I.D. (1990) The dependence of properties of cellulose acetate propionate on molecular weight and the level of plasticizer. *Journal of Applied Polymer Science*, **40**, 943–952.

25. Buchanan, C.M., Gedon, S.C., White, A.W. and Wood, M.D. (1992) Cellulose acetate butyrate and poly(hydroxybutyrate-*co*-valerate) copolymer blends. *Macromolecules*, **25**, 7373–7381.
26. Roche, E., Chanzy, H., Boudeulle, M. *et al.* (1978) Three dimensional crystalline structure of cellulose triacetate II. *Macromolecules*, **11**, 86–94.
27. Zugenmaier, P. (2008) *Crystalline Cellulose and Cellulose Derivatives – Characterization and Structures*, Springer, Berlin.
28. Iwata, T., Fukushima, A., Okamura, K. and Azuma, J.-I. (1997) DSC study on regioselectively substituted cellulose heteroesters. *Journal of Applied Polymer Science*, **65**, 1511–1515.
29. Döpfner, H., Ernegg, M. and Bramsteidl, R. (1996) Process for pruducing work pieces and molded pieces out of cellulose and/or cellulose-containing fiber material. WO 98/11973.
30. Dean, T., Tarvedi, K., Bramsteidl, R. and Achilli, L. (2009) Method and apparatus for defibrillating cellulose fibers. WO 2010/149711.
31. Liebert, T. and Heinze, Th. (2008) Interaction of ionic liquids with polysaccharids 5. Solvents and reaction media for the modification of cellulose. *BioResources* **3**, 576–601.
32. Philipp, B. and Fink, H.-P. (1999) Recent progress in understanding cellulose mercerization. *Polymer News*, **24**, 122–127.
33. Weigel, P. and Fink, H.-P. (1997) Verfahren zur Herstellung von Celloloseblasfolien. *Lenzinger Berichte*, **76**, 115–118.
34. Fink, H.-P., Weigel, P. and Bohn, A. (1997) Struktur-Eigenschaftsbeziehungen von Blasfolien aus Cellulose. *Lenzinger Berichte*, **76**, 119–125.
35. Ebeling, H., Fink, H.-P., Luo, M. and Geus, H.-G. (2006) Cellulose meltblown nonwovens using the lyocell process. *Lenzinger Berichte*, **86**, 124–131.
36. El Seoud, O.A. and Heinze, T. (2005) Organic esters of cellulose: new perspectives for old polymers. *Progress in Polymer Science*, **186**, 103–149.
37. Hummel, A. (2004) Industrial processes, in *Cellulose Acetates: Properties and Applications* (ed. P. Rustemeyer),Wiley-VCH, Weinheim, pp. 61–79.
38. Steinmeier, H. (2004) Acetate manufacturing, process and technology – chemistry of cellulose acetylation, in *Cellulose Acetates: Properties and Applications* (ed. P. Rustemeyer), Wiley-VCH, Weinheim, pp. 49–60.
39. Fordyce, C.R. and Meyer, L.W.A. (1940) Plasticizers for cellulose acetate and cellulose acetate butyrate. *Industrial and Engineering Chemistry*, **32**, 1053–1060.
40. Law, P.W., Longdon, A. and Willins, G.G. (2004) Solvent cast cellulose diacetate film, in *Cellulose Acetates: Properties and Applications* (ed. P. Rustemeyer), Wiley-VCH, Weinheim, pp. 293–322.
41. Malm, C.J., Fordyce, Ch.R. and Tanner, H.A. (1942) Properties of cellulose esters of acetic, propionic, and butyric acids. *Industrial and Engineering Chemistry*, **34**, 430–435.
42. Rustemeyer, P. (2004) CA filter tow for cigarette filters, in *Cellulose Acetates: Properties and Applications* (ed. P. Rustemeyer), Wiley-VCH, Weinheim, pp. 267–291.
43. Law, P.W. (2004) Applications of cellulose acetate – cellulose acetate in textile application, in *Cellulose Acetates: Properties and Applications* (ed. P. Rustemeyer), Wiley-VCH, Weinheim, pp. 255–265.
44. Sata, H., Murayama, M. and Shimamoto, Sh. (2004) Properties and applications of cellulose triacetate film, in *Cellulose Acetates: Properties and Applications* (ed. P. Rustemeyer), Wiley-VCH, Weinheim, pp. 323–333.
45. Carollo, P. and Grospierto, B. (2004) Plastic materials, in *Cellulose Acetates: Properties and Applications* (ed. P. Rustemeyer), Wiley-VCH, Weinheim, pp. 335–351.
46. Ganster, J. and Fink, H.-P. (2006) Novel cellulose fibre reinforced thermoplastic materials. *Cellulose*, **13**, 271–280.
47. Ganster, G., Fink, H.-P., Uihlein, K. and Zimmerer, B. (2008) Cellulose man-made fibre reinforced polypropylenen – correlations between fibre and composite properties. *Cellulose*, **15**, 561–569.
48. Fink, H.-P., Weigel, P., Purz, H.J. and Ganster, J. (2001) Structure formation of regenerated cellulose materials from NMMO-solutions. *Progress in Polymer Science*, **26**, 1473–1524.
49. Koslowski, H. J. (publisher) and Van Bonn, C. (editor-in-chief) (2011) *Man-Made Fiber Year Book 2011*, Chemical Fibers International, Frankfurt am Main, p. 8.

50. Koslowski, H.-J. (publisher) and Van Bonn, C. (editor-in-chief) (2011) *Man-Made Fiber Year Book 2011*, Chemical Fibers International, Frankfurt am Main, p. 11.
51. Synthetic Rayon and Textile Export Promotion Council (India) (2011) Global fibre production. *SRTEPC Newsletter*, **26** (2), 10.
52. Eastman Chemical Company (2005) Publication PPC-101, p. 13.
53. Eastman Chemical Company (1996) Tenite Acetate – Chemical resistance. Publication PP-101B.
54. Northolt, M.G., Boerstoel, H., Maatman, H. *et al.* (2001) The structure and properties of cellulose fibres spun from an anisotropic phosphoric acid solution. *Polymer*, **42**, 8249–8264.
55. Fink, H.-P., Ebeling, H. and Rihm, R. (2006) Fibre Formation from Liquid Crystalline Solutions of Cellulose Carbamate in NMMO. Proceedings of the 7th Int. Symp. 'Alternative Cellulose-Manufacturing, Forming, Properties', September 6–7, 2006, Rudolstadt, pp. 13–23.
56. Feng, L. and Chen, Zh.-I. (2008) Recent progress on dissolution and functional modification of cellulose in ionic liquids. *Journal of Molecular Liquids*, **142**, 1–5.
57. Heinze, T., Dorn, S., Schöbitz, M. *et al.* (2008) Interactions of Ionic liquids with polysaccharides – 2: cellulose. *Macromolecular Symposia*, **262**, 8–22.
58. Endres, H.-J. and Siebert-Rahts, A. (2011) *Engineering Biopolymers – Markets, Manufacturing, Properties, and Applications*, Hanster, Munich.
59. Franko, A., Seavey, K.C., Gumaer, J. and Glasser, W.G. (2001) Continuous cellulose fiber-reinforced cellulose ester composites III. Commercial matrix and fiber options. *Cellulose*, **8**, 171–179.
60. Park, H.-M., Misra, M., Drzal, L.T. and Mohanty, A.K. (2004) 'Green' nanocomposites from cellulose acetate bioplastic and clay: effect of eco-friendly triethyl citrate plasticizer. *Biomacromolecules*, **5**, 2281–2288.
61. Hassan Nejad, M., Ganster, J., Bohn, A. *et al.* (2009) Bio-based nanocomposites of cellulose acetate and nano-clay with superior mechanical properties. *Macromolecular Symposia*, **280**, 123–129.
62. Hassan Nejad, M. (2011) Improving the Mechanical Properties of Polysaccharide Derivatives though Melt Compounding with Nano-Clays. PhD thesis, Technische Universität Berlin. http://opus.kobv.de/tuberlin/volltexte/2011/3042/pdf/hassannejad_mehdi.pdf (accessed 22 June 2013).

4

Materials Based on Chitin and Chitosan

Marguerite Rinaudo
Biomaterials Applications, 6 rue Lesdiguires, France

4.1 Introduction

Chitin (poly-β-(1 → 4)-N-acetyl-D-glucosamine) is a natural, renewable polysaccharide of major importance widely synthesized in a number of living organisms (Figure 4.1). Chitin occurs in nature as semicrystalline microfibrils found as structural components in the exoskeleton of arthropods or in the cell wall of fungi and yeasts. The main commercial source of chitin comes from crab and shrimp shells. Chitin is infusible and difficult to solubilize for processing into different conformations (film, bead, sponge, etc.). The question of its solubility is a major problem in view of the development of processing and uses of chitin as well as for its characterization. Partial deacetylation, under strong alkaline conditions, allows the applications of chitin to be extended by producing chitosan. Chitosan is a random copolymer with a molar fraction DA (degree of acetylation) of β-(1 → 4)-N-acetyl-D-glucosamine and a fraction (1-DA) of β-(1 → 4)-D-glucosamine (Figure 4.1). It is the most important chitin derivative in terms of applications [1,2].

In this chapter, the main characteristics of chitin and chitosan and the more convenient techniques used for their characterization will be presented together with their main physical properties. Furthermore, the materials obtained with these polysaccharides will be described. It is important to recall that chitin is a natural polymer that is also biocompatible and biodegradable, an important advantage for biomedical and pharmaceutical applications. Good film forming properties is valuable for packaging or other applications in the domain of materials.

Bio-Based Plastics: Materials and Applications, First Edition. Edited by Stephan Kabasci.

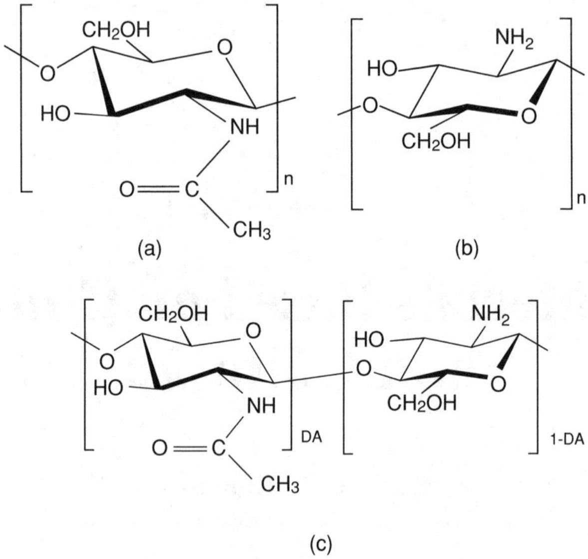

Figure 4.1 Repeat unit of: (a) chitin (poly(N-acetyl-β-D-glucosamine)), and (b) chitosan (poly(D-glucosamine). (c) Structure of partially acetylated chitosan characterized by an average degree of acetylation DA.

4.2 Preparation and Characterization of Chitin and Chitosan

4.2.1 Chitin: Characteristics and Characterization

In industrial processes, chitin is extracted from crustaceans by acid treatment to dissolve calcium carbonate followed by alkaline extraction to solubilize proteins. In addition, a decolourization step is often applied to remove the residual pigments and to obtain a colourless product. These treatments need to be adapted to each chitin source due to differences in the ultrastructure of the initial materials.

In the solid state, chitin mainly occurs as two allomorphs, the α- and β-forms, which are differentiated by infrared and solid state NMR spectroscopies and X-ray diffraction [3, 4]. α-chitin is the most abundant, occurring in krill, lobsters and crab tendons, shrimp and crab shells. It also results from recrystallization of β-chitin in solution treated with strong aqueous HCl (over 7N) and washed with water because the α-form is thermodynamically more stable than β-chitin [5]. β-chitin is found in association with proteins in squid pens [3, 6] or in the tubes synthesized by pogonophoran and vestimetiferan worms [7, 8]. In the solid state, the chains are parallel in β-chitin and antiparallel in α-chitin. Their crystalline structures were reviewed in different papers [1, 4, 9–13]. Chitin is insoluble in all the usual solvents while taking into account their crystalline forms (α or β) because the reactivity of β-chitin is larger than that of the α-isomorph, which is important for enzymatic and chemical transformations of chitin [14].

Only few data are available on the physical properties of chitin in solution due to the difficulties in dissolving this polymer. Solubilization would be necessary to characterize molar mass but also for processing because chitin cannot be processed in molten state. The cohesion of this polymer is too large due to stabilization by a dense hydrogen bond network and thermal decomposition instead of melting occurs upon heating.

One of the most popular systems to solubilize chitin is the following: a complex is formed between chitin and LiCl (at 5 wt% in the solvent used), in which Li is coordinated with the carbonyl group of acetylated units, and this complex becomes soluble in dimethylacetamide (DMAc) and in N-methyl-2-pyrrolidone (NMP) [15–18]. Austin also used formic acid, dichloroacetic and trichloroacetic acids and interpreted the data in terms of solubility parameters [15]. Other solvents were tested: saturated solution of lithium thiocyanate [19,20], phosphoric acid [21], and alkali [22–25]. Solubilization of chitin in aqueous NaOH (16 wt%, [23]) is obtained at temperatures lower than 0 °C. That method allows the preparation of films regenerated in water at 25 °C or in methanol. In these conditions, depolymerization, deacetylation and destructuration of the crystalline material are observed depending on the time of contact between chitin and alkali [23,24, 26]. A review on chitin and chitosan including the question of solubility was recently published in relation to fibre processing [27]. Other solvents such as ($CaCl_2$ 2 H_2O)-saturated MeOH at room temperature were suggested [28,29], and chitin hydrogels are regenerated by dilution in water. Hexafluoroisopropyl alcohol [30] or hexafluoracetone sesquihydrate [31] were also proposed as solvents. The advantage of these solvents is that one can produce fibres by wet and dry spinning [31]. An aqueous mixture of 8 wt% NaOH and 4 wt% urea at −20 °C was shown to be a good solvent in which the chitin solution is stable. Molecular weight and DA are only slightly reduced, but the solution turns into a gel when temperature increases [32]. More recently, ionic liquids, especially 1-butyl-3-methylimidazolinium acetate (BminAc), at room temperature, are proposed as new good solvents for native chitin, which is regenerated in water or methanol [33,34].

As soon as the average fraction of D-glucosamine is high enough (in the range of 0.5), this partially deacetylated chitin (degree of deactylation (DD) ≥ 0.5) is named chitosan and characterized by a DA ≤ 0.5. Chitosan becomes water soluble in acidic conditions due to $-NH_2$ protonation. Under this state, it is possible to obtain a complete characterization of the polymer but it may be considerably different from the starting material especially in terms of molar mass reduction during deacetylation. In these conditions, the existence of free $-NH_2$ groups available on C-2 position of D-glucosamine units along the chains allows to perform specific reactions in homogeneous conditions.

The most convenient technique to determine the DD of native chitin or the DA of chitosan is the solid state ^{13}C or ^{15}N nuclear magnetic resonance (NMR) spectroscopy as discussed recently [35,36]. Figure 4.2 shows the ^{13}C NMR spectra obtained, in solid state, on chitin and chitosan with different DAs.

There is a good agreement between the values of DA obtained in the different experiments demonstrating the advantage of solid state NMR well adapted to all the range of acetylation degrees (see Table 4.1). Infrared spectroscopy was also often used but it is more delicate to interpret due to the difficulty in adopting a convenient base line. This was also discussed previously for samples with different DAs [37].

The second important characteristic of chitin is the molecular weight but its measurement needs polymer solubilization and absence of aggregates in the solution. Few data concern the determination of intrinsic viscosity in different solvents and the relation with molecular weight obtained by light scattering eventually associated with a SEC equipment [38]. These characteristics are given in Table 4.2 in which the parameters K and a used in the Mark–Houwink relationship ($[\eta] = KM^a$) are given as well as the dn/dc useful to analyse light scattering results.

The agreement between K and a parameters does not look very good but this may be related to the quality of the solution. The origin of chitin as well as the conditions of extraction and

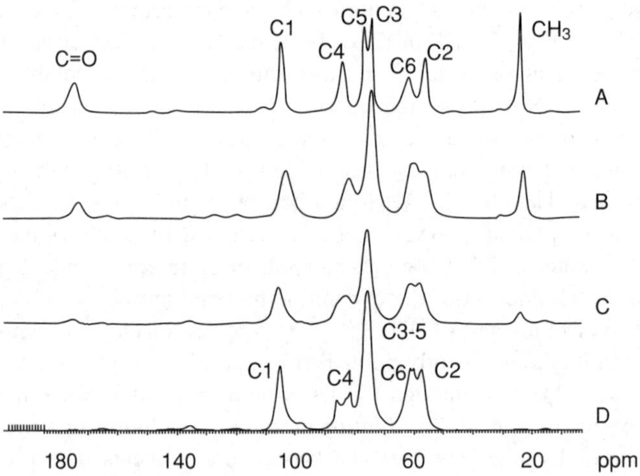

Figure 4.2 ^{13}C *CP-MAS NMR spectra of chitin and chitosan with different degrees of acetylation. A = Chitin; B = chitosan obtained by homogeneous reacetylation to DA~0.6; C = commercial chitosan from Pronova with DA~0.2; D = deacetylated chitosan. (Reproduced with permission from [36]. Copyright © 2000, American Chemical Society.)*

purification are important to solubilize it (in absence of aggregates) and to obtain valuable data in light scattering and viscosity.

4.2.2 Chitosan: Preparation and Characterization

Chitosan results from the deacetylation of chitin and it becomes soluble in aqueous acidic medium when the average DA is lower than 0.5. In fact, this limit depends on the distribution of acetyl groups along the chains. Chitosan fully deacetylated may be reacetylated in homogeneous phase. In these conditions, solubility is observed up to DA~0.6 due to a random distribution of the acetyl groups.

Table 4.1 *Degrees of acetylation of chitin and chitosan obtained by liquid state (^1H) and solid state (^{13}C and ^{15}N) NMR on 4 samples having different degrees of acetylation (same samples as in Figure 4.2). (Reproduced with permission from [36]. Copyright © 2000, American Chemical Society.)*

Samples	A	B	C	D
DA from ^1H NMR (liquid state)	insoluble	0.58	0.21	acetyl traces
DA from ^{13}C NMR (solid state)	0.99	0.61	0.20	0
DA from ^{15}N NMR (solid state)	1	0.63	0.20	0

Table 4.2 *Mark–Houwink parameters for chitin characterization.*

Solvent	K (mL/g)	a	T (°C)	dn/dc	Ref.
2.77 M NaOH	0.1	0.68	20	0.145	[22]
DMAc/LiCl 5%	7.6 × 10−3	0.95	30	0.091	[38]
DMAc/LiCl 5%	2.4 × 10−1	0.69	25	0.1	[39]

Conditions for deacetylation are numerous. It is needed to prepare an alkaline solvent, usually concentrated KOH or NaOH in aqueous solution eventually in the presence of a co-solvent, and to adopt a high temperature, usually from 80 to 140 °C, for a long time, from 2 to 6 hours, in one step or two successive 2-hour reactions [40, 41]. The regeneration of chitosan or derivatives from their solutions takes place in water at pH = 7.5–8 where chitosan is usually insoluble, and allows films, sponges, powders and fibres to be obtained. Chitin, being a semi crystalline polymer, usually insoluble in the medium used for deacetylation, reacts in a heterogeneous way. Therefore, the distribution of the $-NH_2$ groups along the chain may be irregular and give some blockwise arrangement which will control the solubility in aqueous medium even when DA < 0.5 with the risk to favour chain aggregation [42]. Solubilization of chitosan occurs by protonation of the $-NH_2$ function on the C-2 position of the D-glucosamine repeat units (Figure 4.1). This step is necessary for characterization but also to process chitosan for applications.

The first step is to characterize chitosans by their molecular weights, DA and eventually by the distribution of acetyl group along the chain. To characterize chitosan, different solvents based on acetic acid are proposed. It is suggested that 0.3 M acetic acid be adopted with 0.1 or 0.2 M sodium acetate aqueous solution. The presence of an external salt is needed to screen the long-range electrostatic repulsions between the charged chains. Chitosan is also soluble in acetic acid or hydrochloric acid at pH lower than 6 (its intrinsic pK being around 6.5) [1].

The determination of chitosan average DA may be performed by different techniques from which infrared spectroscopy, elementary analysis, and potentiometric titration are mentioned in literature, but ^{1}H liquid state [36, 43] and solid state ^{13}C NMR [36, 44, 45] have to be preferred. Presently, one considers that ^{1}H NMR is the most convenient technique for soluble samples. An example is given in Figure 4.3.

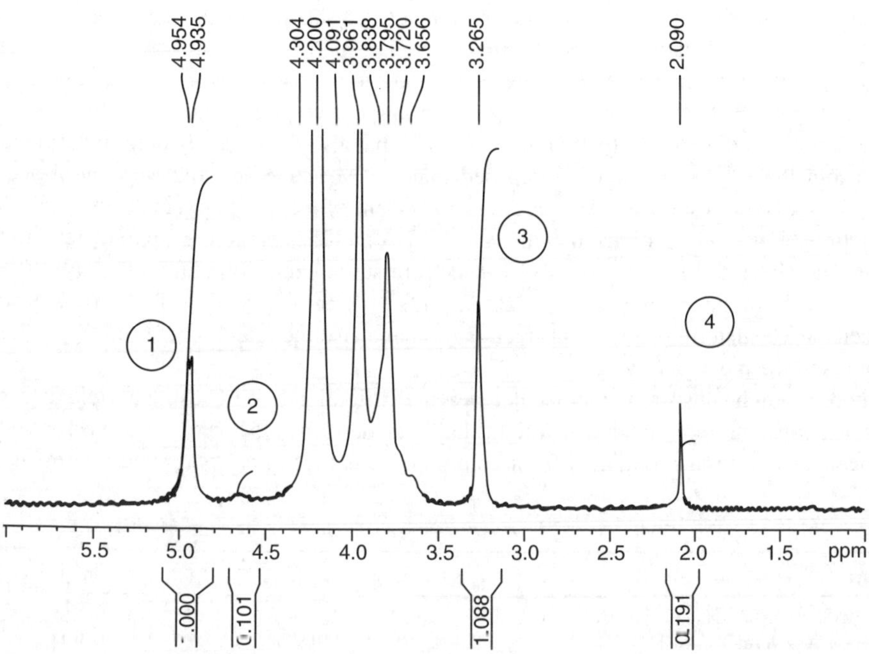

Figure 4.3 *^{1}H NMR spectrum of chitosan in D_2O, pD~4, temperature 85 °C, concentration 5 g/L.*

Figure 4.3 gives the ^{1}H spectrum obtained for chitosan with a low degree of acetylation (DA~0.06). The integral of the signal at 2.09 ppm ④ allows the determination of the acetyl content by reference to the integrals of H-1 signals at 4.94 ppm for the D-glucosamine residue ① and at 4.65 ppm for the H-1 of the N-acetyl-D-glucosamine unit ② at 85 °C; the signal at 3.26 ppm ③ is attributed to H-2. The ^{13}C and ^{15}N solid state NMR were also tested and discussed recently (see Figure 4.2). These techniques were used in the whole range of acetyl content from 0 to 100%. Results are given in Table 4.1: A is an α-chitin, B is a homogeneous reacetylated chitosan and C, D are commercial samples [36]. ^{15}N NMR is especially interesting as it gives only two signals related to the amino group and to the N-acetylated group respectively. It was concluded that a good agreement exists in all the range of DA, whatever the state of the sample and the NMR technique used. As mentioned previously, another important characteristic of chitosan is the distribution (random or blockwise) of acetyl groups along the chain because it may influence the solubility of the polymer and the interchain interactions due to hydrogen bonds and hydrophobic character of the acetyl group. This distribution was evaluated from ^{13}C NMR measurements [46,47]. Diad and triad frequencies were determined for homogeneous and heterogeneous chitosan with different degrees of DA.

The second important characteristic of these polymers is the molecular weight distribution and the average molecular weight as well as the intrinsic viscosity. The first difficulty encountered in the measurement of these values concerns the solubility of the samples and the dissociation of aggregates often present in polysaccharide solutions [48,49].

The question of choice of the solvent for chitosan characterization was examined. Different systems were proposed including an acid at a given concentration for protonation and a salt to screen the electrostatic interactions. The solvent is important also when molecular weight (or the viscometric-average molecular weight with polydispersed samples) has to be calculated from intrinsic viscosity using the Mark-Houwink relationship $[\eta]$ (mL/g) = $K\,M^{a}$. For the determination of the *K* and *a* parameters, an absolute molecular weight detection system is needed. Usually, light scattering is used, which is very sensitive to aggregates, resulting in an overestimated MW value. Some data for the Mark–Houwink parameters are given in Table 4.3. It was demonstrated that the aggregates perturb not only the molecular weight determination, especially by light scattering, but also the viscosity determination [50].

To avoid these artefacts, it was proposed that 0.3 M acetic acid/0.2 M sodium acetate (pH = 4.5) be used as solvent since there was no evidence for aggregation in this mixture [50]. At that time, it was demonstrated that the value of DA modifies the *K* and *a* parameters. In this solvent, the absolute M values were obtained from steric exclusion chromatography (SEC) equipped with online viscometer and light-scattering detector, allowing the Mark–Houwink parameters to be determined without fractionation and also the relation between the radius of gyration and the molecular weight

A more recent analysis of the molecular weight distribution by SEC using a triple detection (viscosity, concentration, molecular weight) has been described [51–53]. It was performed on heterogeneous chitosans, obtained from commercial sources after solid state treatment and

Table 4.3 *Mark–Houwink parameters for chitosan in different conditions.*

Solvent	K (mL/g)	a	T (°C)	dn/dc	Ref.
0.15M AcOH/0.1M AcONa 0.05<DA<0.23	—	—	25	0.192	[52]
0.3M AcOH/0.2M AcONa 0<DA<0.03	7.9×10^{-2}	0.796	25	0.190	[53]
0.02M acetate buffer/0.1M NaCl	8.43×10^{-2}	0.92	25	0.203	[55]

on homogeneous chitosans with different molecular weights obtained by reacetylation of a highly deacetylated chitosan by G. Roberts [53,54]. The DA of these acid-soluble chitosans varied from 0.02 up to 0.61. The data confirm that the Mark–Houwink parameters depend only slightly on the degree of acetylation. This point will be discussed later in relation with the persistence length. It was also shown that the DA has a negligible influence on dn/dc in the 0.3 M acetic acid/0.2 M sodium acetate solvent [53], and a value of dn/dc = 0.190 was obtained. The average values of the Mark–Houwink parameters and dn/dc, given for chitosans characterized by DA over a large range covered, are valid for heterogeneous as well as homogeneous samples. Recent studies propose to use 0.15 M acetic acid/0.1 M Na-acetate as solvent and an average dn/dc = 0.192 [51,52].

The relatively high values obtained for the parameter a are in agreement with the semirigid character of this polysaccharide. To validate this conclusion, chitin and chitosan molecular modelling [56,57] were investigated and the data predicted were compared with experimental results obtained by SEC. The importance of the stiffness of polymers is that their dimensions and their related hydrodynamic volume and viscosimetric contribution depend directly on the semirigid character of the chains reflected by the persistence length. This property is directly related to the performance of chitosan and chitin used in solution. Chitosan in acid medium being a polyelectrolyte, the actual total persistence length Lt at a given ionic concentration contains the intrinsic contribution Lp and the electrostatic contribution Le, calculated following the Odijk treatment. The electrostatic contribution is null in external salt excess [58]. The conformational analysis of chitin with different degrees of deacetylation confirms that chitin and chitosan are semirigid polymers characterized by a persistence length that depends moderately on the degree of acetylation of the molecule. From this analysis, chitosan, free of acetyl groups, has an intrinsic persistence length Lp of 9 nm when the electrostatic repulsions are screened. Lp increases when DA increases up to 12.5 nm for DA = 0.6, above that it remains constant up to pure chitin at 25 °C.

The local stiffness is related to the conformation of the molecule and especially to the intrachain hydrogen bond network stabilizing the chain [1]. Concerning chitosan with different DA, experimental results from steric exclusion chromatography using three detectors on-line give Lp~11nm, Lp being nearly constant for $0 < \text{DA} < 25\%$. Up to 60% acetylation, the stiffness of chitosan is not much influenced by the degree of acetylation and Lp increases to 15 nm in agreement with modelling prediction [53, 56]. The small variation of the persistence length with DA is in direct relation with the small variation observed on the Mark–Houwink parameters. The stiffness of the chain plays a large role in the rheological behaviour of the molecule but also, even in dilute solution, in the existence of interchain hydrogen bonds forming multimers that perturb all the characterization of these polysaccharides. The aggregation has been discussed recently and the causes analysed. It seems that hydrogen bonds as well as hydrophobic attractions play a role, whatever the degree of acetylation [48,49].

4.3 Processing of Chitin to Materials and Applications

4.3.1 Processing of Chitin and Physical Properties of Materials

Chitin, like other polysaccharides derived from cellulose, has good film-forming properties. It is a semicrystalline polymer and the morphology of the materials prepared depends on the conditions of processing. The stability of chitin-based materials is promoted by the establishment of a hydrogen bond network between extended chains. Chitin adds original properties to the new materials as being biocompatible, biodegradable and nontoxic, with antimicrobial activity and low immunogenicity. It is also insoluble in water with some hydrophilic character and moisture controlling. Chitin can be processed in the form of films, beads, sponges and fibres with good physical properties as soon as solutions are able to be prepared in a good solvent. Recently, a short review presented the applications of chitin and chitosan-based nanomaterials [59]. Chitin may also be used in blends with other polymers or as microfibres and incorporated in composites.

Fibres were first developed by Austin [60] and then by Hirano [61–63] in solvents mentioned previously, especially the DMAc/LiCl system. The fibres were obtained by wet-spinning [63]. A recent review presents the different fibres obtained from chitin solution and some of their physical properties [27]. In addition, chitin solutions may be casted to obtain films [64, 65] or regenerated under sponge or bead conformation in dependence of the use. Fibres were often proposed for textile applications [66–68].

Blends were prepared with cellulose or silk as soon as a common solvent was available [63, 69–71]. Recently, ionic liquids were used. The solvent 1-ethyl-3-methyl-imidazolium acetate completely dissolves raw crustacean shells allowing to recover high purity chitin powder or films and fibres by direct spinning [72]. Films of poly(ε-caprolactone) (PCL) blends with α-chitin and chitosan were produced. They are completely biodegradable and the crystallinity of PCL is suppressed in the blends due to hydrogen bond interaction between PCL and polysaccharides [73]. Blends were also realized with poly (3-hydroxybutyric acid) (PHB) and chitin or chitosan. They show faster biodegradation than the pure-state component polymers [74, 75].

Composites were also prepared using chitin fibres (length 5 mm, fineness 2.3 ± 0.5 dtex) dispersed in a chitosan solution and regenerated with alkali in a mould. After washing and drying, the product shows excellent mechanical properties [76]. Another reinforcement of chitosan film was obtained with chitin whiskers (width 5–70 nm; length 150–800 nm). The optimum tensile strength is found for an amount of whiskers around 2.96%, and the percentage of elongation at break decreases compared to that of pure chitosan. Addition of α-chitin whiskers and heat treatment helped to improve the water resistance of the material [77].

4.3.2 Applications of Chitin-Based Materials

The main development is in medical and pharmaceutical applications for wound-dressing material [2, 64] or for controlled drug release [78, 79], taking into account the biodegradability, low toxicity, physiological inertness, antibacterial properties, hydrophilic character, gel forming properties and affinity for proteins. Chitin also accelerates wound-healing [66]. Chitin is inert in the gastrointestinal tract of mammals but biodegradable due to the presence of chitinases widely distributed in nature and found in bacteria, fungi and plants and in the digestive systems of numerous animals. Chitinases are involved in host defence against bacterial invasion. Lysozyme from egg white, fig and papaya plants degrade chitin and bacterial cell walls [80, 81]. The influence of the degree of acetylation was studied and shows that a certain degree of deacetylation is necessary to allow hydrolysis in dependence of the enzyme mechanism.

Highly crystalline chitin nanofibres/chitosan glycolate composite was prepared as biocompatible support of medicaments in spray, gel and gauze giving good results for a variety of wound dressings [82]. A composite porous wound dressing was prepared with β-chitin and chitosan (in a ratio 0.1–1) giving good blood absorbability and yielding strength [83].

Chitin is also used as excipient and drug carrier under film, gel or powder forms for applications involving mucoadhesivity [84]. A recent review describes the use of chitin and chitosan as biomaterials in different biomedical applications [85]. Different applications of chitin and that of few chitin derivatives are described in a recent book devoted to chitosan [86].

Another interesting application is a composite material made of hydroxyapatite-chitin-chitosan as bone filling material for guided tissue regeneration, for treatment of periodontal bone defects; it forms a self-hardening paste [87,88].

Chitin was used also to immobilize enzymes and whole cells. Enzyme immobilization was used for applications in food industries such as clarification of fruit juices and processing of milk with α- and β-amylases and invertase fixed on chitin [89]. Chitin is also used for the treatment of environmental pollutants [90,91], or in biosensors [92,93].

Chitin has been used in columns to isolate pure lectins and determine their structure [94]. Chitin and 6-O-carboxymethylchitin activate peritoneal macrophages *in vivo*, suppress the growth of tumour cells in mice and stimulate nonspecific host resistance against *Escherichia Coli* infection [95]. Chitin is used for biological activity as mentioned previously: oligomers were claimed as anticancer drug and the $DP = 5$ is active to control the photosynthesis of maize and soybean [96].

Regenerated chitin derivative fibres were used as binder in a paper-making process. N-iso-butyl chitin fibres were added at 10% yield to reinforce the breaking strength of paper [97,98]. Chitin fibres and/or chitosan fibres with diameter <100 nm and an aspect ratio >5 were used in pneumatic tires and give a good abrasion resistance [99].

4.4 Chitosan Processing to Materials and Applications

4.4.1 Processing of Chitosan

Often, in the literature, the polymer used is either chitin (high degree of acetylation) or chitosan ($DA < 0.5$). Chitosan is easier to use to prepare hydrogels, films, fibres or sponges as previously mentioned. Most of these materials are used in the biomedical domains where chitosan biocompatibility is needed [85]. Solutions of chitosan are prepared in acidic conditions and processed to the needed conformation (cast for a film, spun for fibres, freeze dried for sponges etc.) and immersed in a precipitation alkaline solution, washed and dried. The processing of chitosan is easier than that of chitin but the stability of the materials is lower due to the larger hydrophilic character and especially the pH sensitivity [100]. For better stability, chitosan may be crosslinked by reagents such as epichlorohydrin [101], diisocyanate [102], 1,4-butanediol diglycidyl ether [103], or glutaraldehyde [104]. Specific crosslinking was performed on a blend made of starch and chitosan. Starch was oxidized to produce a polyaldehyde reacting on the $-NH_2$ group of chitosan in the presence of a reducing agent [105]. The same reaction was realized with oxidized maltodextrins, galactomannans and methylcelluloses [106]. A natural cross-linker named genepin was also used recently [107, 108]. Many chitosan hydrogels are obtained by treatment with multivalent anions as oxalic acid [109,110] or citric acid [111–114]. Tripolyphosphate is used to crosslink chitosan [115, 116].

The following example is particularly interesting: chitosan in the presence of β-glycerophosphate gives a thermoforming gel at temperatures over 37 °C and was developed, on the basis of research by Buschmann and coworkers, by Biosyntec in Canada under the trade name BST-Gel™. These products are well adapted for drug delivery, tissue repair, rheumatology and reconstructive chirurgy. The gel BST-CarGel (MD) is proposed for articular cartilage repair [117, 118].

Blends or composites were also prepared especially by Hirano in the same way as mentioned previously for chitin [63]. Other systems were proposed in the literature and listed previously [1, 2]. Some of them are described in detail, for example, chitosan/polyamide 6 and chitosan/polyamide 66 [119,120], chitosan/cellulose using a common solvent [121, 122], chitosan/poly(oxyethylene) (POE) [123], chitosan/polyvinylpyrrolidone (PVP), chitosan/polyvinyl alcohol (PVA) [123, 124], chitosan/polycaprolactone [124–126], chitosan/collagen [127], chitosan/PHB [128], and chitosan/cellulose fibres [129]. In some case, the polymers are crosslinked in the blends [130]. Hydrogels or films have been prepared by mixing chitosan solution with monomers, initiator and cross-linker followed by copolymerization [131, 132].

Recently, reinforcement of chitosan film with carbon nanotubes was tested. This composite demonstrates a large increase in the tensile modulus with the incorporation of only 0.8% of multiwalled carbon nanotubes [133]. Many papers concern bio-inorganic composites, including hydroxyapatite, which increases the mechanical properties and mimics the nanostructure of bone [88, 134, 135].

Chitosan derivatives. Due to the chemical structure of chitosan, specific reactions can be performed on the amino group in C-2 position of the glucosamine unit, especially by reductive amination reaction. Also two –OH groups are available on C-3 and C-6 positions [1, 2]. The more interesting is to test reactions on the C-2 position. Two different types of derivatives may be recognized: (i) derivatives obtained by chemical modification and introduction of different substituents, (ii) grafted copolymers on C-2 position.

Cationic derivatives (quaternized chitosan like trimethylchitosan ammonium), and anionic derivatives (–N– and/or O– carboxylated chitosans) are charged polysaccharides and water soluble in a wide range of pH, especially in neutral and alkaline pH where chitosan is not soluble [136–138]. Alkylated chitosans behave like polymeric surfactants but stabilize lipidic membranes in contrast to simple charged surfactants. The mechanism suggested is flat adsorption on the external leaflet of the vesicles in a patch-like model [139–144]. But a sulfated alkylated chitosan (S-C10-chitosan) was suggested to be incorporated into the lipidic membrane without toxicity [145]. In addition, alkylated chitosan with a side chain C10 or C12 behaves as an amphiphilic polymer and forms a physical gel in presence of external salt in acidic conditions. The salt screens the electrostatic repulsions between the cationic chains [140]. Chitosan substituted with cyclodextrin (CD) was also prepared in relation with the ability of CD to encapsulate hydrophobic molecules, for example, drugs or lipidic substances [146–148]. It was shown that new physical gels may be obtained by mixing chitosan-CD with chitosan-adamantane [149].

Another class of materials are the thermoreversible gels obtained when LCST polymers such as PNIPAm [150–152], PEOP [153], or methylcellulose [154] are grafted on chitosan. Figure 4.4 shows the variation of mechanical moduli to stepwise periodic changes in temperature from 10 to 30 °C.

For chitosan-g-methylcellulose, whatever the temperature, $G' > G''$ indicating loose interchain interactions, and G' increases strongly when temperature increases just as for initial methylcellulose (with a hysteresis on decreasing temperature). On the other hand, for chitosan, $G' < G''$ and both moduli remain nearly constant when temperature varies (with no hysteresis). Few results are shown in Figure 4.5.

The physical cross-linking in these amphiphilic polymers is based on hydrophobic interactions between side chains which counterbalances the electrostatic repulsions between cationic chitosan chains in acidic condition [154].

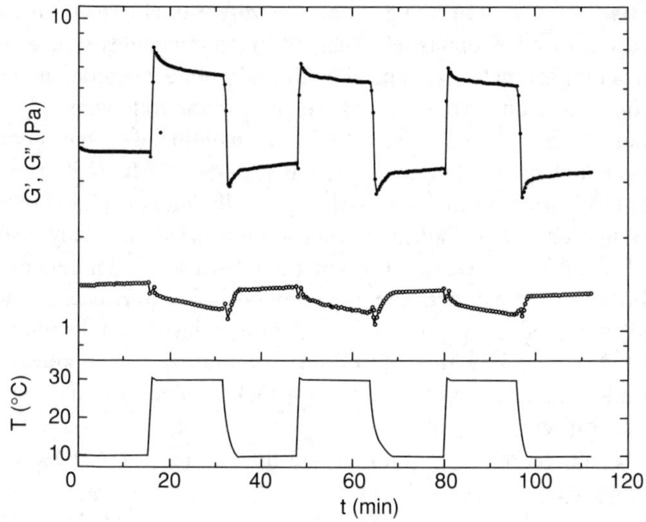

Figure 4.4 *Variations in mechanical moduli, G′ (•) and G″ (○), for 1% (w/w)chitosan-g- PNIPAm solution in 10% aqueous acetic acid to stepwise periodic changes in temperature between 10 and 30 °C (ω = 1 rad s^{-1}; γ = 5%). (Reproduced with permission from [150]. Copyright © 2009, American Chemical Society.)*

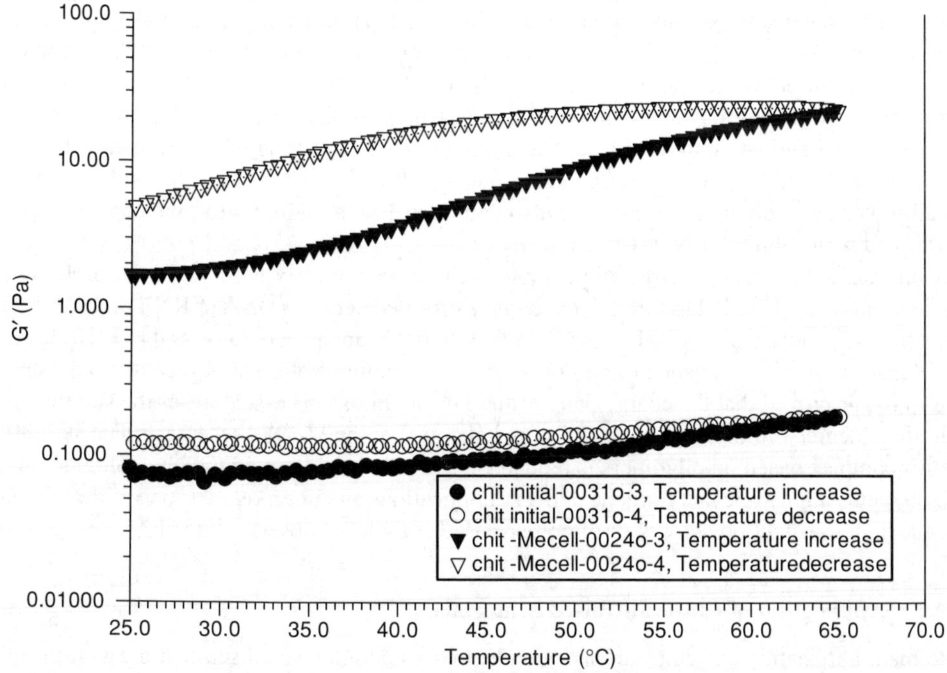

Figure 4.5 *Storage modulus (G′ in Pa.) for chitosan-g-methylcellulose in 0.3 M acetic acid solution as a function of temperature. Comparison with initial chitosan. Filled symbols for temperature increase and open symbols for temperature decrease.*

Electrostatic complex. Chitosan being a polyelectrolyte is able to form interesting electrostatic complexes with other oppositely charged macromolecules to give hydrogels. The properties of these complex materials depend on the polymer concentration, temperature, pH, ionic concentration, and so on. Few examples are given in the following.

Electrostatic polyelectrolyte complexes (PEC) are mentioned in the literature involving chitosan and synthetic or natural polymers such as PAA and CMC [155, 156], xanthan, carrageenan, alginate [157–162], pectin [163, 164], heparin, hyaluronan (HA) [165–169], sulfated cellulose, dextran sulfate, or N-acylated chitosan/chondroitin sulfate. Many systems were cited in the literature [2, 170]. The electrostatic interaction was discussed in relation with the stiffness of the backbone and the nature of the ionic groups involved. Especially, with alginate or HA, a pH-dependent complex is formed, whose stability depends on the ionic concentration, The complex formation was investigated in dilute solution by potentiometry (pH) and conductivity to determine the fraction of ion pairs ($-COO^- \, ^+NH_3-$) formed in dependence of the experimental conditions [164, 166].

Electrostatic interactions between chitosan and lipidic vesicles are also important in the biological and pharmaceutical fields due to chitosan's bioadhesive and permeabilizer roles [171–173]. The coating of liposomes with chitosan increases their biocompatibility and stabilizes the composite membrane against pH and ionic concentration shocks as shown using different techniques from which ^{31}P NMR [172, 174] demonstrates the electrostatic interaction of chitosan and polar phosphate head groups.

One aspect of these electrostatic complexes now in development is the preparation of layer-by-layer polyelectrolyte capsules or films based on charged biocompatible polysaccharides or chitosan/synthetic polyelectrolytes. Core-shell phospholipid nanoparticles were stabilized via layer-by-layer self-assembly of anionic alginate and cationic chitosan and were proposed for protein release [175]. Using chitosan and different anionic polyelectrolytes (hyaluronan, alginate, nucleic acid) to encapsulate *E. coli* cells, it was concluded that the cellular activity was not affected by the multilayer assembly [176, 177]. Complex beads were obtained by dropwise addition of Na-alginate into a chitosan-$CaCl_2$ solution. The swelling of the beads is different from that of Ca-alginate beads, with a maximum swelling at pH 9 [178]. Oligo-chitosans or small molecular weight chitosans were also complexed with alginates to give capsules with controlled permeability [179, 180].

Chitosan and alginate electrostatic complexes have been mostly used so far for biological applications [158, 159, 181] as well as the complex formed between DNA or RNA and chitosan (oligomers or polymers) actually under investigation in many laboratories [182–185]. An investigation on the mechanism and cooperativity of complexation was recently published [186]. It was proved that the charge density and DA of chitosan is essential for the stability. A minimum degree of polymerization (DP) (and charges) around 6–9 is necessary. The stability of this complex based on oligomers is reduced above pH $= 7.4$, around the physiological pH, which seems highly relevant for gene delivery applications and interpreted as one of the reason for the high transfection activity of oligomer-based complexes observed [186].

4.4.2 Application of Chitosan-Based Materials

The main applications of chitosan and the electrostatic complexes discussed are antithrombogenic materials, controlled-release encapsulation of drugs, immobilization of enzymes and cells, and gene carriers. The advantage of chitosan in such materials is its biodegradability but also its hydrophilic property in addition to the polar groups able to form secondary interaction (–OH and –NH_2 groups involved in hydrogen bonds with other polymers) and antibacterial activity. The main promising developments at present are in pharmaceutical and biological or biomedical domains and, at a lower level, in cosmetics and food [187–193].

The main properties and specific qualities of chitosan are:

Biodegradability	Film forming
Biocompatibility	Hydrating agent
Bioadhesivity	Renewable
Polycationic substance	Absorption promoters
Antifungal	Non-toxicity
Antibacterial	Non-allergenic
Immunoadjuvant	Anticholesteremic agent
Antithrombogenic	

(Reproduced with permission from reference [2]. Copyright Elsevier, 2012.) They explain the great interest of many researchers as well as industries in applications of chitosan and specific derivatives.

The cationic character of chitosan in acidic medium is unique. It is the only pseudonatural cationic existing polymer. The film-forming property and biological activity of chitosan also open new applications. In Table 4.4, the main applications of chitosan as biomaterials are recalled. In particular, the drug-delivery applications are available by oral, nasal, parenteral and/or transdermal administration.

Wound-dressing materials and tissue engineering are important domains under development [194–200]. New adhesives were proposed recently [201, 202]. Especially Az-chitosan was shown to be nontoxic, cytocompatible and mechanically suitable in peripheral surgeries [202]. Chitosan is suitable for transmucosomal administration of drugs [203].

Concerning chitosan derivatives, the mucoadhesivity of chitosan and its cationic derivatives is recognized and has been proven to enhance the adsorption of drugs especially at neutral pH. N-trimethyl chitosan chloride interacts with the negatively charged cell membranes [204, 205]. N-lauryl-carboxymethylchitosan being an amphiphilic polymer forms micelles solubilizing taxol, which becomes more efficient. This type of chitosan derivative was found to be safe in terms of membrane toxicity and it could be useful as carrier for hydrophobic cancer drugs [206, 207]. Chitosan or derivatives were used for gene transfection. For N-alkylated chitosan, it was shown that transfection efficiency increases upon elongating the alkyl side chains but levels off after the number of carbons in the side chain exceeds 8 [208]. Quaternized chitosan was also used for the same purpose [209]. Porous chitosan (and derivatives) microspheres were prepared to deliver antigens in a controlled way [210]. This type of particle was loaded with Newcastle disease virus vaccine and tested *in vitro* and *in vivo* [211]. In tumour therapy, delivery of As_2O_3 adsorbed on chitosan was proposed [212].

An interesting application concerns the chitosan- calcium phosphate cement. Chitosan or chitosan glycerophosphate mixed with calcium phosphate and citric acid forms an attractive injectable self-hardening system for bone repair or filling indications [134, 213, 214].

Table 4.4 Main applications of chitin and chitosan as biomaterials. (Reproduced with permission from [2]. Copyright © John Wiley, 2008.)

Forms	Applications	Route of delivery/Properties
Beads	Drug delivery	Oral
Microspheres [76]	Enzyme immobilization	Oral, Implantable, ocular, injectable
Coatings	Surface modification Textile finishes	Chitosan increases mucoadhesivity of alginate capsules
Fibers	Medical textiles	Biodegradable
Nanofibers [75]	Suture Guided bone regeneration Scaffold for nerve tissue regeneration	
Films	Wound care Dialysis membrane	
Powder	Adsorbent for pharmaceutical and medical devices Surgical glove powder Enzyme immobilization	
Sponge [77]	Mucosomal hemostatic dressing Wound dressing Drug delivery Enzyme entrapment	
Shaped objects	Orthopaedics Contact lenses	
Solutions	Cosmetics Bacteriostatic agent Hemostatic agent Anticoagulants Antitumor agent	Moisture holding Oral, nasal Complex formation → Gene delivery
Gels	Delivery vehicle Implants, coating Tissue engineering	
Tablets	Compressed diluent Disintegrating agent	Oral, bucal
Capsules	Delivery vehicle	

Chitosan and chitosan with immobilized enzymes found application in a large range of applications due to the original properties of chitosan itself [190, 215]. A review for applications in food was published previously but it seems that this domain is not in development [216, 217]. Packaging [218] and paper industry are also concerned [98, 219–221]. In paper fabric, it may be used for paper reinforcement or as coagulant of wastewater. In textile industry, some application of chitosan is also mentioned in the literature as it was for chitin [61, 63, 222]. Another field where the specificity of chitosan still is to be recognized is cosmetics (especially for hair treatment in relation with electrostatic interaction) (Table 4.5) [1].

The last point to discuss is biological activity in agriculture because chitosan has antivirus and antiphage activities [223]. It inhibits the growth of bacteria and bacterial infection and stimulates the natural defences in plants. A mechanism was proposed through the octadecanoid pathway [224].

Table 4.5 *Specific characteristics for applications of chitosan in hair treatment. (Reproduced with permission from [1]. Copyright © Elsevier, 2006.)*

Properties	Uses
Aqueous solution interacting with negatively charged hairs (electrostatic interaction)	Shampoos Hair tonics
Antistatic effect (due to hydrophilic character), it maintains moisture in low humidity and hair style in high humidity	Rinses, Permanent wave agents
Removing sebum and oils from hairs (due to hydrophobic character)	Hair colorants Lacquers, Hair sprays Time release deliveries (chitosan beads, gels or granules)
Antibacterial and antifungal activities	
Thickening polymer	
Role on surfactant stability; stabilize emulsion	
Make hair softer, increase mechanical strength	
Protecting elastic film on hairs, increasing their softness	

4.5 Conclusion

In this chapter, the characteristics of chitin were described. This was followed by a discussion on the solubilization necessary to process the polysaccharide to obtain new materials. Its fibre and film-forming performances are recognized and allow application in the textile domain or in packaging. Nevertheless, the more important applications are in the domain of biomaterials due to the hydrophilic character and antimicrobial properties.

The characterization and processing of chitosan were then described. Solubility in acidic media is a valuable property, extending processing methods and the domains of application. The polysaccharide may be processed as fibre, film, sponge, bead, gel or solution. Due to the presence of free $-NH_2$ groups, specific modifications can be performed on the C-2 position of the D-glucosamine unit.

Chitin and chitosan can be blended with synthetic or natural polymers (proteins, DNA, alginate, hyaluronan, etc.) or introduced as fibres in composite materials. The majority of the applications covered in the literature seem to be that in the biomedical and pharmaceutical domains. Many of these applications are mentioned in this review.

References

1. Rinaudo, M. (2006) Chitin and chitosan: properties and applications. *Progress in Polymer Science*, **31**, 603–632.
2. Rinaudo, M. (2008) Main properties and current applications of some polysaccharides as biomaterials. *Polymer International*, **57**(3), 397–430.
3. Rudall, K.M. and Kenchington, W. (1973) The chitin system. *Biological Reviews*, **40**, 597–636.
4. Blackwell, J. (1973) Chitin, in *Biopolymers* (eds A.G. Walton and J. Blackwell), Academic Press, New York, pp. 474–489.
5. Lotmar, W. and Picken, L.E.R. (1950) A new crystallographic modification of chitin and its distribution. *Experientia*, **6**, 58–59.
6. Rudall, K.M. (1969) Chitin and its association with other molecules. *Journal of Polymer Science Part C-Polymer Letters*, **28**, 83–102.

7. Blackwell, J., Parker, K.D. and Rudall, K.M. (1965) Chitin in pogonophore tubes. *Journal of the Marine Biological Association of the United Kingdom*, **45**, 659–661.
8. Gaill, F., Persson, J., Sugiyama, J. *et al.* (1992) The chitin system in the tubes of deep sea hydrothermal vent worms. *Journal of Structural Biology*, **109**, 116–128.
9. Gonell, H.W. (1926) Rötgenographische studien an chitin. *Zeitschrift Fur Physiologische Chemie*, **152**, 18–30.
10. Clark, G.L. and Smith, A.F. (1936) X-ray studies of chitin, chitosan, and derivatives. *Journal of Physical and Chemical*, **40**, 863–879.
11. Gardner, K.H. and Blackwell, J. (1975) Refinement of the structure of β-chitin. *Biopolymers*, **14**, 1581–1595.
12. Minke, R. and Blackwell, J. (1978) The structure of α-chitin. *Journal of Molecular Biology*, **120**, 167–181.
13. Blackwell, J. (1969) Structure of β-chitin or parallel chain systems of poly-β-(1→4)-N-acetyl-D-glucosamine. *Biopolymers*, **7**, 281–298.
14. Kurita, K., Tomita, K., Ishi, S. *et al.* (1993) β-chitin as a convenient starting material for acetolysis for efficient preparation of N-acetylchitooligosaccharides. *Journal of Polymer Science Part A-Polymer Chemistry*, **31**, 2393–2395.
15. Austin, P.R. (1984) Chitin solvents and solubility parameters, in *Chitin and Chitosan and Related Enzymes* (ed. J.P. Zikakis), Academic Press Inc., Orlando, pp. 227–237.
16. Austin, P.R. (1975) Solvents for extraction and purification of chitin. U.S. Patent 3,892,731; 1975 and Purification of chitin. U.S. Patent 3,879, 377.
17. McCormick, C.L., Callais, P.A. and Hutchinson, B.H. (1985) Solution studies of cellulose in lithium chloride N,N-dimethylacetamide. *Macromolecules*, **18**, 2394–2401.
18. Elevan, Y. and Murat, B. (2003) Preparation and characterization of physical gels and beads from chitin solution. *Carbohydrate Polymers*, **54**, 479–488.
19. Gagnaire, D., Saint-Germain, J. and Vincendon, M. (1982) NMR studies of chitin and chitin derivatives. *Makromolekulare Chemie-Macromolecular Chemistry and Physics*, **183**, 593–601.
20. Vincendon, M. (1985) ^{1}H NMR study of the chitin dissolution mechanism. *Makromolekulare Chemie-Macromolecular Chemistry and Physics*, **186**, 1787–1795.
21. Vincendon, M. (1994) Solution of chitin in phosphoric acid, in *Chitin World* (ed. S.Z. Karniki), Wirtschaftsverlag NW, Bremerhaven, Germany, pp. 91–97.
22. Einbu, A., Naess, S.N., Elgsaeter, A. and Vårum, K.M. (2004) Solution properties of chitin in alkali. *Biomacromolecules*, **5**, 2048–2054.
23. Sannan, T., Kurita, K. and Iwakura, Y. (1975) Studies on chitin, 1. *Makromolekulare Chemie-Macromolecular Chemistry and Physics*, **176**, 1191–1195.
24. Sannan, Y., Kurita, K. and Iwakura, Y. (1976) Studies on chitin. *Makromolekulare Chemie-Macromolecular Chemistry and Physics*, **177**, 3589–3600.
25. Argüelles-Monal, W., Goycoolea, F.M., Lizardi, J. *et al.* (2003) Chitin and chitosan in gel network systems, in *Polymer Gels* (eds H.B. Bohidar, P. Dubin and Y. Osada), ACS Symposium Series No. 833, American Chemical Society, Washington D.C., pp. 102–122.
26. Kubota, N. and Eguchi, Y. (1997) Facile preparation of water-soluble N-acetylated chitosan and molecular weight dependence of its water-solubility. *Polymer Journal*, **29**, 123–127.
27. Pillai, C.K.S., Paul, W. and Sharma, C.P. (2009) Chitin and chitosan polymers: chemistry, solubility and fiber formation. *Progress in Polymer Science*, **34**, 641–678.
28. Tamura, H., Hamaguchi, T. and Tokura, S. (2003) Destruction of rigid crystalline structure to prepare chitin solution, in *Advances in Chitin Science*, vol. **7** (eds I. Boucher, K. Jamieson and A. Retnakaran), ISM BioPolymer Inc., Montreal, pp. 84–87.
29. Tokura, S. and Tamura, H. (2003) Regeneration of chitin and chitosan under the influence of calcium ion, in *Advances in Chitin Science*, vol. **7** (eds I. Boucher, K. Jamieson and A. Retnakaran), ISM BioPolymer Inc., Montreal, pp. 62–65.

30. Louvier-Hernandez, J.F., Luna-Barcenas, GT, Thakur, R. and Gupta, R.B. (2005) Formation of nanofibers by supercritical antisolvent. *Journal of Biomedical Nanotechnology*, **1**, 109–114.
31. Carpozza, R.C. (1976) Spinning and shaping poly-(N-acetyl-D-glucosamine). US Patent 3,988, 411; Carpozza, R.C. (1976) Solution of poly-(N-acetyl-D-glucosamine), US Patent 3,989,535.
32. Hu, X., Du, Y., Tang, Y. *et al.* (2007) Solubility and properties of chitin in NaOH/urea aqueous solution. *Carbohydrate Polymers*, **70**, 451–458.
33. Wu, Y., Sasaki, T., Irie, S. and Sakurai, K. (2008) A novel biomass-ionic liquid platform for the utilization of native chitin. *Polymer*, **49**, 2321–2327.
34. Mantz, R.A., Fox, D.M., Green, J.M. *et al.* (2007) Dissolution of biopolymers using ionic liquids. *Zeitschrift Fur Naturforschung Section A-A Journal of Physical Sciences*, **62**, 275–280.
35. Tanner, S.T., Chanzy, H., Vincendon, M. *et al.* (1990) High-resolution solid-state carbon-13 nuclear magnetic resonance study of chitin. *Macromolecules*, **23**, 3576–3583.
36. Heux, L., Brugnerotto, J., Desbrieres, J. *et al.* (2000) Solid state NMR for determination of degree of acetylation of chitin and chitosan. *Biomacromolecules*, **1**, 746–751.
37. Brugnerotto, J., Lizardi, J., Goycoolea, F.M. *et al.* (2001) An infrared investigation in relation with chitin and chitosan characterization. *Polymer*, **42**, 3569–3580.
38. Poirier, M. and Charlet, G. (2002) Chitin fractionation and characterization in N, N-dimethylacetamide/lithium chloride solvent system. *Carbohydrate Polymers*, **50**, 363–370.
39. Terbojevich, M., Carraro, C. and Cosani, A. (1988) Solution studies of the chitin-lithium chloride-N, N-dimethylacetamide system. *Carbohydrate Research*, **180**, 73–86.
40. Rhazi, M., Desbrieres, J., Tolaimate, A. *et al.* (2000) Investigation of different natural sources of chitin: influence of the source and deacetylation process on the physicochemical characteristics of chitosan. *Polymer International*, **49**, 337–344.
41. Tolaimate, A., Desbrieres, J., Rhazi, M. and Alagui, A. (2003) Contribution to the preparation of chitins and chitosans with controlled physico-chemical properties. *Polymer*, **44**(26), 7939–7952.
42. Rinaudo, M. (2012) Physical properties of chitosan and derivatives in sol and gel states, in *Chitosan-Based Systems for Biopharmaceuticals: Delivery, Targeting and Polymer Therapeutics* (eds B. Sarmento and José das Neves), John Wiley & Sons, Ltd, Chichester.
43. Rinaudo, M., Le Dung, P., Gey, C. and Milas, M. (1992) Substituent distribution on O, N-carboxymethylchitosans by ^{1}H and ^{13}C NMR. *International Journal of Biological Macromolecules*, **14**, 122–128.
44. Saito, H., Tabeta, R. and Ogawa, K. (1987) High-resolution solid-state ^{13}C NMR study of chitosan and its salts with acids: conformational characterization of polymorphs and helical structures as viewed from the conformation-dependent ^{13}C chemical shifts. *Macromolecules*, **20**, 2424–2430.
45. Raymond, L., Morin, F.G. and Marchessault, R.H. (1993) Degree of deacetylation of chitosan using conductometric titration and solid-state NMR. *Carbohydrate Research*, **246**, 331–336.
46. Vårum, K.M., Anthonsen, M.W., Grasdalen, H. and Smisrød, O. (1991) Determination of the degree of N-acetylation and the distribution of N-acetyl groups in partially N-deacetylated chitins (chitosans) by high-field n.m.r. spectroscopy. *Carbohydrate Research*, **211**, 17–23.
47. Värum, K.M., Anthonsen, M.W., Grasdalen, H. and Smisrød, O. (1991) ^{13} C-NMR studies of the acetylation sequences in partially N-deacetylated chitins (chitosans). *Carbohydrate Research*, **217**, 19–27.
48. Philippova, O.E., Volkov, E.V., Sitnikova, N.L. *et al.* (2001) Two types of hydrophobic aggregates in aqueous solutions of chitosan and its hydrophobic derivative. *Biomacromolecules*, **2**, 483–490.
49. Philippova, O.E., Korchagina, E.V., Volkov, E.V. *et al.* (2012) Aggregation of some water-soluble derivatives of chitin in aqueous solutions: Role of the degree of acetylation and effect of hydrogen bond breaker. *Carbohydrate Polymers*, **87**(1), 687–694.
50. Rinaudo, M., Milas, M., and Le Dung, P. (1993) Characterization of chitosan. Influence of ionic strength and degree of acetylation on chain expansion. *International Journal of Biological Macromolecules*, **15**, 281–285.

51. Nguyen, S., Winnik, F.M. and Buschmann, M.D. (2009) Improved reproducibility in the determination of the molecular weight of chitosan by analytical size exclusion chromatography. *Carbohydrate Polymers*, **75**, 528–533.
52. Nguyen, S., Sisiger, S., Jolicoeur, M. *et al.* (2009) Fractionation and characterization of chitosan by analytical SEC and 1H NMR after semi-preparative SEC. *Carbohydrate Polymers*, **75**, 636–645.
53. Brugnerotto, J., Desbrières, J., Roberts, G. and Rinaudo, M. (2001) Characterization of chitosan by steric exclusion chromatography. *Polymer*, **42**, 9921–9927.
54. Maghami, G.G. and Roberts, G.A.F. (1988) Evaluation of the viscometric constants for chitosan. *Makromolekulare Chemie-Macromolecular Chemistry and Physics*, **189**, 195–200.
55. Berth, G. and Dautzenberg, H. (2002) The degree of acetylation of chitosans and its effect on the chain conformation in aqueous solution. *Carbohydrate Polymers*, **47**, 39–51.
56. Mazeau, K., Perez, S. and Rinaudo, M. (2000) Predicted influence of N-acetyl group content on the conformational extension of chitin and chitosan chains. *Journal of Carbohydrate Chemistry*, **19**, 1269–1284.
57. Mazeau, K. and Rinaudo, M. (2004) The prediction of the characteristics of some polysaccharides from molecular modeling. Comparison with effective behaviour. *Food Hydrocolloids*, **18**(6), 885–898.
58. Odijk, T. (1979) On the ionic-strength dependence of the intrinsic viscosity of DNA. *Biopolymers*, **18**, 3111–3113.
59. Jayakumar, R., Menon, D., Manzoor, K. *et al.* (2010) Biomedical applications of chitin and chitosan based nanomaterials. A short review. *Carbohydrate Polymers*, **82**(2), 227–232.
60. Austin, P.R. and Brine, J. (1977) Chitin films and fibers. U.S. Patent 4,029,727.
61. Hirano, S. and Midorikawa, T. (1998) Novel method for the preparation of N-acylchitosan fiber and N-acylchitosan-cellulose fiber. *Biomaterials*, **19**, 293–297.
62. Hirano, S., Nakahira, T., Nakagawa, M. and Kim, S.K. (1999) The preparation and applications of functional fibres from crab shell chitin. *Journal of Biotechnology*, **70**, 373–377.
63. Hirano, S. (2001) Wet-spinning and applications of functional fibers based on chitin and chitosan. *Macromolecular Symposia*, **168** (1), 21–30.
64. Yusof, N.L., Wee, A., Lim, L.Y. and Khor, E. (2003) Flexible chitin films as potential wound-dressing materials: wound model studies. *Journal of Biomedical Materials Research-Applied Biomaterials*, **66A**, 224–232.
65. Yusof, N.L., Lim, L.Y. and Khor, E. (2004) Flexible chitin films: structural studies. *Carbohydrate Research*, **339**(16), 2701–2711.
66. Hudson, S.M. and Jenkins, D.W. (2003) Chitin and chitosan, in *EPST*, 3rd edn, vol. **1** (ed. H.F. Mark), John Wiley & Sons, Inc., New York, pp. 569–580.
67. Hudson, S.M. (1999) The applications of chitin and chitosan to fiber and textile products, in *Advances in Chitin Science*, vol. **3** (eds R.H. Chen and H.C. Chen), National Taiwan Ocean University, Taiwan, pp. 80–87.
68. Rathke, T.D. and Hudson, S.M. (1994) Review of chitin and chitosan as fiber and film formers. *Reviews in Macromolecular Chemistry & Physics-C*, **34**, 375–437.
69. Wu, J., Liang, S., Dai, H. *et al.* (2010) Structure and properties of cellulose/chitin blended hydrogel membranes fabricated via a solution pre-gelation technique. *Carbohydrate Polymers*, **79**(3), 677–684.
70. He, C., Ma, B. and Sun, J. (2009) The preparation and properties of cellulose/chitin blend filaments. *Journal of Applied Polymer Science*, **113**(5), 2777–278.
71. He, C. and Ma, B. (2010) The spinning, structure, and properties of cellulose/ chitin blend filaments through HWM method. *Polymers For Advanced Technologies*, **21**(7), 496–505.
72. Qin, Y., Lu, X., Sun, N. and Rogers, R.D. (2010) Dissolution or extraction of crustacean shells using ionic liquids to obtain high molecular weight purified chitin and direct production of chitin film and fibers. *Green Chemistry*, **12**, 968–971.
73. Senda, T., He, Y. and Inoue, Y. (2002) Biodegradable blends of poly(ε-caprolactone) with α-chitin and chitosan: specific interactions, thermal properties and crystallisation behaviour. *Polymer International*, **51**, 33–39.

74. Ikejima, T. and Inoue, Y. (1999) Crystallisation behaviour and environmental biodegradability of the blend films of poly (3-hydroxybutyric acid) with chitin and chitosan. *Carbohydrate Polymers*, **41**, 351–356.
75. Ikejima, T., Yagi, K. and Inoue, Y. (1999) Thermal properties and crystallization behaviour of poly(3-hydroxybutyric acid) in blends with chitin and chitosan. *Macromolecular Chemistry and Physics*, **200**, 413–421.
76. Hu, Q. and Wang, Z. (2010) Preparat ion of chitin fiber-reinforced three –dimensional chitosan rod materials. Faming Zhuanli Shenqing Gogkai Shuomingshu CN101665589 A 20100310.
77. Sriupayo, J., Supaphol, P., Blackwell, J. and Rujiranavit, R. (2005) Preparation and characterization of α-chitin whiskers-reinforced chitosan nanocomposite films with and without heat treatmernt. *Carbohydrate Polymers*, **62**, 130–136.
78. Kanke, M., Katayama, H., Tsuzuki, S. and Kuramoto, H. (1989) Application of chitin and chitosan to pharmaceutical preparations. *Chemical and Pharmaceutical Bulletin*, **37**, 523–525.
79. Kato, Y., Onishi, H., Machida, Y. (2003) Application of chitin and chitosan derivatives in the pharmaceutical field. *Current Pharmaceutical Biotechnology*, **4**, 303–309.
80. Rinaudo, M. (2007) Properties and degradation of selected polysaccharides: hyaluronan and chitosan. *Corrosion Engineering, Science and Technology*, **42**(4), 324–334.
81. Sashiva, H., Saimoto, H., Sgigemasa, Y. *et al.* (1990) Lysozyme susceptibility of partially deacetylated chitin. *International Journal of Biological Macromolecules*, **12**, 295–296.
82. Muzzarelli, R.A.A., Morganti, G., Palombo, P. *et al.* (2007) Chtin nanofibrils/chitosan glycolate composites as wound medicaments. *Carbohydrate Polymers*, **70**, 274–284.
83. Lee, J-S., Shiu, J-Y. and Lin, W-H. (2003) Wound dressing made of chitin and chitosan composite porous body and manufacturing method thereof, TW 244397 B 20051201, Taiwan 2005 and Composite sponge wound dressing made of β-chitin and chitosan, US Patent Appl. Publ. 2003, US20030190346 A120031009.
84. Bernkop-Schnurch, A. (2002) Mucoadhesive polymers, in *Polymeric Biomaterials* (ed. S. Dumitriu), Marcel Dekker, New York, pp. 147–165.
85. Khor, E. and Lim, LY. (2003) Implantable applications of chitin and chitosan. *Biomaterials*, **24**, 2339–2349.
86. Muzzarelli, R.A.A. and Muzzarelli, C. (2002) *Chitosan in Pharmacy and Chemistry*, Atec, Grottammare, Italy.
87. Ito, M., Matahira, Y. and Sakai, K. (1998) The application of chitin-chitosan to bone filling materials. *Kichin, Kitosan Kenkyu*, **4**, 142–143.
88. Swetha, M., Sahithi, K., Moorthi, A. *et al.* (2010) Biocomposites containing natural polymers and hydroxyapatite for bone tissue engineering. *International Journal of Biological Macromolecules*, **47**, 1–4.
89. Krajewska, B. (2004) Application of chitin- and chitosan-based materials for enzyme immobilizations: a review. *Enzyme and Microbial Technology*, **35**, 126–139.
90. Songkroah, C., Nakbanpote, W. and Thiravetyan, P. (2004) Recovery of silver-thiosulfate complexes with chitin. *Process Biochemistry*, **39**, 1553–1559.
91. Kosyakov, V.N., Yakovlev, N.G. and Veleshko, I.E. (2002) Application of chitin-containing fiber material 'mycoton' for actinide adsorption. *Journal of Nuclear Science and Technology*, **39** (suppl. 3), 508–511.
92. Bouvree, A., Feller, J.F., Castro, M. *et al.* (2009) Conductive polymer nano-biocomposites (CPC): chitosan-carbon nanoparticle a good candidate to design polar vapour sensors. *Sensors and Actuators B-Chemical*, **138**, 138–147.
93. Actis, P., Vilozny, B., Seger, R.A. *et al.* (2011) Voltage-controlled metal binding on polyelectrolyte-functionalized nanopores. *Langmuir*, **27**(10), 6528–6533.
94. Datta, P.K., Basu, PS. and Datta, TK. (1984) Isolation and characterization of *Vicia faba* lectin affinity purified on chitin column. *Preparative Biochemistry*, **14**, 373–387.
95. Nishimura, M., Nishimura, S., Nishi, N. *et al.* (1984) Immunological activity of chitin and its derivatives. *Vaccine*, **2**, 93–94.

96. Khan, W., Prithiviraj, B. and Smith, DL. (2002) Effect of foliar application of chitin and chitosan oligosaccharides on photosynthesis of maize and soybean. *Photosynthetica*, **40**, 621–624.
97. Tokura, S. and Nishi, N. (1982) Preparation and properties of alkyl-chitin fibers, in *Chitin Chitosan* (eds S. Hirano and S. Tokura), Proceedings of the Second International Conference on Chitin and Chitosan, Japanese Society of Chitin Chitosan, Tattori, Japan, pp. 244–247.
98. Kobayashi, Y., Nishiyama, M., Matsuo, R. *et al.* (1982) Application of chitin and its derivatives to paper industry, in *Chitin Chitosan* (eds S. Hirano and S. Tokura), Proceedings of the Second International Conference on Chitin and Chitosan, Japanese Society of Chitin Chitosan, Tattori, Japan, pp. 244–247.
99. Imoto, Y., Ichikawa, N., Chirachanchai, S. and Rartwatanaseri, T. (2009) Chitin/chitosan fiber-reinforced diene rubber compositions, their manufacture, and pneumatic tires using the compositions. Jpn. Kokai Tokkyo Koho, JP 2009001671 A 20090108.
100. Rinaudo, M. (2011) Biomaterials based on chitin and chitosan. *Bioplastics*, **6**, 48–50.
101. Wei, Y.C., Hudson, S.M., Mayer, J.M. and Kaplan, D.L. (1992) The crosslinking of chitosan fibers. *Journal of Polymer Science Part A-Polymer Chemistry*, **30**, 2187–2193.
102. Welsh, E.R. and Price, R.R. (2003) Chitosan cross-linking with a water-soluble, blocked diisocyanate.2. Solvates and hydrogels. *Biomacromolecules*, **4**, 1357–1361.
103. Roy, S.K., Todd, J.G. and Glasser, W.G. (1998) Crosslinked hydrogel beads from chitosan. US Patent 5,770,712.
104. Arguelles-monal, W., Goycoolea, FM, Peniche, C. and Higuera-Ciapara, I. (1998) Rheological study of the chitosan/glutaraldehyde chemical gel system. *Polymer Gels and Network*, **6**, 429–440.
105. Baran, E.T., Mano, J.F. and Reis, R.L. (2004) Starch-Chitosan hydrogels prepared by reductive alkylation cross-linking. *Journal of Materials Science-Materials in Medicine*, **15**, 759–765.
106. Rinaudo, M. (2010) New way to crosslink chitosan in aqueous solution. *European Polymer Journal*, **46**, 1537–1544.
107. Mi, F.L., Sung, H.W. and Shyu, S.S. (2000) Synthesis and characterization of a novel chitosan-based network prepared using naturally occurring crosslinker. *Journal of Polymer Science Part A-Polymer Chemistry*, **38**, 2804–2814.
108. Espinosa-Garcia, B.M., Arguelles-Monal, W.M., Hernandez, J. *et al.* (2007) Molecular imprinted chitosan-genepin hydrogels with recognition capacity toward o-xylene. *Biomacromolecules*, **8**, 3355–3364.
109. Hirano, S., Yamaguchi, R., Fukui, N. and Iwata, M. (1990) A chitosan oxalate gel: its conversion to an N-acetylchitosan gel via a chitosan gel. *Carbohydrate Research*, **201**, 145–149.
110. Yamaguchi, R., Hirano, S., Arai, Y. and Ito, T. (1978) Chitosan salt gels thermally reversible gelation of chitosan. *Agricultural and Biological Chemistry*, **42**, 1981–1982.
111. Hsieh, S.H., Huang, Z.K., Huang, Z.Z. and Tseng, Z.S. (2004) Antimicrobial and physical properties of woollen fabrics cured with citric acid and chitosan. *Journal of Applied Polymer Science*, **94**, 1999–2007.
112. Yokoyama, A., Yamamoto, S., Kawasaki, T. *et al.* (2002) Development of calcium phosphate cement using chitosan and citric acid for bone substitute materials. *Biomaterials*, **23**, 1091–1101.
113. Shen, X., Tong, H., Jiang, T. *et al.* (2007) Homogeneous chitosan/carbonate apatite/citric acid nanocomposites prepared through a novel *in situ* precipitation. *Composites Science and Technology*, **67**, 2238–2245.
114. Hsieh, S.H., Chen, W.H. and Wei, L.L. (2003) A spectroscopic analysis of the reaction mechanism of polycarboxylic acid crosslinking with chitosan and cotton fabric. *Cellulose Chemistry and Technology*, **37**, 359–369.
115. Desai, K.G.H. and Park, H.J. (2005) Encapsulation of vitamin C in tripolyphosphate cross-linked chitosan microspheres by spray drying. *Journal of Microencapsulation*, **22**, 179–192.
116. Lee, S-T., Mi, F-L., Shen, Y-J. *et al.* (1998) The copper (II) uptake by chitosan-tripolyphosphate chelating resin and related kinetic studies, in *Advances in Chitin Science*, vol. **3** (eds R.H. Chen and H.C. Chen), National Taiwan Ocean University, Taiwan, pp. 475–480.

117. Chenite, A., Chaput, C., Wang, D. *et al.* (2000) Novel injectable neutral solutions of chitosan form biodegradable gels in situ. *Biomaterials*, **21**, 2155–2161.
118. Chenite, A., Buschmann, M., Wang, D. *et al.* (2001) Rheological characterization of thermogelling chitosan/glycerol-phosphate solutions. *Carbohydrate Polymers*, **46**:39–47.
119. Ko, M.J., Jo, W.H., Kim, H.C. and Lee, S.C. (1997) Miscibility of chitosans/polyamide 6 blends. *Polymer Journal*, **29**, 997–1001.
120. Gonzalez, V., Guerrero, C. and Ortiz, U. (2000) Chemical structure and compatibility of polyamide-chitin and chitosan blends. *Journal of Applied Polymer Science*, **78**(4), 850–857.
121. Hosokawa, J., Nishiyama, M., Yoshihara, K. and Kubo, T. (1990) Biodegradable film derived from chitosan and homogenized cellulose. *Industrial and Engineering Chemistry Research*, **29**, 800–805.
122. Hasegawa, M., Isogai, A., Kuga, S. and Onabe, F. (1994) Preparation of cellulose-chitosan blend film using chloral/dimethylformamide. *Polymer*, **35**, 983–987.
123. Mucha, M., Piekielna, J. and Wieczorek, A. (1999) Characterization and morphology of biodegradable chitosan/synthetic polymer blends. *Macromolecular Symposia*, **144**, 391–412.
124. Abou-Aiad, T.H.M., Abd-El-Nour, K.N., Hakim, I.K. and Elsabee, M.Z. (2005) Dielectric and interaction behavior of chitosan/polyvinyl alcohol and chitosan/polyvinyl pyrrolidone blends with some antimicrobial activities. *Polymer*, **47**, 379–389.
125. Honma, T., Zhao, L., Asakawa, N. and Inoue, Y. (2006) Poly(ε-caprolactone)/chitin and poly(ε-caprolactone)/chitosan blend films with compositional gradients: fabrication and their biodegradability. *Macromolecular Bioscience*, **6**(3), 241–249.
126. Honma, T., Senda, T. and Inoue, Y. (2003) Thermal properties and crystallization behaviour of blends of poly(ε-caprolactone) with chitin and chitosan. *Polymer International*, **52**(12), 1839–1846.
127. Hou, C., Gu, Q. and Jiang, L. (2008) Manufacture of composite nerve conduit containing collagen and chitosan for repairing peripheral nerve defect. Faming Zhuanli Shenqing, CN 101138656 A 20080312.
128. Cardenas, T.G., Sanzana, L.J. and Mei, L.H.I. (2002) Synthesis and characterization of chitosan-PHB blends. *Boletin De La Sociedad Chilena De Quimica*, **47**(4), 529–535.
129. Zhang, L., Yang, L-B., Sui, S-Y. *et al.* (2010) Study on the preparation and performance of the antibacterial cellulose/chitosan derivative composite film. *Yinran Zhuji*, **27**(6), 20–22.
130. Ferdous, S., Mustafa, A.I. and Khan, M. (2003) A Study on mechanical properties of photo-cured films of chitosan/PVA and PEO/PVA blend with acrylic monomers. *Journal of Macromolecular Science, Part A. Pure and Applied Chemistry*, **A40**(8), 817–832.
131. Martinez-Ruvalcaba, A., Sanchez-Diaz, J.C., Becerra, F. and Cruz-Barba, L.E. (2009) Swelling characterization and drug delivery kinetics of polyacrylamise-co-itaconic acid/chitosan hydrogels. *eXPRESS Polymer Letters*, **3**, 25–32.
132. Alam, R., Khan, M.A., Khan, R.A. *et al.* (2008) Study on the physico-mechanical properties of photocured chitosan films with oligomer and acrylate monomer. *Journal of Polymers and the Environment*, **16**, 213–219.
133. Wang S-F., Shen L., Zhang W-D. and Tong, Y-J. (2005) Preparation and mechanical properties of chitosan/carbon nanotubes composites. *Biomacromolecules*, **6**, 3067–3072.
134. Vankatesan, J. and Kim, S-K. (2010) Chitosan composites for bone tissue engineering- an overview. *Marine Drugs*, **8**, 2252–2266.
135. Volkmer, E., Hoffmann, B., Kokott, A. *et al.* (2009) A short-term bioresorbable bone filling material based on hydroxyapatite, chitosan, and oxidized starch tested in a novel orthotopic meta physeal mouse model. *Advanced Engineering Materials*, **4**, B114–B121.
136. Domard, A., Rinaudo, M. and Terrassin, C. (1986) A new method for the quaternization of chitosan. *International Journal of Biological Macromolecules*, **8**, 105–107.
137. Domard, A., Gey, C., Rinaudo, M. and Terrassin, C. (1987) ^{13}C and 1H NMR spectroscopy of chitosan and N-trimethyl chloride derivatives. *International Journal of Biological Macromolecules*, **9**, 233–237.
138. Pham, L.D., Milas, M., Rinaudo, M. and Desbrieres, J. (1994) Water soluble derivatives obtained by controlled chemical modification of chitosan. *Carbohydrate Polymers*, **24**, 209–214.

139. Holme, K.R. and Hall, L.D. (1991) Chitosan derivatives bearing C-10–alkyl glycoside branches: a temperature-induced gelling polysaccharide. *Macromolecules*, **24**, 3828–3833.
140. Desbrières, J., Martinez, C. and Rinaudo, M. (1996) Hydrophobic derivatives of chitosan: characterization and rheological behaviour. *International Journal of Biological Macromolecules*, **19**, 21–28.
141. Muzzarelli, R.A.A., Frega, N., Miliani, M. *et al.* (2000) Interactions of chitin, chitosan, N-lauryl chitosan and N-dimethylaminopropyl chitosan with olive oil. *Carbohydrate Polymers*, **43**, 263–268.
142. Babak, V.G., Rinaudo, M., Desbrières, J. *et al.* (1997) The effect of alkyl chain length of a polysoap on the surface activity of its complexes with cationic surfactants. *Mendeleev Communications*, **4**, 149–151.
143. Quemeur, F., Rinaudo, M. and Pepin-Donat, B. (2008) Influence of molecular weight and pH on adsorption of chitosan at the surface of large and giant vesicles. *Biomacromolecules*, **9**(1), 396–402.
144. Rinaudo, M., Auzely, R., Vallin, C., Mullagaliev, I. (2005) Specific interactions in modified chitosan systems. *Biomacromolecules*, **6**, 2396–2407.
145. Nonaka, K-I., Kazama, S., Goto, A. *et al.* (2002) Spin probe study on the interaction of chitosan-derived polymer surfactants with lipid membrane. *Journal of Colloid and Interface Science*, **246**, 288–295.
146. Sakairi, N., Nishi, N. and Tokura, S. (1999) Cyclodextrin-linked chitosan: synthesis and inclusion complexation ability, in *Polysaccharide Applications: Cosmetics and Pharmaceuticals* (eds M.A. El-Nokaly and H.A. Soini), American Chemical Society, Washington, pp. 68–84.
147. Auzely-Velty, R. and Rinaudo, M. (2001) Chitosan derivatives bearing pendant cyclodextrin cavities: synthesis and inclusion performance. *Macromolecules*, **34**, 3574–3580.
148. Auzely-Velty, R. and Rinaudo, M. (2002) New supramolecular assemblies of a cyclodextrin grafted chitosan through specific complexation. *Macromolecules*, **35**, 7955–7962.
149. Charlot, A., Auzely-Velty, R. and Rinaudo, M. (2003) Specific interactions in model charged polysaccharide systems. *Journal of Physical Chemistry B*, **107**, 8248–8254.
150. Recillas, M., Silva, L.L., Peniche, C. *et al.* (2009) Thermoresponsive behavior of chitosan-g-*N*-isopropylacrylamide copolymer solutions. *Biomacromolecules*, **10**(6), 1633–1641.
151. Prabaharan, M. and Mano, J.F. (2007) A novel pH and thermosensitive N,O-carboxymethylchitosan-graft-poly(N-isopropylacrylamide) hydrogel for drug release. *e-Polymers*, **043**, 1–14.
152. Chen, J.P. and Cheng, T.H. (2006) Thermo-responsive chitosan-grafted-poly(N-isopropylacrylamide) injectable hydrogel for cultivation of chondrocytes and meniscus cells. *Macromolecular Bioscience*, **6**(12), 1026–1039.
153. Creuzet, C., Auzely-Velty, R. and Rinaudo, M. (2006) Synthèse et étude d'hydrogels thermosensibles obtenus par modification chimique contrôlée du chitosane. *Actualite Chimique*, **294**, 34–38.
154. Rinaudo, M. (2011) New amphiphilic graft copolymers based on polysaccharides. *Carbohydrate Polymers*, **83**, 1338–1344.
155. Peniche, C. and Arguelles-Monal, W. (2001) Chitosan based polyelectrolyte complexes. *Macromolecular Symposia*, **168**, 103–116.
156. Arguelles-Monal, W. and Peniche, C. (1988) Study of the interpolyelectrolyte reaction between chitosan and carboxymethylcellulose. *Makromolekulare Chemie-Rapid Communications*, **9**: 693–697.
157. Daly, M.M. and Knorr, D. (1988) Chitosan-alginate complex coacervate capsules: effects of calcium chloride, plasticizers, and polyelectrolytes on mechanical stability. *Biotechnology Progress*, **4**, 76–81.
158. Majima, T., Funakosi, T., Iwasaki, N. *et al.* (2005) Alginate and chitosan polyion complex hybrid fibers for scaffolds in ligament and tendon tissue engineering. *Journal of Orthopaedic Science*, **10**, 302–307.
159. Iwasaki, N., Yamane, S-T., Majima, T. *et al.* (2004) Feasibility of polysaccharide hybrid materials for scaffolds in cartilage tissue engineering: evaluation of chondrocyte adhesion to polyion complex fibers prepared from alginate and chitosan. *Biomacromolecules*, **5**, 828–823.
160. Chung, T.W., Yang, J., Akaike, T. *et al.* (2002) Preparation of alginate/galactosylated chitosan scaffold for hepatocyte attachment. *Biomaterials*, **23**, 2827–283.
161. Donati, I., Haug, I.J., Scarpa, T. *et al.* (2007) Synergetic effects in semidilute mixed solutions of alginate and lactose-modified chitosan (Chitlac). *Biomacromolecules*, **8**, 957–962.

162. Donati, I., Borgogna, M., Turello, E. *et al.* (2007) Tuning supramolecular structuring at the nanoscale level: nonstoichiometric soluble complexes in dilute mixed solutions of alginate and lactose-modified chitosan (Chitlac). *Biomacromolecules*, **8**, 1471–1479.
163. Diaz-Rojas, E.I., Pacheco-Aguilar, R., Lizardi, J. *et al.* (2004) Linseed pectin: gelling properties and performance as an encapsulation matrix for shark liver oil. *Food Hydrocolloids*, **18**(2), 293–304.
164. Arguelles-Monal, W., Cabrera, G., Peniche, C. and Rinaud, M. (2000) Conductimetric study of the interpolyelectrolyte reaction between chitosan and polygalacturonic acid. *Polymer*, **41**, 2373–2378.
165. Demartini, A., Alloisio, M. and Cuniberti, C. (2004) Preparation and characterization of chitosan and chitosan/hyaluronic acid hydrogels for biomedical applications. *Journal of Applied Biomaterials & Biomechanics*, **2**, 208.
166. Vasiliu, S., Popa, M. and Rinaudo, M. (2005) Polyelectrolyte capsules made of two biocompatible natural polymers. *European Polymer Journal*, **41**, 923–932.
167. Kim, S.J., Shin, S.R., Lee, K.B. *et al.* (2004) Synthesis and characteristics of polyelectrolyte complexes composed of chitosan and hyaluronic acid. *Journal of Applied Polymer Science*, **1**, 2908–2913.
168. Yamane, S., Iwasaki, N., Majima, N.T. *et al.* (2005) Feasibility of chitosan-based hyaluronic acid hybrid biomaterial for a novel scaffold in cartilage tissue. *Biomaterials*, **26**, 611–619.
169. Rusu-Balaita, L., Desbrières, J. and Rinaudo, M. (2003) Formation of a biocompatible polyelectrolyte complex: chitosan-hyaluronan complex stability. *Polymer Bulletin*, **50**, 91–98.
170. Kubota, N. and Kikuchi, Y. (1998) Macromolecular complexes of chitosan, in *Polysaccharides. Structural Diversity and Functional Versatility* (ed. S. Dumitriu), Marcel Dekker, Inc., New York, pp. 595–628.
171. Quemeneur, F., Rinaudo, M., Maret, G. and Pepin-Donat, B. (2010) Decoration of lipid vesicles by polyelectrolytes: mechanism and structure. *Soft Matter*, **6**, 4471–4481.
172. Rinaudo, M., Quemeneur, F. and Pepin-Donat, B. (2009) Stabilization of liposomes against stress using polyelectrolytes: interaction mechanisms, influence of pH, molecular weight, and polyelectrolyte structure. *International Journal of Polymer Analysis and Characterization*, **14**(8), 667–677.
173. Bordi, F., Sennato, S. and Truzzolillo, D. (2009) Polyelectrolyte-induced aggregation of liposomes: a new cluster phase with interesting applications. *Journal of Physics-Condensed Matter*, **21**, 203102/1–203102/26.
174. Mertins, O., Schneider, P.H., Pohlmann, A.R. and de Silveira, N.P. (2010) Interaction between phospholipids bilayer and chitosan in liposomes investigated by 31P NMR spectroscopy. *Colloids and Surfaces B*, **75**, 294–299.
175. Haidar, Z.S., Hamdy, R.C. and Tabrizian, M. (2008) Protein release kinetics for core-shell hybrid nanoparticles based on the layer-by-layer assembly of alginate and chitosan on liposomes. *Biomaterials*, **29**, 1207–1215.
176. Boddohi, S., Killingsworth, C.E. and Kipper, M.J. (2008) Polyelectrolyte multilayer assembly as a function of pH and ionic strength using the polysaccharides chitosan and heparin. *Biomacromolecules*, **9**, 2021–2028.
177. Hillberg, A.L. and Tabrizian, M. (2006) Biorecognition through layer-by-layer polyelectrolyte assembly: in situ hybridation on living cells. *Biomacromolecules*, **7**, 2742–2750.
178. Lee, K.Y., Park, W.H. and Ha, W.S. (1997) Polyelectrolyte complexes of sodium alginate with chitosan or its derivatives for microcapsules. *Journal of Applied Polymer Science*, **63**, 425–432.
179. Bartkowiak, A. and Hunkeler, D. (1999) Alginate-oligochitosan microcapsules: a mechanistic study relating membrane and capsule properties to reaction conditions. *Chemistry of Materials*, **11**, 2486–2492.
180. Bartkowiak, A. and Hunkeler, D. (2000) Alginate-oligochitosan microcapsules. II. Control of mechanical resistance and permeability of the membrane. *Chemistry of Materials*, **12**, 206–212.
181. Suh, J.K.F. and Matthew, H.W.T. (2000) Application of chitosan-based polysaccharide biomaterials in cartilage tissue engineering: a review. *Biomaterials*, **21**, 2589–2598.
182. Lavertu, M., Méthot, S., Tran-Khanh, N. and Buschmann, M.D. (2006) High efficiency gene transfer using chitosan/DNA nanoparticles with specific combinations of molecular weight and degree of deacetylation. *Biomaterials*, **27**, 4815–4824.

183. Jean, M., Smaoui, F., Lavertu, M. *et al.* (2009) Chitosan-plasmid nanoparticle formulations for IM and SC delivery of recombinant FGF-2 and PDGF-BB or generation of antibodies. *Gene Therapy*, **16**, 1097–1110.
184. Alameh, M.Z., Jean, M., Dejesus, D. *et al.* (2010) Chitosanase-based method for RNA isolation from cells transfected with chitosan/siRNA nanocomplexes for real-time RT-PCR in gene delivery. *International Journal of Nanomedicine*, **5**, 473–481.
185. Thibault, M., Nimesh, S., Lavertu, M. and Buschmann, M. (2010) Intracellular trafficking and decondensation kinetics of chitosan-pDNA polyplexes. *Molecular Therapy*, **18**, 1787.
186. Strand, S.P., Danielsen, S., Christensen, B.E. and Varum, K.M. (2005) Influence of chitosan structure on the formation and stability of DNA-Chitosan polyelectrolyte complexes. *Biomacromolecules*, **6**, 3357–3366.
187. Kadajji, V.G. and Betageri, G.V. (2011) Water soluble polymers for pharmaceutical applications. *Polymers*, **3**, 1972–2009.
188. Kumirska, J., Weinhold, M.X., Thöming, J. and Stepnowski, P. (2011) Biomedical activity of chitin/chitosan based materials – influence of physicochemical properties apart from molecular weight and degree of N-acetylation. *Polymers*, **3**, 1875–1901.
189. Jayakumar, R., Menon, D., Manzoor, K. *et al.* (2010) Biomedical applications of chitin and chitosan based nanomaterials-A short review. *Carbohydrate Polymers*, **82**, 227–232.
190. Aranaz, I., Mengíbar, M., Harris, R. *et al.* (2009) Functional characterization of chitin and chitosan. *Current Chemical Biology*, **3**, 203–230.
191. Dutta, P.K., Dutta, J. and Tripathi, V.S. (2004) Chitin and chitosan: chemistry, properties and applications. *Journal of Scientific and Industrial Research*, **63**, 20–31.
192. Shi, C., Zhu, Y., Ran, X. *et al.* (2006) Therapeutic potential of chitosan and its derivatives in regenerative medicine. *Journal of Surgical Research*, **133**, 185–192.
193. Ravi Kumar, M.N.V., Muzzarelli, R.A.A., Muzzarelli, C. *et al.* (2000) Chitosan chemistry and pharmaceutical perspectives. *Reactive and Functional Polymers*, **46**, 1–27.
194. Liu, J. and Zhu, L. (2009) Method for preparing chitin composite artificial skin that can be used as woundplast. Faming Zhuanli Shenqing, CN 101411897 A 20090422.
195. Wongpanit, P., Sanchavanakit, N., Pavasant, P. *et al.* (2005) Preparation and characterization of microwave-treated carboxymethylchitin and carboxymethylchitosan films for potential use in wound care application. *Macromolecular Bioscience*, **5**(10), 1001–1012.
196. Freier, T., Montenegro, R., Koh, H.S. and Shoichet, M.S. (2005) Chitin tubes for tissue engineering in the nervous system. *Biomaterials*, **26**(22), 4624–4632.
197. Guo, J.X. and Gregory, K.W. (2010) Medical devices for wound dressing and hemorrhage control, and methods. PCT Int. *Appl.*, WO 2010042540 A1 20100415.
198. Foda, N.H., El-Iaithy, M. and Tadros, I. (2007) Implantable biodegradable sponges: effect of interpolymer complex formation of chitosan with gelatin on the release behavior of tramadol hydrochloride. *Drug Development and Industrial Pharmacy*, **33**(1), 7–17.
199. Loke, W-K., Lau, S-K., Yong, L.L. *et al.* (2000) Wound dressing with sustained anti-microbial capability. *Journal of Biomedical Materials Research*, **53**(1), 8–17.
200. Murakami, K., Aoki, H., Nakamura, S. *et al.* (2009) Hydrogel blends of chitin/chitosan, fucoidan and alginate as healing-impaired wound dressings. *Biomaterials*, **31**(1), 83–90.
201. Farzaneh, N-H., Soheila, S-K., Faramarz, A-T. and Zahra, S. (2011) A novel topical biocompatible tissue adhesive based on chitosan-modified urethane pre-polymer. *Iranian Polymer Journal*, **20**, 671–680.
202. Rickett, T.A., Amoozgar, Z., Tuchek, C.A. *et al.* (2011) Rapidly photo-cross-linkable chitosan hydrogel for peripheral neurosurgeries. *Biomacromolecules*, **12**, 57–65.
203. Illum, L. and Davis, S. (2005) Chitosan as a delivery system for the transmucosal administration of drugs, in *Polysaccharides. Structural Diversity and Functional Versatility*, 2nd edn (ed. S. Dumitriu), Marcel Dekker Publ., New York, pp. 643–660.
204. Kotzé, A.F., Hamman, J.H., Snyman, D. *et al.* (2002) Mucoadhesive and absorption enhancing properties of N-trimethyl chitosan chloride, in *Chitosan in Pharmacy and Chemistry* (eds R.A.A. Muzzarelli and C. Muzzarelli), ATEC, Grottammare, Italy, pp. 31–40.

205. Hamman, J.H. and Kotzé, A.F. (2002) Paracellular absorption enhancement across intestinal epithelia by N-trimethyl chitosan chloride, in *Chitosan in Pharmacy and Chemistry* (eds R.A.A. Muzzarelli and C. Muzzarelli), ATEC, Grottammare, Italy, pp. 41–50.
206. Miwa, A., Ishibe, A., Nakano, M. *et al.* (1998) Development of novel chitosan derivatives as micellar carriers of taxol. *Pharmaceutical Research*, **15**, 1844–1850.
207. Liu, W., Sun, S.J., Zhang, X. and De Yao, K. (2003) Self-aggregation behaviour of alkylated chitosan and its effect on the release of a hydrophobic drug. *Journal of Biomaterials Science, Polymer Edition*, **14**, 851–859.
208. Liu, W., Zang, X., Sun, S.J. *et al.* (2003) N-alkylated chitosan as a potential nonviral vector for gene transfection. *Bioconjugate Chemistry*, **14**, 782–789.
209. Ouchi, T., Murata, J-I. and Ohya, Y. (1999) Gene delivery by quaternary chitosan with antennary galactose residues, in *Polysaccharide Applications: Cosmetics and Pharmaceuticals* (eds M.A. El-Nokaly and H.A. Soini), American Chemical Society, Washington DC, pp. 15–23.
210. Mi, F.L., Shyu, S.S., Chen, C.T. and Schoung, J.Y. (1999) Porous chitosan microspheres suitable for controlling the antigen release of Newcastle disease vaccine: preparation of antigen-adsorbed microsphere and in vitro release. *Biomaterials*, **20**, 1603–1612.
211. Park, I.K., Jiang, H.L., Yun, C.H. *et al.* (2004) Release of newcastle disease virus vaccine from chitosan microspheres in vitro and in vivo. *Asian – Australasian Journal of Animal Sciences*, **17**, 543–547.
212. Sacco, L.D. and Masotti, A. (2010) Chitin and chitosan as multipurpose natural polymers for groundwater arsenic removal and As2O3 delivery in tumor therapy. *Marine Drugs*, **8**, 1518–1525.
213. El Zein, A.R., Dabbarh, F. and Chaput, C. (2002) Injectable self-setting calcium phosphate cement, in *Chitosan in Pharmacy and Chemistry* (eds R.A.A. Muzzarelli and C. Muzzarelli), ATEC, Grottammare, Italy, pp. 365–370.
214. Yi, H., Wu, L-Q., Bentley, W.E. *et al.* (2005) Biofabrication with chitosan. *Biomacromolecules*, **6**, 2881–2894.
215. Krajawska, B. (2004) Application of chitin-and chitosan-based materials for enzyme immobilizations: a review. *Enzyme and Microbial Technology*, **35**, 126–139.
216. Fereidoon, S., Janak, K.V.A. and You-Jin, J. (1999) Food applications of chitin and chitosans. *Trends in Food Science & Technology*, **10**, 37–51.
217. Agullo, E., Rodriguez, M.S., Ramos, V. and Albertengo, L. (2003) Present and future role of chitin and chitosan in food. *Macromolecular Bioscience*, **3**(10), 521–530.
218. Srinivasa, P.C. and Tharanathan, R.N. (2007) Chitin/chitosan-safe, ecofriendly packaging materials with multiple potential uses. *Food Reviews International*, **23**(1), 53–72.
219. Kablov, V.F. and Ioshchenko, Y.P. (2009) Modified paper with improved gas barrier, heat/fire resistance and thermal oxidation stability for use in construction. Russ. RU 2355842 C1 20090520.
220. Rodrigues, A.C., Boroski, M., Shimada, N.S. *et al.* (2008) Treatment of paper pulp and paper mill wastewater by coagulation-flocculation followed by heterogeneous photocatalysis. *Journal of Photochemistry and Photobiology A-Chemistry*, **194**(1), 1–10.
221. Ganjidoust, H., Tatsumi, K., Yamagishi, T. and Gholian, R.N. (1997) Effect of synthetic and natural coagulant on lignin removal from pulp and paper wastewater. *Water Science and Technology*, **35**(2), 291–296.
222. Masuda, M. and Yokota, N. (2006) Mixture yarns for clothing with antibacterial properties and tender handle, comprising yarns containing chitin fibers or chitosan fibers and antibacterial woven and knitted fabrics therefrom. Jpn. Kokai Tokkyo Koho, JP 2006225785 A 20060831.
223. Pospieszny, H., Struszczyk, H., Chirkov, S.N. and Atabekov, J.G. (1994) New applications of chitosan in agriculture, in *Chitin World* (ed. S.Z. Karniki), Wirtschaftsverlag NW, Bremerhaven, Germany, pp. 246–254.
224. Doares, S.H., Syrovets, T., Weiler, E.W. and Ryan, C.A. (1995) Oligogalacturonides and chitosan activate plant defensive genes through the octadecanoid pathway. *Proceedings of the National Academy of Sciences of the United States of America*, **92**, 4095–4098.

5

Lignin Matrix Composites from Natural Resources – ARBOFORM®

Helmut Nägele[1], Jürgen Pfitzer[1], Lars Ziegler[1], Emilia Regina Inone-Kauffmann[2], Wilhelm Eckl[2], and Norbert Eisenreich[2]
[1] *Tecnaro GmbH, Germany*
[2] *Fraunhofer Institute for Chemical Technology ICT, Germany*

5.1 Introduction

Most of the energy, organic chemicals and synthetic organic polymers in the world are obtained from fossil resources such as petroleum and natural gas. To preserve resources, the utilization of biomass must be an obligatory topic for the future. It requires ecological innovations in future product design and has to initiate new approaches in material development based on raw material from renewable resources [1–22]. Current bio-based materials do not fully meet industrial and customers' requirements in all cases. For example, wood, a traditional construction material, remains competitive in a challenging environment due to high resistance against impact, thermal loading and styling opportunities. Disadvantages are the high costs, splintering on breakage and the use of environmentally harmful glues when shaping complex profiles via layered configurations. Plastics including all types of composites show a broad range of properties that meet differing needs for products in various industrial branches and mass consumer applications [23]. Disadvantages are softening at higher temperatures, shrinkage on processing and environmental problems on recycling – especially in case of multicomponent compounds, composites and sandwich structures. Driving forces of the research for these environmentally friendly materials are the branches of mass consumer, automotive, construction and electronic industries which have to aim towards a zero-waste strategy, CO_2 saving in manufacturing and product responsibility at long term to keep competitiveness in a global market [1–22] (see also http://www.biocomp.eu.com/uploads/Final_Summary_Report.pdf, accessed 22 June 2013). A new substitute should meet both requirements: saving wood resources for processed

Bio-Based Plastics: Materials and Applications, First Edition. Edited by Stephan Kabasci.

wooden parts of complex shapes, and saving crude oil for synthetic plastics [24–37]. The lignin matrix composite family ARBOFORM [24–34] uses lignin as main component, which is superabundantly available from the worldwide output of pulp mills in masses surpassing 50×10^6 tons every year. The use of lignin in ARBOFORM needs no further chemical processing in contrast to extracts from biorefinery techniques. Mechanically treated natural fibres can be incorporated without any additional processing in chemical engineering.

In the following, results are reported from research into and development of novel engineering plastics with properties as expected from classical engineering plastics or wood. These comprise high stiffness and impact strength, surface smoothness, various functionalities like flame retarding, thermal and electrical conductivity, various colours, full absence of processing chemicals, and modifications of formulations. Approaches for advanced bioinspired materials are outlined. Selected examples provide an overview on diversified applications to mass consumer goods and industrial parts, developed earlier or being under detailed investigation currently. They demonstrate the relations of formulation and material properties, as well as appropriateness for the targeted applications.

5.2 Approaches for Plastics Completely Made from Natural Resources

Much work has been done on using chemical raw materials from biomass to produce materials similar to synthetic plastics, especially when considering high-quality engineering ones. An enormous amount of money was spent on these developments but there was only limited success. The reasons were that the resulting materials had to compete with the mass production synthetic polymers like polyethylene and polystyrene. Past and recent reviews and textbooks (see e.g. references [1–22]) give a detailed overview on the current development and the progress achieved [15–17]. The approach to biocomposites is shown schematically in Figure 5.1. Some examples of these materials at a mature state of development are:

- Starch: only useful in low-grade applications; if higher quality material is made then admixtures of synthetic polymers are required.
- Poly (lactic acid) PLA, poly (hydroxybutyrates) PHB: mainly useful in the packaging industry for bottles, beakers, and so on where these materials have to compete with cheap synthetic plastics. In addition, there are no established production and processing facilities in operation. Therefore, the acceptance at the market is difficult.
- Natural fibres incorporated in synthetic polymeric matrices (mainly polypropylene): had a real industrial success; natural additives like lignin were also used.

Natural fibres show many advantages over glass fibres when used as reinforcement of synthetic polymers (see Table 5.1): the relatively high density of glass fibres (≈ 2.5 g/cm^3) compared to cellulose or ligno-cellulose fibres of 1.5 g/cm^3 makes lightweight applications possible.

Composites from natural fibres have not yet been fully established in high-tech industry because a sufficient quality for engineering applications is not adequately offered (e.g. by Natural Fiber Composites Inc. and North Wood Plastics). Fibres are available from many plants (highlighted of high quality): for example, wood, abaca, coir, cotton, flax, hemp, henequen, istle, jute, kenaf, ramie, sisal or sunn.

An approach to use also biopolymers, like PLA, PHB, starch, resins from furfuryl alcohol and vegetable oils in wood like plastics has often been considered in the past [2, 8, 9, 14–22, 35–44]. Compounding of the fibres with the thermoplastic biopolymers provides granules

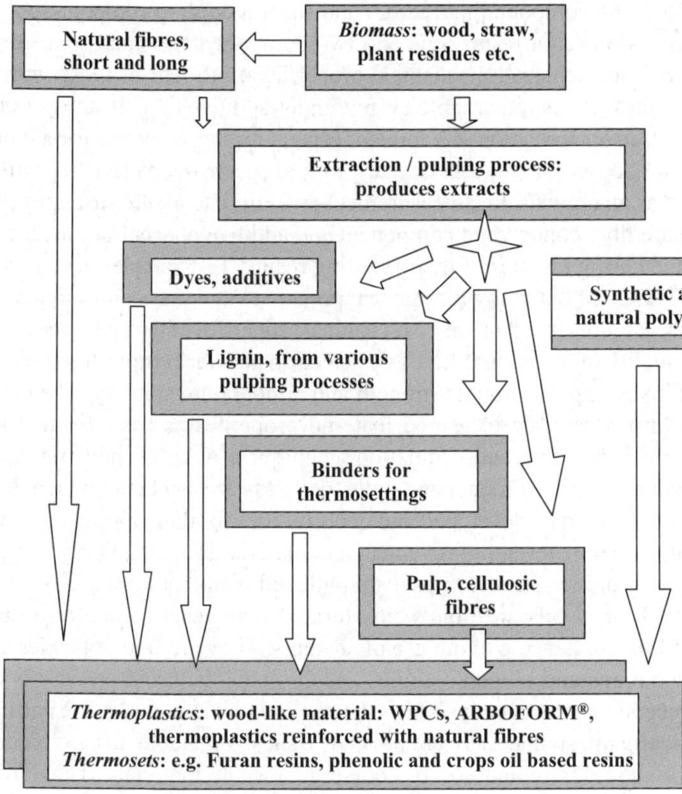

Figure 5.1 *Plastic-like materials from biomass. (© Fraunhofer ICT, 2013.)*

for injection moulding or extrusion of model products and profiles. The research has already achieved improvement of temperature stability, a 50% increase in impact strength and a flame resistance fulfilling classifications appropriate for extensive use in electronic products. An example of a successful broad research approach gave the Biocomp project (see e.g. Biocomp project final report [14–17] and recent reviews [19–22]). Advanced processing methods involve cone extrusion, film stacking and long fibre direct processing. Thermoset processing uses all important types of related methods like vacuum infusion, compression

Table 5.1 *Properties of natural fibre reinforced synthetic plastics materials.*

Material	Density (g/cm^3)	Tensile strength (MPa)	Failure strain (%)	Bending modulus (GPa)
E-Glass fibres	2.40			
Natural fibres	1.15–1.50			
PP[a]	0.91	26	>50	1.55
PP + Glass[b]	1.14	80	7	4.10
PP + Woodfibres[c]	1.04	26	5.4	3.30

[a] Stamylan (DSM).
[b] Stamylan (30% glass fibres).
[c] WPC (PP + 40% wood fibres).

moulding, bulk moulding compounding (BMC) and sheet moulding compounding (SMC). It is based on low emission furan resins, one and two component versions, and samples with hemp and flax fibre mats achieved mechanical properties up to half of those with polyester resins. The flame retardancy is remarkable even without additives. In all steps of processing, standard methods of characterization are applied, especially those needed for a database and the specific characterization for end users. Near-infrared spectroscopy (NIR-spectroscopy) is included to be used as an online/inline measurement procedure to assure sustained qualities of materials, to measure fibre content and component and additive concentrations. It enables the determination of the curing kinetics of thermosetting resins. Test samples are investigated by standard test procedures for obtaining mechanical properties and thermal behaviour as well as by methods to analyse failure mechanisms. Mechanical properties of the materials are improved by increasing the quality of the natural fibres by selection and various pre-treatment methods. Modifiers and additives improve impact strength and temperature stability. The development of parts for applications uses the measured material properties as basis for the design and tool construction, the processing and simulation techniques. A lot of industrial applications require flame retardancy. Various model and industrial parts are available, which demonstrate the performance of the materials developed and opportunities for their application. Work in the near future aims at further material improvement and material functionalizing using additives to fulfil high classes of fire resistance, impact strength and temperature stability.

Raw and refined linseed oils are highly unsaturated (and reactive) and are obtained by simple pressing of linseed, hence without use of solvents. They are free of pesticide residues, heavy metals, mycotoxins and biogenic amines, as the manufacturing process is constantly monitored by corresponding trace analysis tests. Under ambient conditions, the partial pressure of the oils is practically nil, so that there are no VOC issues. Vegetable oils are biodegradable, have no toxicity or eco-toxicity and are, of course, renewable materials. They are available worldwide in very large amounts.

Biomass-based furan resins use furfuryl alcohol (FA) [17]. It has affected the success of the rather old biomass-based chemical industry. This reactive alcohol plays a vital role in the production of foundry sand binders. For over 30 years these furan resins have been extensively used in the foundry industry to produce cores that yield high-quality castings. Despite the good chemical, mechanical and thermal properties of FA resins/polymers such as corrosion resistance, flame retardancy, low smoke emission and excellent char formation, the consumption of these bio-based resins in other markets is of minor significance:

- This type of bio-based adhesive can be used as sustainable substitute for petroleum-based resins. Moreover, furan resins are price competitive with synthetic counterparts.
- Furfural industry has been neglected as a source for bio-based adhesives.
- Excellent mechanical properties of fibre-reinforced plastics (epoxy, PUR, phenolics) are necessary when a bio-based thermoset system is used.
- Low VOC furan resins will be available.

5.3 Formulation of Lignin Matrix Composites (ARBOFORM)

5.3.1 Lignin

The name lignin comes from the Latin word *lignum* which means wood. The discovery of lignin and its exploitation began over 150 years ago, when scientists considered wood to be a definite chemical substance based on three main components: cellulose, lignin and hemicellulose.

Table 5.2 Lignin content in some biomass resources.

Biomass	Cellulose (%)	Hemicellulose (%)	Lignin (%)
Softwood	40–50	20–30	25–35
Hardwood	40–50	30–40	15–20
Hemp	60–70	10–15	3–10
Hemp shives	45–55	15–20	20–30
Wheat straw	20–30	45–55	10–20
Miscanthus	35–40	20–30	10–15
Bagasse	30–40	20–30	10–20

Lignin is a kind of binder between the wood cells, which imparts rigidity, creating a very strong composite material, resistant to impact, compression and bending [38–40].

The content of lignin in different types of biomass resources varies strongly; it is especially high in wood (see Table 5.2).

In addition, many residues of agricultural products that are currently unused contain substantial amounts of lignin (see Table 5.3, [41]).

The natural polymer lignin is made up of phenylpropane units and, next to the polysaccharides, it is the most abundant organic polymer in the plant world, probably being the second most abundant naturally occurring material on earth.

The lignin macromolecule has a variety of functional groups that have an impact on its reactivity. Lignin mostly contains methoxyl groups, phenolic hydroxyl groups and a few terminal aldehyde groups. Only a small proportion of the phenolic hydroxyl groups are free because most are occupied in linkages to neighbouring phenylpropane units. Carbonyl and alcoholic hydroxyl groups are incorporated into the lignin structure during enzymatic dehydrogenation. Tables 5.4 and 5.5 illustrate the linkages of some common functional groups found in lignin [42]. Some monomers of lignin are plotted in Figure 5.2.

In recent years, additional incentives to exploit the more efficient use of the lignin-containing wastes from the chemical pulping and wood-using industries have been established. Sulfite liquors and their sulfonated lignin components (lignosulfonates) still represent the most widely

Table 5.3 Content of lignin, hemicellulose, cellulose, nitrogen and ash (given in %) referring different rests and residues from plants and fruits.

	Lignin	Cellulose	Hemi-cellulose	Nitrogen	Ash
Meal from maize (hominy)	2.0	11.0	42.0	1.84	1.0
Malt sprouts hulls	3.0	15.0	28.0	4.50	7.0
Brewers grains, dehydrate	6.0	18.0	22.0	4.16	4.0
Grape stems	12.0	–	–	<0.5	–
Straw from rice	12.5	32.1	24.0	–	17.5
Coffee ground	15.0	–	9.0	0	2.0
Hay, weathered	15.0	30.0	13.0	1.60	8.0
Straw from Soybean	16.0	38.0	16.0	0.83	6.0
Bamboo	20.1	–	19.6	–	3.3
Hulls from peanuts	23.0	42.0	9.0	1.25	4.0
Cocoa bean	25.0	–	–	2.56	–
Olive waste	28.0	–	–	0	–
Grape pomace	35.0	19.0	1.0	0	10.0
Fruit pits	40.0	–	–	–	–

Table 5.4 Linkages found in softwood lignin.

Linkage type	Dimer structure	Percentage of total linkage (%)
β-O-4	Phenylpropane β-aryl ether	45–50
5-5	Biphenyl and Dibenzodioxocin	18–25
β -5	Phenylcoumaran	9–12
β -1	1,2-Diaryl propane	7–10
α-O-4	Phenylpropane α-aryl ether	6–8
4-O-5	Diaryl ether	4–8
β-β	β-β-linked structures	3

Table 5.5 Functional groups in softwood lignin per 100 phenyl propane units.

Functional group	Abundance per 100 C9 units
Carbonyl	10–15
Benzyl	15–20
Phenolic hydroxyl (free)	15–30
Methoxyl	92–96

used waste products of the pulp and paper industry. The relatively large-scale use of lignosulfonates is based on their physical and chemical properties. The dispersing properties of the lignosulfonates, combined with their ability to sequester metallic ions, have contributed to their successful use as additives in the preparation of oil-well drilling mud. Other application fields of lignosulfonates include road binders, animal feed pelletizers, cement products, industrial cleaners, trace element chelation, leather, tanning, metallic ion complexation for agricultural use, and boiler-scale inhibition. The expected complete utilization of these products in the future has resulted in continued physical and chemical studies of lignosulfonates in solution.

These are mainly processes of the pulp and paper industry, which consumes most of the biomass, nearly exclusively wood, to produce cellulose for paper, packing and tissues. Most applied chemical processes for the breaking down ligno-cellulose are shown in Figure 5.3.

Figure 5.3, also shows that chemicals like $[SO_4]^{2-}$ and $[SO_3]^{2-}$ as well as $[CO_3]^{2-}$ are always needed.

A variety of methods exits to extract native (nonmodified) lignin totally or partially from lignocellulose biomass. These methods use organic solvents like methanol or ethanol (Organosolv process [43]) or even pure water (steam explosion process, Aquasolv process [41, 44–47]), but

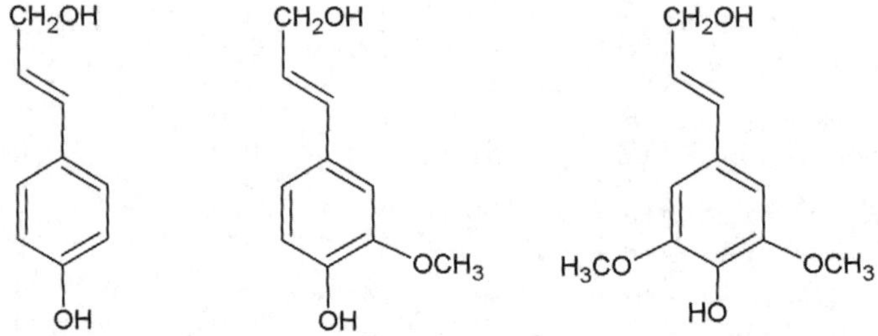

Figure 5.2 Lignin monomers: p-coumaryl alcohol, coniferyl alcohol, sinapyl alcohol. (© Fraunhofer ICT, 2013.)

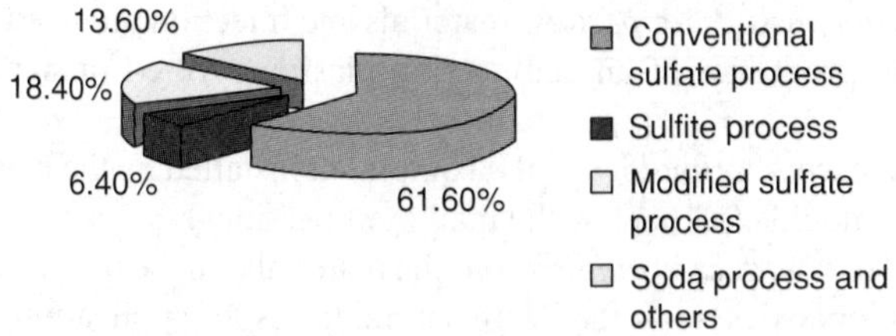

Figure 5.3 *Conventional chemical processes to receive delignificated cellulose for subsequent treatments by paper industry to produce paper, packing and tissues. (Data taken from ref. [41].)*

need special separation techniques to recover the lignin. Only a few of these new extraction processes, such as Organocell, an alkaline sulfite anthraquinone methanol, (ASAM), pulping process consisting of three main sections: digestion and brownstock washing, screening and four-stage bleaching, evaporation and methanol rectification), Formacell and Milox were developed in the past for the cellulose industry to include some extended ecological aspects [43].

Alkaline (both sulfate, commonly known as Kraft, and soda) pulping processes are the most dominant methods for the production of chemical wood pulps. The economics of these processes require the burning of the lignin in the spent liquor, both for recovering the pulping chemicals and for providing process energy. As a result, the Kraft and soda pulp mills in the United States burn some million tons of lignin annually. Although this transformation of lignin into energy is an economical method for the use of the spent pulping liquor, it is not an ideal way to dispose such a potentially valuable raw material.

Substantial efforts to use lignin for material development have already been undertaken. However, the applications for thermoset binders or fillers in composites were only of limited success (see e.g. [48–52]).

5.3.2 Basic Formulations and Processing of ARBOFORM

The approach to use lignin as prime matrix component resulted in a new family of a plastic-like and also wood-like biocomposites meeting relevant industrial engineering challenges [24–34]. One important aspect of a success is in the fact that the material can be processed like a thermoplastic by injection and compression moulding using standard technologies. In order to start developing such a material, model mouldings were designed for various industrial branches, which were, after meeting the requirements, transferred to industrial mouldings for mass manufacturing. The development aimed at substituting plastics and wood in selected applications. Following the 'green' vision the material has to consist only of natural renewable constituents with the main components being lignin, natural fibres and additives to fully meet 'green' technology requirements. As many types of lignin are precipitated by chemicals man-made chemicals (sulfur, sodium or chlorine trace residues as in the case of current types of lignin) are present in the resulting material. A purely native lignin (e.g. from an Aquasolv process, see below) might substitute the current lignin types in the future, as early experiments proved successfully.

Basic objectives of the actual manufacturing and the future developments include:

- *Biocomposite completely from natural resources*: only natural components without man-made chemicals and additives should be used in the materials.

- *Achieve engineering plastics properties:* materials and injection moulded model component parts are to be of properties of an engineering plastic, verified in standard property test procedures.
- *Use standard processing technologies:* the equipment installed by the manufacturers should be used, even the moulds (perhaps with small modifications).
- *Identify technical advantages in application:* the materials can be used to obtain high quality engineering thermoplastics reinforced with natural fibres for application in industrial goods of mass consumer, electronic, automotive industry, and so on.
- *Save energy:* processing at low temperatures of injection moulding (<160 °C), mechanical properties similar to those of polyamide, thermally stable at least till 95 °C or even 105 °C.

A composite material of this type fulfilling these criteria was made of lignin, which can be processed as a thermoplastic material. It has been developed by the company Tecnaro GmbH together with Fraunhofer Institute for Chemical Technology ICT [24–34]. Based on sales of a family of compounds of material by Tecnaro GmbH, both partners continue its development to broaden applicability by modifying composition to achieve system compatible properties. The trade name of this material is ARBOFORM® (Latin: *arbor* – tree). The components are mixed and compacted/agglomerated without heating them to manufacture granules as a compound. This compound can be fed to standard injection moulding and compression moulding machines widely used in industrial plastics manufacturing. On heating, the mixture can be thermo-plastically processed to moulded parts or panels even using moulds similar to synthetic plastic manufacturing. If sophisticated parts have to be designed, however, the mould design is recommended to be conform to the special properties of the compound.

Lignin acts as binding matrix in this natural fibre reinforced composite, resulting in a biocomposite. The processing of the lignin-based material according to industrial needs implies the interesting advantage compared to a wood, while being composed of the main constitutes of wood or other biomass. Its appearance, however, and some of its properties, like density, haptic quality, and so on come close to those of wood. Standard surface treatment like polishing and/or varnishing can generate a noble wood-like appearance. The thermal expansion, which is similar to that of wood, makes the material compatible with wood veneers, which can be introduced as inlays into the moulds. Exposure to water for weeks does not lead to strong swelling, but rather does not change the shape in any way. In general, the following variations in composition are possible (Table 5.6):

Table 5.6 *Range of compositions.*

Matrix:		
Lignin	30%	60%
Fibre reinforcement (loose fibres):		
hemp (e.g. H1)	10%	60%
flax (e.g. F1)	10%	60%
Additives		
processing aids (PrAi)	0%	10%
impact modifier	0%	20%
Flame retardants	0%	15%

Some general guidelines for optimized lignin composite processing (ARBOFORM F-grades) on injection moulding machines are summarized below:

- Do not predry lignin-compounds, normal storage conditions (dry, not more than 25 °C).
- Achieve the manufacturing temperature of the injection moulding machine with low-melting PE-LD.
- Do not keep the material longer than 15 minutes with manufacturing temperature in the injection moulding machine, in the case of longer dwell time clean intermediately with LD-PE. Clean the injection moulding machine with PE-LD or cleaning granules after manufacturing.
- Do not keep lignin-compounds together with other hot polymers; cool down immediately, for example, using water.
- It is normal for the material to develop fumes and smell when discharged into the open air.
- Smell of mouldings will be considerably less after 2 days and change its characteristics (to vanilla-like) after 2 weeks.

Screw design:	Standard three-zone-screw (no mixing elements), standard non-return valve
Nozzle:	open (2.5 to 5.0 mm)
Screw feed rate:	Depending on frictional heat and cycle time requirement (feed rate slowed down if necessary)
Back pressure:	High enough to prevent the screw running jerkily
Injection pressure:	More than 50% of machine capacity in most cases
Injection speed:	50–90% of the maximum rate possible
Holding pressure:	20–60% of the machine capacity
Time of holding pressure:	about 0.5–3 s (very low shrinkage)
Mould temperature:	20–55 °C (depending on the flow distance)
Hot runner:	155–165 °C
Cooling periods:	Up to a maximum of 20% higher than for unfilled thermoplastics

Lignin, extracted by various processes and stemming from various sources, is useful to generate such a biocomposite; the choice of the lignin depends on the target application required by the final product. For applications where the sulfur content is not important, lignosulfonates can be included. For applications where no man-made chemicals are allowed, native lignin types obtained by a chemical-free process (e.g. Aquasolv or Organosolv) can be used. Additives define the properties, especially for functionalities like fire retardancy, conductivity, and so on in a similar way to standard plastics and predominantly engineering plastics for high tech utilization. However, the additives for ARBOFORM® generating the desired functions are also of natural origin. A sample of granules and some model parts are shown in Figure 5.4.

5.3.3 The Influence of the Fibre Content

ARBOFORM formulations show a dominant dependence of the mechanical properties on the fibre content (especially hardwood fibres), thus enabling material properties of the injection

Figure 5.4 Granules of a lignin-based composite and injection moulded sample. (© Tecnaro GmbH, 2013.)

moulded parts to be varied in a broad range. Figure 5.5 exhibits a strong increase of the tensile strength related to the content of fibres in the formulations using hardwood fibres. At about 45% fibre content a good balance between stiffness and strength is obtained. For the other fibre types a substantial high modulus is reached even at lower fibre contents.

The processing leads to an effective bonding between fibres and lignin matrix with a homogeneous distribution of the fibres without bonding agents. It is evident from analysing zones of break of the material under linear tension experiments that the fibres are not being pulled out of the matrix, which confirms the good bonding behaviour (Figure 5.6).

Optical microscopy shows that a higher amount of hardwood can be integrated into the lignin matrix whereas the fibres of softwood need more lignin to be fully covered (Figures 5.7 and 5.8). Therefore, better mechanical properties are already obtained at lower fibre contents and cannot be further increased by higher amounts of fibres than 45–50%.

Further data and ranges of ARBOFORM to be covered by the standard version of the material are published in various data sheets of Tecnaro GmbH, Ilsfeld-Auenstein (see an example

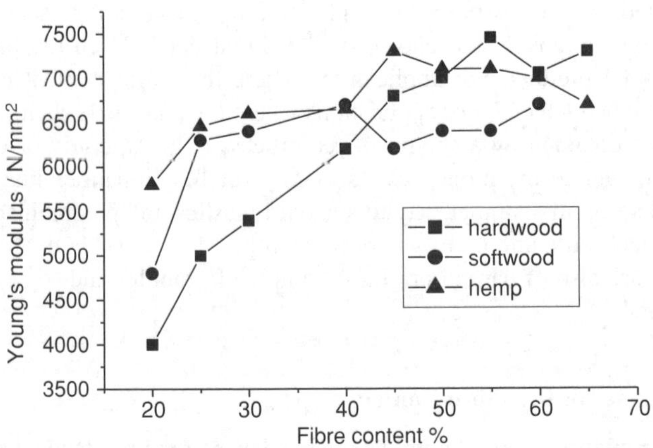

Figure 5.5 Dependence of Young's Modulus on the fibre content. (© Fraunhofer ICT, 2013.)

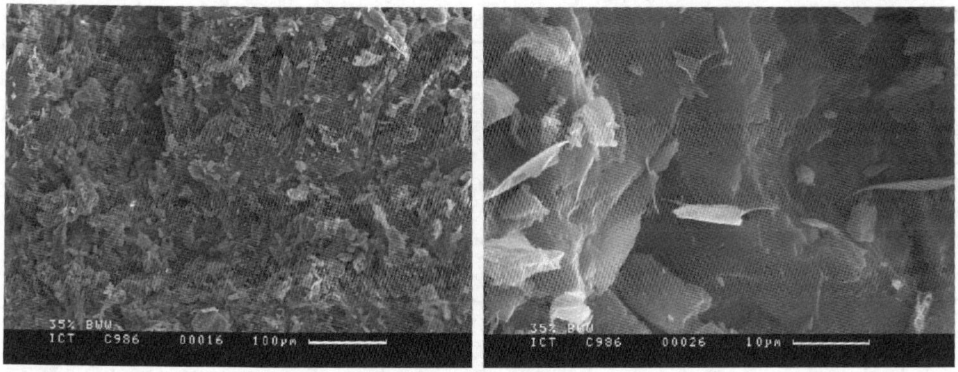

(a) (b)

Figure 5.6 *Breaking zone on tensile tests, where no de-bonding of fibre/matrix can be observed. (© Fraunhofer ICT, 2013.)*

(a) (b) (c)

Figure 5.7 *Hardwood fibres in various amounts by optical microscopic pictures. (© Fraunhofer ICT, 2013.)*

(a) (b) (c)

Figure 5.8 *Softwood fibres in various amounts by optical microscopic pictures. (© Fraunhofer ICT, 2013.)*

Table 5.7 Data sheet of a standard ARBOFORM material.

Mechanical qualities	Norm	Unit	Mean[a]
Tension at break	DIN 53 455	N/mm^2	18
Elongation at tear	DIN 53 455	%	0.3
Modulus of elasticity in tension	DIN 53 457	N/mm^2	6000
Bending stress	DIN EN ISO 178	N/mm^2	40
Impact strength (Charpy)	DIN EN ISO 179	kJ/m^2	2
Ball indentation hardness	DIN 53 456	N/mm^2	122
Thermal qualities	**Norm**	**Unit**	**Mean[a]**
Linear expansion coefficient	DIN 53 752	1/°C	1.8.10E-5
Vicat temperature (A)	DIN 53 460	°C	82.5
Martens temperature	DIN 53462	°C	54
Heat conductivity	DIN 52 612	W/(m.K)	0.384
Glow wire test	DIN IEC 60695-2-1	—	650 °C passed
Electrical qualities	**Norm**	**Unit**	**Mean[a]**
Conductivity, surface	DIN 53 482	G Ω	5
Conductivity, passage	DIN 53482	G Ω	3
Other qualities	**Norm**	**Unit**	**Mean[a]**
Decrease in size	—	%	0.2
Density (in moulded parts)	ISO 1183	g/cm3	1.3
Water content (at 105 °C)	—	%	6
Migration of certain elements	DIN EN 71-3	—	passed
Resistance to saliva and sweat	DIN 53 160	—	passed
Reaction to fire/Flammability	DIN 4102-B2	—	passed
Flow length (spiral test, 4 * 4 mm)	—	cm	60

[a] Depending on material composition.

in Table 5.7). Materials with special properties are developed according to the customer's specifications, if possible.

5.4 Chemical Free Lignin from High Pressure Thermo-Hydrolysis (Aquasolv)

5.4.1 Near Infrared Spectroscopy of Lignin Types

The natural polymer lignin is extracted from biomass, mainly wood by various technologies as described above. It accumulates in masses up to more than 50×10^6 tons at chemical pulp mills every year, worldwide, as a by-product of the pulp and paper industry [48]. When using it for material development, the most abundantly available types of lignin are modified by chemicals on extraction depending on the type of process and these might be used for their identification [53–55].

In order to use biomass in an environmentally friendly manner, various chemical-free approaches for lignin extraction have been applied. Pure subcritical water and supercritical water technologies (Aquasolv) have recently received increasing attention based on earlier research [41, 44–47, 56–65]. Subcritical water and supercritical water behave very differently

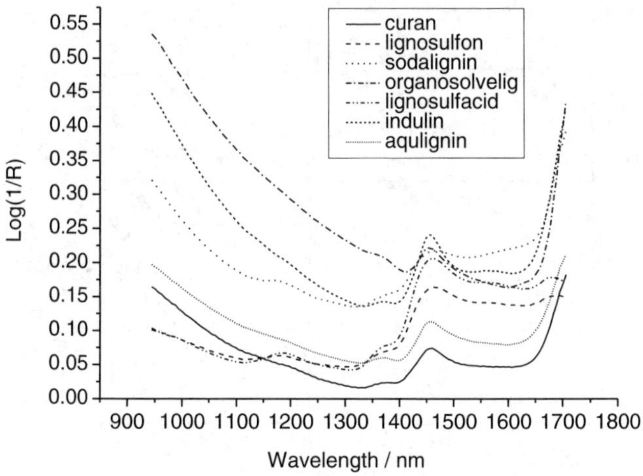

Figure 5.9 *NIR reflectance spectra of various types of lignin. (© Fraunhofer ICT, 2013.)*

from water under normal conditions, especially when considering the ability to dissolve or to extract substances. The ionic product of subcritical and supercritical water, ranging between 10^{-12} and 10^{-10} mol^2/l^2, is about three to four orders of magnitude higher than the ionic product under normal conditions. Therefore, supercritical water can work as a hydrolysis fluid. The dielectric constant of water in these treatments is in a range between 15 and 20, which is close to that of organic solvents like acetone under normal conditions. The dielectric constant of water under normal condition is about 80, so the properties of water drastically change when increasing pressure and temperature. Furthermore, in the supercritical state the dielectric constant decreases to about 2.

These conditions lead to interesting extraction yields, but residence times in reactors have to be kept short in order to not degrade the biomass components substantially. In the following, lignin from lignocellulosic biomass was separated by subcritical water ($<$220 °C, $<$5 MPa) and was characterized by various analytical methods. Its efficient utilization as a matrix for thermoplastic materials could be proved. It is important in material manufacturing to have a simple method for quality control of the raw materials. One of them can be Near Infrared Spectroscopy (NIRS), for example [53–55].

There are many types of lignin available; more than 20 are frequently produced. It is important for a reliable lignin-matrix material like ARBOFORM to control the input type and quality [55]. Near-infrared spectroscopy permits online control of thermoplastic processing. It had been used to measure lignin in the past [53, 54]. Reflectance spectra of some types of lignin are shown in Figure 5.9. Whereas differences are not obvious in the main band between 1400 and 1500 nm, the other structure is caused by physical properties like particle size. A multivariate statistical analysis [66, 67] by principal component analysis (PCA, e.g. Unscrambler®), however, can clearly distinguish between the lignin types investigated as plotted in Figure 5.10.

5.4.2 Lignin Extraction by High-Pressure Hydrothermolysis (HPH)

Figure 5.11 shows the principles of the high-pressure hydrothermolysis (HPH) process for lignin extraction and the concept to realize it for experiments with various types of biomass

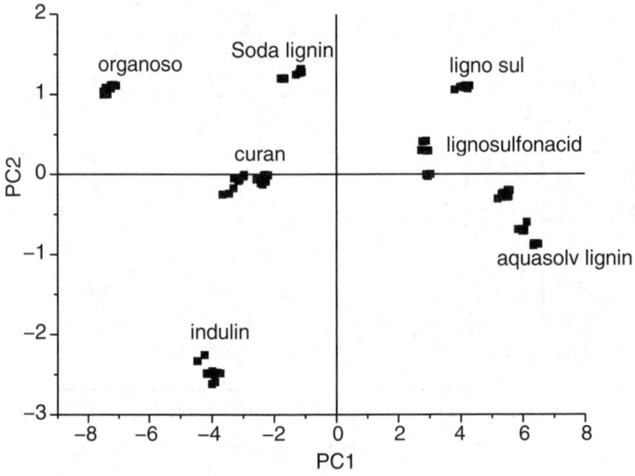

Figure 5.10 *Principal component analysis shows that the different kinds of lignin can be clearly distinguished by NIRS. (© Fraunhofer ICT, 2013.)*

including the further processing to obtain, at the end, native Aquasolv-lignin [41, 46, 47]. Summarizing the activities:

- Various materials, wood residues, straw, hemp shives, and so on, were treated in a laboratory-scale reactor recording details of heating and cooling phases.
- Various separation methods were tested to obtain solid lignin for analysis and for the supply to the injection moulding of material samples.
- Analysis of the lignin by various analytical methods.

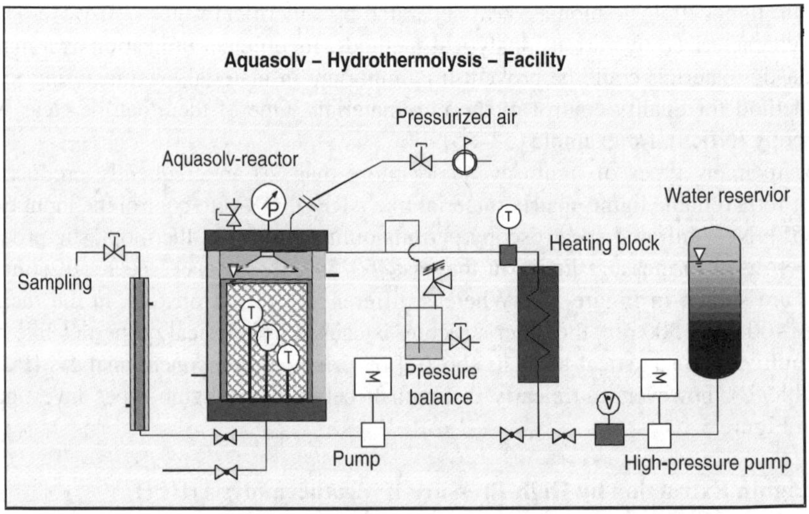

Figure 5.11 *Scheme of the equipment to extract lignin at sub-critical conditions, the Aquasolv process. (© Fraunhofer ICT, 2013.)*

One aspect of the complete HPH philosophy is to combine an ecological approach with an economical extraction process to use all the constituents from inexpensive basic natural (waste) materials. In this case, biological rests and residues out of agriculture and food industry are considered for use. The selection of biomass also depends on the end-products that shall be gained from the biomass, because the content of lignin, hemicellulose, cellulose and even all other organic components varies strongly between different types of biomass. The process was realized at a scale of 12 L, which was sufficient for separation raw materials to enable the generation of granulates for injection moulding and to obtain test samples for material testing. It operates at conditions of:

Volume: 12 L
Temperature range: 100–220 °C, preferable 180–220 °C
Pressure range: Up to max. 4 MPa
Solvent: pure water, no additives
Solid biomass feed: circa 800 g
Water volume: 3–10 L
Pump feed: 3 L per min
Residence time: 10–25 min

The separation of the lignin from the eluate is not trivial. One option is the successive application of ultrafiltration, rotational evaporation, and vacuum drying.

The lignin obtained via the ultrafiltration consists of fine powders which are composed of spheres of about 200 nm diameter (Figure 5.12). Bigger and smaller particles are included.

Results were obtained from experiments using Japanese wood chips treated with HPH at 190 °C and 20 min residence time. For the rotational evaporation a temperature of 35 °C and a pressure of 40 mbar proved to be of highest effectiveness. High-pressure hydrothermolysis gives lignin particle sizes below 300 nm when ultrafiltration is used. The membrane which was applied had a nominal cut-off of 100 kDa (100 000 g/mol). The lignin was characterized by NIR/IR-spectroscopy, molecular weight analysis (Gel Permeation Chromatography – GPC),

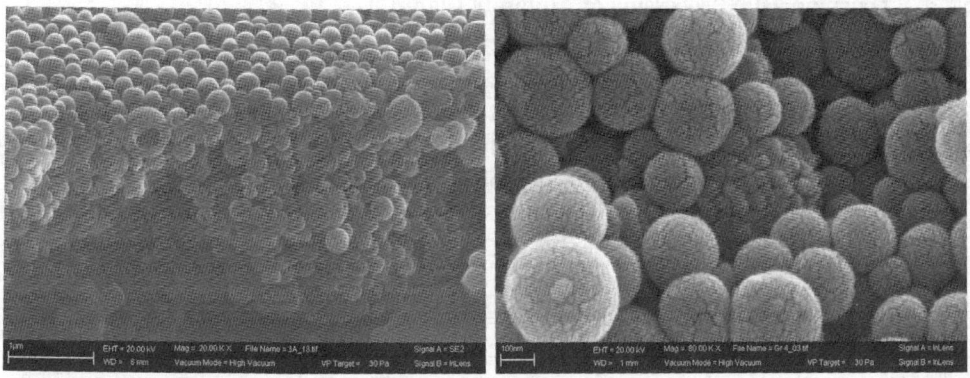

(a) (b)

Figure 5.12 agglomerated HPH-lignin particles (|———| scale 1000 nm, left hand; |—| scale 100 nm, right hand) extracted at a temperature of 190 °C and pressures between 2 to 4 MPa. (© Fraunhofer ICT, 2013.)

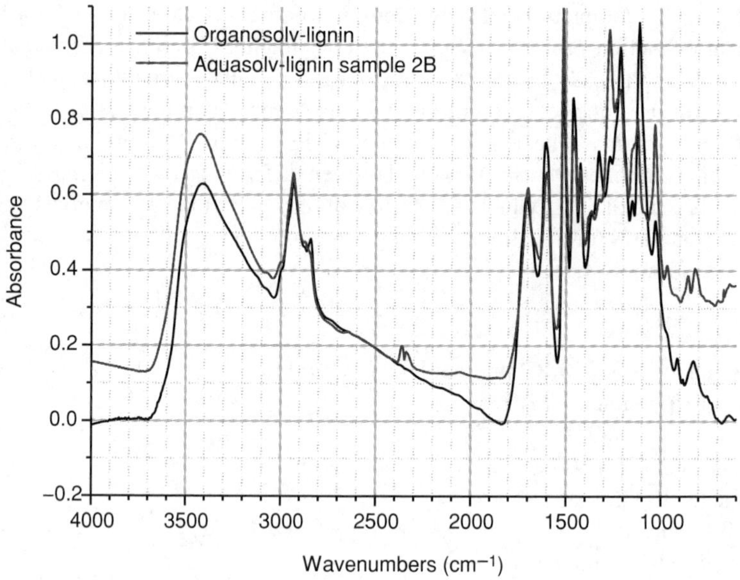

Figure 5.13 IR- Analysis of two samples comparing a sample of HPH Aquasolv lignin with Organosolv lignin. (© Fraunhofer ICT, 2013.)

methods of thermo-analysis and scanning electron microscopy (SEM). IR/NIR spectra are similar to those of other lignin types (see Figure 5.13), however chemometric methods (principal components analysis, PCA, and partial least square, PLS) can clearly distinguish HPH-lignin from the other types as shown above. The molecular weight by GPC is approximately half of that of Kraft or Organosolv lignins, the polydispersity about 2/3 (see Table 5.8).

The molecular weight distribution (data plotted in Figure 5.14) shows that the HPH process under the condition described above extracts the lower molecular weight fraction of lignin present in biomass.

5.4.3 Thermoplastic Processing of Aquasolv Lignin

The separated Aquasolv-lignin was further processed [68, 69]. It started with the standard mixing and pelletizing procedures for the integration of natural fibres and additives to form a compound (like ARBOFORM, see above). Despite the unfavourable conditions and the small amounts of Aquasolv-lignin, the resulting granules were injection moulded using a Babyplast® injection moulding machine to obtain test samples (see Figure 5.3). The application temperature for lignin should not surpass 200 °C to avoid degradation. However, the

Table 5.8 Molecular weight of some lignins.

Sample	Mn	Mw	MP	Mz + 1	Polydispersity
Kraft-Lignin I	623	1240	933	2954	1.99
Kraft-Lignin C	664	1345	1059	3404	2.02
Organosolv Lignin	666	1467	1135	4531	2.20
Aquasolv- / HPH-Lignin from experiments described	505	750	525	1554	1.49

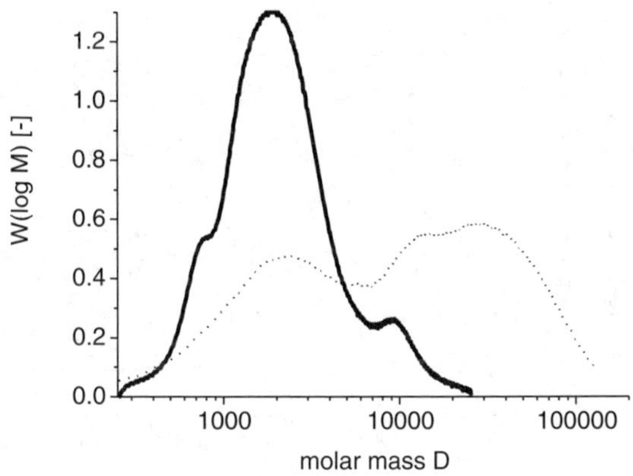

Figure 5.14 The molecular weight distributions for an exemplary HPH lignin of the experiments described in this paper (solid line) in comparison with Curan (broken line). (© Fraunhofer ICT, 2013.)

methods of thermal analysis showed that the thermal stability of HPH-lignin is sufficient for this process and even better suited than most other lignin types which were earlier used in a great variety in ARBOFORM. The process caused a strong fibre matrix bonding and fully covered the embedded fibres by the lignin up to 60% for hard wood and to 15% for softwood. The processed samples are shown in Figure 5.15. The resulting properties were similar to ARBOFORM with other lignin types (see Table 5.7): Yield stress 18.03 N/mm^2, Yield strain 0.28%, Stress at break 18.03 N/mm^2, Strain at break 0.28%, Young's-Modulus 6483 N/mm^2.

5.5 Functionalizing Lignin Matrix Composites

Development of ARBOFORM started in the late 1990s, the material being composed only of the components lignin, reinforcing natural fibres and some natural additives. The material

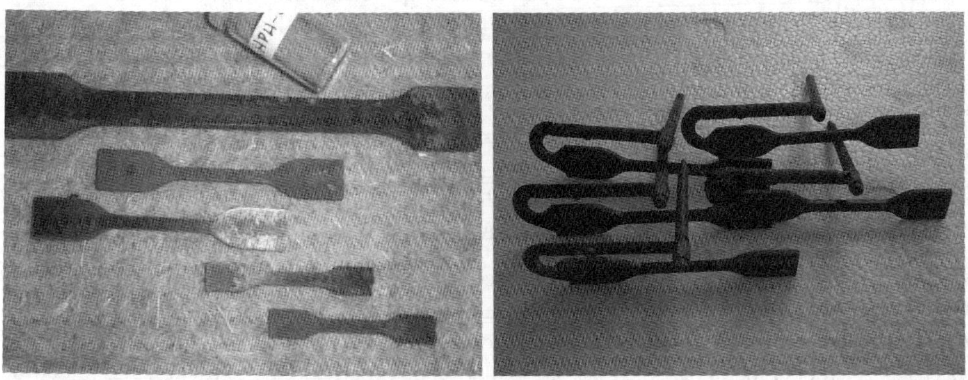
(a) (b)

Figure 5.15 HPH-lignin and test material samples composed of HPH-lignin, natural fibres and natural additives obtained by injection moulding. (© Fraunhofer ICT, 2013.)

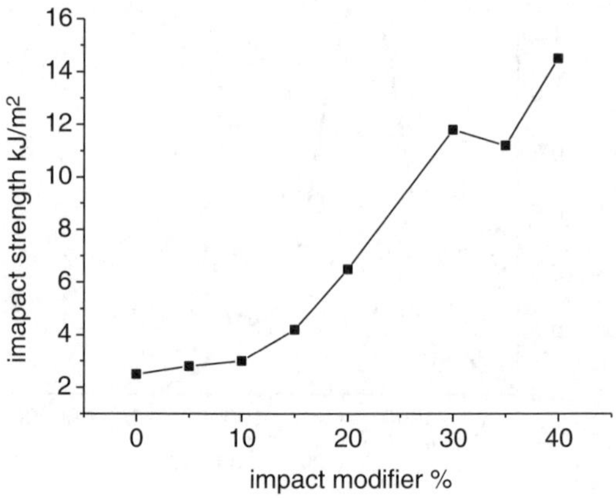

Figure 5.16 Dependence of the impact strength on the amount of impact modifier used. (© Fraunhofer ICT, 2013.)

already fulfilled important criteria of easy processing on standard machines and of engineering performance – with promising mechanical properties, especially high Young's modulus.

5.5.1 Impact Strength

Improved processing could already achieve an increase of impact strength well above 2 kJ/m^2. A further increase of impact strength required the integration of a substantial amount of impact modifiers, which at least are bio-compatible and biodegradable. However, in the end a broad range can now be covered as shown in Figure 5.16.

5.5.2 Flame Retardancy

To achieve certain classes of flame retardancy is a challenging task when using only bio-based or at least biocompatible substances. The amounts of additives to obtain relevant values are high and often may negatively influence mechanical properties. Halogen-free additives with a justifiable environmental impact are used, only. In synthetic plastics like polypropylene high amounts of these types of additives have to be used to fulfil some basic criteria. The lignin matrix composite ARBOFORM needs lower amounts of flame retardant to achieve the fire classifications UL V1 and UL HB and the mechanical properties are merely reduced by 10–20%. Obtaining the UL V0 and UL HB classifications is possible, too.

5.5.3 Electrical Conductivity with Nanoparticles

Various approaches exist to increase the electrical conductivity of plastic materials with nanoparticles [70, 71]. The precipitation of nanometal particles to the reinforcing fibres is a very effective approach to achieve a good electrical conductivity at low concentrations. Pt was used as a model metal particle to avoid oxidation on processing by extrusion and injection moulding. Stable suspensions of Pt (supplied by Tsutomu Sakai, KRI Inc., Japan) [72] were applied in two concentrations: suspension 1: 150 mg Pt/L, suspension 2: 1500 mg Pt/L, both containing particles of the size between 2 and 5 nm (Figure 5.17). The fibres were immersed

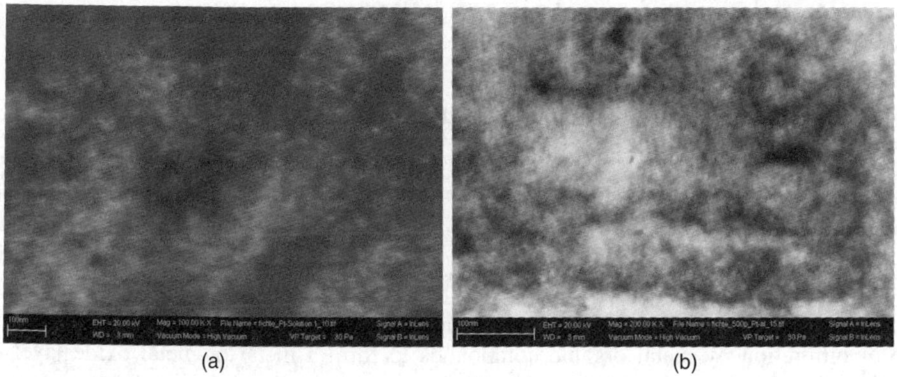

Figure 5.17 Deposition of Pt on natural fibres; left picture (a): with suspension 1 (150 mg Pt/L) on soft wood fibres; right picture (b): suspension 2: (1500 mg Pt/L) on hemp fibres. (© Fraunhofer ICT, 2013.)

in the suspensions and the particles precipitated in a vacuum rotation evaporator by removing the residual liquid medium at 70 °C. Various types of fibres were investigated including those from glass, PA, PET, CRF and the natural fibres hemp and wood used in lignin matrix composites. The fibres were incorporated into materials to obtain composites by typical processing methods. SEM and EDAX were applied to analyse the deposited Pt. The fibres were pressed to strands for measuring the electrical resistance of samples with 5 mm thickness and electrical contacts at a distance of 12 mm. Homogeneous layers are formed on the fibres with Pt concentrations between 0.01 and 1%. The surface configuration of the fibres governs the thickness of the metal layers and the concentration. The electrical resistance of pressed fibre strands varies strongly with pressing conditions and can decrease from 10 MΩ to 0.1 MΩ. The electrical resistance of the processed composites was more reproducible. Higher amounts of Pt are deposited on the natural hemp fibres with their rough surface structure (Figures 5.17 and 5.18). A composite with the bio-polymeric matrix lignin was processed to standard test strands. The electrical resistance of that material was essentially reduced in the configuration

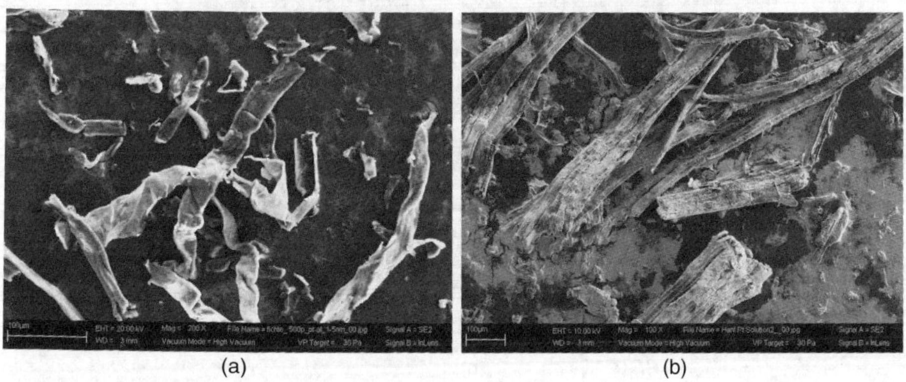

Figure 5.18 Deposition of Pt on natural fibres with Pt-suspension 2 (1500 mg Pt/L): left picture (a): on soft wood fibres; right picture (b): on hemp fibres. (© Fraunhofer ICT, 2013.)

described from about 10^4 MΩ to 10^1 MΩ. These are values of interest for surfaces to handle electronic circuits without the risk of damaging the semiconductor components.

5.5.4 Pyrolysis to Porous Carbonaceous Structures

Pyrolysis of ARBOFORM parts leads to highly porous carbonaceous structures, which might be used for various applications [73]. The pyrolysis of injection moulded parts might enable the formation of complex shapes designed by the mould. The resulting part exhibits a graphite-like structure (see Figure 5.19) with a high porosity (see Figure 5.20). This high porosity would also allow the pyrolysed ARBOFORM to be used as isolating or wear-resistant material thermally stable at high temperatures. Ceramics (e.g. metal carbide) could be obtained in case of infiltration by metal organic compounds to form a metal or metal oxide layer on decomposition. A new opportunity might be the deposition of Si from cyclopentasilane as

(a)

(b)

Figure 5.19 Pyrolysis of ARBOFORM leads to a stable part, which might be used for various carbon applications. (© Fraunhofer ICT, 2013.)

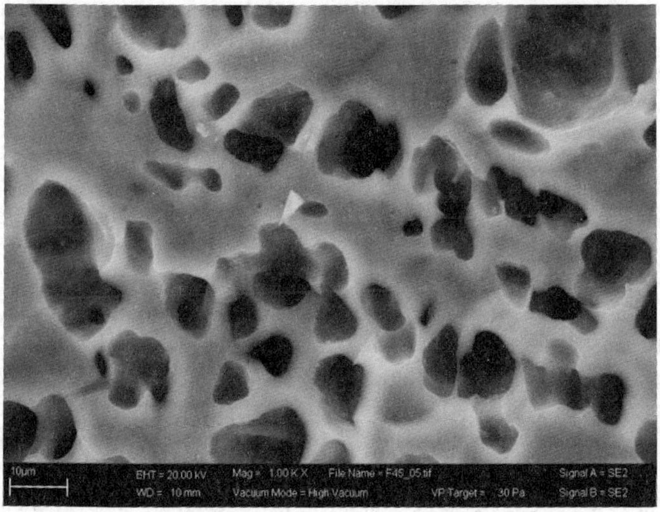

Figure 5.20 Porous structure of pyrolysed ARBOFORM. (© Fraunhofer ICT, 2013.)

described with glass fibres [74, 75]. The electrical conductivity of the carbonaceous material varies with the pyrolysis temperature and can be modified by additives.

The presence of carbon fibres in ARBOFORM can also stabilize the structure and act as precursor. Unpyrolysed ARBOFORM F45 is a biocomposite with a low electrical conductivity – this means that it is an isolating material. The conductivity of the material can be adjusted on one hand by precipitation of metal particles onto the reinforcing natural fibres (see above) but on the other hand also by varying the pyrolysis temperature. When pyrolysed at 600 °C conductivity is still low, but pyrolysed at 1000 °C, a reasonable conductivity value is obtained, which substantially decreases when the material is pyrolysed at 1200 °C. A report on more detailed investigations can be found in ref. [76].

5.6 Injection Moulding of Parts – Case Studies

The lignin based bio-composite ARBOFORM reinforced by natural fibres exhibits a series of interesting properties, which beneath ecological benefits also reveal technical advantages in processing and in finished products. These include:

- *Saving weight (10–20%):* Because the density of natural fibres (1.0–1.4 g/cm^3) is reduced compared to glass fibres (2.5 g/cm^3).
- On processing they are *not abrasive* especially when passing nozzles.
- *Nearly no shrinkage* of finished products versus mould dimensions and no dips arise even for thick walled parts.
- Strong *resistance against water* and humidity uptake.
- *Acoustics:* strong damping of resonance frequencies, even enhanced by isotropic fibre configuration (beneficial for loudspeaker applications).
- *Shock absorbing* also at low temperatures, more flexible when forming, no splintering on crashes.
- *Thermal expansion* similar to wood, enabling easy adaptation of noble wooden veneers.

- *A lower amount of flame retardant* is needed to achieve the desired classification.
- Processing requires no extruder for compounding and occurs at relatively *low temperatures and saves energy*.

The manufacturing of component parts and panels from this material occurs at lower temperatures and the resulting parts show a lower shrinkage than those made from synthetic plastics. The mechanical behaviour, however, exhibits some similarities with wood and the measured properties depend strongly on the fibre type and content. The model product loudspeaker box combines an interesting noble wood like design with outstanding acoustic properties. They were injection moulded in different complex shapes at various wall thicknesses.

Accordingly, a large series of products made of ARBOFORM has accumulated up to now.

Existing tools can be used; however, they normally should be constructed for the serial production of material according to the criteria of the ARBOFORM family (see above). In addition, numerous parts for various applications are currently available from trials with existing moulds and those specially adapted following the requirements. The number of realized products running in continuous industrial manufacturing for domestic and industrial use is currently growing strongly. Some examples with special technical challenges that could only be met with this lignin matrix composite are discussed below.

5.6.1 Loudspeaker Boxes

A favourite application is the substitution of materials for boxes in sound generation because the material is optimal for suppressing unwanted resonances and damping oscillations.

Figures 5.21 and 5.22 show a loudspeaker boxes frame made from ARBOFORM® granules on a standard injection moulding machine. The thickness of the spherical boxes is about 10 mm and represents an example of extremely low shrinkage. The injection moulded parts are of high precision, having stable dimensions without dips and tolerances less than 0.1 mm.

The spherical loud speakers were also manufactured with a tool that allowed the manufacturing within one single mould.

5.6.2 Precision Parts

As already outlined, the low shrinkage of ARBOFORM on demoulding enables high-precision fittings to be manufactured. This cannot be achieved by normal synthetic plastics.

(a) (b) (c)

Figure 5.21 Tools and parts for a loudspeaker for use of ARBOFORM, injection moulded part. (© Fraunhofer ICT, 2013.)

(a) (b)

Figure 5.22 *Loudspeaker boxes for surround systems (for one type see mould and raw parts in Figure 5.21), the texture on the parts is a feature of ARBOFORM products, which makes each a unique one. (© Fraunhofer ICT, 2013.)*

Figure 5.23 gives an illustration for two fittings from the lignin matrix composites meeting strict specifications of the user.

5.6.3 Thin Walled and Decorative Gift Boxes and Toys

The special texture of ARBOFORM parts results in decorative gift boxes usable for example for precious jewellery and so on (Figure 5.24a). The challenge in their production is the thin walls below 1 mm thickness and below 0.5 mm at the overlapping closures. Toys and figures can also be made of ARBOFORM (Figure 5.24b), which demonstrates the manufacturing

(a) (b)

Figure 5.23 *Technical parts made from ARBOFORM. (© Tecnaro GmbH, 2013.)*

(a) (b) (c)

Figure 5.24 *Gift boxes, Christmas figures for a crib and 'wood watch'. (© Fraunhofer ICT, 2013.)*

of filigree parts. Resistance to saliva makes the material suited for applications in toys for children; here a chemical-free lignin type should be used.

Acknowledgements

The work was partially supported by the Commission of the European Union, General Directorate XII, within BIOCOMP, project N° NMP2-CT-2005-515769; the NEDO grant Nr. 2002GP008, Japan; the German Federal Ministry for Agriculture with the German Agency for Renewable Resources (Fachagentur Nachwachsende Rohstoffe FNR); the German Federal Ministry for Economics with the Arbeitsgemeinschaft Industrielle Forschungsvereinigungen AIF; the state of Baden-Württemberg, Ministry of Education and Research with the Project Management Jülich (PTJ); and by the Deutsche Bundesstiftung Umwelt, Germany.

References

1. Carrasco, F. and Pages, P. (1997) Propiedades mecanicas de materiales compuestos de polietileno/celulosa sometidos a envecimiento natural. *Información Tecnológica*, **8**, 31–35.
2. Kaplan, D.L. (ed.) (1998) *Biopolymers from Renewable Resources*, Springerverlag, Heidelberg.
3. Gauthier, R., Joly, C., Coupas, A.C. *et al.* (1998) Interfaces in polyolefin/cellulosic fiber composites: chemical coupling, morphology, correlation with adhesion and aging in moisture. *Polymer Composite*, **19**, 287–300.
4. Bledzki, A.K. and Gassan, J. (1999) Composites reinforced with cellulose based fibres. *Progress in Polymer Science*, **24**, 211–274.
5. Coutinho, F.M.B. and Costa, T.H.S. (1999) Performance of polypropylene-wood fiber composites. *Polymer Testing*, **18**, 581–587.
6. Snijder, M.H.B. and Bos, H.L. (2000) Reinforcement of polypropylene by annual plant fibers: optimisation of the coupling agent efficiency. *Composites Interfaces*, **7**, 69–75.
7. Nabi Saheb D. and Jog, J.P. (1999) Natural fiber polymer composites: a review. *Advances in Polymer Technology*, **18**, 295–382.
8. Mohanty A.K., Misra M. and Hinrichsen G. (2000) Biofibres, biodegradable polymers and biocomposites: an overview. *Macromolecular Materials and Engineering*, **276/277**, 24–1.

9. Mohanty A.K., Misra M. and Drzal L.T. (2002) Sustainable bio-composites from renewable resources: opportunities and challenges in the green materials world. *Journal of Polymers and the Environment*, **10**, 19–26.
10. Riedel, U. and Nickel, J. (2003) Activities in biocomposites. *MaterialsToday*, **6** (10), 44–48.
11. Madsen, B. (2004) Properties of Plant Fibre Yarn Polymer Composites. Ph.D. thesis, Technical University of Denmark.
12. Wool, R.P. and Sun, X.S. (eds) (2005) *Bio-Based Polymers and Composites*, Elsevier, Inc., Amsterdam.
13. Bledzki, A.K., Faruk, O. and Sperber, V.E. (2006) Cars from bio-fibres. *Macromolecular Material Engineering*, **291**, 449–457.
14. Fowler, P.A., Hughes, J.M. and Elias R.M. (2006) Review biocomposites: technology, environmental credentials and market forces. *Journal of Food and Agriculture*, **86**, 1781–1789.
15. Inone-Kauffmann, E. and Eisenreich, N. (2007) Biocomp - composites from renewable resources - a European project. Proceedings of The Polymer Processing Society, Annual Meeting, PPS23, San Salvador, Brazil, May, 2007.
16. Nägele, H. (2007) Engineering thermoplastic bio-composites. JEC Composites Forum Proceedings, Biocomposites, Paris, 3 April 2007.
17. Hoydonckx, H. (2007) Biomass based thermoset furan resin systems. JEC Composites Forum Proceedings, Biocomposites, Paris, 3 April 2007.
18. John, M. and Thomas, S. (2008) Biofibres and biocomposites. *Carbohydrate Polymers*, **71**, 343–364.
19. Christian, S. and Billington, S. (2009) Sustainable Biocomposites for Construction. Composites and Polycon 2009, American Composites Manufacturers Association, Jan. 15–17, 2009, Tampa, Fl.
20. Westman, M.P., Laddha, S.G., Fifield, L.S. *et al.* (2010) Natural Fiber Composites: a Review. National Technical Information Service, US Department of Commerce, Springfield, VA.
21. Tripathi, G., Choudhury, P. and Basu, B. (2010) Development of polymer based biocomposites: a review. *Materials Technology: Advanced Performance Materials*, **25**, 158–176.
22. Pilla, S. (ed.) (2011) *Handbook of Bioplastics and Biocomposites Engineering Applications*, John Wiley & Sons, Inc., Hoboken, and Scrivener Publishing LLC, Salem.
23. Elsner, P., Eyerer, P. and Hirth, Th. (eds) (2012) *DOMININGHAUS – Kunststoffe, Eigenschaften und Anwendungen*, Springer, Heidelberg.
24. Nägele, H., Pfitzer, J., Inone, E. *et al.* (1999) Flüssiges Holz - durch Wärme verformbarer thermoplastischer Werkstoff aus nachwachsenden Rohstoffen. *HK Holz- und Kunststoffverarbeitung*, **11**, 44–46.
25. Eisenreich, N., Eckl, W., Inone E.R. *et al.* (2000) ARBOFORM® – a thermoplastic made of renewable resources. Proceedings of the Electronic Goes Green 2000, Vol. 1, Technical Lectures (Joint International Congress and Exhibition, Berlin, Germany, Sept. 2000), p. 625.
26. Nägele, H., Pfitzer, J., Nägele, E. *et al.* (2002) ARBOFORM® - A thermoplastic, processable material from lignin and natural fibers, in *Chemical Modification, Properties, and Usages of Lignin* (ed. Th.Q. Hu), Kluwer Academic / Plenum Publisher, New York, pp. 101–120.
27. P. Eyerer, N. Eisenreich, E. Inone *et al.* (2001) Faserverstärkter Werkstoff auf der Basis duroplastischer Polymere. Patent DE 2001-10132749 A: 2001 -07-03.
28. P. Eyerer, N. Eisenreich, P. Elsner *et al.* (2000) Production of natural material-based fiber-reinforced. Patent WO 2000027925.
29. P. Eyerer, N. Eisenreich, P. Elsner *et al.* (2000) Composites with decorative layers. Patent WO 2000027926.
30. P. Eyerer, N. Eisenreich, P. Elsner *et al.* (2000) Verfahren zur Herstellung eines faserverstaerkten Kunststoff-Werkstoffs. Patent DE198552036.
31. P. Eyerer, N. Eisenreich, P. Elsner *et al.* (2000) Kunststoff-Werkstoff aus einem Polymer-Blend und dessen Verwendung. Patent DE19852081C1.
32. P. Eyerer, N. Eisenreich, E. Inone *et al.* (2000) Herstellung sowie Verarbeitung pulverförmiger, rieselfähiger Schüttgüter oder zu Pelletts agglomerierter Mischungen aus Naturfasern, Schellack und Lignin. Patent DE19852034C1.
33. P. Eyerer, N. Eisenreich, E. Inone *et al.* (2004) Agglomeration pulverförmiger Schüttgüter zu thermoplastisch verarbeitbaren Briketts, Pelletts oder Tabletten. Patent PCT/EP01/10006.

34. Eyerer, N. Eisenreich, E. Inone *et al.* (2000) Verbesserung der Eigenschaften von naturfaserverstärkten Compounds auf Basis von Lignin bzw. synthetischen Polymeren. Patent DE19852067A1.
35. Kamm, B., Gruber, P.R. and Kamm, M. (eds) (2006) *Biorefineries – Industrial Processes and Products: Status Quo and Future Directions*, Wiley-VCH, Weinheim, Germany.
36. Plackett, D., Logstrup Andersen, T., Batsberg Pedersen, W. and Nielsen, L. (2003) Biodegradable composites based on l-polylactide, and jute fibres. *Composites Science and Technology*, **63**, 1287–1296.
37. Mukherjee, T. and Kao, N. (2011) PLA based biopolymer reinforced with natural fibre: a review. *Journal of Polymers and the Environment*, **19**, 714–725.
38. Brauns, F.E. (1952) *The Chemistry of Lignin*, Academic Press, New York.
39. Glasser, W.G., Northey, R.A. and Schultz, T.P. (1999) *Commercial Use of Lignin-Based Materials, Lignin: Historical, Biological, and Materials Perspectives*, American Chemical Society, Washington DC, pp. 304–320.
40. Glasser, W.G., Northey, R.A. and Schultz, T.P. (eds) (2000) Lignin: Historical, Biological, and Materials Perspectives, American Chemical Society, Washington DC, pp. 219–239.
41. Jedicke, O., Eisenreich, N. and Dümpert, H. (2000) Aquasolv® - Hydrothermolyse -The Development of a Process for Completely Use of Biomass. Proceedings 1st Biomass World Conference, Sevilla, Spain, 2000.
42. Sjöström, E. (1993) *Wood Chemistry: Fundamentals and Application*, Academic Press, Orlando.
43. Muurinen, E. (2000) Organosolv Pulping: A Review and Distillation Study Related to Peroxyacid Pulping, Department of Process Engineering, University of Oulu, http://herkules.oulu.fi/isbn9514256611/isbn9514256611.pdf (accessed 22 June 2013).
44. Bobleter, O. (1994) Hydrothermal degradation of polymers derived from plants. *Progress in Polymer Science*, **19**, 797–841.
45. Bobleter, O. (1998) Hydrothermal degradation and fractionation of saccharides and polysaccharides, in *Polysaccharides, Structural Diversity and Functional Versatility*, (ed. M. Dimitriu), Marcel Dekker, Inc., New York, pp. 775–833.
46. Jedicke, O., Eisenreich, N., Steurenthaler, E. and Dümpert, H. (1999) Aquasolv® – Verfahren – Ein Konzept zur ganzheitlichen Verwendung von Biomasse in Verbindung mit 'grüner Chemie'. Proceedings of 2nd Scientific Meeting on Zero Emissions, Frohnleiten; Technische Universität Graz, Austria, 1999.
47. Jedicke, O., Eisenreich, N. and Dümpert, H. (2002) Aquasolv-Process as a Multifunctional Process for Isolated Areas. International Conference on New and Renewable Technologies for Sustainable Development; 24–26 June 2002, Ponta Delgada; Azores; Portugal.
48. Roffael, E. and Dix, B. (1991) Lignin and ligninsulfonate in non-conventional bonding systems – an overview. 1st European Workshop on Lignocellulosics and Pulp, Utilization and Analysis of Lignins – Hamburg-Bergedorf, Federal Republic of Germany, Sept. 18–20, 1991.
49. Rozman, H.D., Tan, K.W., Kumar, R.N. *et al.* (2000) The effect of lignin as compatibilizer on the physical properties of coconut fiber-polypropylene composites. *European Polymer Journal*, **26**, 1483–1494.
50. Etün, N.S. and Zmen, N.I. (2003) Studies on lignin-based adhesives for particleboard panels. *Turkish Journal of Agriculture and Forestry*, **27**, 183–189.
51. Thielemans, W. and Wool, R.P. (2004) Butyrated kraft lignin as compatibilizing agent for natural fiber reinforced thermoset composites. *Composites Part A: Applied Science and Manufacturing*, **35**, 327–338.
52. Satheesh Kumar, M.N., Mohanty, A.K., Erickson, L. and Misra, M. (2009) Lignin and its applications with polymers. *Journal of Biobased Materials and Bioenergy*, **3**, 1–24.
53. Oldham, P.B., Wang, J., Conners, T.E. and Schultz, T.P. (1993) Rapid analysis of pulp lignin: a review of NIR and FTIR and preliminary investigation of multidimensional fluorescence spectroscopy. Pulping Conference 1993, pp. 653.
54. Porter, J., Sands, T. and Trung, T. (2009) Understanding Kraft liquor cycle: a need for online measurement and control. TAPPI Engineering, Pulping and Environmental Conference October 11–14, 2009 Memphis Tennessee, pp. 1–12.

55. Becker, W., Inone-Kauffmann, E., Eckl, W. and Eisenreich, N. (2007) Near infrared spectroscopy for in-line control of biopolymer Processing. Proceedings of the Polymer Processing Society, Annual Meeting, PPS23, San Salvador, Brazil, May, 2007.
56. Antal, M.J., Allen, S.G., Schulman, D. *et al.* (2000) Biomass gasification in supercritical water. *Industrial and Engineering Chemistry Research*, **39**, 4040–4053.
57. Antal, M.J., Manarungson, S. and Mok, W.S. (1994) Hydrogen production by steam reforming glucose in supercritical water. *Advances in Thermochemical Biomass Conversion*, **3**, 1367–1377.
58. Antal, M.J. and Xu X. (1999) Hydrogen production from high moisture content biomass in supercritical water. Proceedings of the 1999 U.S DOE Hydrogen Program Review NREL/CP-570-26938.
59. Modell, M. (1977) Reforming of glucose and wood at critical conditions of water. *Mechanical Engineering*, **99**, 108–108.
60. Modell, M. (1980) Reforming of organic-substances in supercritical water. *Journal of the Electrochemical Society*, **127**, C139–C139.
61. Saka, S. and Ueno, T. (1999) Chemical conversion of various celluloses to glucose and its derivatives in supercritical water. *Cellulose*, **6**, 177–191.
62. Saka, S. and Konishi, R. (2001) Chemical conversion of biomass resources to useful chemicals and fuels by supercritical water treatment, in *Progress in Thermochemical Biomass Conversion* (ed. A.V. Bridgwater), Blackwell, Oxford, pp. 1338–1348.
63. Ehara, K., Saka, S. and Kawamoto, H. (2002) Characterization of the lignin-derived products from wood as treated in supercritical water. *Journal of Wood Science*, **48**, 320–325.
64. Saka, S. (2005) Comparative study on lignin-derived products from softwood and hardwood as treated in supercritical water. *Cellulose*, **81**, 4738–4738.
65. Saka, S. and Konishi, R. (2008) Chemical conversion of biomass resources to useful chemicals and fuels by supercritical water treatment, in *Progress in Thermochemical Biomass Conversion* (ed. A.V. Bridgwater), Blackwell Science Ltd, Oxford, UK. doi: 10.1002/9780470694954.ch110
66. Rohe, T., Becker, W., Krey, A. *et al.* (1998) In-line monitoring of polymer extrusion processes by NIR spectroscopy. *Journal of Near Infrared Spectroscopy*, **6**, 325–332.
67. Martens, H. and Naes, T. (1989) *Multivariate Calibration*, John Wiley & Sons, Ltd, Chichester.
68. Eisenreich, N., Sakai, T., Eckl, W. *et al.* Practical research of wood-like thermoplastic using lignin extracted by high pressure hydrolysis process, NEDO project G-02 Registration number 2002GP008, http://www.nedo.go.jp/content/100084713.pdf (accessed 22 June 2013).
69. Inone-Kauffmann, E., Eisenreich, N., Eckl, W. and Plitzko, Y. (2007) Bio-Composites Based on Lignin Matrix from High Pressure Hydrolysis. Proceedings of the Polymer Processing Society, Annual Meeting, PPS23, San Salvador, Brazil, May, 2007.
70. Jordan, J., Jacob, K.I., Tannenbaum, R. *et al.* (2005) Experimental trends in polymer nanocomposites – a review. *Materials Science and Engineering A*, **393**, 1–11.
71. Droval, G., Feller, J.F., Salagnac, P. and Glouannec, P. (2008) Conductive polymer composites with double percolated architecture of carbon nanoparticles and ceramic microparticles for high heat dissipation and sharp PTC switching. *Smart Materials and Structures*, **17** (2), 1–10.
72. Devi, G. and Rao, V.J. (2000) Room temperature synthesis of colloidal platinum nanoparticles. *Bulletin of Materials Science*, **23**, 467–470.
73. Kauffmann, A., Eisenreich, N., Weiser, V. *et al.* (2005) Leitfähiges Formteil, Verfahren zu seiner Herstellung und Verwendung desselben, DE 103 47 701 A1, 19. 05. 2005.
74. Hadjiev, H., Frey, H., Khan, H.R. *et al.* (2011) Decompositon and oxidation of cyclopentasilane diluted in toluene. Proceeding of 42nd International Annual Conference of Fraunhofer ICT, Karlsruhe, Germany, 28 June – 01 July, 2011, p. 103.
75. Hadjiev, H., Khan, H.R., Frey, H. *et al.* (2012) Decomposition and oxidation of cyclopentasilane. Proceeding of 43rd International Annual Conference of the Fraunhofer ICT, Karlsruhe, Germany, 26 June – 29 June 2012.
76. Haensel, T., Comouth, A., Zydziak, N. *et al.* (2010) Pyrolysis of wood-based polymer compounds. *Journal of Analytical and Applied Pyrolysis*, **87**, 124–128.

6

Bioplastics from Lipids

Stuart Coles
International Digital Laboratory, WMG, University of Warwick, United Kingdom

6.1 Introduction

The prominence of lipids in the chemical industry is growing year on year and they are now one of the major renewable feedstocks used in a chemist's synthetic toolbox. They are readily obtained in different parts of the planet and can be easily modified to produce a wide range of chemical precursors for polymer production. However, many associate the commercial use of lipids in areas such as biodiesel, lubricants and paint formulations and not with the largely investigated use of polymers. This chapter aims to define loosely what a lipid is, and highlight the main areas where they have been employed in bioplastics production to date.

6.2 Definition and Structure of Lipids

It is challenging to give an accurate, succinct definition of what a lipid is. Most commonly, lipids are thought of as fatty acids and the related triglycerides, but even this does not include the mono- and di-substituted glycerol compounds. In general terms, lipids can be considered to be fatty acids, their derivatives and any other compounds that may have a similar behaviour, functionality or biosynthetic pathway. They are often hydrophobic, soluble in hydrocarbon or aromatic solvents and insoluble in water.

6.2.1 Fatty Acids

In order to define clearly what a lipid is, it is essential to know the structure of a fatty acid. A fatty acid is a carboxylic acid with an aliphatic carbon chain attached to it. The carbon chain generally has a length of anywhere between four and 22 carbons (Figure 6.1). This carbon

Bio-Based Plastics: Materials and Applications, First Edition. Edited by Stephan Kabasci.

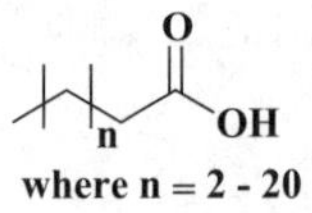

Figure 6.1 Generic structure of a fatty acid.

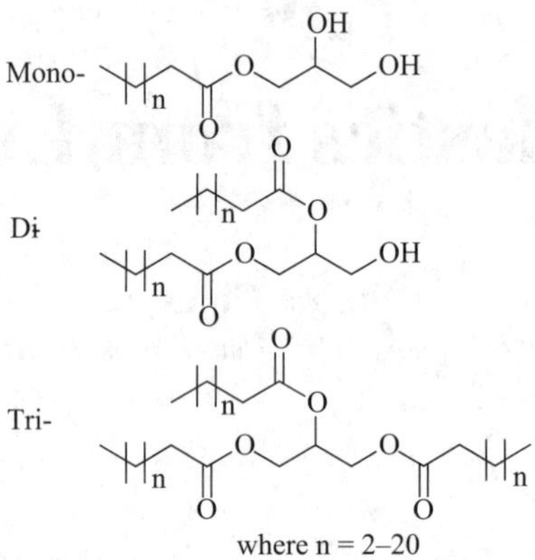

Figure 6.2 Generic structures for mono-, di- and triglycerides.

chain can have no (saturated), one (monounsaturated) or multiple (polyunsaturated) double bonds in its structure.

6.2.2 Mono-, Di- and Tri-Substituted Glycerols

Fatty acids form esters with alcohols and some of the most commonly found esters are those with glycerol. As glycerol has three alcohol groups, up to three ester linkages can be formed (Figure 6.2).

6.2.3 Phospholipids

Phospholipids have a negatively charged, hydrophilic phosphate group, often referred to as the 'head'. This is attached to a glycerol group which is di-substituted with fatty acids, forming a hydrophilic 'tail'. The phosphate head group can have several different types of group attached to it. The example shown in Figure 6.3 is that of a phosphatidylcholine compound, a type of phospholipid.

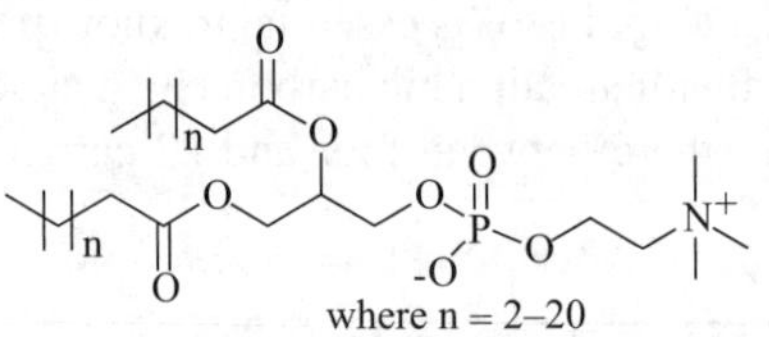

Figure 6.3 Generic structure for the phosphatidylcholine class of phospholipids.

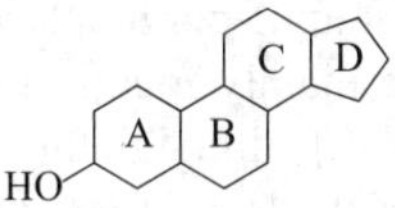

Figure 6.4 Generic sterol structure.

6.2.4 Other Compounds

There are many other forms of lipids but these are generally not used to form bioplastics. A wax is a general term for a group of chemical compounds that are malleable around ambient temperature and melt above 45 °C. Some fatty acids and their ester derivatives fit into this category (covered in sections 6.2.1 and 6.2.2) but other compounds include esters with long-chain alcohols and hydrocarbons. They are always insoluble in water.

Sterols, or steroid alcohols, are a subgroup of the steroid class of compound. Steroids have a particular arrangement of three cyclohexyl rings and one cyclopentyl ring (Figure 6.4), and sterols have an alcohol group in the 3-position of the A ring. Specifically, phytosterols come from plants and zoosterols come from animals. Examples of these compounds include cholesterol and ergosterol.

Other lipid compounds include prenol lipids, saccharolipids, sphingolipids and polyketides (Figure 6.5). Each of these has a slightly different structure, but they have the same indicative properties of the other compounds in the lipid class.

6.3 Sources and Biosynthesis of Lipids

6.3.1 Sources of Lipids

All naturally occurring lipids from botanical sources (plants) or zoological sources (animals and humans) could potentially be used in the production of bioplastics. They have the same chemical structure and function in a very similar manner. Cholesterol (and its related esters) and fatty acids are commonly found in zoological sources, although the fats tend to be of a heavily saturated nature. It is feasible to create other products such as biodiesel from extreme sources including liposuction fat. However, a lack of volume and more importantly ethical approval have prevented their uptake further. The remainder of this chapter will focus on the use of triglycerides from plants and their related compounds for bioplastics production.

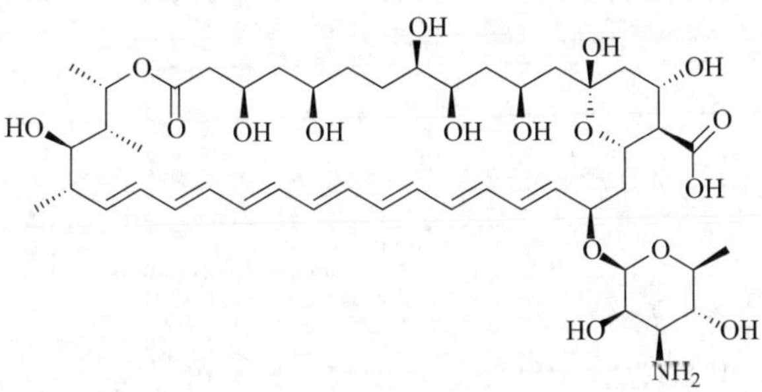

Figure 6.5 Structure of amphotericin B, a polyketide.

Plants are used as the main sources of lipids for bioplastics production. Often used as energy storage in plants [1], triglycerides are mainly found in the seeds although they can also be observed in other parts, such as tubers [2], although this is a rare case. Some bioplastics have been synthesized from fish oils [3] but the lack of reliability in supply, cost and questions of the true sustainability of this route have restricted investigations in this area.

In other cases, it is possible to find lipids as a waxy coating of leaves for example, cutin or in peanuts, where the entire plant is approximately 50% peanut oil [4].

6.3.2 Biosynthesis of Lipids

In animals and plants, the main pathway for synthesis of lipids is for them to metabolize carbohydrates into triglycerides for the purposes of energy storage [5]. When the energy is required, the triglycerides are then converted back to their original carbohydrate form. It is important for crop scientists to understand this biosynthetic cycle as the optimal point of harvesting for oilseed crops will be at the point when energy storage is at its highest.

The main difference between plants and animals is that in animals a single multifunctional protein known as fatty acid synthase carries out all of the synthetic steps [5]. In plants, the steps are broken down and carried out by a number of different enzymes [5]. The introduction of unsaturation into a triglyceride structure is generally carried out by a separate protein, such as the conversion of stearic acid to oleic acid by stearoyl-CoA desaturase (see Figure 6.6) [6].

6.3.3 Composition of Triglycerides

The composition of triglycerides varies from plant to plant, and also varies depending on where the plant was grown, what the conditions were during growth and so on. There has been a significant body of work looking at different oil sources and this is summarized in Table 6.1.

As a result of this natural variation, it has become very important to choose a plant oil carefully for the specific application in mind. Saturated fatty acids and triglycerides will perform well in some applications whereas polyunsaturated fatty acids allow for more chemistry and functionalization to take place, allowing for greater adaptability (see sections 6.5 and 6.6).

6.4 Extraction of Plant Oils, Triglycerides and Their Associated Compounds

The triglycerides found in plants are mainly contained within the seeds and must be obtained from within them. Unfortunately, it is often not simply a case of crushing the seeds and

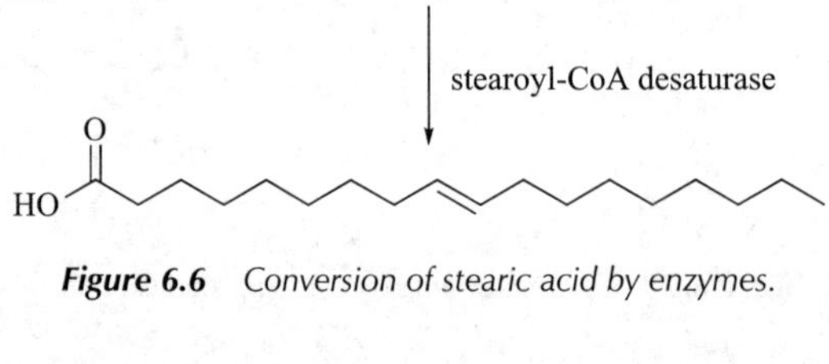

Figure 6.6 Conversion of stearic acid by enzymes.

Table 6.1 Selected fatty acid composition of triglycerides from plant oils.

Source	Palmitic	Stearic	Oleic	Linoleic	Linolenic	Ref.
Castor[a]	1	1	3	4	0	[7–9]
Coconut	7.5–10.5	1–3.5	5–8	1–2.6	0	[7–9]
Cottonseed	22–26	2–3	15–22	47–58	0	[7–9]
Hemp	5–7	1–3	8–13	52–62	12–23	[10,11]
Linseed	6.5–7.5	3–5	18–22	14–17	52–55	[12]
Palm	40–47	3–6	36–44	6–12	0	[7–9]
Peanut	6–14	2–6	36–67	13–43	0	[7–9]
Rapeseed	2–6	4–6	52–65	18–25	10–11	[7–10]
Soybean	10–12	3–5	18–26	49–57	6–9	[7–9]
Sunflower	5–7	3–6	14–40	48–74	0	[7–9]

[a]Castor oil contains approximately 90% ricinoleic acid.

drawing off the oil. There are several stages required to obtain good quality oil suitable for use in bioplastics production [13] and the route chosen depends on the scale of the operation. Liquid extraction is only economically viable on a large scale; whereas the more traditional pressing methods work well on a small scale.

As a result of this natural variation, it has become very important to choose a plant oil carefully for the specific application in mind. Saturated fatty acids and triglycerides will perform well in some applications whereas polyunsaturated fatty acids allow for more chemistry and functionalization to take place, allowing for greater adaptability (see sections 6.5 and 6.6).

6.4.1 Seed Cleaning and Preparation

If a small-scale process is going to be used, the seeds must be cleaned and well prepared before being pressed to remove the oil. Any loose dirt, sand, and so on, will damage the mechanical pressing machines and must be cleared from the seeds. The seeds must also be dry, particularly if the seeds will be stored for a significant period of time. Moisture content levels of 10% or less are desirable in this instance [14].

6.4.2 Seed Pressing

Once effectively prepared, the seeds are mechanically pressed and the oil is removed. The seeds can be warm or cold before extraction [14] although warm oils are more readily extracted. The press designs themselves are often simple and similar to that of a conventional extruder; a screw is driven around a cylinder and the seeds are loaded from a hopper before being crushed. The base of the cylinder will have some small holes drilled into it to allow the oil to escape from the cylinder whilst much of the seed material remains behind and is expelled through a die at the end of the screw.

6.4.3 Liquid Extraction

If the process requires several thousand tons of seeds per day, liquid extraction of the oilseeds becomes the more preferable option. Since triglycerides are soluble in nonpolar organic solvents, often used are liquids such as hexane or petroleum ether [15,16]. Given the quantities involved this presents a significant environmental hazard, not only in the production of the solvents but also in the surrounding area should quantities of solvent end up leaking into water courses and so on.

Investigations into avoid this danger have taken place. Extended anionic surfactants have been employed to extract corn oil instead of hexane [17]. These extractions were shown to extract 83% of the oil available and also maintained the composition of the corn oil when compared to extractions carried out using hexane. A mixture of enzymes to extract oil from *Irvingia gabonensis* kernels was employed; controlled, ordered addition of the enzymes optimized the extraction of the oil up to 90% efficiency [18]. Similar protocols were applied to the extraction of canola oil [19] and showed that the enzymatically assisted extractions were more efficient than by using water alone and produced better oil quality than both aqueous and solvent extractions. The efficiency of the process in this instance was significantly increased from those previously stated (circa 25%).

6.4.4 Post Extraction Processing

After extraction, the oil must be purified for further use. Clarification is an essential step in this procedure. Any residual water is removed by allowing the oil to settle and subsequently stripping off the water layer. If this is insufficient, filtration through a fine material is employed to remove any insoluble, fine particles. In some cases, the oil can be heated to destroy any residual bacteria as well. Settling of the oil will also help remove any phospholipids present [20].

Refining is then carried out to degum and remove any free fatty acids from the oil using an alkali compound, often sodium hydroxide [20]. This material can often be sold as 'once refined' [20] or bleached and deodorized. Often for food applications, the bleaching is performed to give a product an acceptable appearance whereas the deodorization process is designed to remove small volatile molecules that have an effect on flavour or taste, using activated charcoal [20].

6.5 Biopolymers from Plant Oils, Triglycerides and Their Associated Compounds

Triglycerides that occur in nature have a wide variety of chemical reactivity because of the variation in the fatty acid content. As such, each feedstock can be transformed into a variety of different polymer architectures. Many of the polymer architectures have slight and subtle variances, and it becomes far easier to classify these types of bioplastics by the feedstock from which they were generated.

6.5.1 Generic Triglycerides

Whilst not strictly sourced from plants, generic triglyceride structures have also been used in the synthesis of bioplastics. They may have been isolated from plants originally (as many commercially available fatty acids are) but have since been recombined to give model triglycerides that present idealized behaviour and allow chemists to determine reactivity in a relatively simple way. There have been many reports in the literature of triglyceride-based polymers [21–24]. Initial studies focused on the use of triglycerides in photoinitiated cationic polymerizations [25] or where triglycerides were derivatized using acrylonitrile to produce acrylamides [26] (see Figure 6.7).

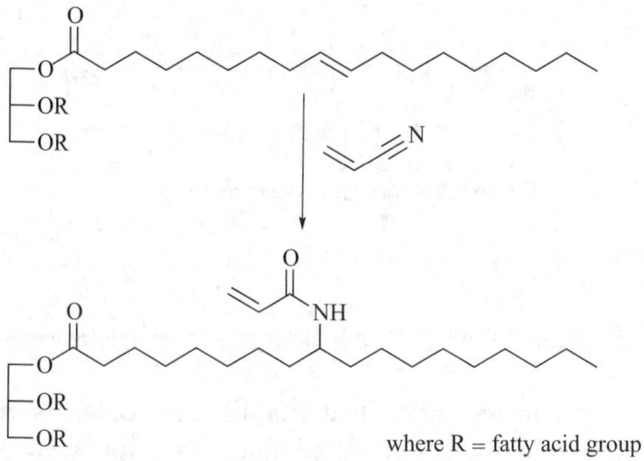

Figure 6.7 Acrylonitrile derivatization of triglycerides.

6.5.2 Common Manipulations of Triglycerides

The structure of triglycerides dictates that synthetic procedures must either be carried out on the glyceride ester linkages or on the unsaturated double bonds that exist on the carbon chain. Transesterification is a common reaction for the breakdown of the triglyceride structure into its constituent fatty acid alkyl esters. These can then be used as biodiesel fuels or manipulated further. Whether the triglycerides have been initially transesterified or not, the double bonds of the fatty acids remain able to be reacted and potentially incorporated in a polymerization reaction. One common manipulation is to epoxidize the double bond, often using a peroxide of some description, generating the reactive epoxide group (see Figure 6.8). Further manipulations

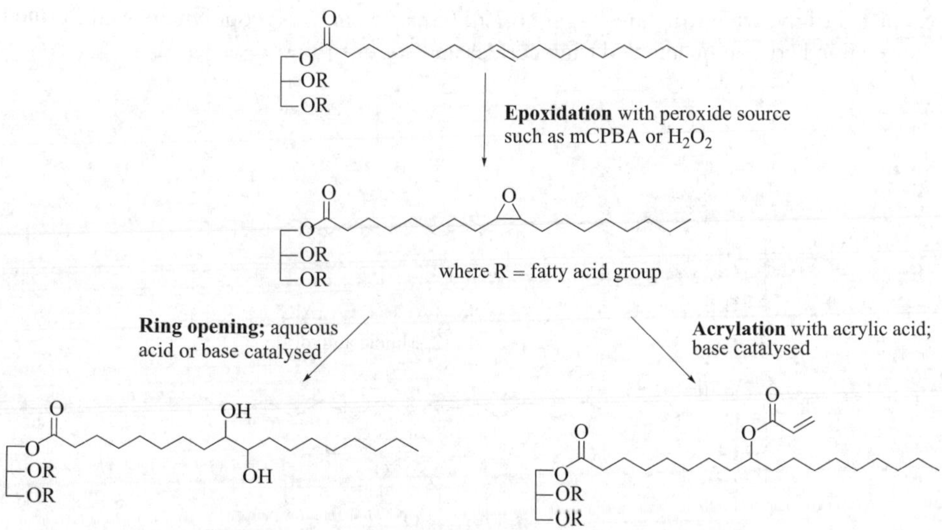

Figure 6.8 Epoxidation, acrylation and ring opening of triglycerides.

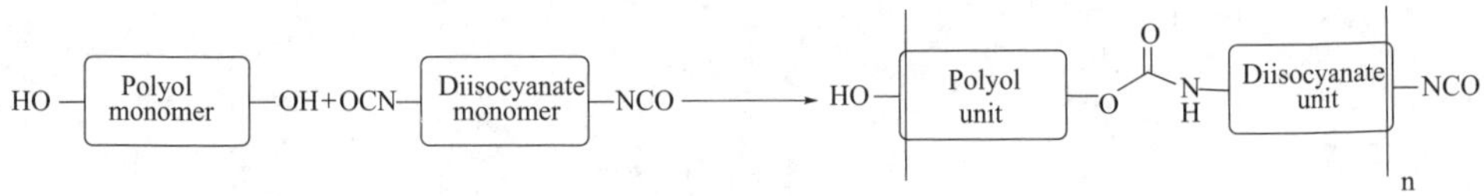

Figure 6.9 Formation of polyurethanes.

have been studied such as acrylation [27], allowing the materials to get incorporated into a styrene-based copolymer.

Another postepoxidation manipulation that can be carried out is ring opening, generating a diol, in the case of monounsaturated fatty acids, or a polyol, in the case of polyunsaturated fatty acids (see Figure 6.8). The generation of polyols is essential for the incorporation of triglycerides into polyurethanes. Polyurethanes from triolein were originally studied to synthesize polyurethanes as the heterogenecity of plant oils prevented a structure-property relationship being identified [28]. Other ways of generating a polyol include ozonolysis, which was studied on triolein and trilinolein; the resulting terminal hydroxyl group was then used to produce polyurethanes [29]. Triolein was epoxidized and ring opened to give triols before being reacted with bromine and sodium azide to give trisamines. These were crosslinked with 1,4-phenylene diisocyanates to give thermally stable polymers [30].

Polyurethanes are produced by an addition polymerization reaction (Figure 6.9) between diols (or polyols) and diisocyanates. These diisocyanates can also be produced from triglyceride structure. Triglycerides were brominated using *N*-bromosuccinimide before being converted to isocyanates with AgNCO; converting approx. 70% of the bromine groups without disrupting the double bonds. Polymerization to form polyurethanes and polyureas was then carried out with polyols (including castor oil) [31].

The reaction with maleic anhydride can also follow epoxidation reactions in order to give a maleinated triglyceride structure (Figure 6.10). In particular, rigid copolymers were formed by the reaction between maleinated triglycerides and styrene [32].

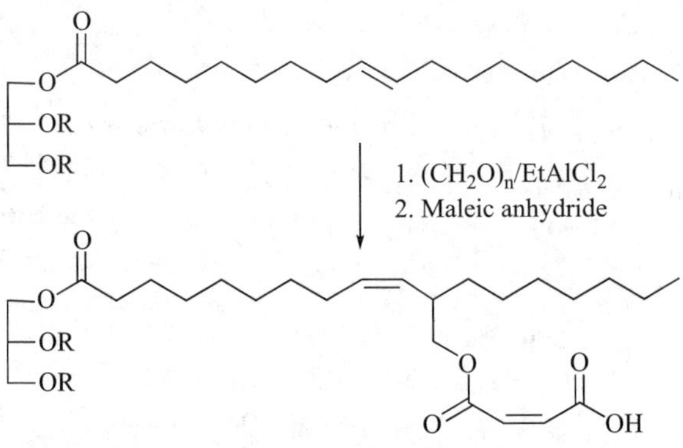

Figure 6.10 Maleination of triglycerides.

6.5.3 Soybean Oil-Based Bioplastics

One of the most obvious sources for triglycerides in bioplastics production is soy beans. Soybeans accounted for over half of the world production of oilseed crops in 2009/10 [33] and are readily available since they grow in many different areas of the planet. It is possible to polymerize the soybean oil directly, although the residual phytosterol content was shown to inhibit the polymerization by up to 20% [34]. Following some of the simpler manipulations, epoxidized soybean oil was maleinated and then polymerized using UV light by itself [35]. Rigid thermoset copolymers of maleinated or epoxidized soybean oil have also been formed with styrene, vinyl acetate, methyl methacrylate [35], maleinated polypropylenes [36] and maleinated butadiene [37]. With the latter, a ratio of 1:1 of epoxy to anhydride could not be increased as it negatively impacted on the mechanical properties of the polymers when compared with the butadiene homopolymer.

Styrene is commonly used as a copolymer with acrylated, epoxidized soybean oil; styrene polymerization can be easily initiated in either a free radical or cationic fashion (Figure 6.11). These polymerizations will incorporate the double bonds of the acrylate groups should they be present. Examples of this include the reaction of epoxidized soybean oil or soybean oil diglyceride with a styrene-maleic anhydride comonomer [38] or acrylated epoxidized soybean oil with styrene to produce thermoset polymers [39]. The latter showed increased tensile and impact strength as the iodine value of the epoxidized soybean oil decreased. With rigid thermoset polymers, an increased level of crosslinking often generated by highly unsaturated triglycerides produces a brittle polymer often unsuitable for structural applications. However, rigidity is not always present in these types of polymers. Hydrolysis at 80 °C occurred in a copolymer of soybean oil and styrene; the oil was epoxidized and reacted with 4-vinyl benzenesulfonic acid before polymerization took place [40].

In the same fashion, methacrylate groups have also been incorporated in soybean oils to generate similar polymer architectures. Acrylated epoxidized soybean oil and maleinated acrylated epoxidized soybean oil were reacted with methacrylated lauric acid as a method of reducing styrene input [41]. This increased both the visocosity and the glass transition temperature when compared with the styrene analogue as well as having the benefits of improving the environmental properties of the polymer even further. Acrylated, epoxidized soybean oil was copolymerized with methyl methacrylate in a polymerization initiated by benzoyl peroxide to produce a clear bioplastic [42].

Also similar to styrene and methyacrylates is the use of divinylbenzene as a comonomer. Acrylated, epoxidized soybean oil was polymerized with divinylbenzene or modified with phthalic anhydride before polymerization to give materials with increased glass transition temperatures [43]. Modification of the soybean oil with cinnamate esters and subsequent

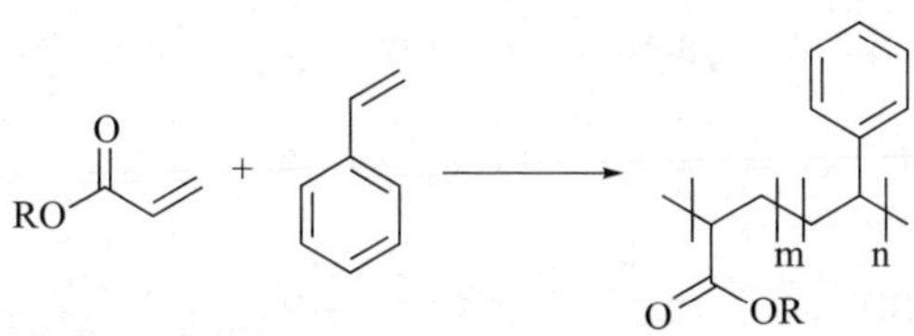

Figure 6.11 *Reaction of acrylated soybean oil with styrene.*

free radical copolymerization with styrene, vinyl acetate or methyl methacrylate produced a range of bio-based polymers [44]. The homopolymer was also be synthesized via photopolymerization.

Maleination of soybean oil is interesting as it has been shown to produce two wildly different polymer structures. Maleic anhydride-modified soybean oil was polymerized with styrene to give a hard, rigid polymer; modification improved modulus, strength and glass transition values when compared to the unmodified polymerized oil [45]. These also showed comparable strengths to commercial unsaturated polyester resins. In contrast, by using diols as the comonomer, maleinated soybean oil was used to form soft, resilient rubbers at room temperature [46].

Polyurethane-type structures are incredibly common with soybean oil (see section 6.5.2). Soybean oil, esterified with sucrose to create a sucrose soyate, was epoxidized and then reacted with carboxylic acids or alcohols to create esters/ethers (polyols), which can further be reacted with aliphatic NCO to give PU [47]. Soybean oil was functionalized to form an isocyanate adduct which was then reacted with various polyols (including castor oil) to give polyurethane structures [48]. However, the polymer exhibited poor thermal stability, with degradation onset around 150 °C. Allylically brominated soybean oil was converted into isothiocyanated soybean oil using ammonium thiocyanate [49]. Subsequent reaction with polyols (including castor oil) produced polythiourethanes which had higher tensile strength lower elongation than their urea analogues.

Rigid thermoset polymers have also been studied using other crosslinking agents. The reaction of epoxidized soybean oil was investigated with a series of pyridine derivatives [50] or fatty acid methyl ester monomers [51]. *In situ* crosslinking of these formulations produced rigid polymers with the ability to be used in composite structures, although the latter required blending with styrene to achieve a suitable viscosity.

6.5.4 Castor Oil-Based Bioplastics

Castor oil has been heavily investigated for the preparation of polyurethane structures due to the presence of ricinoleic acid in the oil harvested from the plant (Figure 6.12). Ricinoleic acid constitutes around 90% of the fatty acids in castor oil and has a naturally occurring hydroxyl group, eliminating the need for chemical modification of the triglyceride structure. However, harvesting castor oil is difficult due to the production of the nerve gas ricin as a by-product and as such its production is being scaled back whilst investigation takes place to modify the plant and inhibit the ricin production.

Direct reaction of castor oil with 4,4′-methylenebisphenyl diisocyanate gave standard polyurethane material [52]. In efforts to modify the physical properties of the polymer, castor oil was mixed with trimethylol propane (a common industrial building block) and the stoichiometric ratios of the diisocyanates involved were varied [53]. The reaction in this instance displayed second order kinetics. Polycarbonate-urethanes were also produced from

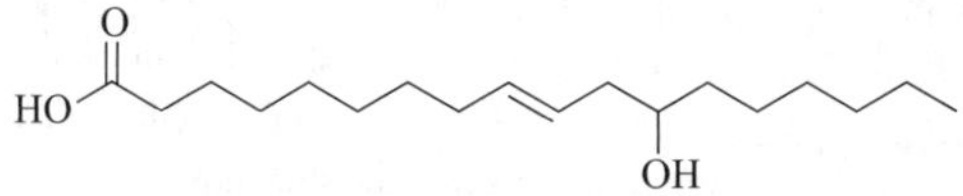

Figure 6.12 Structure of ricinoleic acid.

castor oil, although the incorporation of the oil reduced the glass transition temperature and increased the softening temperature, limiting its industrial potential [54]. Suspension-addition techniques were employed to generate polyurethane nanoparticles and microparticles capable of encapsulating levofloxacin, an antibacterial drug [55].

Other modifications, similar to those carried out with soybean oil, have also been investigated. Maleinated castor oil was used a precursor for low viscosity composite resins by crosslinking styrene or fatty acid methyl esters [51]; similar high performance applications were demonstrated with the use of maleinated dehydrated castor oil, which formed a crosslinked polymeric product for high-performance applications [56]. Castor oil was split using ozonolysis and polymerized to give high molecular weight polyesters with better thermal stability, higher melting point and glass transition when compared with the analogous polyester polycaprolactone [57]. Comonomers such as styrene-maleic anhydride [38] or methyacrylated lauric acid [41] have been investigated for the polymerization with castor oil; in both instances a rigid thermoset polymer was formed. Blending of epoxidized castor oil with polyvinyl alcohol and subsequent microwave irradiation improved the mechanical properties of the blend by promoting hydrogen bond formation [58].

6.5.5 Linseed Oil-Based Bioplastics

Linseed oil presented another interesting opportunity for the production of bioplastics. The highly polyunsaturated nature of the oil allowed for a large number of reactive sites to be introduced onto the material for subsequent polymerization. As has previously been mentioned, this increased reactivity has been shown to limit the physical properties of the materials and as such there has been little commercial interest in taking these materials further. Epoxidation, acrylation and maleinization of linseed oil before polymerization produced a heavily crosslinked polymer [59]. A formulation for potential in infusion resins was developed with a mixture of epoxidized linseed oil and phthalic anhydride with the curing catalysed by 2-methylimidazole [60].

Bioprocessing has also been investigated with this oil, with epoxidation using *Candida antartica* lipase B taking place [61]. Mixing with maleinated linseed oil and commercial polyamide resin produced a polymer with controllable curing properties. Increasing the level of epoxidation was shown to lead to a higher enthalpy in the curing process by DSC.

6.5.6 Other Plant Oil-Based Bioplastics

Sunflower oil has been studied since it allows for fairly uniform triglycerides to be reacted due to its high oleic acid content. Branched polyesters were produced using high oleic sunflower oil via acyclic triene methathesis [62, 63]. This route used commercial Hoveyda-Grubbs second generation catalysts; first generation Grubbs catalysts produced no cross linking between polymer chains (Figure 6.13). High oleic sunflower oil was converted to a tertiary phosphine oxide and acrylated before being polymerized in the presence of pentaerythritol [64].

Palm oil was used as a model feedstock for the formation of copolymers with cyclooctene by metathesis; with most of the products investigated being in the liquid state [65]. Tung oil was copolymerized with divinylbenzene in cationic fashion to give stable materials exhibiting a high thermal stability and a modulus >1 GPa at room temp [66]. A rare oil that has been used in bioplastics synthesis is vernonia oil, of which the main constituent is vernolic acid, the only naturally occurring epoxidized fatty acid (Figure 6.14). This was incorporated in styrene/acrylate

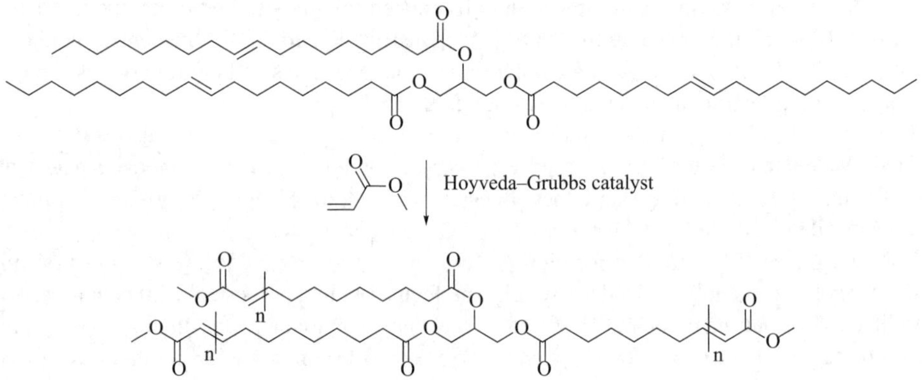

Figure 6.13 Metathesis of triglycerides.

copolymerizations to form films with high solvent resistance [67] or elastomers with varied properties [68].

6.5.7 Biological Synthesis of Polymers

Triglycerides have not only been used as feedstocks for chemical production of bioplastics, but also the enzymatic production of polymers. Polyhydroxyalkanoates were produced by using *Brassica carrinata* oil with high erucic acid content (35–48%) as a feedstock for *Pseudomonas aeruginosa* [69]. Fatty acids with a chain length greater than 20 carbon atoms were shown to be beneficial in terms of increasing the molecular weight, in some cases more than doubling the molecular weight from 56 kDa to 126 kDa.

6.6 Applications

The perception of generating naturally sourced plastic materials may be that of an academic curiosity but there are many industrial applications that have been investigated. The readily available, low-cost feedstock is attractive to companies looking to get a competitive edge over their rivals. The inherent varied reactivity allows manipulations to be carried out efficiently to optimize the properties of the desired application.

6.6.1 Mimicking to Reduce R&D Risk

One of the main issues with the use of plant oils has been the inferior physical properties of the polymers that they are used to make. This is largely due to the fact that they have been bred and optimized for food usage over the past several hundred years and are not fit for purpose.

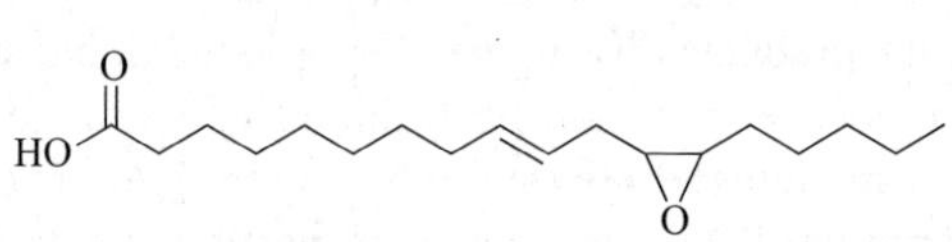

Figure 6.14 Structure of vernolic acid.

Breeding of specialized plants in crop trials would take 3–5 years in order to extract enough oil to investigate its properties fully; this is too long for industry to take the risk of a plant oil not being fit for purpose. In order to mitigate this, rapeseed oil profiles were mimicked in the laboratory and then converted to polyurethanes and compared with their natural analogues [10]. The mimicking process was shown to be accurate and this could be used as a method for reducing R&D times in developing biopolymer materials further.

6.6.2 Composites

One of the largest areas of applications for triglyceride-based bioplastics has been in the area of composite structures. Formulations have been developed and bio-based resins present obvious benefits in terms of greenhouse gas emissions and reduction of the environmental impact of manufacturing [70]. Canola oil and other triglycerides have been modified in several ways (acrylated epoxidized, maleinated acrylated epoxidized and maleinated hydroxylated) before being mixed with styrene or acetic anhydride for use a low viscosity resin for composite manufacture [71, 72]. Sheet-moulding compounds were made from linseed oil where the increased level of unsaturation created strong materials with 2.8 GPa flexural modulus and 100 MPa flexural strength with a high thermal degradation onset at 300 °C [59].

The combination with fibres has proved difficult however. Often there are issues with compatibility between bio-resins and fibres (both natural and synthetic), which cause defects in the composite structure and ultimately poorer physical properties. Castor oil polyurethane was compared with phenolic resins when infused over sisal fibres; however, the phenolic resins showed better structural performance when compared with the castor oil-based material [52]. This is not always the case, as some improvements have been made. Soybean oil thermoset polymers were used in a glass/flax hybrid composite resulting in improved mechanical performance [73]. Thermoset resins were produced from triglyceride oils with a wide range of properties (tensile modulus ~ 1–2 GPa, glass transition temperature T_g ~ 70–120 °C) and glass- and hemp- fibre composites were manufactured [74, 75].

Composites that have been reinforced using organoclays have also been studied with general improvements to the physical properties noted. Reaction of a functionalized soybean oil matrix with montmorillonite produced a polymeric composite with improved thermal and mechanical characteristics than the cured soybean matrix [76]. Clay nanocomposites with a triglyceride matrix were manufactured and treated at 900 °C to leave a porous ceramic material [77]. Toughening agents such as poly(butadiene-co-acrylonitrile), epoxidized soybean liquid rubber and organo-clay were used to improve both the mechanical and thermal properties of the polymers [78, 79]. The clay was the most effective in this instance; the polymer and liquid rubber have detrimental effects in either the mechanical or thermal stability respectively. Acrylated oleic acid methyl ester was polymerized and mixed into an organoclay to form an elastic nanocomposite with vastly improved properties over the original natural polymer [80].

6.6.3 Coatings

Soybean oil has been used as a drying oil in paint formulations for many years, so it was obvious that the application of plant oils polymers in coatings would be investigated. Crosslinked triglycerides have been used as coatings for metal substructures since they have been shown to have beneficial properties when compared to the commercial standards. Triglycerides crosslinked with either 4,4′-methylenedianiline or phthalic anhydride gave glassy polymers, which both had a lower modulus than commercial epoxy resins but adhered more readily to steel or aluminium substructures with increased ductility [81]. Polyurethanes were used as a coating for steel based on a soybean oil starting material and modification to give a sucrose-soyate ester polyol before polymerization was shown to improve the hardness of the coating due to the extensive cross-linking present [47]. Other industrial coatings studied included tack-free coatings, synthesized using an air-catalysed free-radical crosslinked soybean oil polymer [82] and a polyester-polyurethane industrial coating akin to an alkyd resin made from lesquerella oil, a potential castor oil substitute [83].

6.6.4 Packaging Materials

Biopolymers have been used for packaging materials recently, particularly with the growth of poly(lactic acid) as a feedstock. The incorporation of plant oils has shown some benefit, with poly(lactic acid) being strengthened by blending with free-radical polymerized soybean oil, resulting in significantly tougher tensile strength [84]. Similarly, polymer films were developed by incorporating triglycerides with poly(vinyl acetate) to give increased water and solvent resistance due to hydrophobic nature of the triglyceride [85].

6.6.5 Foams

The extensive research into polyurethanes has naturally driven foams to be investigated as they are a major output of the polyurethane industry. Rigid and flexible polyurethane foams were created from epoxidized soybean oil with the oil used as a replacement for the petrochemical polyols [86]. These water-blown foams had up to 50% bio-based content, although it was shown that over 20% oil incorporation had inferior physical properties when compared with the straight petrochemical counterpart. A variety of plant oils were also studied to make polyurethane foams with densities between 27–43 kg m^{-3} [87]. Industrial foams have been generated by both Dow [88] and Lear, with the latter producing a soybean polyurethane foam seat for the WorldFirst F3 racing car (http://www.worldfirstracing.co.uk/, accessed 22 June 2013).

These foams can then be extended into the area of flame-retardant materials, where methyl oleate-polyesters were used as polyols in the synthesis of silicon-containing polyurethanes [89]. Despite not strictly being foams, methyl oleate, soybean and sunflower oils have also been investigated to produce semi-rigid flame retardant materials [90]. In this instance, they were brominated, acrylated and then radically copolymerized with styrene to form the polymeric material.

6.6.6 Biomedical Applications

The explosion in our population density has driven the medical research community to look for innovative, low-cost solutions to today's medical problems. It was suggested that polyurethane based on triglycerides reacting with diisocyanates could be used in wound-dressing applications by analysis of their gas permeability, film and mechanical properties [91,92]. Nanocomposite films based on free-radical polymerized triglycerides and silver nanoparticles were shown to have an antibacterial effect versus Gram-positive, Gram-negative and Spore-forming bacteria [93]. Castor oil polymers were studied and shown to be effective at controlled drug release in *in vitro* studies [55, 94]. In particular, copolymers with lactic acid showed that control of the L- or D- conformation of the lactic acid or increasing the hydrophobicity of the castor oil content were the dominant effects on effectively controlling the release.

6.6.7 Other Applications

Other industrial applications include the fabrication of two-part epoxy resins (similar to those commonly found in household maintenance stores) [95–97]. These were synthesized using triglycerides and diamines. These resins are often used as adhesives; these have also been studied using soybean oil, which provided beneficial properties in terms of fast curing, thermal stability and ease of removal (peel strength) [98]. A blend of divinylbenzene/styrene/tung oil mix gave a polyurethane-based material which behaved like a smart polymer with shape memory behaviour [66].

6.7 Conclusions

Lipids present an excellent opportunity for the production of bioplastics. They can be directly polymerized using free-radical or cationic processes with the addition of some additives. The reactivity present in many triglycerides (hydroxyl groups, double bonds) allows for the facile manipulation to whatever the desired functional group for polymerization may be. Triglycerides are an excellent, sustainable platform chemical for polymer production; the explosion in their use since the early 2000s will only continue as concerns over petrochemical feedstocks and prices grow.

References

1. Murphy, D.J. and Vance, J. (1999) Mechanisms of lipid-body formation. *Trends in Biochemical Sciences*, **24**(3), 109–115.
2. Eteshola, E. and Oraedu, A. (1996) Fatty acid compositions of tigernut tubers (*Cyperus esculentus* L.), baobab seeds (*Adansonia digitata* L.), and their mixture. *Journal of the American Oil Chemists' Society*, **73**(2), 255–257.
3. Benaniba, M.T., Belhaneche-Bensemra, N. and Gelbard, G. (2001) Stabilizing effect of epoxidized sunflower oil on the thermal degradation of poly(vinyl chloride). *Polymer Degradation and Stability*, **74**(3), 501–505.
4. Grosso, N.R. and Guzman, C.A. (1995) Chemical composition of aboriginal peanut (*Arachis hypogaea L.*) seeds from Peru. *Journal of Agricultural and Food Chemistry*, **43**(1), 102–105.
5. Chirala, S. and Wakil, S. (2004) Structure and function of animal fatty acid synthase. *Lipids*, **39**(11), 1045–1053.
6. Zhang, L., Ge, L., Parimoo, S. *et al.* (1999) Human stearoyl-CoA desaturase: alternative transcripts generated from a single gene by usage of tandem polyadenylation sites. *The Biochemical Journal*, **340**(1), 255–264.
7. Knothe, G., Krahl, J. and Gerpen, J.V. (2005) *The Biodiesel Handbook*, AOCS Press, Urbana, IL.
8. Gunstone, F.D., Harwood, J.L. and Dijkstra, A.J. (2007) *The Lipid Handbook*, 3rd edn, CRC Press, Boca Raton, FL.
9. Knothe, G. (2008) 'Designer' biodiesel: optimizing fatty ester composition to improve fuel properties. *Energy & Fuels*, **22**(2), 1358–1364.
10. Coles, S.R., Barker, G., Clark, A.J. *et al.* (2008) Synthetic mimicking of plant oils and comparison with naturally grown products in polyurethane synthesis. *Macromolecular Bioscience*, **8**(6), 526–532.
11. Leizer, C., Ribuicky, D., Poulev, A. *et al.* (2000) The composition of hemp seed oil and its potential as an important source of nutrition. *Journal of Nutraceuticals, Functional & Medical Foods*, **2**(4), 35–53.

12. Vereshchagin, A. and Novitskaya, G. (1965) The triglyceride composition of linseed oil. *Journal of the American Oil Chemists' Society*, **42**(11), 970–974.
13. Romanik, G., Gilgenast, E., Przyjazny, A. and Kamiński, M. (2007) Techniques of preparing plant material for chromatographic separation and analysis. *Journal of Biochemical and Biophysical Methods*, **70**(2), 253–261.
14. Kurki, A., Bachmann, J. and Hill, H. (2008) Oilseed Processing for Small Scale Producers, https://attra.ncat.org/attra-pub/summaries/summary.php?pub=102 (accessed 7 July 2013).
15. Barthet, V.J., Chornick, T. and Daun, J.K. (2002) Comparison of the methods to measure the oil contents in oilseeds. *Journal of Oleo Science*, **51**(589–597).
16. Baümler, E., Crapiste, G. and Carelli, A. (2010) Solvent extraction: kinetic study of major and minor compounds. *Journal of the American Oil Chemists' Society*, **87**(12), 1489–1495.
17. Kadioglu, S., Phan, T. and Sabatini, D. (2011) Surfactant-based oil extraction of corn germ. *Journal of the American Oil Chemists' Society*, **88**(6), 863–869.
18. Womeni, H.M., Ndjouenkeu, R., Kapseu, C. *et al.* (2008) Aqueous enzymatic oil extraction from *Irvingia gabonensis* seed kernels. *European Journal of Lipid Science and Technology*, **110**(3), 232–238.
19. Latif, S., Diosady, L.L. and Anwar, F. (2008) Enzyme-assisted aqueous extraction of oil and protein from canola (*Brassica napus L.*) seeds. *European Journal of Lipid Science and Technology*, **110**(10), 887–892.
20. Proctor, R. (1997) Soybean oil extraction and processing, in *Soybeans: Chemistry, Technology and Utilization* (ed. K. Li), Chapman & Hall, New York, pp. 297–346.
21. Hazimah, A.H., Kosheela Devi, P.P., Tuan Noor Maznee, T.I. *et al.* (2011) Palm-based polyols and polyurethanes: two decades of experience. *Oil Palm Bulletin*, **63**, 33–42.
22. Ronda, J.C., Lligadas, G., Galia, M. and Cadiz, V. (2011) Vegetable oils as platform chemicals for polymer synthesis. *European Journal of Lipid Science and Technology*, **113**(1), 46–58.
23. Gallezot, P. (2007) Catalytic routes from renewables to fine chemicals. *Catalysis Today*, **121**(1–2), 76–91.
24. Seniha Guener, F., Yagci, Y. and Tuncer Erciyes, A. (2006) Polymers from triglyceride oils. *Progress in Polymer Science*, **31**(7), 633–670.
25. Crivello, J.V. and Narayan, R. (1992) Epoxidized triglycerides as renewable monomers in photoinitiated cationic polymerization. *Chemistry of Materials*, **4**(3), 692–699.
26. Eren, T. and Kusefoglu, S.H. (2005) Synthesis and polymerization of the acrylamide derivatives of fatty compounds. *Journal of Applied Polymer Science*, **97**(6), 2264–2272.
27. La Scala, J. and Wool, R.P. (2005) Rheology of chemically modified triglycerides. *Journal of Applied Polymer Science*, **95**(3), 774–783.
28. Zlatanic, A., Petrovic, Z.S. and Dusek, K. (2002) Structure and properties of triolein-based polyurethane networks. *Biomacromolecules*, **3**(5), 1048–1056.
29. Petrovic, Z.S., Zhang, W. and Javni, I. (2005) Structure and properties of polyurethanes prepared from triglyceride polyols by ozonolysis. *Biomacromolecules*, **6**(2), 713–719.
30. Zhao, H.-P, Zhang, J.-F, Sun, X.S. and Hua, D.H. (2008) Syntheses and properties of cross-linked polymers from functionalized triglycerides. *Journal of Applied Polymer Science*, **110**(2), 647–656.
31. Cayli, G. and Kusefoglu, S. (2008) Biobased polyisocyanates from plant oil triglycerides: synthesis, polymerization, and characterization. *Journal of Applied Polymer Science*, **109**(5), 2948–2955.
32. Eren, T. and Kuesefoglu, S.H. (2004) Hydroxymethylation and polymerization of plant oil triglycerides. *Journal of Applied Polymer Science*, **91**(6), 4037–4046.
33. United States Department of Agriculture (2011) Oilseeds: World Markets and Trades, http://www.fas.usda.gov/oilseeds/Current/ (accessed 7 July 2013).
34. Winkler, J.K. and Warner, K. (2008) Effect of phytosterol structure on thermal polymerization of heated soybean oil. *European Journal of Lipid Science and Technology*, **110**(11), 1068–1077.
35. Esen, H., Kusefoglu, S. and Wool, R. (2007) Photolytic and free-radical polymerization of monomethyl maleate esters of epoxidized plant oil triglycerides. *Journal of Applied Polymer Science*, **103**(1), 626–633.
36. Ozturk, C. and Kusefoglu, S.H. (2010) Polymerization of epoxidized soybean oil with maleinized soybean oil and maleic anhydride grafted polypropylene mixtures. *Journal of Applied Polymer Science*, **118**(6), 3311–3317.

37. Oeztuerk, C. and Kuesefoglu, S.H. (2011) Polymerization of epoxidized soybean oil with maleinized polybutadiene. *Journal of Applied Polymer Science*, **120**(1), 116–123.
38. Oeztuerk, C. and Kuesefoglu, S.H. (2010) New polymers from plant oil derivatives and styrene-maleic anhydride copolymers. *Journal of Applied Polymer Science*, **116**(1), 355–365.
39. Fu, L., Yang, L., Dai, C. *et al.* (2010) Thermal and mechanical properties of acrylated expoxidized-soybean oil-based thermosets. *Journal of Applied Polymer Science*, **117**(4), 2220–2225.
40. Oztuerk, C., Mutlu, H., Meier, M.A.R. and Kusefoglu, S.H. (2011) 4-Vinylbenzenesulfonic acid adduct of epoxidized soybean oil: synthesis, free radical and ADMET polymerizations. *European Polymer Journal*, **47**(7), 1467–1476.
41. Campanella, A., La Scala, J.J. and Wool, R.P. (2011) Fatty acid-based comonomers as styrene replacements in soybean and castor oil-based thermosetting polymers. *Journal of Applied Polymer Science*, **119**(2), 1000–1010.
42. Saithai, P., Tanrattanakul, V., Chinpa, W. *et al.* (2011) Synthesis and characterization of triglyceride-based copolymer from soybean oil. *Materials Science Forum*, **695**, 320–323.
43. Zhan, M. and Wool, R.P. (2010) Biobased composite resins design for electronic materials. *Journal of Applied Polymer Science*, **118**(6), 3274–3283.
44. Esen, H. and Kuesefoglu, S.H. (2003) Photolytic and free-radical polymerization of cinnamate esters of epoxidized plant oil triglycerides. *Journal of Applied Polymer Science*, **89**(14), 3882–3888.
45. Can, E., Wool, R.P. and Kusefoglu, S. (2006) Soybean- and castor-oil-based thermosetting polymers: mechanical properties. *Journal of Applied Polymer Science*, **102**(2), 1497–1504.
46. Eren, T., Kuesefoglu, S.H. and Wool, R. (2003) Polymerization of maleic anhydride-modified plant oils with polyols. *Journal of Applied Polymer Science*, **90**(1), 197–202.
47. Pan, X. and Webster, D.C. (2012) New biobased high functionality polyols and their use in polyurethane coatings. *ChemSusChem*, **5**(2), 419–429.
48. Cayli, G. and Kuesefoglu, S. (2010) A simple one-step synthesis and polymerization of plant oil triglyceride iodo isocyanates. *Journal of Applied Polymer Science*, **116**(4), 2433–2440.
49. Cayli, G. and Kuesefoglu, S. (2010) Isothiocyanate derivatives of soybean oil triglycerides: synthesis, characterization, and polymerization with polyols and polyamines. *Journal of Applied Polymer Science*, **116**(1), 125–131.
50. Oztuerk, C. and Kusefoglu, S.H. (2011) New polymers from epoxidized soybean oil with pyridine derivatives. *Journal of Applied Polymer Science*, **121**(5), 2976–2984.
51. Campanella, A., La Scala, J.J. and Wool, R.P. (2009) The use of acrylated fatty acid methyl esters as styrene replacements in triglyceride-based thermosetting polymers. *Polymer Engineering & Science*, **49**(12), 2384–2392.
52. Milanese, A.C., Cioffi, M.O.H. and Voorwald, H.J.C. (2011) Mechanical behavior of natural fiber composites. *Procedia Engineering*, **10**, 2022–2027.
53. Kaushik, A. and Singh, P. (2012) Castor oil/trimethylol propane-based polyurethane reactions: modeling in a batch reactor. *Journal of Applied Polymer Science*, **125**, E51–E60.
54. Szelest-Lewandowska, A., Masiulanis, B., Szymonowicz, M. *et al.* (2007) Modified polycarbonate urethane: synthesis, properties and biological investigation in vitro. *Journal of Biomedical Materials Research, Part A*, **82A**(2), 509–520.
55. Zanetti-Ramos, B., Soldi, V., Lemos-Senna, E. and Borsali, R. (2005) Use of natural monomer in the synthesis of nano- and microparticles of polyurethane by suspension-polyaddition technique. *Macromolecular Symposia*, **229**, 234–245.
56. Hirayama, K.-I, Irie, T., Teramoto, N. and Shibata, M. (2009) High-performance bio-based thermosetting resins composed of dehydrated castor oil and bismaleimide. *Journal of Applied Polymer Science*, **114**(2), 1033–1039.
57. Petrovic, Z.S., Milic, J., Xu, Y. and Cvetkovic, I. (2010) A chemical route to high molecular weight vegetable oil based polyhydroxyalkanoate. *Macromolecules (Washington, DC, United States)*, **43**(9), 4120–4125.

58. Sharif, A., Nijas, P.K. and Ufana, R. (2011) Effect of microwave processing on the spectral, mechanical, thermal, and morphological characteristics of sustainable resource based castor oil epoxy/PVA blends. *Advances in Polymer Technology*, **30**(2), 96–109.
59. Lu, J. and Wool, R.P. (2006) Novel thermosetting resins for SMC applications from linseed oil: synthesis, characterization, and properties. *Journal of Applied Polymer Science*, **99**(5), 2481–2488.
60. Boquillon, N. and Fringant, C. (2000) Polymer networks derived from curing of epoxidised linseed oil: influence of different catalysts and anhydride hardeners. *Polymer*, **41**(24), 8603–8613.
61. Mahendran, A.R., Aust, N., Wuzella, G. and Kandelbauer, A. (2012) Synthesis and characterization of a bio-based resin from linseed oil. *Macromolecular Symposia*, **311**, 18–27.
62. Biermann, U., Metzger, J.O. and Meier, M.A.R. (2010) Acyclic triene metathesis oligo- and polymerization of high oleic sun flower oil. *Macromolecular Chemistry and Physics*, **211**(8), 854–862.
63. Fokou, P.A. and Meier, M.A.R. (2008) Acyclic triene metathesis polymerization with chain-stoppers: molecular weight control in the synthesis of branched polymers. *Macromolecular Rapid Communications*, **29**(19), 1620–1625.
64. Montero De Espinosa, L., Ronda, J.C., GaliA M. and Cadiz, V. (2009) A straightforward strategy for the efficient synthesis of acrylate and phosphine oxide-containing vegetable oils and their crosslinked materials. *Journal of Polymer Science, Part A: Polymer Chemistry*, **47**(16), 4051–4063.
65. Buang, N.A., Nordin, N.A.M., Kamarudin, R.A. and Muda, A. (1997) Production of a low molecular weight polymer from the cometathesis reaction of palm oil and cyclooctene. *Journal of Molecular Catalysis A: Chemical*, **115**(1), 95–99.
66. Meiorin, C., Aranguren, M.I. and Mosiewicki, M.A. (2012) Smart and structural thermosets from the cationic copolymerization of a vegetable oil. *Journal of Applied Polymer Science*, **124**(6), 5071–5078.
67. Casebolt, E.D., Mote, B.E. and Trumbo, D.L. (2002) Applications of Vernonia oil in thermoset coatings II. *Progress in Organic Coatings*, **44**(2), 147–151.
68. Fernandez, A.M., Murphy, C.J., DeCosta, M.T. *et al.* (1983) Vernonia oil characterization and polymerization and simultaneous interpenetrating polymer networks based on vernonia oil-sebacic acid/polystyrene-DVB compositions. *Polymer Science and Technology (Plenum)*, **17**, 273–288.
69. Impallomeni, G., Ballistreri, A., Carnemolla, G.M. *et al.* (2011) Synthesis and characterization of poly(3-hydroxyalkanoates) from *Brassica carinata* oil with high content of erucic acid and from very long chain fatty acids. *International Journal of Biological Macromolecules*, **48**(1), 137–145.
70. Wool, R.P., Khot, S.N., LaScala, J.J. *et al.* (2002) Affordable composites and plastics from renewable resources: part II: manufacture of composites. *ACS Symposium Series*, **823**, 205–224.
71. Okieimen, F.E., Bakare, I.O. and Pavithran, C. (2009) Network structures, flexibility and stability in thermal and chemical environments of polyurethane prepared from rubber seed oil. *Advanced Materials Research (Zuerich, Switzerland)*, **62–64**, 324–334.
72. Adekunle, K., Aakesson, D. and Skrifvars, M. (2008) Synthetic modification of reactive soybean oils for use as biobased thermoset resins in structural natural fiber composites. *Polymer Preprints (American Chemical Society, Division of Polymer Chemistry)*, **49**(1), 279–280.
73. Morye, S.S. and Wool, R.P. (2005) Mechanical properties of glass/flax hybrid composites based on a novel modified soybean oil matrix material. *Polymer Composites*, **26**(4), 407–416.
74. Khot, S.N., Lascala, J.J., Can, E. *et al.* (2001) Development and application of triglyceride based polymers and composites. *Journal of Applied Polymer Science*, **82**(3), 703–723.
75. Williams, G.I. and Wool, R.P. (2000) Composites from natural fibers and soy oil resins. *Applied Composite Materials*, **7**(5–6), 421–432.
76. Sen, S. and Cayli, G. (2010) Synthesis of bio-based polymeric nanocomposites from acrylated epoxidized soybean oil and montmorillonite clay in the presence of a bio-based intercalant. *Polymer International*, **59**(8), 1122–1129.
77. Tsujimoto, T., Kuwabara, M., Uyama, H. *et al.* (2010) Preparation and properties of green nanocomposites from epoxidized triglycerides and a modified clay. *Nippon Setchaku Gakkaishi*, **46**(4), 131–136.
78. Lu, J. and Wool, R.P. (2008) Additive toughening effects on new bio-based thermosetting resins from plant oils. *Composites Science and Technology*, **68**(3–4), 1025–1033.

79. Lu, J., Hong, C.K. and Wool, R.P. (2004) Bio-based nanocomposites from functionalized plant oils and layered silicate. *Journal of Polymer Science, Part B: Polymer Physics*, **42**(8), 1441–1450.
80. Zhu, L. and Wool, R.P. (2005) Acrylate rubber/organoclay hybrid nanocomposites from renewable resources. Proceedings of the American Society for Composites, Technical Conference, 2005; 20th, 148/1-/18.
81. Earls, J.D., White, J.E., Lopez, L.C. *et al.* (2007) Amine-cured ω-epoxy fatty acid triglycerides: fundamental structure-property relationships. *Polymer*, **48**(3), 712–719.
82. Fornof, A.R., Onah, E., Ghosh, S. *et al.* (2006) Synthesis and characterization of triglyceride-based polyols and tack-free coatings via the air oxidation of soy oil. *Journal of Applied Polymer Science*, **102**(1), 690–697.
83. Thames, S.F., Bautista, M.O., Watson, M.D. and Wang, M.D. (1994) Application of lesquerella oil in industrial coatings. *ACS Symposium Series*, **575**, 212–222.
84. Robertson, M.L., Chang, K., Gramlich, W.M. and Hillmyer, M.A. (2010) Toughening of polylactide with polymerized soybean oil. *Macromolecules (Washington, DC, United States)*, **43**(4), 1807–1814.
85. Salvini, A., Saija, L.M., Lugli, M. *et al.* (2010) Synthesis of modified poly(vinyl acetate) adhesives. *Journal of Adhesion Science and Technology*, **24**(8–10), 1629–1651.
86. Tu, Y.-C, Suppes, G.J. and Hsieh, F.-H. (2008) Water-blown rigid and flexible polyurethane foams containing epoxidized soybean oil triglycerides. *Journal of Applied Polymer Science*, **109**(1), 537–544.
87. Prociak, A. (2007) Properties of polyurethane foams modified with natural oil-based polyols. *Cellular Polymers*, **26**(6), 381–392.
88. Sanders, A., Babb, D., Prange, R. *et al.* (2007) Producing polyurethane foam from natural oil. *Chemical Industries (Boca Raton, FL, United States)*, 115, 377–384.
89. Galia, M., Montero de Espinosa, L., Ronda, J.C. et al. (2009) Vegetable oil-based thermosetting polymers. *European Journal of Lipid Science and Technology*, **112**(1), 87–96.
90. Eren, T. and Kuesefoglu, S.H. (2004) Synthesis and polymerization of the bromoacrylated plant oil triglycerides to rigid, flame-retardant polymers. *Journal of Applied Polymer Science*, **91**(4), 2700–2710.
91. Madra, H., Tantekin-Ersolmaz, S.B. and Guner, F.S. (2009) Monitoring of oil-based polyurethane synthesis by FTIR-ATR. *Polymer Testing*, **28**(7), 773–779.
92. Ozkaynak, M.U., Atalay-Oral, C., Tantekin-Ersolmaz, S.B. and Guner, F.S. (2005) Polyurethane films for wound dressing applications. *Macromolecular Symposia*, **228**, 177–184.
93. Eksik, O., Erciyes, A.T. and Yagci, Y. (2008) In situ synthesis of oil based polymer composites containing silver nanoparticles. *Journal of Macromolecular Science, Part A: Pure and Applied Chemistry*, **45**(9), 698–704.
94. Sokolsky-Papkov, M. and Domb, A.J. (2008) Stereoisomeric effect on in vitro drug release from injectable poly(lactic acid co castor oil) polyesters. *Polymers for Advanced Technologies*, **19**(6), 671–679.
95. Dirlikov, S., Frischinger, I. and Chen, Z. (1996) Phase separation of two-phase epoxy thermosets that contain epoxidized triglyceride oils. *Advances in Chemistry Series*, **252**, 95–109.
96. Frischinger, I., Muturi, P. and Dirlikov, S. (1994) Two-phase interpenetrating epoxy thermosets that contain epoxidized triglyceride oils. Part II. Applications. *Advances in Chemistry Series*, **239**, 539–556.
97. Frischinger, I. and Dirlikov, S. (1994) Two-phase interpenetrating epoxy thermosets that contain epoxidized triglyceride oils. Part I. Phase separation. *Advances in Chemistry Series*, **239**, 517–538.
98. Ahn, B.K., Kraft, S., Wang, D. and Sun, X.S. (2011) Thermally stable, transparent, pressure-sensitive adhesives from epoxidized and dihydroxyl soybean oil. *Biomacromolecules*, **12**(5), 1839–1843.

7

Polyhydroxyalkanoates: Basics, Production and Applications of Microbial Biopolyesters

Martin Koller[1,2], Anna Salerno[1], and Gerhart Braunegg[2]
[1] *Graz University of Technology, Institute of Biotechnology and Biochemical Engineering, Austria*
[2] *ARENA Arbeitsgemeinschaft für Ressourcenschonende und Nachhaltige Technologien, Austria*

7.1 Microbial PHA Production, Metabolism, and Structure

7.1.1 Occurrence of PHAs

Polyhydroxyalkanoates (PHAs) are a family of biopolyesters, synthesized as intracellular products by prokaryotic genera, encompassing Gram-positive and Gram-negative eubacteria [1, 2], oxygen-evolving cyanobacteria [3–5], and even archaea, often occupying extreme environments [6–8].

In addition to prokaryotic PHA accumulation, low-molecular mass poly-hydroxybutyrate complexed to other macromolecular cell components like proteins or polyphosphates (*c*PHB) can be found in the tissue of eukaryotic organisms. *c*PHB classically occurs in low quantities in the organisms and does not serve as storage material; not all of the details of its metabolic functions have been elucidated [9].

Regarding genetically modified eukaryotic organisms, PHA production is reported in plants, yeasts [10] and other eukaryotic organisms harbouring PHA synthesis genes. Especially in the field of genetically engineered crops, PHA synthesis starting from photosynthetic CO_2-fixation has already been well studied [11].

Table 7.1 provides an up-to-date compilation of microbial genera that are reported to encompass PHA producing species.

Bio-Based Plastics: Materials and Applications, First Edition. Edited by Stephan Kabasci.

Table 7.1 *Microbial PHA producing genera.*

Eubacteria		
Gram negative	Gram positive	Archaea
(A) *Accumulibacter, Acidovora, Acinetobacter, Actinobacillus, Aeromonas, Alcaligenes, Alcanivorax, Alkalilimuicola, Allochromatium, Amphritea, Anabaena*[a]*, Anacystis*[a]*, Aphanocapsa*[a]*, Aphanothece*[a]*, Aquaspirillum, Aromatoleum, Asticcaulus, Aulosira*[a]*, Aureobasidium, Axobacter, Azomonas, Azospirillum, Azotobacter*	*Actinomycetes*	**(C)** *Cenarchaenum* **(H)** *Halobacterium* *Haloarcula* *Halobiforma* *Halococcus* *Haloferax*
(B) *Beggiatoa, Beijerinckia, Beneckea, Brachymonas, Bradyrhizobium, Brevundimonas, Burkholderia*	*Bacillus*	*Halogeometricum* *Halopiger*
(C) *Caenibacterium, Caldimonas, Calothrix*[a]*, Caryophanon, Caulobacter, Chelatococcus, Chloroflexus*[a]*, Chlorogloea*[a]*, Chrococcus*[a]*, Chromatium, Chromobacterium, Chromohalobacter, Cobetia, Comamonas Competibacter, Cupriavidus* (formal name of genera: *Ralstonia, Wetursia*), *Cyanobacterium*[a]*, Cyanothece*[a]	*Clostridium* *Corynebacterium*	*Haloquadratum* *Haloterrigena* *Halorhabdus* *Halorhodospira* *Halorubrum* **(K)** *Kushneria*
(D) *Dechloromonas, Defluviicoccus, Delftia, Desulfococcus, Desulfobacterium*	*Derxia*	**(N)** *Natrialba* *Natrinema*
(E) *Ectothiorhodospira, Ensifer* (formal name of genera: *Sinorhizobium*), *Erwinia, Escherichia* (wilde type and recombinant strain)		*Natronobacterium-* *Natronococcus* *Nitrosopumilus*
(F) *Ferrobacillus, Fischerella*[a]*, Fundibacter*		
(G) *Gamphosphaeria, Gloeocapsa*[a]*, Gloeothece*[a]		
(H) *Haemophilus, Hahella, Halomonas, Hydrogenophaga, Herbaspirillum, Hyphomicrobium*		
(I, J, K) *Ideonella, Janthinobacterium, Klebsiella* (recombinant strain)		
(L) *Labrenzia, Lampropedia, Leptothrix, Legionella*		
(M) *Magnetospirillum, Marinobacter, Marinospirillum, Mesorhizobium, Methanomonas, Methylarcula, Methylibium, Methylobacterium, Methylosinus, Methylocystis, Methylomonas, Methylovibrio, Microcoleus, Microcystis*[a]*, Microvoleus, Moraxella, Mycoplana*	*Micrococcus* *Microlunatus*	
(N) *Neptunomonas, Nitrobacter, Nitrococcus, Nodularia*[a]*, Novosphingobium, Nostoc*[a]	*Nocardia*	
(O) *Oceanospirillum, Oligotropha, Oscillatoria*[a]		
(P) *Paracoccus, Pedomicrobium, Pelomonas, Photobacterium, Protomonas, Plasticicumulans, Pleurocapsa*[a]*, Pseudoanabaen*[a]*, Pseudomonas*	*Paucispirillum*	
(R) *Rhizobium, Rhodobacter*[a]*, Rhodopseudomonas, Rhodospirillum, Rivularia*[a]*, Rubrivivax*[a]*, Roseateles*[a]*, Ruegeria*	*Rhodococcus*	
(S) *Saccharophagus, Schlegelella, Scytonema*[a]*, Sphaerotilus, Sphingomonas, Sphingopyxis, Spirillum, Spirulina*[a]*, Stella, Synechococcus*[a]*, Synechocystis*[a]	*Staphylococcus* *Streptomyces* *Syntrophomonas*	
(T) *Thauera, Thermus, Thiobacillus, Thiocapsa, Thiococcus, Thiocystis, Thiodictyon, Thiopedia, Thiosphaera, Tolypothrix*[a]		
(V) *Variovorax, Vibrio*		
(W, X, Z) *Westielopsis*[a]*, Xanthobacter, Zobellella, Zoogloe*		

Note: [a] Cyanobacteria.

Polyhydroxyalkanoate biosynthesis starts from renewable feedstocks like hexoses, pentoses, starch, sucrose, lactose, maltose, lipids, alcohols, organic acids, or gases like methane or CO_2, classically under unfavourable growth conditions due to imbalanced nutrient supply. In particular, a high intracellular energy charge, characterized by an elevated pool of Acetyl-CoA, NAD(P)H, ATP and others, promotes PHA accumulation. Such conditions result from a sufficient availability of external carbon source and restricted supply with growth-essential substrates like nitrogen, phosphate, or dissolved oxygen, or certain micro-components like sulfur and certain metals with essential enzymatic functions [9, 10]. Among all bio-based plastics, PHAs are unique by being the only biopolymer family entirely produced and degraded by living cells.

7.1.2 *In Vivo* Characteristics and Biological Role of PHAs

If PHA-rich cells are investigated using a light-optical microscope, the biopolymer is visible as insoluble, inert cytoplasmic inclusions. Figure 7.1 provides a scanning transmission electron microscopy (STEM) picture of PHA-rich *Cupriavidus necator* cells. Polyhydroxyalkanoate inclusions are observable as bright, refractive granules and, in the figure, amount to 52 wt.-% in cell mass. Under optimized conditions, this intracellular PHA content can exceed 90% [12]. Polyhydroxyalkanoate granules possess a typical diameter of 0.2 to 0.7 μm and consist of 97.7% PHA, 1.8% protein and 0.5% lipids. Native PHA granules can be stained with Sudan Black B and more specifically with Nile Blue A or Nile Red. Further studies showed that the strongly fluorescent dye Nile Red can be used to directly monitor PHA inclusions in growing colonies [13, 14]. Proteins and lipids form a membrane coat surrounding the core region and stabilize the transition between the hydrophobic core region and its hydrophilic surroundings. Every granule consists of at least a thousand polymer chains forming right-handed 2^1-helices

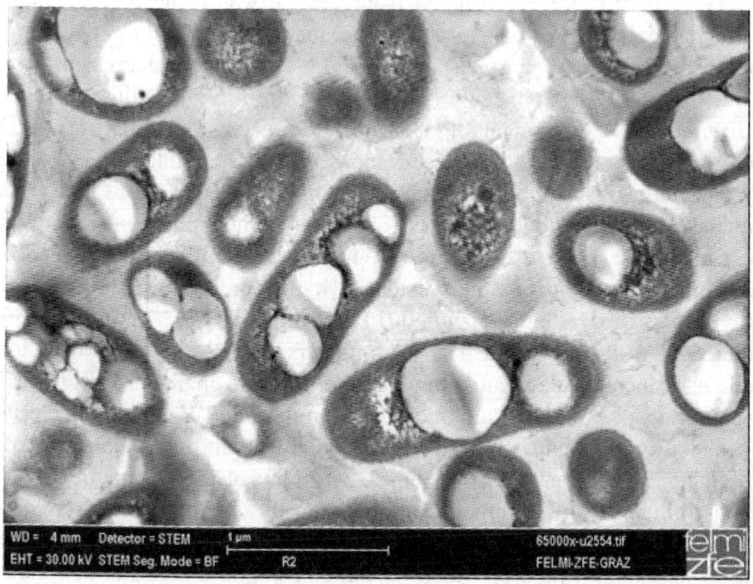

Figure 7.1 STEM picture of *Cupriavidus necator* cells harbouring granules of PHA biopolymers (magnification 1/65000).

with a twofold screw axis and a fibre repeat of 0.596 nm, stabilized by hydrogen bonds and van der Waals intramolecular forces [15]. Decades ago it was generally believed that the core of the inclusion bodies was almost crystalline, explaining the brittle character of PHB. X-ray diffraction analysis underlined this theory. Later it was shown with ^{13}C NMR spectroscopy that the bulk of PHB and poly(3HB-*co*-3HV), a copolyester of 3-hydroxybutyrate (3HB), and 3-hydroxyvalerate (3HV), in intact cells is not crystalline, but can be regarded as a mobile amorphous state [16].

Polyhydroxyalkanoates fulfil important biological tasks, mainly as energy and carbon reserve materials for the microbial cells [17]. Under conditions lacking extracellular carbon sources, they can be remobilized and utilized as carbon and energy substrates, thus providing the cell with an advantage for surviving during starvation periods. Polyhydroxyalkanoates are key-compounds for the regulation of intracellular energy flow – for example, for cell motility or osmoregulation – and contribute to distribution and routing of carbon reserves to different metabolic pathways. Further, they play a major role as endogenous carbon- and energy source for the formation of spores (*Bacillus* sp.) and cysts (*Azotobacter* sp.) [18]. In addition, several important functions of PHAs in various ecosystems were elucidated, such as protection against environmental stress conditions like osmotic shock, UV irradiation, desiccation, or thermal and oxidative stress. More recently, the role of PHA in the maintenance of the intracellular redox state has been discussed [19]. Further, microbial PHA synthesis and degradation plays a crucial role for syntrophic interactions between consortia of different organisms in aquatic or terrestrial ecosystems [2].

7.1.3 Structure and Composition of PHAs

Polyhydroxyalkanoates mainly consist of 3-hydroxyalkanoates (3HAs) as monomeric building blocks. Until the 1960s, poly(3-hydroxybutyrate) (PHB) was the only reported representative of PHAs. In 1974, microbes were isolated from sewage water; these organisms contained PHA consisting of 3HB, 3HV and, to a minor extent, building blocks with more than 5 carbon atoms [20]. The number of discovered PHA components increased tremendously with the extraction of PHAs from microbes isolated from marine sediments that contained at least 11 different PHA components [21]. Until 1995, 91 components forming thioesters that can act as substrates for PHA synthases, and hence that can undergo polymerization to PHAs, were detected. This number has increased until today it is more than 150. Most of these constituents (including lactic acid) have only been demonstrated *in vitro* to be potential PHA building blocks [22]. Most of them are optically active (important exceptions are 4-hydroxybutyrate, 4HB, and, to a minor extent, 5-hydroxyvalerate, 5HV) and, due to the stereospecifity of the biosynthetic enzymes (PHA synthases), *R*-(-)-configured; reports of any *S*-configured HAs do not exist. These monomers can display saturated, unsaturated, straight, cyclic or branched side-chains. In general, these side chains are aliphatic, but, when cultivated on unusual substrates, certain microbes (mainly found among the bacterial group of fluorescent pseudomonades) can synthesize PHAs with pendant functionalities like aromatic, halogenic, pseudohalogenic or alkoxy groups [23]. Many of these 'functional' side chains provide the possibility for post-synthetic modification by chemical or enzymatic methods. This way, functional biopolymers are accessible with properties tailor-made for special niche applications. Prime examples are provided by the conversion of unsaturated side chains of monomeric PHA building blocks to carboxylate, epoxide, or hydroxyl groups that provide the polyester with enhanced hydrophilicity and a variety of additional unusual physical properties

[24]. Based on these functionalities, linkage groups can be added to the polyester for carrying active compounds [25]. The general structure of PHA and its follow-up products after post-synthetic modification are provided in Figure 7.2.

Depending on various factors like the type of microbial production strain, the nutrient supply regime and the process parameters during biosynthesis, PHA polymer chains contain a magnitude of 10^2 to 10^5 3HA monomers. As described before, 3HAs are normally enantiomerically pure, *R*-(-)-configured chiral compounds. Among all known PHAs, poly([*R*]-3-hydroxybutyrate) (PHB) is the best characterized one. Poly([*R*]-3-hydroxybutyrate) is a homopolyester consisting merely of 3HB building blocks; this material features rather high crystallinity and restricted processability. The low difference between the decomposition temperature (around 270 °C) and the high melting point (around 180 °C) provides a window of processability that is too small for many processing techniques, for example, in melt extrusion or production of polymeric films. These obstacles can be overcome by interrupting the crystalline PHB matrix by additional building blocks like 3HV or the achiral building block 4HB, resulting in co-polyesters with enhanced material properties and a broader range of applications.

The exact material properties are strongly dependent on the monomeric composition of the copolyesters; this composition can be triggered during PHA biosynthesis by feeding precursor substrates to obtain the desired building blocks. For example, 3HV building blocks are produced by many PHA accumulating strains by supplying 3HV related precursors, namely odd-numbered fatty acids, such as propanoic or pentanoic acid, whereas 4HB units are incorporated into PHAs by many microbial species after feeding them with the 4HB-related precursors γ-butyrolactone, 4-hydroxybuytric acid, or 1,4-butanediol [27].

7.1.4 Metabolic Aspects

Among all the steps involved in PHA biosynthesis and degradation, three enzymatic reactions are the central switch points where activation or inhibition on the enzyme activity shifts the carbon flux towards protein formation, PHA biosynthesis or PHA degradation.

Polyhydroxyalkanoates biosynthesis is regulated, on one hand, by the activity of *3-ketothiolase* (EC 2.3.1.16), and on the other hand of *acetoacetyl-CoA reductase* (EC 1.1.1.36); intracellular PHA breakdown is dependent on the activity of *3-hydroxybutyrate dehydrogenase* (EC 1.1.1.30). Besides these three enzymes, the following compounds can be pointed out as major factors responsible of the activities of the key enzymes: acetyl-CoA, free CoA, $NAD(P)^+$ (or $NAD(P)H_2$, respectively) and, to a lower extent, ATP, pyruvate and oxalacetate. In any case, acetyl-CoA can be considered as the central metabolite both for biomass formation and PHB biosynthesis. This compound stems from the catabolic break down of carbon substrates like sugars (mainly catabolized by the 2-Keto-3-desoxy-6-phosphogluconate pathway) or fatty acids (converted by ß-oxidation).

In general, high yields of PHA are achieved under conditions of high intracellular energy charge. This can be a consequence of an excess of carbon- and energy source together with a depletion of nitrogen [28] or oxygen [29]. Under these conditions, NAD(P)H is piled up. Limitation of sulfur (stop of protein synthesis due to inability of production of sulfur-containing amino acids such as methionin or cystein) and of phosphates (stop of nucleic acid biosynthesis) also increase the NAD(P)H and acetyl-CoA levels. This results in inhibition of the tricarboxylic acid (TCA) cycle enzymes *citrate synthase* and *isocitrate dehydrogenase*, hampering the entrance of acetyl-CoA into the TCA cycle. As a consequence, the intracellular acetyl-CoA

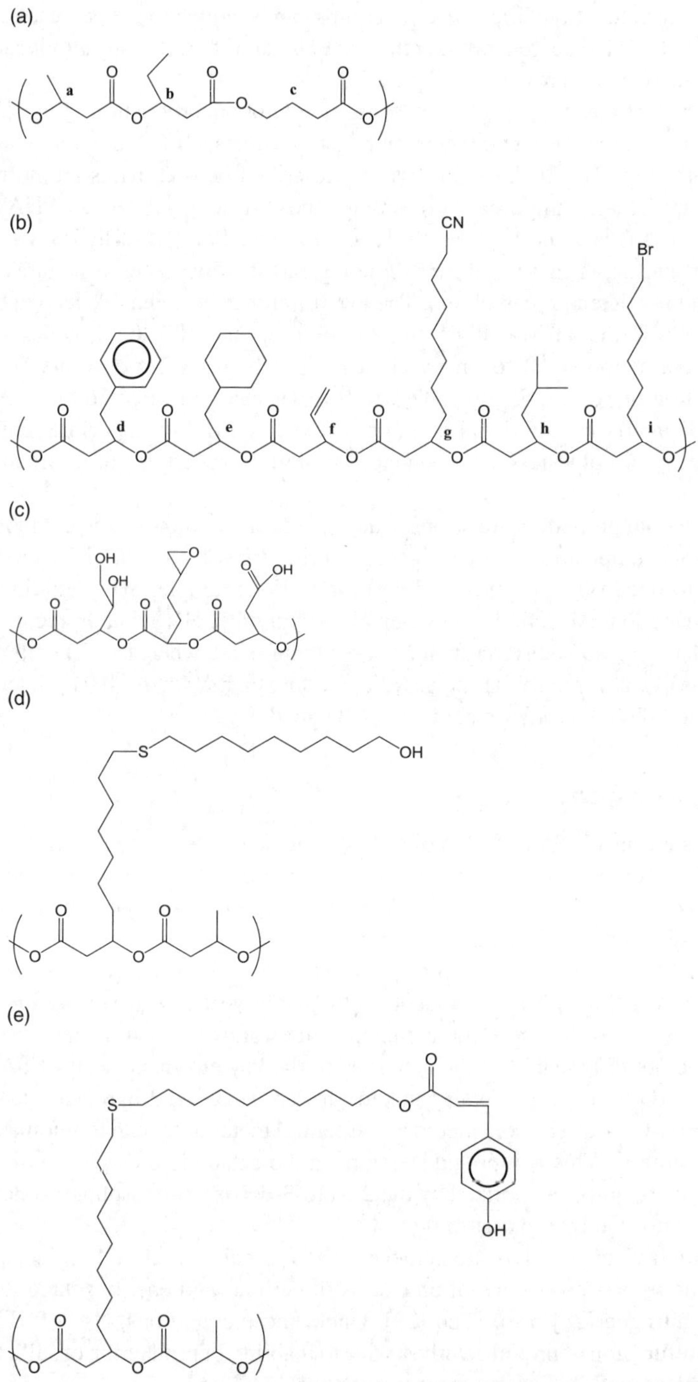

Figure 7.2 *Chemical structures of biobased polyesters. (a) General structure of PHA; (b) example of a functional PHA harbouring 5-Phenylvaleryl-, Cyclohexylbutyryl-, 4-Pentenoyl-, 9-Cyanononanoyl-, 5-Methylhexanoyl-, and 8-Bromooctanoyl groups; (c) PHA after post-synthetic functionalization; (d) PHA with linker group; (e) functional PHA active against biofouling [25]. Chiral centres exist at the central carbon atoms indicated by a, b and d-i; c indicates the non-chiral center of 4HB [26].*

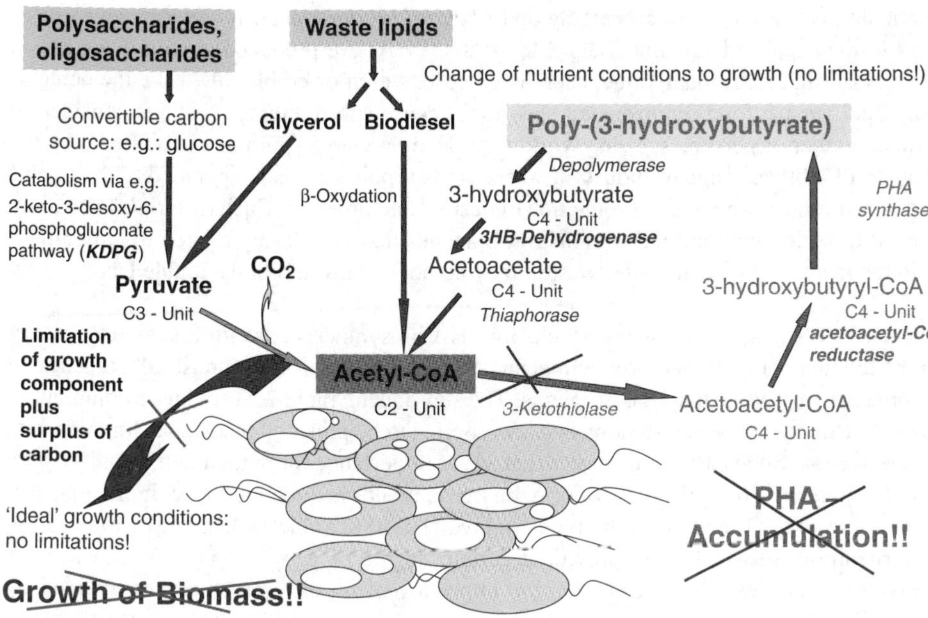

Figure 7.3 Cyclic nature of the basic metabolic routes from carbon substrates to biomass and PHB.

is piled up. By conversion to acetoacetyl-CoA (via *3-ketothiolase*), the metabolic flux can be channelled towards PHA synthesis. Under oxygen-limiting condition, PHA synthesis provides the cell a possibility to regenerate reducing power, also known as 'pseudo-fermentation'. In contrast, if sufficient oxygen is available, free co-enzyme A (CoA) increases due to the acetyl-CoA entering the TCA cycle.

Free CoA is inhibiting *3-ketothiolase*; as a consequence, the carbon flux towards PHAs is decreased. To a lower extent, *3-ketothiolase* is also inhibited by NAD^+ and ATP. Additionally, high amounts of NAD^+ slow down the reduction steps from acetoacetyl-CoA to 3-hydroxybutyryl-CoA by *acetoacetyl-CoA reductase*.

Poly([*R*]-3-hydroxybutyrate) degradation is regulated by inhibition of the oxidative conversion of 3–hydroxybutyrate to acetoacetate. Here, $NADH_2$, pyruvate and oxalacetate act as inhibitors. The last step, the splitting of acetoacetyl-CoA into two acetyl-CoA molecules by *3-ketothiolase*, is product inhibited by high concentrations of acetyl-CoA. High concentrations of free CoA, the second substrate of this cleaving reaction, relieve this product inhibition [28].

Figure 7.3 illustrates the basic metabolic route from carbon substrates to PHB biosynthesis and its subsequent remobilization.

7.2 Available Raw Materials for PHA Production

Using 'white biotechnology', PHAs can be produced starting from renewable resources. The implementation of 'white biotechnology' can be regarded as really promising for sustainable development, although for a range of industrial products, especially bulk products, biotechnological production strategies have not yet passed the threshold for economic feasibility. This is mainly caused by the expenses on the raw materials. The economics of PHA production, in particular, is to a large extent, namely up to 50% of production costs, determined by costs related to the supply of raw materials. Classically, PHAs are produced starting from prized substrates of high nutritional value, such as glucose, starch or edible oils. Like the contemporary 'fuel versus. food' controversy, this might cause severe ethical conflicts considering the huge number of persons starving worldwide. Here, a viable solution can be found by the utilization of a broad range of industrial waste- and surplus materials upgraded to the role of feedstocks for the bio-mediated production of desired end products. Such materials are mainly produced in agriculture and those industrial branches that are closely related to agriculture. Due to the fact that they constitute 'waste', they do not occupy additional arable land [17, 29, 30].

The use of waste materials as feedstocks for PHA biosynthesis constitutes a viable strategy for cost-efficient biopolymer production and supports various agro-industrial branches to overcome existing waste disposal problems. The subsequent Table 7.2 provides a compilation of selected carbon-rich waste streams that are reported to be potential feedstocks for microbial PHA production. Such carbon substrates that are of importance for human nutrition, like pure starch or edible oils from olives, soya or palm trees, are not included in this compilation.

In many regions of the world, the industrial realization of value-added conversion of low-cost agricultural feedstocks can provide a certain degree of geopolitical independence for the involved countries. This means that the importing of fossil reserves can be avoided by replacing fossil feedstocks with efficient implementation of renewable resources available in the respective countries. In addition, the creation of different qualified jobs in these regions in the field of biopolymer production from local resources can provide a socio-economic benefit.

The selection of the appropriate waste stream as feedstocks for biotechnological purposes mainly depends on the region of the world where a production plant is to be constructed. In Europe or North America, whey or waste from the animal processing and biodiesel industry are available at huge quantities. In other regions, lignocelluloses of diverse origin (e.g. rice straw or sugar cane bagasse), molasses, various lipid-rich side streams from plant oil or olive oil manufacturing and so on can be converted by adequate means (hydrolysis) to substrates for biopolymer production [26, 68].

In order to save costs for transportation, facilities for the production of biopolymers, biofuels and biochemicals should be integrated into existing production lines, where the feedstocks accrue directly as waste streams. In addition, the availability of convertible feedstocks all over the year has to be ascertained. This rises up the question of the material's appropriateness for long-term storage, enabling the creation of stocks of raw materials for 'off-season' operation, where no harvest takes place (e.g. in the case of waste streams from the cane sugar industry).

7.3 Recovery of PHA from Biomass

7.3.1 General Aspects of PHA Recovery

Downstream processing constitutes a key part of the entire PHA production process. After biosynthesis of the polyester and separation of the bacterial biomass (normally by well-established techniques like centrifugation, sedimentation, flocculation or filtration), the process required for PHA recovery from the microbial cell mass must be considered as an important cost factor, especially considering large scale production. Choosing the adequate method for separating PHAs from residual biomass is dependent on several factors: the production strain, the required product purity that is determined by the final application of the biopolymer, the

Table 7.2 *Available carbon-rich waste streams for PHA production.*

Carbon rich waste stream	Occurrence	Main convertible carbon source after upstream processing	Microbial production strain	References
Lignocellulosics	Sugar bagasse, rice straw, fruit crops, corn straw, fruit seed production	Hexoses, pentoses, sugar alcohols	*Pseudomonas pseudoflava, Burkholderia sacchari* IPT 101 and *Burkholderia cepacia* IPT 048	[31–33]
Whey	Dairy industry	Lactose and its hydrolysis products (glucose and galactose)	*Methylobacterium* sp. ZP24, *Hydrogenophaga pseudoflava* DSM1034, *Haloferax mediterranei* DSM 1411, *Pseudomoans hydrogenovora* DSM 1749, rec. *Escherichia coli*	[34–41]
Crude glycerol phase	Biodiesel production	Glycerol	*Haloferax mediterranei, Bacillus sphaericus, Methylobacterium rhodesianum Halorcula* sp., *Methylobacterium extorquens, Novosphingobium capsulatum, Cupriavidus necator, Pseudomonas oleovorans, Pseudomonas corrugata*	[35, 42–46]
Saturated biodiesel fraction	Biodiesel production from animal lipids	Methyl esters of fatty acids	*Cupriavidus necator*	[47,48]
Alpechìn	Olive oil production	Triacyl glycerides	*Azotobacter chroococcum* H23	[49–51]
Waste oils	Gastronomy	Triacyl glycerides	*Pseudomonas aeruginosa* 42A2	[52]
Tallow	Rendering and slaughtering industry	Triacyl glycerides	*Pseudomonas resinovorans*	[53]
Molasses	Sugar industry	Sucrose and its hydrolysis products (glucose and fructose)	*Azotobacter beijerinckii, Bacillus megaterium,* rec. strains of *Escherichia coli, Klebsiella oxytoca* and *Klebsiella aerogenes*	[54–56]
Starchy waste	Potato processing	Starch and its hydrolysis products (maltose, glucose)	*Ralstonia eutropha* NCIMB 11 599, *Alcaligenes eutrophus; Bacillus cereus; Haloferax mediterranei*	[57–60]
CO_2	Effluent gases	CO_2	*Alcaligenes eutrophus; Ideonella* sp.	[61–66]
CH_4	Biogas production, waste water treatment plants, landfills	CH_4	*Methylosinus trichosporium, Methylocystis parvus*	[67]

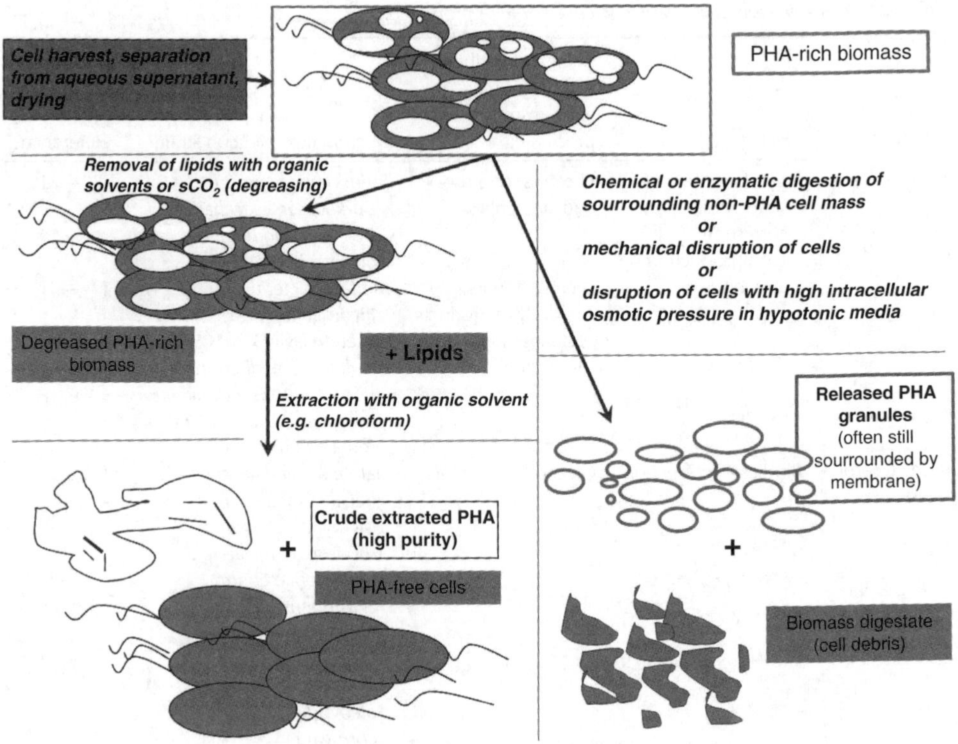

Figure 7.4 Schematic of downstream processing for PHA recovery from biomass.

in-house availability of isolation agents, and the acceptable impact of the isolation method on PHAs final molecular mass. Principally, three different strategies are described for PHA isolation:

- Direct extraction of PHA from biomass (solvent – antisolvent methods; PHA is a dissolved intermediary).
- Chemical or enzymatic digestion, or mechanical disruption of the non-PHA cellular material (PHA granules are set free, no intermediary dissolving of PHA).
- Disruption of cells of osmophilic microbes in hypotonic medium (PHA granules are set free, no intermediary dissolving of PHA).

Figure 7.4 illustrates the different strategies for PHA recovery from biomass.

7.3.2 Direct Extraction of PHA from Biomass

Cost effectiveness of PHA isolation does not only depend on equipment and chemicals needed, but, most of all, on the yields for product recovery and the possibility to reutilize the compounds needed for the isolation. For direct extraction of PHA from biomass extraction solvents that can easily be recycled will be of interest [69]. Solubility of PHAs, especially pure crystalline PHB, is a complex topic. Unlike most low molecular mass compounds and noncrystallizing amorphous polymers, the solubility of crystalline polymers such as PHB cannot be predicted from simple criteria, such as the similarity of chemical architecture, matching of refractive indices, dielectric constants or even solubility parameters, although such attempts are described in literature [70]. For example, hexane is a nonsolvent for linear poly(propylene) (PP), although both compounds are made of the same repeating units.

Typical halogenated extraction solvents like chloroform or, to a minor extent, dichloromethane or 1,2-dichloroethane at room temperature show excellent performance in isolation of short chain length (*scl*)- and medium chain length (*mcl*)-PHA in terms of extraction yields and product purity [71]. After the polyester is extracted from the biomass, its solubility is reduced by the addition of a PHA antisolvent – typically ethanol, methanol or acetone, resulting in the precipitation of highly pure PHA. Until some years ago, the remaining mixture of ethanol and chloroform had to be disposed of, because a reutilization of the solvents after separation via distillation is highly energy demanding. This is clearly in contrast to the aims of the biotechnological production of PHAs to be a sustainable, 'green' technology. Based on simple physico-chemical considerations, a simple and effective method for separation of the mixture by the addition of water was developed. The three-component-system water-chloroform-ethanol consists, in a specific relative concentration range of the components, of two phases. The lower phase (95% chloroform, residues of ethanol and water) can directly be reused for PHA extraction; the upper phase contains only negligible amounts of halogenated solvents [72].

These halogenated solvents, especially chloroform, pose a high risk not only to the environment, but also to the personnel working with them. In order to avoid leaving the patterns of sustainability in biopolymer production, it will be indispensable to concentrate the development of new extraction processes on such recyclable solvents that are also of an environmentally benign nature, such as lactic acid esters [73].

In Brazil, PHB is produced on pilot plant scale from sucrose, integrated into the multi-dimensional production line of a sugarcane mill, where bioethonol is produced from sugar cane sucrose as well as PHB. By-products of the bioethanol production are fusel alcohols like iso-pentanol, which are used as extraction solvents for PHB. After cell harvest, separation and concentration of cells results in a sludge containing 25–30% of solids. This sludge is subjected to a multistage extraction process in continuous stirred tank reactors. Subsequently, cell debris is removed and the remaining solution is cooled down, resulting in a gel. From this gel, the major part of the solvent is removed by mechanical and thermal treatment. The remaining solvent is removed by mixing the wet PHA with water and finally co-distilling the alcohols and the washing water. The extraction yield of the described process is 95%. Although the purity of the material is reported to be higher than 98%, the product displays a yellowish to brownish colour together with an unpleasant odour. A significant loss of molecular mass is observed due to the thermal treatment [92].

7.3.3 Digestion of the non-PHA Cellular Material

An enzymatic digestion method has been developed by Imperial Chemical Industries (London) to recover PHB from *C. necator* by proteases. This process includes thermal treatment of PHA-rich biomass, enzymatic disintegration of the cell material, and washing with an anionic surfactant to dissolve residual biomass [74]. Here, the costs for enzymes and the requirements of steps for increase of product purity are quite high.

Pseudomonas putida cells can be disrupted by a combination of heat shock and treatment of the cells with the enzyme alcalase, sodium dodecyl sulfate, and finally EDTA. By this procedure, the PHA granules remain unscathed [75].

Alternatively, after pretreatment with the surfactant Triton X-100, the PHA-rich biomass can be disrupted with sodium hypochlorite under strong alkaline conditions. This method turned out to be quite simple and effective, but results in an enormous reduction of molecular mass of the polyester of approximately 50% [76, 77].

Later, sodium hypochlorite was applied to obtain ultra-pure, endotoxin-free PHA for potential medical application. The cell pellets were lysed in hypochlorite solution, centrifuged and washed successively with water, acetone, and ethanol. Chloroform was added to dissolve the polymer pellet. By adding water, it was possible to separate cell debris from the solvent phase. The chloroform phase was evaporated and the remaining PHA films were used for polymer characterization [45].

Using recombinant *Escherichia coli* strains for PHA production, one can take profit from the fact that such cells become very fragile at high PHA contents. This is due to the fact that *E. coli* is naturally not equipped for storing large amounts of reserve materials. This encouraged Choi and Lee to develop a simple recovering method without freezing and drying steps. A variety of surfactants, acids and hydroxides were tested and the recovery yields and product purities were determined at different temperatures. Among the chemicals investigated, NaOH, KOH and sodium dodecyl sulfate (SDS) turned out to have sufficient recovery yields and product purities. In direct comparison, SDS results in the highest purities, but the recovery yields were superior using the alkaline compounds. The authors also stated that the alkaline substances are cheaper and ecologically more benign than SDS, and their impact on the molecular mass of PHB was almost negligible [78].

A simple, cost-efficient and effective method for the recovery of PHB directly from high cell-density culture broth with no pretreatment steps has been developed by Kim and colleagues. This method consists of direct addition of SDS to the culture broth, shaking, heat treatment, and washing steps. When the SDS/biomass ratio exceeded 0.4, the purity of recovered PHB was over 95% for various cell concentrations. The recovery yield of PHB was over 90% regardless of cell density and SDS concentration; the reduction in molecular mass was negligible [79].

In 1978, a patent was published by the AGROFERM AG claiming that cyclic carbonic esters such as ethylene carbonate and 1,2-propylene carbonate are particularly suitable solvents for PHB, especially for the extraction of PHA from moist microbial biomass. Here, PHA-rich biomass is suspended in the cyclic ester and heated under stirring. After removal of the hot extract, PHA is precipitated from the liquid by simple cooling down or addition of small amounts of water. As a drawback it turned out that the extraction with both cyclic esters, but especially with ethylene carbonate, results in a high degree of depolymerization, hence in inferior molecular masses. This effect is very dependent on temperature and time of the extraction [80].

A mechanical, nonsolvent-based method using a high-pressure homogenizer in the presence of SDS is also described. A maximum recovery yield of 98% was achieved by homogenizing the cells at 400 kg/cm^2 in 5% SDS solution [81].

7.3.4 Disruption of Cells of Osmophilic Microbes in Hypotonic Medium

Using highly osmophilic PHA production strains like *Haloferax mediterranei* or *Halomonas* sp., one can profit from the fragility of their cell envelopes when they are exposed to hypotonic media, classically distilled water. Under these conditions, the osmophilic cells burst due to high intracellular pressure, releasing all the cell components into the medium. The considerable size and density of the PHA granules means that they can easily be recovered from this cocktail by low-speed centrifugation, sedimentation or filtration. If higher purity is demanded, the whitish crude sediment of the PHA granules obtained has to be washed several more times with detergents (e.g. SDS) which can break down impurities consisting of proteins and lipids. After drying, a fine powder made up of PHA with superior purity is obtained, which can be used directly for polymer processing [81].

7.4 Different Types of PHA

7.4.1 Short Chain Length vs. Medium Chain Length PHAs

Depending on their monomeric composition, short chain length (*scl*) and medium chain length (*mcl*) PHAs are distinguished. Especially low glass transition temperatures (classically around −30 to −40 °C), a broad melting range at rather low peaks of melting temperatures (around 60 °C), and a typically low degree of polymerization (molecular masses classically below 100 000 Da) discriminate *scl*-PHAs from *mcl*-PHAs. In the latter case, high-glass transition temperatures around 0 °C, sharp melting points of up to 180 °C and more, and high molecular masses (up to 4 MDa) are reported [82, 83].

Regarding the physical properties, *mcl*-PHA feature an incredibly high elongation at break of some 100%; in the case of PHB, the most important representative of *scl*-PHAs, this value amounts to about 6%. Typical characteristics of *scl*-PHAs can also be visualized regarding their tensile strength; here, values for PHB and its co-polyesters, harbouring 3HV building blocks, are reported with 30–40 MPa, for *mcl*-PHAs with only about 10 MPa. The homopolyester of 4HB, poly(4HB), constitutes an important exception. This flexible material is the only naturally occurring *scl*-PHA not consisting of chiral building blocks; its elongation at break, about 1000%, by far surpasses the reported values for *mcl*-PHAs; its glass transition temperature of about −50 °C is in the same range as shown for *mcl*-PHAs [82, 83].

In the case of *scl*-PHAs, the monomers consist of three to five carbon atoms, and mainly constitute *R*-configured chiral 3-hydroxyalkanoates. *scl*-PHAs mainly feature properties of classical thermoplasts; hence, on the market, they compete with poly(ethylene) (PE) or poly(propylene) (PP), and, to a certain extent, also with bio-based poly(*L*-lactic acid). The strain *Cupriavidus necator*, a member of the Burkholderiaceae family, can be regarded as the best investigated bacterial *scl*-PHA producer.

In contrast, *mcl*-PHAs are by far less crystalline than their *scl*-relatives; their monomeric building blocks are mainly *R*-configured chiral 3-hydroxyalkanoates and, to a minor extent, *R*-configured chiral 3-hydroxyalkenoates consisting of 6 to 14 carbon atoms. Sometimes, these *mcl*-building blocks possess functionalities that allow postsynthetic chemical modification of *mcl*-PHA for fine-tuning the material properties. Characteristics of *mcl*-PHA resemble those of elastomers and latexes that, due to their low glass transition temperature, do not become brittle even at temperatures far below the freezing point. This makes them interesting as biological rubberlike materials and latexes. As most prominent microbial *mcl*-PHA producer, *Pseudomonas putida* is well-described by the scientific community [84]. The subsequent Table 7.3 provides an overview of material characteristics of selected *scl*- and *mcl*-PHAs.

7.4.2 Enzymatic Background: PHA Synthases

The type of PHA produced biotechnologically is highly dependent on the microbial production strain. *Cupriavidus necator* can only polymerize 3HAs consisting of 3–5 C-atoms

Table 7.3 *Characteristics of representative PHAs [82, 83, 137].*

	PHB	Poly(3HB-*co*-3% 3HV)	Poly(3HB-*co*-20% 3HV)	Poly(4HB)	Poly(3HB-*co*-3% 4HB	Poly(3HB-*co*-16% 4HB)	Poly(3HB-*co*-64% 4HB	Poly(3HO-*co*-12% 3HH)
Melting temperature (Tm, °C)	177	170	145	60	166	152	50	61
Glass transition temperature (Tg, °C)	4	–	−1	−50	–	−8	–	−35
Tensile strength (MPa)	40	38	32	104	28	26	17	9
Young's modulus (GPa)	3.5	2.9	1.2	0.149	–	–	0.03	0.008
Elongation at break (%)	6	–	50	1000	45	444	591	380
Characteristic	Thermoplastic	Thermoplastic	Thermoplastic – elastomeric	Elastomeric	Thermoplastic – elastomeric	Thermoplastic – Elastomeric	Elastomeric	Latex
Type of PHA	*scl*-PHA	*scl*-PHA	*scl*-PHA	*scl*-PHA	*scl*-PHA	*scl*-PHA	*scl*-PHA	*mcl*-PHA

Notes: PHB: Poly-3-hydroxybutyrate; 3HB: 3-hydroxybutyrate; 3HV: 3-hydroxyvalerate; 4HB: 4-hydroxybutyrate; 3HH: 3-hydroxyhexanoate; 3HO: 3-hydroxyoctanoate; *scl*-PHA: short chain length PHA; *mcl*-PHA: medium chain length PHA.

producing *scl*-PHAs. *Pseudomonas putida*, a representative of the so-called pink or fluorescent pseudomonades, only accepts 3HAs of 6–14 C-atoms, resulting in *mcl*-PHAs. This different behaviour is caused by the specificity of the PHA synthase (EC 2.3.1.-). This enzyme, encoded by the phbC gene, polymerizes *R*-3-hydroxybutyryl-CoA monomers stemming from the prior metabolic steps. The growing polymer chain is elongated by incorporation of additional monomers [85]. PHA synthases can be classified in three different classes:

- **Class I** PHA synthases, (prototype organism *C. necator)*, are encoded by phaC genes and are relatively large (approximately 64 kDa). The polymerization of three- to five-carbon backbones hydroxyacyl-CoAs is catalysed (*scl*-PHA) [87].
- **Class II** PHA synthases (prototype organism: *Ps. putida*) are also encoded by phaC genes and are also relatively large (approximately 63 kDa). The polymerization of 6- to 14-carbon backbones hydroxyacyl-CoAs is catalysed (*mcl*-PHA) [87].
- **Class III** PHA synthases (prototype organism: *Allochromatium vinosum*), are heteromeric, requiring two subunits of app. 40 kDa each, encoded by phaC and phaE. The polymerization of *scl*-hydroxyacyl-CoAs is catalysed [87].

Due to this high specificity of PHA synthases, it was believed for a long time that one strain can either produce *scl*-PHAs or *mcl*-PHAs. Some exceptions were recently reported [86–88]. Today it is known that PHA synthases of the strains *Thiocapsia pfennigii* and *Aeromonas caviae* constitute exceptions within class III; they are able to incorporate as well *scl*- as *mcl*-hydroxyacyl-CoAs [87]. Another exception of the three-class system described above seems to be the polymerizing enzyme isolated from a *Bacillus megaterium* strain, which, by some scientists, is proposed as a fourth class of PHA synthases [89]. This enzyme type requires phaC genes that are similar to those in class III, but demands phaR instead of phaE [87].

At the moment, PHA synthesis genes of extremophilic archaea are under detailed investigation. *Haloferax mediterranei*, a representative of the extreme halophilic group of the archaea branch, displays the rare ability to synthesize copolyesters of 3HB and 3HV (PHBHV) from unrelated carbon sources like sugars, hence without the need of a structurally related 3HV precursor [36, 90]. This ability saves costs normally needed for 3HV precursors.

Polyhydroxyalkanoate synthases can use many different substrates and the number of viable pathways for the biosynthesis of PHAs is huge. When new substrates are used, novel PHAs can be produced. In aiming to reduce production costs, it will be useful to search for bacteria that can synthesize precursor substrates from simple and cheap carbon sources [36, 92]. It has also been revealed that, when a PHA synthase enzyme is expressed in a different host microbe, it may result in different substrate specificity and therefore new PHAs with new chemical and physical properties are accessible [11].

7.5 Global PHA Production

It is challenging to collect reliable data for the actual quantities of PHA that are currently produced on (semi)industrial scale. This is due to an obvious gap between the often highly enthusiastic public announcements made by the manufacturers, and the *de facto* realized PHA production. It took quite a long time and generations of scientists and engineers from the first well grounded report describing the occurrence and chemical structure of PHB by Lemoigne in the 1920s, until the first serious attempts for applications of these materials as bioplastics were accomplished in the 1970s. The oil crisis in the early 1970s resulted in quadruplicating of the crude oil price within only three months. This was the decisive episode regarding the public sensitization for problems resulting from established methods of classical industrial end-of-pipe production strategies that are based on robber economy on limited fossil resources. It finally brought forth the awareness of researchers and political and industrial decision makers that liberation of the global economy from the dependence on fossil resources has to be forcefully aspired. Due to these economic aspects and increasing ecological awareness, alternatives to petrol-based products for energy and plastic production were suddenly vehemently investigated, also resulting in strong activities in PHA research.

Table 7.4 provides prime examples for successful PHA production on a larger scale

7.6 Applications of PHAs

7.6.1 General

Concerning potential applications, PHAs display very versatile materials that raise the attention of different industrial branches. Considering the increasing global demands for packaging for safe covering and transportation of goods, this field certainly constitutes the most important one in terms of quantity. Here, it has to be considered that in 2010, the world-wide quantity of plastics amounted to 250×10^6 tons with a clear upwards trend. Nevertheless, the subsequent paragraphs compile various additional areas were PHAs can be implemented.

7.6.2 Packaging and Commodity Items

As the best known and least pretentious application, PHAs are of interest for packaging purposes. Especially, in such areas where compostable packaging is wanted like for easily spoiling food and feed and the utilization of classical packaging materials results in enormous piles of waste. In this field, PHAs can act as a real alternative. The high barrier of PHAs towards oxygen permeation is of enormous interest for producing packaging materials, which prevent the oxidative spoiling of products. Compared to the classical packaging material HDPE, food quality is at least as good when PHA is used [93]. Poly(3-hydroxybutyrate-co-3-hydroxyvalerate) (PHBHV) received European approval for food contact use in 1996. For very simple packaging materials such as plastic bags, no high purity demands are required. Here, the pretensions for substrate purity on the one hand, and for isolation and refining on the other hand, are lower than for higher sophisticated applications.

In addition, bottles for shampoos made of PHAs were already commercially available in the past based on Biopol™ products, encompassing PHB and PHBHV. PHA can be used for production of daily commodity items like razors, diapers, hygiene products, or cups and dishes. In agriculture, mulch films can be produced from PHAs. For these applications, PHAs can be processed by techniques of injection moulding or film blowing by using the same equipment as known from the well-established processing of petrochemical plastics. As an example, an ongoing Austrian research project develops biodegradable carriers for electrical spools made of PHAs. Tests carried out on laboratory scale have demonstrated the suitability of PHA for this special application.

In the paper industry there is an increasing interest in coating paper in order to improve its properties. Polyhydroxyalkanoates can be used as coating latex resulting in a paper material which has strength, gloss, brightness and overall performance comparable to other coated paper products [94]. However, when latex films of PHA are made, heat and pressure is required to accomplish satisfying properties. In this niche of applications the costs could be lowered by

Table 7.4 Selected examples for (semi)industrial PHA production.

Company	Country	Production since/until	Microbial production strain	Substrate(s)	Type of PHA	Trade mark	Annual production scale
Chemie Linz (technology transferred to Urs Hänggi, Biomer)	Austria	Late 1980s to early 1990s	*Alcaligenes latus* DSM 1124 (today: *Azohydromonas lata*)	Glucose from carbohydrate feedstocks	PHB	–	< 50 t
Biomer	Germany	1993	*Cupriavidus necator*	Glucose from corn starch	PHB	Biomer™	500 t–1000 t
Imperial Chemical Industries (ICI) (Later: Zeneca, Monsanto)	UK	1976–1998	*Alcaligenes eutrophus* (today *Cupriavidus necator*)	Glucose from carbohydrate feedstocks	PHB, later also PHBHV	BIOPOL®	800 t (later phase under Monsanto)
Metabolix (with Monsanto technology)	USA	1980-ongoing	n.r.		PHBHV and others	–	n.r.
Telles (joint venture of Metabolix and ADM)	USA	2007–2012	n.r.		PHBHV and others	Mirel™ and Mvera™	planned 50 000 t
Bio-On	Italy		n.r.	Sugar co-products or sugar waste material (molasses)	n.r.	MINERV-PHA™	10 000 t
Tepha Inc.	USA	2007-ongoing	n.r.		P4HB, P(3HB-co-4HB), *mcl*-PHA	TephaFLEX™	n.r.
Polyferm Canada	Canada	ongoing	wild type bacteria, e.g. *Ps. putida, Aeromonas hydrophila, C. necator*	Vegetable oils, sugars	*mcl*-PHA (even and odd numbered, saturated and unsaturated building blocks)	VersaMer™	kg scale
Tianjin Green Bioscience & DSM	PR China	2004- ongoing	n.r.		P(3HB-co-4HB)	GreenBio™	10 000 t
Tianan	PR China		*Cupriavidus necator*		PHBHV	Enmat™	100 t–1000 t
PHB Industrial/ Copersucar (PHB SA)	Brazil	1995-ongoing	*Alcaligenes eutrophus* (today *Cupriavidus necator*); *Burkholderia sacchari*	Cane sugar	PHB and PHBHV	BIOCYCLE™	100 t (capacity 5000 t)

Note: n.r. not reported (information not provided by manufacturer).

using PHA blends with starch. The product would still be totally biodegradable but the costs would be significantly lower [95].

7.6.3 Medical Applications

As a result of their biocompatibility, PHAs are tailor-made materials for application in the medical field. The ideal biocompatibility of PHAs is underlined by the natural occurrence of 3-hydroxybutyric acid (3HB) and its low molecular weight oligo- and polymers (cPHB) in human blood and tissue. For *in vivo* applications, PHAs were successfully investigated as bone implant materials, for tissue engineering, as implants, surgical pins, screws, meshes and sutures, and as carrier matrices for controlled drug release [83]. Furthermore, the production of highly sophisticated surgical articles such as artificial blood vessels and vein valves, spinal fusion cages, bone marrow scaffolds, joints, and meniscus regeneration devices is reported [96–98].

At the moment, implants for children femoral fracture healing are developed by the project Consortium of BRIC ('BioResorbable Implants for Children') coordinated by the Medical University Graz (www.medunigraz.at/bric, accessed 22 June 2013). Additional partners are distributed all over Austria: Graz University of Technology, University of Natural Resources and Life Sciences Vienna, and Vienna University of Technology. This research is carried out in a Laura Bassi Center of Excellence, financed by the Austrian Research Promotion Agency FFG and the industrial partners AT&S Austria Technologie und Systemtechnik AG, and Hereus Medical. Strong support in material development is also provided by ETH Zürich. Beside the development of PHA-based implants, BRIC also focuses on novel biodegradable metal alloys for the purpose of femoral fracture healing. In principle, the applied implants can be degraded and absorbed quickly by children, thus avoiding a physically and psychologically wearing secondary surgery to remove the implant.

For *in vivo* applications, high purity of the items is a prerequisite in order to prevent negative impacts on the human organism. Considering the fact that natural PHA production predominantly occurs by Gram-negative microorganisms, endotoxins (lipopolysaccharides) are frequently occurring and unwanted byproducts. Together with the shape of the materials and the structure of the surface, inflammatory reactions of cells and tissue to biomaterials are mainly determined by the contamination by endotoxins. After extraction and precipitation of PHAs from microbial biomass, the product should therefore be redissolved and precipitated several times to ensure a sufficient removal of impurities [99]. Postsynthetic treatment for removal of endotoxins uses H_2O_2, sodium hypochlorite or sodium hydroxide [84, 100]. The resulting purities are not sufficient for medical applications; further, the treatment can cause degradation of the biopolyesters [84]. Investigating *mcl*-PHA poly(3-hydroxyoctanoate-co-3-hydroxyhexanaote) (PHO) production by *Pseudomonas putida* GPo1, an efficient extraction method was recently developed. Here, it was demonstrated that n-hexane and 2-propanol are optimal solvents for achieving highly pure PHO. Extraction with n-hexane was done at 50 °C, precipitation occurred between 0 and 5 °C. Following this protocol, an endotoxicity level between 10 and 15 endotoxin units (EU) per g PHO was reached. Additional redissolving in 2-propanol at 45 °C and precipitation at 10 °C resulted in minimal endotoxicity of 2 EU/g PHO [84]. Recently, endotoxin-free PHA was produced by Gram-negative *Novosphingobium* sp. from glycerol. Here, cells were disrupted by sodium hypochlorite solution. The remaining PHA granules were subsequently washed by water, acetone and ethanol. Finally, the remaining polymer was dissolved in chloroform [45]. At the moment, efforts are devoted to genetically modify Gram-positive organisms like *Corynebacterium glutamicum* to produce endotoxin-free polymers by inserting the required PHA synthesis genes [101].

Prior to *in vivo* application, the surgical items have to be sterilized. First attempts in the past used steam sterilization; this procedure is only viable for devices displaying melting points above 140 °C as is the case for PHB and PHBHV, but not for *mcl*-PHA and most copolyesters containing 4HB building blocks. As an alternative, ethylene oxide was intensively used to sterilize PHA-based implants, sutures, meshes and so on. This treatment does not cause any negative effect on the material properties, but residual ethylene oxide might be present at or in the device even after a rather long period. Many studies report the use of γ–irradiation as an efficient sterilization method. However, it features the drawback that high doses of irradiation can negatively impact the material properties of PHB and PHBHV, mainly regarding molecular mass reduction due to polymer chain scission.

It is important to distinguish between such medical applications where high hydrolytic stability of the material is desired, and applications demanding fast degradation. The variable composition of PHA, in particular, allows the manufacturing of materials with tailor-made mechanical properties and a fine-tuned degradation rate under *in vivo* conditions. Considering PHB and PHBHV, *in vivo* degradation normally takes a rather long time; in contrast, P(3HB4HB) displays accelerated biodegradation. In fact, one can expect a quick increase of the number of medical applications for PHAs and its composites; a first step in this direction was done by the Food and Drug Administration (FDA) approval of the TephaFLEX™ P4HB as implant material.

7.6.4 Application of the Monomeric Building Blocks

Hydrolysis of PHA to the monomers generally results in a rich source of chiral, optically pure *R*(-)-configured bifunctional hydroxy acids that can be used as starting materials (synthons) for syntheses of fine chemicals and marketable products such as pheromones, aromatics, vitamins or antibiotics, or can even be used as pharmaceutically active compounds [102]. Most prominently, 3HB and its oligomers have therapeutic effects. They promote cell proliferation and prevent necrotic cell death. As one of the well-known ketone bodies, 3HB is a naturally occurring metabolite in human blood plasma. It was successfully tested as carbon and energy source in order to prevent major loss of proteins, for example, for obese patients experiencing provoked therapeutic starvation [83, 103, 104]. A current field of research is the application of 3-hydroxyalkanoate methyl esters as an effective drug in *Morbus Alzheimer* therapy [105]. Considering the vast number of reported PHA building blocks, a variety of such chiral compounds is accessible.

These follow-up compounds of PHAs typically possess higher market values than the polymers themselves. Some of them also display biological activity against pathogenic bacteria or viruses [106]. Their production is complex and not economical; classical PHA depolymerization by acidic alcoholysis after recovery from bacterial biomass is highly solvent demanding and energy demanding [102, 107, 108]. Efficient methods were therefore developed for *in vivo* depolymerization of intracellular PHA by triggering the activities of the enzymes involved in intracellular PHA degradation via changing the process conditions. After PHA is synthesized by the micro-organisms, the polyester is rapidly hydrolysed by increasing activity of PHA depolymerase following a rigorous lowering in pH value. Under such acidities, the activity of *R(-)*-3HB dehydrogenase is zero, hence, the unwanted reaction of *R(-)*-3HB towards acetoacetate is totally blocked and *R(-)*-3HB can be collected in a pure form. Instead, *R(-)*-3HB can be released in the medium by PHA-rich cells, when external carbon source is depleted. Here, the subsequent metabolizing of *R(-)*-3HB towards acetoacetate can be avoided by interruption of the oxygen supply [109].

7.6.5 Smart Materials

Further, PHAs harbouring special building blocks can be applied as so-called 'functional materials' for different niche applications. Here, they can act as heat sensitive adhesives, latex materials, or smart gels [110]. Of special importance is the utilization of PHA as basic material to be connected by linkers with biologically active substances. As an example, Hany and colleagues produced an *mcl*-PHAs-based carrier matrix linked with zosteric acid; this material can be used on the surface of ships and boats to avoid bio-fouling [25].

7.6.6 Controlled Release of Active Agents

Carrier materials and degradable matrices can be produced from PHAs and are desired in many fields, such as agriculture, food technology, or pharmacy. Compounds that can be released from these matrices at controlled rates include drugs, hormones, pesticides, antibiotics, dyestuffs, or flavours [111].

7.7 Economic Challenges in the Production of PHAs and Attempts to Overcome Them

7.7.1 PHA Production as a Holistic Process

Despite all efforts globally devoted to biopolymer research, PHAs are still not really competitive to petrochemical plastics mainly considering production costs and, to a certain extent, also regarding the material properties [27, 71, 112]. If the realization of a substantially cost-effective process for biopolymer production is the intention, one has to consider every single process step [23, 71]. The subsequent stages in biotechnological PHA production can be decisive for the economic assessment: purchasing, installing and operation of the production plant; raw materials (carbon and nitrogen sources); precursor costs for copolyester production; and downstream processing for product recovery. Figure 7.5 shows a schematic of the entire PHA production process (Figure 7.5).

7.7.2 Substrates as Economic Factor

Up to half of the entire production costs is related to the carbon substrates. The difficult nutritional situation for mankind in many global regions strictly prevents the use of various renewable feedstocks for production of chemicals, plastics or fuels from an ethical point of view. This goes especially for the application of carbohydrates, high-value proteinaceous materials or edible oils. However, diverse waste materials exist that represent severe disposal problems for the concerned industrial branches, and, at the same time, their use does not interfere with the nutrition chain. Their utilization is a viable strategy to overcome this ethical conflict; it can be considered the most promising approach in making PHAs economically competitive. Such materials are mainly produced in agriculture and those industrial branches that are closely related to agriculture [17, 29, 30, 82].

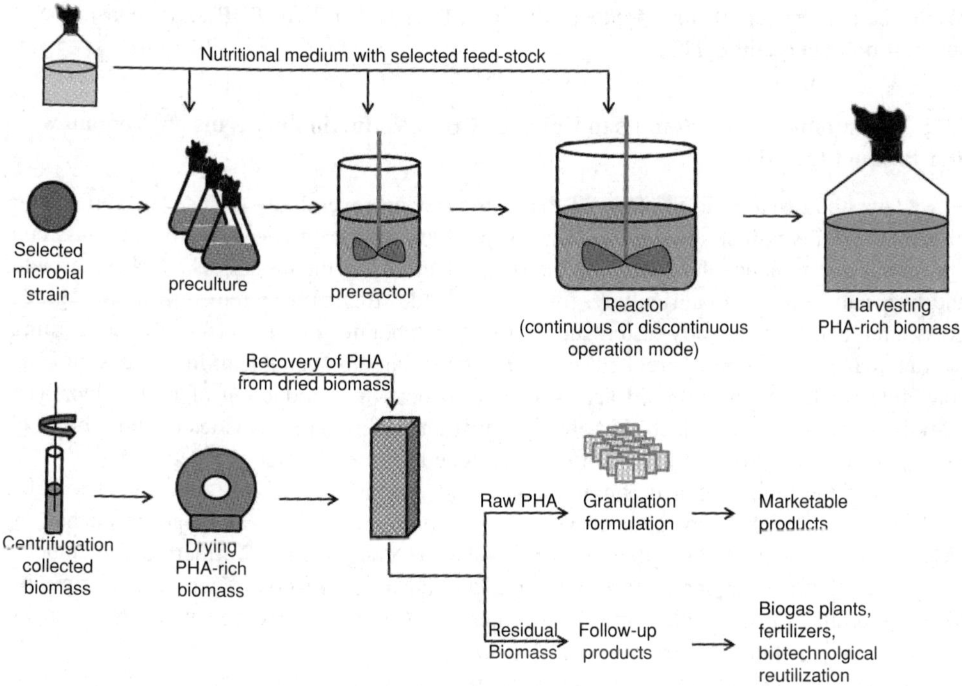

Figure 7.5 Schematic illustration of the entire PHA production process.

7.7.3 Downstream Processing

In addition to the substrate expenses, costs have to be saved by optimizing the downstream processing for PHA recovery after cell harvest. As intracellular products, PHAs have to be separated from the surrounding non-PHA cell mass (NPCM), mainly consisting of proteins, lipids, nucleic acids and polysaccharides. Here, high input of often hazardous solvents and enormous energy demand still constitute the state-of-the art in PHA recovery, antagonizing the claims of these bioplastics to be ecologically benign materials [26, 113].

7.7.4 Process Design

Apart from the selected raw materials and the downstream processing, the increase in productivity by applying an optimal engineering process design is indispensable for the final break-through of PHAs on the market. Batch and fed-batch discontinuous fermentation mode are currently the most common techniques for microbial PHA production [38, 91, 114]. In contrast, continuous biotechnological production mode is a well-known tool for achieving high productivities, lower production costs and constant product quality. Due to these facts, a growing number of research activities were accomplished recently, investigating and assessing the potential of continuous PHA production processes [115–117]. Recently, the continuous production of PHB in a five-stage cascade of bioreactors was investigated. Here, the biopolymer was produced from glucose by the eubacterial strain *Cupriavidus necator*. As a main result, the authors report high productivities of 1.85 g L^{-1} h^{-1} for PHB and a constant and superior product quality [117].

7.7.5 Contemporary Attempts to Enhance PHA Production in Terms of Economics and Product Quality

Nowadays, the research in the field of PHAs focuses on several key topics. The application of growth additives that shorten the time for production of catalytically active biomass is a pre-requisite to enhance the entire volumetric productivity of the process. Such cheap growth additives can be found in agriculture, for example, side streams from the cultivation of green grass land, and were already tested successfully on laboratory scale [118, 119]. Concerning the raw materials, simple 'unrelated' carbon sources that are available at low prices, or even constitute waste streams, should act as sole feedstocks for production of high-value PHA co-polyesters using specialized microbial production strains. This provides the possibility to save costs for precursor compounds normally needed for copolyester production.

Efforts in the field of genetic engineering mainly intend the increase of volumetric PHA productivity and higher molecular masses of the biopolymers. This can be accomplished by knocking-out enzymes responsible for intracellular PHA degradation. Furthermore, metabolic bottlenecks that can hamper a fast and complete substrate conversion can be overcome by genetic modifications, as well as the enhancement of the microbial oxygen uptake by inserting genes encoding catalase or peroxidase [120, 121].

In the field of downstream processing, environmentally safe and efficient solvents are investigated for enhanced recovery of PHA from the cells. This goes in parallel with the examination of novel biological lysis methods and enhanced strategies for mechanical cell disruption. In any case, improved downstream processing has to manage with lower energy consumption compared with contemporary methods. For efficient polymer recovery, the increase of the intracellular polymer content as well as the increase of the PHA granule size is beneficial; these factors are determined during PHA bio-production [110]. The remaining non-PHA cell mass (NPCM) has to be converted in a sustainable, value-adding way. Currently, research in this direction is devoted to the anaerobic digestion of NPCM in biogas plants, or to the chemical or enzymatic hydrolysis of NPCM to a carbon- and nitrogen-rich source for subsequent microbial cultivations. As an alternative, NCPM can be applied in agriculture as a 'green fertilizer'. Downstream processing can be facilitated by genetic modifications, too: an excretion of high quantities of nuclease enzymes after cell disruption results in decreased amounts of nucleic acids in the medium, leading to lower viscosities that facilitate the separation of PHA granules from the surrounding liquid phase by centrifugation or flocculation.

The technological drawbacks of the bio-production itself can be handled by the application of robust microbial production strains that remain genetically stable for a long time period under continuous cultivation conditions, and, at the same time, can resist the risk of microbial contamination by microbial competitors that can endanger whole fermentation batches, generating a high economic loss [2]. Here, extremophilic species like the salt-requiring archaeon *Haloferax mediterranei* might be a viable solution in order to minimize the normally indispensable, highly energy demanding sterility requirements for PHA production setups [86]. Detailed data for this approach were recently presented by Tan and colleagues [122]. An unsterile and continuous fermentation process was developed based on a halophilic bacterium termed *Halomonas* TD01 isolated from a salt lake in Xinjiang, China. The strain reached 80 g/L cell dry weight containing 80% PHB on glucose salt medium during a 56 h fed-batch process. In a 14-day open unsterile and continuous process, the cells grew to an average of 40 g/L cell dry weight containing 60% PHB in the first bioreactor with glucose salt medium. Continuous pumping of cultures from the first bioreactor to the second stage containing the nitrogen-deficient glucose salt medium diluted the cells but allowed them to maintain a PHB level of between 65 and 70% of cell dry mass.

In future, continuous PHA production should not only aim at the increase of volumetric productivity but should also open the door for tailor-made material properties by fine-tuning the polyester composition. This can be accomplished by the formation of block-copolymers, where the sequential arrangement of softer and harder polymer parts can result in well-adjusted novel polymeric materials. Here, a multistage bioreactor cascade for PHA production as presented by Atlić and colleagues [117] might be the adequate process engineering equipment. Synthesis of diblock copolymer of 3-hydroxybutyrate covalently bound to 3-hydroxyhexanoate [PHB-b-PHHx] by a beta-oxidation weakened *Pseudomonas putida* strain was recently demonstrated by Tripathi and colleagues. This material was produced to overcome the negative characteristics of the homopolymer PHB, a highly crystalline material, and the homopolymer PHHx, an amorphous sticky material; and is expected to resist post-synthetic polymer aging that negatively affects polymer properties. The authors state that block copolymerization opens the door to some new and useful properties in comparison to homo-, random and blend polymers [123].

During the last two to three decades, the preparation of composites and blends has become one of the key research fields in biopolymer science. For enhancement of the material properties, PHAs can be processed together with a variety of compatible matters, resulting in the creation of novel PHA-based blends and composites. For this purpose, the utilization of polymeric materials like poly(vinyl alcohol) (PVA), poly(lactic acid) (PLA), poly(ε-caprolactone) (PCL) and so on, including synthetic analogues of PHA (e.g. atactic PHB), inorganic fillers (clays, sepiolites, Montmorillonite, or calcium carbonate), and organic fillers of agricultural origin was already tested [124, 125]. Concerning fillers from agriculture, the application of surplus materials like lignocelluloses like sugar cane bagasse, wheat flour, fruit peels, crop fruit fibres, saw dust and wheat straw is reported in literature [124]. In general, nanocomposites and natural fibres composites can be distinguished. Nanocomposites have the potential to improve special polymer properties, such as gas permeability and thermal and mechanical characteristics. For the creation of nanocomposites, rather small amounts of filler, commonly an organophilic modified clay, are needed for efficient enhancement of the properties. Natural fibres composites often display excellent mechanical properties, and, as desired for many applications, they lower the density of the final product. Due to the fact that in most cases fibres constituting agricultural residues are used as fillers, the biodegradability of the final product is enhanced. Because such fibres do not feature a considerable price or even are waste materials, this normally goes in parallel with a reduction of the entire production cost of the marketable product [125].

Entirely new PHA-based materials, namely segmented polyester-urethanes (PEUs) can be prepared from diol-terminated blocks ('macrodiols') of PHB and blocks of diol-terminated biodegradable co-components like α, ω-dihydroxypoly(butylene glycol adipate). These diols are joined by chemical reaction with hexamethylenediisocyanate [126]. The obtained materials can be processed with various fibres like, for example, cellulose acetate recycled material, resulting in compostable, cost-efficient composites; processing of such composites was already successfully tested on a kg scale using an injection moulding machine with a screw conveyor [127].

7.8 Process Design

Classically, PHA production is accomplished under fed-batch feeding conditions, where substrates are supplied to the fermentation broth when required. Today, this addition can be automated to a high degree, for example, by coupling of the substrate addition to process parameters like the dissolved oxygen tension, pH-value, the CO_2 concentration in the air leaving the bioreactor, or by automatic feeding. Due to their higher productivity, however, continuous production processes appear of special commercial interest, especially for fast-growing microbial strains [128, 129]. On laboratory scale, continuous PHA production has been demonstrated as well for *scl*-PHAs as for *mcl*-PHAs; single-stage, two-stage and multi-stage processes are reported [117, 130–134].

To choose the adequate bioreactor design for continuous PHA production, kinetics for both biomass and PHA production by the microbial strain should be considered. In the case of PHA production directly associated with microbial growth as it is found in *Alcaligenes latus* DSM 1122 on sucrose [128], or for *Pseudomonas putida* ATCC 29147 on fatty acids [97, 98], a one-step continuous process using a continuous stirred tank reactor (CSTR) is a viable solution.

The situation changes significantly in the case of strains like *C. necator*, where autocatalytic growth of biomass is followed by a phase of linear PHA production. In this case, biomass production should occur in the first step in a CSTR which is coupled to a subsequent plug flow reactor (PFR). The combination CSTR-PFR not only ensures higher productivity, but also minimizes the loss of substrates and co-substrates. Furthermore, product quality can be enhanced by the fact that the PFR features a narrow residence time distribution, leading to higher uniformity of cell populations. This should also have positive impacts on the distribution of the PHA molecular masses and the composition of polyesters [128].

Continuous processes are operated as so called 'chemostat', hence, the concentrations of as well substrates as products stay constant during the entire process duration. Such conditions of constant concentrations at constant process parameters (pH-value, temperature, dissolved oxygen concentration etc.) are known as 'steady state conditions' in literature [128]. Besides the significance for industrial process development, continuous studies in chemostats are also a precious tool for elucidating the relationships between cells and their environment. For example, the optimization of the composition of nutritional media can be accomplished this way. In addition, continuous processes enable the fine-tuned supply of growth inhibiting substrates as it is often the case in *mcl*-PHA production.

The first studies on one-stage chemostat continuous PHA production were published by Ramsay and colleagues [130]. *C. necator* was cultivated on glucose at a dilution rate of 0.15 h^{-1} and produced 5 g/L of biomass with a PHB content of 33%. Similar results were obtained with *A. latus* using sucrose at the same dilution rate, biomass concentration and PHB content were 4 g/L and 40%, respectively. When grown on glucose and various concentrations of propionic acid up to 5 g/L, *A. latus* produced PHBHV with a content range of 3HV monomer in the PHA of up to 20% [130].

Several other studies were conducted to produce PHBHV with *C. necator* in one-stage continuous culture. This copolyester was produced from fructose and propionic acid with a maximum PHA productivity of 0.31 g L^{-1} h^{-1} and a 3HV content in the range of 11 to 79%. It turned out that the molecular mass increased with the dilution rate [129]. When propionate was employed as 3HV precursor, the continuous culture systems did not reach steady states when propionate concentration exceeded 7 g/L [131]. Zinn and colleagues [133] grew *C. necator* in a chemostat under the conditions of simultaneous limitations by carbon (butyric and/or valeric acid) and nitrogen source. No concentrations of biomass or polymer were reported, only the absorbance of the culture = 34 ± 4 and the maximum 3HV content in the copolymer of 62%. The study was rather focused on material properties, with the melting temperature of the material decreasing from 178 °C for PHB to about 80 °C for PHA with the highest 3HV content. High relative molecular mass of the obtained polymers was between 0.9 and 1.2 × 10^6 was reported together with a polydispersity index of around 3 [133].

In the case of *mcl*-PHAs, a Swiss group reports the continuous, growth-associated production of poly(3-hydroxyalkanoate-*co*-3-hydroxyalkenoates) in one-stage chemostat cultures of *Pseudomonas putida* ATCC 29147 in a single CSTR. The applied substrates encompassed 5-phenylvalerate, octanoate and 10-undecenoate. Multiple nutrient limited growth conditions were chosen at a dilution rate of $D = 0.1\ h^{-1}$. Different mixtures of the substrates in the feed resulted in the formation of copolyesters with varying compositions and different amounts of aromatic and unsaturated side chains that make the products accessible for further modification. Based on the results, the authors conclude that the steady state conditions in a continuous culture provide a strategy especially suited for the production of tailored PHA copolymers [117].

All single-stage experiments with *C. necator* resulted in rather low biomass concentrations and intracellular PHA contents, which is an expected outcome, since *C. necator* produces PHAs that are not growth associated. For such organisms, optimal conditions for both cell growth and PHA accumulation cannot be maintained in a single-stage system. For that reason, a multistage system should be more suitable, and in fact, the results from a two-stage chemostat set-up were superior to those from one-stage experiments. Using a system of two CSTRs, continuous PHB production with *C. necator* was examined by Yu and colleagues [131]. Investigating the first stage, best results were obtained at a dilution rate of $D = 0.21\ h^{-1}$, when 27 g/L of cell dry mass with about 11 wt.-% PHB were produced. As expected, the authors report high specific cell growth rates in the first stage and low PHB synthesis rates under carbon-limited and nitrogen–rich conditions. In the second stage, maximum PHB production was observed at a dilution rate of $D = 0.14\ h^{-1}$, giving a PHB concentration of 48 g/L. The maximal PHB productivity was reported with $1.43\ g\ L^{-1}\ h^{-1}$ at a dilution rate of $0.12\ h^{-1}$ but with relatively low PHB content of 47.6% [131].

A few other studies in two-stage continuous systems were conducted with other microorganisms. In the case of P(3HB4HB) production, *Delftia acidovorans* P4a was cultivated on mixtures of acetic acid and γ-butyrolactone; here, P(3HB4HB) copolymers with a molar fraction of 2.7–19% 4HB were obtained. The authors state that, especially in the case of toxic substrates like acetic acid and γ-butyrolactone, the two-stage continuous production strategy is very convenient [132].

Based on literature data, Atlić and colleagues [117] concluded that in the case of nongrowth-associated PHA producers and with totally or partially growth-associated PHA producing micro-organisms, optimal conditions for cell growth as well as for PHA accumulation could more easily be achieved in continuous multistage bioreactor systems than in a single-stage or in a two stage setup. The authors aimed for high productivity as well as excellent PHA quality. Applying previously published engineering approaches [128, 135], which considered a bioreactor cascade of at least five reactors in series as a process-engineering substitute for a continuous PFR, a five-stage bioreactor cascade system for PHA production was tested [117]. The potential of such systems is based on the fact that different controlled nutrient conditions can be applied along the reactor cascade in order to produce PHB with high molecular weight and low polydispersity index. The five-stage system was designed to provide balanced biomass growth regarding the complete nutrient supply in the first reactor. Subsequently, the second vessel was intended to finalize consumption of nitrogen source so that in this stage the growth-associated PHA synthesis should occur. Further on, in the last three reactors, during the non growth-associated synthesis of PHA, the aim was to increase molecular mass of PHA and to reach a suitable low polydispersity index. Under operating conditions (dilution rate for the first reactor $D1 = 0.139\ h^{-1}$; an overall dilution rate $D\ TOTAL = 0.034\ h^{-1}$), a final concentration of 81 g/L of cell dry mass (CDM) containing 77% w/w of PHB and overall PHB productivity $2.14\ g\ L^{-1}\ h^{-1}$ were obtained.

In the case of *mcl*-PHAs, *Pseudomonas oleovorans* was cultivated on octane with a volumetric PHA productivity of $1.06\ g\ L^{-1}\ h^{-1}$ and PHA content of 63% in the second bioreactor. Dilution rates of $D = 0.21\ h^{-1}$ in the first stage and $0.16\ h^{-1}$ in the second stage were reported to result in the highest PHA productivity [136].

Similar to the production of P(3HB4HB), continuous modes provide a viable strategy for the production of *mcl*-PHAs in order to overcome the high sensitivity of the production strains against higher concentration of the required substrates, mainly fatty acids and their derivatives. Here, continuous operation allows for permanent feeding of the organisms with the exact amounts of substrates covering their metabolic requirements, but without reaching inhibiting concentrations [135, 136].

7.9 Conclusion

This chapter highlighted the huge diversity of microbial PHAs, demonstrating that these versatile biopolyesters possess the potential to replace various contemporary fossil-based plastics. Drawbacks in the large-scale production of PHAs were discussed, and economic feasibility was considered.

It was shown how industrial waste like surplus whey, crude glycerol phase, lignocelluloses, molasses and residues from the slaughtering and biodiesel industry can be upgraded to substrates for biopolymer production. Applying such waste streams as carbon source can be regarded as the most promising route to make the entire PHA biopolymer production process economically competitive; this is valid for bulk plastics made of petrochemical competitors as well as for special polymers currently used for niche products.

It is generally undisputed that, within the next couple of years, the global demand in bio-plastics will rapidly increase. In this context, it has to be considered that, especially in many emerging and developing countries, huge efforts are devoted to research and also industrial production of biopolymers presently. Here, biopolymer production is often based on raw materials with high nutritional value produced under doubtful working conditions for the involved employees. Applying feedstocks that are available as waste streams is a viable route to avoid that. Independent from the selected microbial production strain, facilities for PHA production from whey and animal processing industry should be integrated into existing process lines of large dairies, or biodiesel companies, where the raw materials directly accrue. By taking profit of synergistic effects, this can be considered as a viable strategy to minimize production costs using other by-products as raw materials, too.

Choosing the adequate bioreactor design and mode of operation constitutes an additional process-engineering challenge on one side, but provides an important tool to achieve higher productivity and constant and tailor made product quality on the other side. Attempts for single-step and multistep continuous production of PHAs of different quality on laboratory scale are compared with data for classical discontinuous processes.

Beside the raw material supply and the fermentation process itself, downstream processing for polymer recovery from the surrounding cells is another cost-determining factor in biopolymer production. Especially for this process step, additional research progress is required in order to decrease the demands for energy and chemicals that severely antagonize the sustainability of PHA production.

Uniting the potential enhancements of each process step, one can definitely make substantial progress towards an environmentally benign and cost-efficient technology. In any case, the development of really efficient biopolymer production processes needs the narrow cooperation of experts from industry and different scientific fields; hence, a multidisciplinary approach is required. Decision makers from companies, chemical engineers, microbiologists, enzymologists, polymer scientists, genetic engineers, and experts for LCA and Cleaner Production studies have to concentrate their special knowhow. This can close existing gaps between the meritorious successes on laboratory scale that in most cases affect only singular aspects of biopolymer production, and the final market success of competitive bioplastics.

Regarding the high number of available data from literature, one can conclude that important progress has already been achieved in terms of combining the environmental benefit of future-oriented bio-polyesters with economic viability of their production. This should finally facilitate the decision of responsible policymakers from various waste-generating industrial branches and from polymer industry to break new ground in sustainable production.

References

1. Mallick, N., Sharma, L. and Singh, A.K. (2007) Poly-b-hydroxybutyrate accumulation in Nostoc muscorum: effects of metabolic inhibitors. *Journal of Plant Physiology*, **164**, 312–317.
2. Koller, M., Gasser, I., Schmid, F. and Berg, G. (2011) Linking ecology with economy: insights into polyhydroxyalkanoate-producing microorganisms. *Engineering in Life Sciences*, **11**(3), 222–237.
3. Stal, L.J. (1992) Poly (hydroxyalkanoate) in cyanobacteria: an overview. *FEMS Microbiology Reviews*, **103**, 169–180.
4. Ducat, D.C., Way, J.C. and Silver, P.A. (2011) Engineering cyanobacteria to generate high-value products. *Trends in Biotechnology*, **29**, 95–103.
5. Balaji, S., Gopi, K., Lavanya, B. and Muthuvelan, B. (2012) Isolation and optimization of poly β hydroxybutyrate producing cyanobacterial strains. *International Journal of Applied Biology and Pharmaceutical Technology*, **3**(1), 137–145.
6. Lillo, J.A.G. and Rodriguez-Valera, F. (1990) Effects of culture conditions on poly (β-hydroxybutyric acid) production by *Haloferax mediterranei*. *Applied Environmental Microbiology*, **56**, 2517–2521.
7. Quillaguamán, J., Guzmán, H., Van-Thuoc, D. and Kaul, R.H. (2010) Synthesis and production of polyhydroxyalkanoates by halophiles: current potential and future prospects. *Applied Microbiology and Biotechnology*, **85**, 1687–1696.
8. Poli, A., Di Donato, P., Abbamondi, G.R. and Nicolaus, B. (2011) Synthesis, production, and biotechnological applications of exopolysaccharides and polyhydroxyalkanoates by Archaea. *Archaea*, doi:10.1155/2011/693253.
9. Reusch, N., Rosetta, N., Shabalin, O. *et al.* (2002) Posttranslational modification of *E. coli* histone-like protein H-NS and bovine histones by short-chain poly-(R)-3-hydroxybutyrate (cPHB). *FEBS Letters*, **527**(1–3), 319–322.
10. Haddouche, R., Poirier, Y., Delessert, S. *et al.* (2011) Engineering polyhydroxyalkanoate content and monomer composition in the oleaginous yeast *Yarrowia lipolytica* by modifying the ß-oxidation multifunctional protein. *Applied Microbiology and Biotechnology*, **91**(5), 1327–1340.

11. Poirier, Y., Nawrath, C. and Somerville, C. (1995) Production of polyhydroxy-alkanoates, a family of biodegradable plastics and elastomers, in bacteria and plants. *Biotechnology*, **13**, 142–150.
12. Anderson, A.J. and Dawes, E.A. (1990) Occurrence metabolism, metabolic role, and industrial uses of bacterial polyhydroxyalkanoates. *Microbiological Reviews*, **54**, 450–472.
13. Gorenflo, V., Steinbüchel, A., Marose, S. *et al.* (1999) Quantification of bacterial polyhydroxyalkanoic acids by Nile red staining. *Applied Microbiology and Biotechnology*, **51**, 765–772.
14. Spiekermann, P., Rehm, B.H.A., Kalscheuer, R. *et al.* (2006) Biosynthesis of medium-chain-length poly(hydroxyalkanoates) from soy molasses. *Biotechnology Letters*, **28**, 157–162.
15. Cornibert, J. and Marchessault, R.H. (1975) Conformational isomorphism. A general 2^1 helical conformation for poly(β-alkanoates). *Macromolecules*, **8**, 296–305.
16. Barnard, G.N. and Sanders, J.K.M. (1988) Observation of mobile poly(beta -hydroxybutyrate) in the storage granules of *Methylobacterium* AM1 by in vivo super(13)C-NMR spectroscopy. *FEBS Letters*, **231**, 16–18.
17. Braunegg, G., Lefebvre, G. and Genser, K.F. (1998) Polyhydroxyalkanoates, biopolyesters from renewable resources: physiological and engineering aspects. *Journal of Biotechnology*, **65**(2–3), 127–161.
18. Kim, Y.B. and Lenz, R.W. (2001) Polyesters from microorganisms. *Advances in Biochemical Engineering Biotechnology*, **71**, 51–79.
19. Ayub, N., Paula, A.E., Tribelli, M. and Lopez, N. (2009) Polyhydroxyalkanoates are essential for maintenance of redox state in the Antarctic bacterium *Pseudomonas* sp. 14–3 during low temperature adaptation. *Extremophiles*, **13**, 59–66.
20. Wallen, L.L. and Rohwedder, W.K. (1974) Poly-beta-hydroxyalkanoate from activated sludge. *Environmental Science and Technology*, **8**, 576–579.
21. Findlay, R.H. and White, D.C. (1983) Polymeric beta-hydroxyalkanoates from environmental samples and *Bacillus megaterium*. *Applied Environmental Microbiology*, **45**, 71–78.
22. Steinbüchel, A. and Valentin, H.E. (1995) Diversity of bacterial polyhydroxyalkanoic acids. *FEMS Microbiology Letters*, **128**, 219–228.
23. Kim, Y.B. and Lenz, R.W. (2001) Polyesters from microorganisms. *Advances in Biochemical Engineering/Biotechnology*, **71**, 51–79.
24. Lee, M.Y. and Park, W.H. (2000) Preparation of bacterial copolyesters with improved hydrophilicity by carboxylation. *Macromolecular Chemistry and Physics*, **201**, 2771–2774.
25. Hany, R., Böhlen, C., Geiger, T. *et al.* (2004) Toward non-toxic antifouling: synthesis of hydroxycinnamic acid-, sulfate-, and zosteric acid-labeled Poly[3-hydroxyalkanoates]. *Biomacromolecules*, **5**, 1452–1456.
26. Koller, M., Salerno, A., Miranda de Sousa Dias, M. *et al.* (2010) Modern biotechnological polymer synthesis: a review. *Food technology and Biotechnology*, **48**(3), 255–269.
27. Doudoroff, M. and Stanier, R.Y. (1959) Role of poly-b-hydroxybutyric acid in the assimilation of organic carbon by bacteria. *Nature*, **183**, 1440–1442.
28. Oeding, V. and Schlegel, H.G. (1973) b-Ketothiolase from *Hydrogenomonas eutropha* H16 and its significance in the regulation of poly-b-hydroxybutyrate metabolism. *Biochemistry Journal*, **134**, 239–248.
29. Solaiman, D.K.Y., Ashby, R.D., Foglia, T.A. and Marmer, W.N. (2006) Conversion of agricultural feedstock and co-products into poly(hydroxyalkanoates). *Applied Microbiology and Biotechnology*, **71**, 783–789.
30. Khardenavis, A.A., Kumar, M.S., Mudliar, S.N. and Chakrabarti, T. (2007) Biotechnological conversion of agro-industrial wastewaters into biodegradable plastic, poly –ßhydroxybutyrate. *Bioresource Technology*, **98**, 3579–3584.
31. Bertrand, J.-L., Ramsay, B.A., Ramsay, J.A. and Chavarie, C. (1990) Biosynthesis of Poly-ß-Hydroxyalkanoates from Pentoses by *Pseudomonas pseudoflava*. *Applied and Environmental Microbiology*, **56**(10), 3133–3138.

32. Young, F.K., James, R., Sheldon, K. and May, W. (1994) Microbial production of poly-β-hydroxybutyric acid from d-Xylose and lactose by *Pseudomonas cepacia*. *Applied and Environmental Microbiology*, **60**(11), 4195–4198.
33. Ramsay, J.A., Hassan, M.-C.A. and Ramsay, B.A. (1995) Hemicellulose as a potential substrate for production of poly(.beta.-hydroxyalkanoates). *Canadian Journal of Microbiology*, **41**(Suppl. 1), 262–266.
34. Yellore, V. and Desai, A. (1998) Production of poly-3-hydroxybutyrate from lactose and whey by *Methylobacterium* sp. ZP24. *Letters in Applied Microbiology*, **25**, 391–394.
35. Koller, M., Braunegg, G., Bona, R. *et al.* (2005) Production of polyhydroxyalkanoates from agricultural waste and surplus materials. *Biomacromolecules*, **6**, 561–565.
36. Koller, M., Hesse, P.J, Bona, R. *et al.* (2007) Various archae- and eubacterial strains as potential polyhydroxyalkanoate producers from whey lactose. *Macromolecular Bioscience*, **7**, 2218–2266.
37. Povolo, S. and Casella, S. (2003) Bacterial production of PHA from lactose and cheese whey permeate. *Macromolecular Symposia*, **197**, 1–9.
38. Ahn, W.S., Park, S.J. and Lee, S.Y. (2000) Production of poly(3-hydroxybutyrate) by fed-batch culture of recombinant *Escherichia coli* with a highly concentrated whey solution. *Applied and Environmental Microbiology*, **11**, 3624–3627.
39. Ahn, W.S., Park, S.J. and Lee, S.Y. (2001) Production of poly(3-hydroxybutyrate) from whey by cell recycle fed-batch culture of recombinant *Escherichia coli*. *Biotechnological Letters*, **23**, 235–240.
40. Schubert, P., Steinbüchel, A. and Schlegel, H.G. (1988) Cloning of the *Alcaligenes eutrophus* poly-b-hydroxybutyrate synthetic pathway and synthesis of PHB in *Escherichia coli*. *Journal of Bacteriology*, **170**, 5837–5847.
41. Wong, H.H. and Lee, S.Y. (1998) Poly(3-hydroxybutyrate) production from whey by high density cultivation of recombinant *Escherichia coli*. *Applied Microbiology and Biotechnology*, **50**, 30–33.
42. Bormann, E.J. and Roth, M. (1999) The production of polyhydroxybutyrate by *Methylobacterium rhodesianum* and *Ralstonia eutropha* in media containing glycerol and casein hydrolysates. *Biotechnology Letters*, **5**, 1059–1063.
43. Ashby, R.D., Solaiman, D.K.Y. and Foglia, T.A. (2004) Bacterial Poly(hydroxyalkanoate) polymer production from the biodiesel co-product stream. *Journal of Polymers and the Environment*, **12**(3), 105–112.
44. Ashby, R.D., Solaiman, D.K.Y. and Foglia, T.A. (2005) Synthesis of short-/medium-chain-length poly(hydroxyalkanoate) blends by mixed culture fermentation of glycerol. *Biomacromolecules*, **6**, 2106–2112.
45. Teeka, J., Imai, T., Reungsang, A. *et al.* (2012) Characterization of polyhydroxyalkanoates (PHAs) biosynthesis by isolated *Novosphingobium* sp. THA_AIK7 using crude glycerol. *Journal of Industrial Microbiology and Biotechnology*, **39**, 749–758.
46. Cavalheiro, J.M.B.T., de Almeida, M.C., Grandfils, C. and da Fonseca, M.M.R. (2009) Poly(3-hydroxybutyrate) production by *Cupriavidus necator* using waste glycerol. *Process Biochemistry*, **44**, 509–515.
47. Titz, K.H., Kettl, K., Shahzad, M. *et al.* (2012) Process optimization for efficient biomediated PHA production from animal-based waste streams. *Clean Technologies and Environmental Policy*, **14**, 495–503.
48. Koller, M., Salerno, A., Muhr, A. *et al.* (2013) Polyhydroxyalkanoates: biodegradable polymeric materials from renewable resources. *Materiali and Technologije*, **47**, (1), 5–12.
49. Ribera, R.G., Monteoliva-Sanchez, M. and Ramos-Cormenzana, A. (2001) Production of polyhidroxyalkanoates by *Pseudomonas putida* KT2442 harboring pSK2665 in wastewater from olive oil mills (alpechín). *Electronic Journal of Biotechnology*, **4**, doi: 10.2225/vol4-issue2-fulltext-6.
50. Gonzalez-Lopez, J., Pozo, C., Martinez-Toledo, M.V. *et al.* (1996) Production of polyhydroxyalkanoates by *Azotobacter chroococcum* H23 in wastewater from olive oil mills (alpechin). *International Biodeterioration and Biodegradation*, **38**, 271–276.

51. Pozo, C., Martínez-Toledo, M.V., Rodelas, B. and González-López, J. (2002) Effects of culture conditions on the production of polyhydroxyalkanoates by *Azotobacter chroococcum* H23 in media containing a high concentration of alpechín (wastewater from olive oil mills) as primary carbon source. *Journal of Biotechnology*, **97**, 125–131.
52. Fernández, D., Rodríguez, E., Bassas, M. *et al.* (2005) Agro-industrial oily wastes as substrates for PHA production by the new strain *Pseudomonas aeruginosa* NCIB 40045: Effect of culture conditions. *Biochemical Engineering Journal*, **26**, 159–167.
53. Cromwick, A.M., Foglia, T. and Lenz, R.W. (1996) The microbial production of poly (hydroxyalkanoates) from tallow. *Applied Microbiology and Biotechnology*, **46**(5–6), 464–469.
54. Page, W.J. (1992) Production of polyhydroxyalkanoates by *Azotobacter vinelandii* UWD in beet molasses culture. *FEMS Microbiology Letters*, **103**, 149–157.
55. Zhang, H., Obias, V., Gonyer, K. and Dennis, D. (1994) Production of polyhydroxyalkanoates in sucrose-utilizing recombinant *Escherichia coli* and *Klebsiella* strains. *Applied and Environmental Microbiology*, **60**, 1198–1205.
56. Purushothaman, M., Anderson, R., Narayana, S. and Jayaraman, V. (2001) Industrial byproducts as cheaper medium components influencing the production of polyhydroxyalkanoates (PHA) – biodegradable plastics. *Bioprocess and Biosystems Engineering*, **24**, 131–136.
57. Haas, R., Jin, B. and Tobias, F. (2008) Production of Poly(3-hydroxybutyrate) from Waste Potato Starch. *Bioscience, Biotechnology, and Biochemistry*, **72**(1), 253–256.
58. Halami, P.H. (2008) Production of polyhydroxyalkanoate from starch by the native isolate *Bacillus cereus* CFR06. *World Journal of Microbiology and Biotechnology*, **24**, 805–812.
59. Chen, C.W., Don, T.-M. and Yen, H.-F. (2006) Enzymatic extruded starch as a carbon source for the production of poly(3-hydroxybutyrate-co-3-hydroxyvalerate) by *Haloferax mediterranei*. *Process Biochemistry*, **41**, 2289–2296.
60. Rusendi, D. and Sheppard, J.D. (1995) Hydrolysis of potato processing waste for the production of poly-ß-hydroxybutyrate. *Bioresource Technology*, **54**, 191–196.
61. Ishizaki, A. and Tanaka, K. (1990) Batch culture of Alcaligenes eutrophus ATCC 17697T using recycled gas closed circuit culture system. *Journal of Fermentation and Bioengineering*, **69**, 170–174.
62. Ishizaki, A. and Tanaka, K. (1991) Production of poly-β-hydroxybutyric acid from carbon dioxide by *Alcaligenes eutrophus* ATCC 17697T. *Journal of Fermentation and Bioengineering*, **4**, 254–257.
63. Sugimoto, T., Tsuge, T., Tanaka, K. and Ishizaki, A. (1999) Control of acetic acid concentration by pH-stat continuous substrate feeding in heterotrophic culture phase of two-stage cultivation of *Alcaligenes eutrophus* for production of P(3HB) from CO_2, H_2, and O_2 under non-explosive conditions. *Biotechnology and Bioengineering*, **62**, 625–631.
64. Taga, N., Tanaka, K. and Ishizaki, A. (1997) Effects of rheological change by addition of carboxymethylcellulose in culture media of an air-lift fermentor on poly-D-3-hydroxybutyric acid productivity in autotrophic culture of hydrogen-oxidizing bacterium, *Alcaligenes eutrophus*. *Biotechnology and Bioengineering*, **53**, 529–533.
65. Ishizaki, A., Tanaka, K. and Taga, N. (2001) Microbial production of poly-D-3-hydroxybutyrate from CO_2. *Applied Microbiology and Biotechnology*, **57**, 6–12.
66. Tanaka, K., Miyawaki, K., Yamaguchi, A. *et al.* (2011) Cell growth and P(3HB) accumulation from CO_2 of a carbon monoxide-tolerant hydrogen-oxidizing bacterium, *Ideonella* sp. O-1. *Applied Microbiology and Biotechnology*, **92**, 1161–1169.
67. Pieja, A.J., Rostkowski, K.H. and Criddle, C.S. (2011) Distribution and selection of poly-3-hydroxybutyrate production capacity in methanotrophic proteobacteria. *Microbial Ecology*, **62**, 564–573.
68. Koller, M., Hesse, P.J, Kutschera, C. *et al.* (2010) Sustainable embedding of the bioplastic poly-(3-hydroxybutyrate) into sugarcane industry: principles of a future-oriented technology in Brazil, in *The Handbook of Environmental Chemistry 12: Polymers - Opportunities and Risks II* (eds P. Eyerer, M. Weller and C. Hübner), Springer, Berlin, pp. 81–96.
69. Chen, G.Q., Zhang, G., Park, S.J. and Lee, S.Y. (2001) Industrial scale production of poly(3-hydroxybutyrat-co-3-hydroxyhexanoate). *Applied Microbiology and Biotechnology*, **57**, 50–55.

70. Terada, M. and Marchessault, R.H. (1999) Determination of solubility parameters for poly(3-hydroxyalkanaotes). *International Journal of Biological Macromolecules*, **25**, 207–215.
71. Choi, J. and Lee, S.Y. (1999) Factors affecting the economics of polyhydroxyalkanoate production by bacterial fermentation. *Applied Microbiology and Biotechnology*, **51**, 13–21.
72. Braunegg, G., Bona, R., Schellauf, F. and Wallner, E. (2002) Polyhydroxyalkanoates (PHAs): Sustainable Biopolyester Production. *Polimery*, **47**, 13–18.
73. Metzner, K., Sela, M. and Schaffer, J. (1997) Agents for extracting polyhydroxyalkane acids. European-Patent 9708931.
74. Holmes, P.A. and Lim, G.B. (1990) Separation process. US Patent 4910145.
75. Yasotha, K., Aroua, M.K., Ramachandran, K.B. and Tan, I.K.P. (2006) Recovery of medium-chain-length polyhydroxyalkanoates (PHAs) through enzymatic digestion treatments and ultrafiltration. *Biochemical Engineering Journal*, **30**, 260–268.
76. Berger, E., Ramsay, B.A., Ramsay, J.A. *et al.* (1989) PHB recovery by hypochlorite digestion of non-PHB biomass. *Biotechnology. Techniques*, **3**, 227–332.
77. Ramsay, B.A., Ramsay, J., Berger, E. *et al.* (1992) Separation of poly-beta-hydroxyalkanoic acid from microbial biomass. US Patent 5110980.
78. Choi, J.I. and Lee, S.Y. (1999) Efficient and economical recovery of poly (3-hydroxybutyrate) from recombinant *Escherichia coli* by simple digestion with chemicals. *Biotechnology and Bioengineering*, **62**, 546–553.
79. Kim, M., Cho, K.S., Ryu, H.W. *et al.* (2003) Recovery of poly(3-hydroxybutyrate) from high cell density culture of *Ralstonia eutropha* by direct addition of sodium dodecyl sulphate. *Biotechnology Letters*, **25**, 55–59.
80. Lafferty, R.M. and Heinzle, E. (1978) Cyclic carbonic acid esters as solvents for poly-(β-hydroxybutyric acid. US patent 4101533.
81. Ghatnekar, M.S., Pa, J.S. and Ganesh, M. (2002) Production and recovery of poly-3-hydroxybutyrate from *Methylobacterium* sp. V49. *Journal of Chemical Technology & Biotechnology*., **77**, 444–448.
82. Khanna, S. and Srivastava, A.K. (2005) Recent advances in microbial polyhydroxyalkanoates. *Process Biochemistry*, **40**, 607–619.
83. Williams, S.F. and Martin, D.P. (2002) Applications of PHAs in medicine and pharmacy, in *Biopolymers polyesters III* (eds Y. Doi and A. Steinbüchel), Wiley-VCH, pp. 91–127.
84. Furrer, P., Panke, S. and Zinn, M. (2007) Efficient recovery of low endotoxin medium-chain-length poly([*R*]-3-hydroxyalkanoate) from bacterial biomass. *Journal of Microbiological Methods*, **69**, 206–213.
85. Madison, L.L. and Huisman, G.W. (1999) Metabolic engineering of poly(3-hydroxyalkanoates): from DNA to plastic. *Microbiology and Molecular Biology Reviews*, **63**, 21–53.
86. Kobayashi, G., Shiotani, T., Shima, Y. and Doi, Y. (1994) Biosynthesis and characterization of poly-(3-hydroxybutyrate-co-3-hydroxyhexanoate) from oils and fats by *Aeromonas* sp. OL-338 and Aeromonas sp. FA-440, in *Biodegradable Plastics and Polymers* (eds Y. Doi and K. Fukuda), Elsevier, Amsterdam, pp. 410–416.
87. McCool, G.J. and Cannon, M.C. (2001) PhaC and PhaR are required for polyhydroxyalkanoic acid synthase activity in *Bacillus megaterium. Journal of Bacteriology*, **183**(14), 4235–4320.
88. Abe, H., Matsubara, I., Doi, Y., Hori, Y. and Yamaguchi, A. (1994) Physical properties and enzymatic degradability of poly(3-hydroxybutyrate) stereoisomers with different stereoregularities. *Macromolecules*, **27**, 6018–6025.
89. Steinbüchel, A. and Lütke-Eversloh, T. (2003) Metabolic engineering and pathway construction for biotechnological production of relevant polyhydroxyalkanoates in microorganisms. *Biochemical Engineering Journal*, **16**, 81–89.
90. Koller, M., Hesse, P.J., Bona, R. *et al.* (2007) Biosynthesis of high quality polyhydroxyalkanoate xo- and terpolyesters for potential medical application by the archaeon *Haloferax mediterranei*. *Macromolecular Symposia*, **253**, 33–39.

91. Nonato, R.V., Mantelatto, P.E. and Rossell, C.E.V. (2001) Integrated production of biodegradable plastic, sugar and ethanol. *Applied Microbiology and Biotechnology*, **57**, 1–5.
92. Chen, C.W., Trong-Ming, D. and Hsiao-Feng, Y. (2006) Enzymatic extruded starch as a carbon source for kiothe production of poly(3-hydroxybutyrate-co-3-hydroxyvalerate) by *Haloferax mediterranei*. *Process Biochemistry*, **41**, 2289–2296.
93. Haugaard, V.K., Udsen, A.M., Hoegh, L. *et al.* (2001) Potential Food Applications of Biobased Materials an EU-concerted Action Project. *Starch*, **53**, 189–200.
94. Marchessault, R.H., Bluhm, T.L., Deslandes, Y. *et al.* (1988) Poly(b-hydroxyalkanoates): biorefinery polymers in search of applications. *Makromolekulare Chemie. Macromolecular Symposia*, **19**, 235–254.
95. Lauzier, C.A. and Marchessault, R.H. (1994) Polyhydroxyalkanoates: morphology of tender morsels for microbes, in *Proceedings of The Symposium on Physiology, Kinetics, Production and Use of Biopolymers, Seggau, Austria, May 13–15* (ed. G. Braunegg), Technical University Graz, pp. 59–69.
96. Chen, C.Q. and Wu, Q. (2005) The application of polyhydroxyalkanoates as tissue engineering materials. *Biomaterials*, **26**(33), 6565–6578.
97. Zinn, M., Witholt, B. and Egli, T. (2001) Occurrence, synthesis and medical application of bacterial polyhydroxyalkanoate. *Advanced Drug Delivery Reviews*, **53**, 5–21.
98. Valappil, S.P., Misra, S.K., Boccaccini, A.R. and Roy, I. (2006) Biomedical applications of polyhydroxyalkanoates, an overview of animal testing and in vivo responses. *Expert Review of Medical Devices*, **3**(6), 853–868.
99. Volova, T., Shishatskaya, E., Sevastianov, V. *et al.* (2003) Results of biomedical investigations of PHB and PHB/PHV, fibers. *Biochemical Engineering Journal*, **16**, 125–133.
100. Sevastianov, V.I., Perova, N.V., Shishatskaya, E.I. *et al.* (2003) Production of purified polyhydroxyalkanoates (PHAs) for applications in contact with blood. *Journal of Biomaterials Science Polymer*, **14**, 1029–1042.
101. Song, Y., Matsumoto, K., Yamada, M. *et al.* (2012) Engineered *Corynebacterium glutamicum* as an endotoxin-free platform strain for lactate-based polyester production. *Applied Microbiology and Biotechnology*, **93**, 1917–1925.
102. Ren, Q., Grubelnik, A., Hoerler, M. *et al.* (2005) Bacterial poly(hydroxyalkanoates) as a source of chiral hydroxyalkanoic acids. *Biomacromolecules*, **6**, 2290–2298.
103. Pawan, G.L.S. and Semple, S.J.G. (1983) Effect of 3-hydroxybutyrate in obese subjects in very-low-energy diets and during therapeutic starvation. *Lancet*, **15**, 1–8.
104. Mizobata, Y., Hiraide, A., Katayama, M. *et al.* (1996) Oxidation of D(−)3-hydroxybutyrate administered to rats with extensive burns. *Surgery Today*, **26**, 173–178.
105. Cao, Q. and Zhang, J. (2112) 3-hydroxyalkanoate methyl esters as Alzheimer disease drugs. 13th International Symposium on Biopolymers ISBP2012, Cairns, Queensland (Australia).
106. Ruth, K., Grubelnik, A., Hartmann, R. *et al.* (2007) Efficient production of (R)-3-hydroxycarboxylic acids by biotechnological conversion of polyhydroxyalkanoates and their purification. *Biomacromolecules*, **8**, 279–286.
107. Seebach, D. and Züger, S. (1982) Über die Depolymerisierung von Poly-(*R*)-3-hydroxy-buttersäureester (PHB). *Helvetica Chimica Acta*, **65**, 495–503.
108. Seebach, D., Brunner, A., Bürger, H.M. *et al.* (1994) Isolation and 1H-NMR spectroscopic identification of poly(3-hydroxybutanoate) from prokaryotic and eukaryotic organisms. *European Journal of Biochemistry*, **224**, 317–328.
109. Lee, S.Y. and Lee, Y. (2003) Metabolic engineering of Escherichia coli for production of enantiomerically pure (R)-(−)-hydroxycarboxylic acids. *Applied and Environmental Microbiology*, **69**, 3421–3426.
110. Chen, G.Q. (2010) Plastics completely synthesized by bacteria: polyhydroxyalkanoate, in *Plastics from Bacteria: Natural Functions and Applications* (ed. C.-Q. Chen), Springer, Berlin, pp. 17–37.
111. Steinbüchel, A. and Fuchtenbusch, B. (1998) Bacteria and other biological systems for polyester production. *Trends in Biotechnology (TIBTECH)*, **16**, 419–427.
112. Sudesh, K. and Iwata, T. (2008) Sustainability of biobased and biodegradable plastics. *Clean*, **36**(5–6), 433–442.

113. Kunasundari, B. and Sudesh, K. (2011) Isolation and recovery of microbial polyhydroxyalkanoates. *eXPRESS Polymer Letters*, **5**(7), 620–634.
114. Kim, B.S., Lee, S.C., Lee, S.Y. *et al.* (1994) Production of poly(3-hydroxybutyric-*co*-3-hydroxyvaleric acid) by fed-batch culture of *Alcaligenes eutrophus* with substrate control using on-line glucose analyser. *Enzyme Microbiology and Technology*, **16**, 556–561.
115. Zinn, M., Weilenmann, H.U., Hany, R. *et al.* (2003) Tailored synthesis of poly([R]-3-hydroxybutyrate-co-3-hydroxyvalerate) (PHB/HV) in *Ralstonia eutropha* DSM 428. *Acta Biotechnology*, **23**, 309–316.
116. Sun, Z., Ramsay, J.A., Guay, M. and Ramsay, B.A. (2007) Fermentation process development for the production of medium-chain-length poly-3-hydroxyalkanoates. *Applied Microbiology and Biotechnology*, **75**, 475–485.
117. Atlić, A., Koller, M., Scherzer, D. *et al.* (2011) Continuous production of poly([(R]-3-hydroxybutyrate) by *Cupriavidus necator* in a multistage bioreactor cascade. *Applied Microbiology and Biotechnology*, **91**, 295–304.
118. Koller, M., Horvat, P., Hesse, P.J. *et al.* (2006) Assessment of formal and low structured kinetic modeling of polyhydroxyalkanoate synthesis from complex substrates. *Bioprocess and Biosystems Engineering*, **29**(5–6), 367–377.
119. Koschuh, W., Koller, M., Essl, R. and Kromus, S. (2012) Green Biorefinery - Chloroplast Rich Biomass (CRB) as alternative Bio Feedstock. 25th VH Yeast Conference, Hamburg, 2012.
120. Ouyang, S.P., Liu, Q., Fang, L. and Chen, G.Q. (2007) Construction of phaoperon-defined knockout mutants of Pseudomonas putida KT2442 and their applications in poly(hydroxyalkanoate) production. *Macromoles Biosciences*, **7**, 227–233.
121. Ouyang, S.P., Luo, R.C., Chen, S.S. *et al.* (2007) Production of polyhydroxyalkanoates with high 3-hydroxydodecanoate monomer content by fadB and fadA knockout mutant of *Pseudomonas putida* KT2442. *Biomacromolecules*, **8**, 2504–2511.
122. Tan, D., Xue, Y.S., Aibaidula, G. and Chen, G.Q. (2011) Unsterile and continuous production of polyhydroxybutyrate by *Halomonas* TD01. *Bioresource Technology*, **102**, 8130–8136.
123. Tripathi, L., Wu, L.P., Chen, J. and Chen, G.Q. (2012) Synthesis of Diblock copolymer poly-3-hydroxybutyrate -block-poly-3-hydroxyhexanoate [PHB-b-PHHx] by a β-oxidation weakened *Pseudomonas putida* KT2442. *Microbial Cell Factories*, **44**, 1–11.
124. Chiellini, E. and Grillo Fernandes, E. (2004) Thermo-mechanical and morphological characterization of plasticized poly(R - 3-hydroxybutyric acid). *Macromolecular Symposia*, **218**, 157–164.
125. Pietrini, M., Roes, L., Patel, M.K. and Chiellini, E. (2007) Comparative life cycle studies on poly(3-hydroxyalkanoate)-based composites as potential replacement for conventional petrochemical plastics. *Biomacromolecules*, **8**, 2210–2218.
126. Saad, G.R., Lee, Y.J. and Seliger, H. (2001) Synthesis and characterization of biodegradable poly (ester-urethanes) based on bacterial poly (R-3-hydroxybutyrate). *Journal of Applied Polymer Science*, **83**(4), 703–718.
127. Häberlein, H., Seliger, H., Kohler, R. and Sulzberger, P. (2005) Cost-effective synthesis of environmentally benign materials on the basis of poly-3-hydroxybutyrate. *Polímeros*, **15**(2), 122–126.
128. Braunegg, G., Lefebvre, G., Renner, G. *et al.* (1995) Kinetics as a tool for polyhydroxyalkanoate production optimization. *Canadian Journal of Microbiology*, **41**, 239–248.
129. Koyama, N. and Doi, Y. (1995) Continuous production of poly(3-hydroxybutyrate-co-3-hyhroxyvalerate) by *Alcaligenes eutrophus*. *Biotechnology Letters*, **17**, 281–284.
130. Ramsay, B.A., Lomaliza, K., Chavarie, C. *et al.* (1990) Production of poly-(β-Hydroxybutyric-co-β-Hydroxyvaleric) acids. *Applied Environmental Microbiology*, **56**, 2093–2098.
131. Yu, S.T., Lin, C.C. and Too, J.R. (2005) PHBV production by *Ralstonia eutropha* in a continuous stirred tank reactor. *Process Biochemistry*, **40**, 2729–2734.
132. Mothes, G. and Ackermann, J.U. (2005) Synthesis of Poly(3-hydroxybutyrate-co-4- hydroxybutyrate) with a target mole fraction of 4-hydroxybutyric acid units by two-stage continuous cultivation of *Delftia acidovorans* P4a. *Engineering in Life Sciences*, **5**, 58–62.

133. Zinn, M., Weilenmann, H.U., Hany, R. *et al.* (2003) Tailored synthesis of poly([R]-3-hydroxybutyrate-co-3-hydroxyvalerate) (PHB/HV) in *Ralstonia eutropha* DSM 428. *Acta Biotechnology*, **23**, 309–316.
134. Hartmann, R., Hany, R., Geiger, T. *et al.* (2004) Tailored biosynthesis of olefinic medium-chain-length poly[(R)-3-hydroxyalkanoates] in *Pseudomonas putida* GPo1 with improved thermal properties. *Macromolecules*, **37**, 6780–6785.
135. Moser, A. (1988) *Bioprocess Technology: Kinetics and Reactors*, Springer, New York.
136. Jung, K., Hazenberg, W., Prieto, M.A. and Witholt, B. (2001) Two stage chemostat process development for the effective production of medium-chain-length poly(3-hydroxyalkanoates). *Biotechnology and Bioengineering*, **72**, 19–24.
137. Saito, Y. and Doi, Y. (1994) Microbial synthesis and properties of poly (3-hydroxybutyrate-co-4-hydroxybutyrate) in *Comamonas acidovorans*. *International Journal of Biological Macromolecules*, **16**(2), 99–104.

8

Poly(Lactic Acid)

Hideto Tsuji
Department of Environmental and Life Sciences, Graduate School of Engineering, Toyohashi University of Technology, Japan

8.1 Introduction

Over the past several decades, polylactide – i.e. poly(lactic acid) (PLA) – and its copolymers have attracted significant attention in environmental, biomedical, and pharmaceutical applications as well as alternatives to petro-based polymers [1–18]. Plant-derived carbohydrates such as glucose, which is derived from corn, are most frequently used as raw materials of PLA. Among their applications as alternatives to petro-based polymers, packaging applications are the primary ones. Poly(lactic acid)s can be synthesized either by direct polycondensation of lactic acid (IUPAC name: 2-hydroxypropanoic acid) or by ring-opening polymerization (ROP) of lactide (LA) (IUPAC name: 3,6-dimethyl-1,4-dioxane-2,5-dione). Lactic acid is optically active and has two enantiomeric forms, that is, L- and D- (S- and R-). Lactide is a cyclic dimer of lactic acid that has three possible stereoisomers: (i) L-lactide (LLA), which is composed of two L-lactic acids, (ii) D-lactide (DLA), which is composed of two D-lactic acids, and (iii) meso-lactide (MLA), which is composed of an L-lactic acid and a D-lactic acid. Due to the two enantiomeric forms of lactic acids, their homopolymers are stereoisomeric and their crystallizability, physical properties, and processability depend on their tacticity, optical purity, and molecular weight; the latter two are dominant factors.

Poly(lactic acid) has three typical optical isomeric forms: (i) optically active and crystallizable poly(L-lactide) (i.e. poly(L-lactic acid) (PLLA)), (ii) optically active and crystallizable poly(D-lactide) (i.e. poly(D-lactic acid) (PDLA)), and (iii) optically inactive and noncrystallizable poly(DL-lactide) [i.e. poly(DL-lactic acid) (PDLLA)]. Of these isomeric polymers, PLLA is most frequently used because its production cost is lower due to its joint mass production of 1.4×10^5 metric tons per year by NatureWorks LLC, which is owned by Cargil

Bio-Based Plastics: Materials and Applications, First Edition. Edited by Stephan Kabasci.

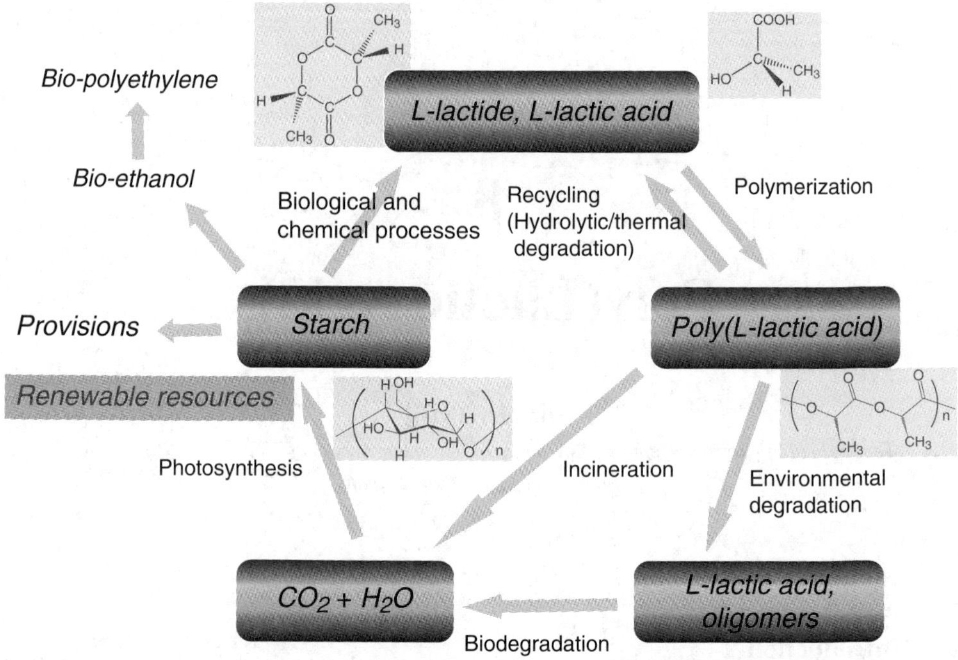

Figure 8.1 *Synthesis, recycling, and degradation of PLLA [13]. (Reproduced with permission from Yoneda Shuppan, Japan.)*

Inc., and Teijin Ltd., which is known as Ingeo™ in the USA and Biofront® in Japan. Another company, Purac, also produces biomedical application-oriented PLA-based materials under the PURASORB(R) brand name. Figure 8.1 shows the synthesis, recycling, and degradation of PLLA [13]. However, due to the recent initiation of the production of biobased polyethylene (PE) from bio-ethanol, PLLA is not the sole mass-produced biobased polymer. To forestall biobased PE and other biobased polymers that will be produced in the near future, high performance PLA-based materials must be developed to suppress their hydrolytic/thermal degradability and increase their mechanical performance.

Poly(lactic acid) is a biobased polymer that can be produced from renewable resources including starch from corn and potatoes, sugar from beets and sugar cane, and so on. The carbon in PLA originates from atmospheric carbon dioxide, which is immobilized in glucose by photosynthesis; therefore, the carbon dioxide formed by its disposal, incineration, or biodegradation does not increase the total amount of atmospheric carbon dioxide. As a result, its impact on the environment during production and disposal, that is, its carbon footprint, is low compared to other petro-based polymers. Poly(lactic acid) can also be recycled via hydrolytic or thermal degradation to its lactic acid and LA monomers, which further reduces its negative impact on the environment. Poly(lactic acid)-based materials, especially crystalline PLLA or PDLA, have mechanical properties that are comparable with those of petro-based commercial polymers such as polystyrene. Poly(lactic acid) is also degradable in the human body and in the environment. The toxicity of its degradation products, which are lactic acid and its oligomers, is very low. The functionality of hydrolytic degradability and the very low toxicity of the resultant products are beneficial for environmental, biomedical, and pharmaceutical applications.

Most commercially available PLLA is used for packaging, automobile interiors, electronics chassis, and other consumer products. However, some applications require a higher mechanical performance and resistance to hydrolytic/thermal degradation. In addition to composite or fibre-reinforced plastic formation, stereocomplexation between enantiomeric PLLA and PDLA is a promising method for producing high-performance PLA-based materials because it has been shown to enhance the mechanical performance and resistance to hydrolytic/thermal degradation of PLA-based materials [11]. Stereocomplexation occurs due to the strong interactions between the two different enantiomeric polymer chains (i.e. PLLA and PDLA) compared to the weaker interactions between the same enantiomeric polymer chains (i.e. PLLA and PLLA, or PDLA and PDLA). The physical properties, hydrolytic degradation, and biodegradation of PLAs can be controlled by altering their molecular and higher ordered structures and so on. This chapter outlines the basic aspects of synthesis, processing, structures, physical properties, degradation, and applications of PLAs.

8.2 Historical Outline

In 1845, a synthetic route to the linear dimer of lactic acid, that is, lactoyllactic acid, through the esterification of lactic acid by removal of water at high temperature (130 °C) under atmospheric pressure was discovered [19]. In 1914, lactic acid oligomers from trimers to heptamers were reported to form when lactic acid was dehydrolysed at an elevated temperature (90 °C) under reduced pressure (15 mmHg) [20]. In 1932 and 1954, a two-step polymerization procedure via the cyclic dimer of lactic acid, LA, was suggested and developed [21, 22] for the synthesis of high-molecular-weight (HMW) PLA. Since the late 1960s, because of the breakthrough work by Kulkarni [23], PLA and its copolymers have been widely investigated for their biomedical applications such as sutures and prostheses [24]. In the 1970s, commercial biodegradable sutures called Vicryl or polyglactin 910 were produced from the copolymers of LA and glycolide (GA) [25]. Since the late 1970s, PLAs and their copolymers have been studied for pharmaceutical applications such as matrices for drug delivery systems [24]. The crystalline structure of the α-form of PLLA was first analysed by De Santis and Kovacs in 1968 [26] and then by Hoogsteen *et al.* in 1990 [27], Kobayashi *et al.* in 1995 [28], Sasaki and Asakura in 2003 [29], and Wasanasuk and Tashiro *et al.* in 2011 [30]. Later, the crystalline structure of the β-form of PLLA was reported by Hoogsteen *et al.* in 1990 [27] and Puiggali *et al.* in 2000 [31], while the γ-form of PLLA was analysed by Cartier *et al.* in 2000 [32]. The crystalline structure of the PLLA/PDLA stereocomplex was reported by Okihara *et al.* in 1991 [33].

Through the 1980s and 1990s, basic information concerning the synthesis, physical properties, crystallization behaviour, and medical and pharmaceutical applications of PLAs and their copolymers was obtained. A large amount of research has been targeted towards the synthesis and molecular characterization of LA homo- and co-polymers [34–36]. Moreover, the crystallization, spinning, and physical properties of PLA-based materials have been intensively investigated [37–39]. Studies have also focused on medical and pharmaceutical applications and the *in vitro* and *in vivo* degradation of PLAs and their copolymers [2, 40, 41]. The stereocomplex between PLLA and PDLA was reported in 1987 [42] and studies on its crystallization, physical properties, and hydrolytic degradation and biodegradation revealed its high mechanical performance, resistance to hydrolytic/thermal degradation, and gas barrier properties [11, 43–45]. Since the second half of the 1990s, PLAs have attracted significant attention in terms of ecological applications and have been recognized as substitutes to commercial petro-based polymers because of their cost-effective (<$4/kg) production from renewable plant resources.

In the 1990s, Baker *et al.* synthesized substituted racemic PLA compounds including poly(2-hydroxybutyrate), that is, poly(2-hydroxybutanoic acid) (P(2HB)), poly(2-hydroxypentyrate), that is, poly(2-hydroxypentanoic acid), and poly(2-hydroxyhexyrate), that is, poly(2-hydroxyhexanoic acid), and discovered their enhanced thermal resistance [46–48]. In 2009 and 2010, Tsuji *et al.* synthesized optically pure and crystallizable L- and D-forms of P(2HB) (i.e. P(L-2HB) and P(D-2HB), respectively), which comprise substituted PLLA and PDLA. Stereocomplexation between enantiomeric P(2HB)s [49] and hetero-stereocomplexation between P(2HB) and PLA with opposite configurations were also revealed [50, 51]. The latter finding may enable the production of biodegradable materials that have a wide variety of physical properties and hydrolytic degradation and biodegradation behaviours and rates. Similar to PLA stereocomplexation, stereocomplexation between P(L-2HB) and P(D-2HB) or enantiomeric blends accelerates crystallization and enhances hydrolytic/thermal degradation resistance. Jung *et al.* employed metabolically engineered *Escherichia coli* to synthesize PLA and its copolymers [52], whereas Han *et al.* synthesized poly(3-hydroxybutyrate-co-2-hydroxybutyrate) with wild-type class I PHA synthase from *Ralstonia eutropha* [53].

8.3 Synthesis of Monomer

Lactic acid, which is the monomer of PLA, is a well known 2-(α-)hydroxycarboxylic acid with an asymmetric carbon atom that is synthesized by the bacterial fermentation of carbohydrates such as glucose. Of the sources of carbohydrates, corn is the most effective and is frequently utilized to produce glucose. However, the use of corn has issues such as limitations to the potential production of corn and the effect of supply on its price. Lactic acids include optically active compounds with L- and D-forms. The melting temperature (T_m) of a 1 : 1 racemic mixture (compound) of L- and D-lactic acids, that is, DL-lactic acid, is 52.8 °C, which is higher than that of the respective enantiomers (16.8 °C). The optical purity, yield, and dominant enantiomeric form of lactic acid is determined by the type of *Lactobacillus* used and the fermentation conditions such as the kind and concentration of nutrition, aeration, pH, temperature, and so on [1, 3]. From an industrial perspective, it is crucial to produce lactic acid with high optical purity and yield. The removal of impurities from the fermentation broth of the lactic acid production is important in order to increase the molecular weight and optical purity of PLA. These impurities include carbohydrates, organic acids, alcohols, amino acids, and proteins. LAs, which are the cyclic dimers of lactic acids, have three different forms: (i) L-lactide (LLA), which comprises two L-lactyl units, (ii) D-lactide (DLA), which comprises two D-lactyl units, and (iii) meso-lactide (MLA), which comprises one L-lactyl unit and one D-lactyl unit. Racemic lactide or DL-lactide (DLLA) is a 1 : 1 physical mixture or a 1 : 1 racemic compound (stereocomplex) of LLA and DLA and has a T_m of 124 °C, which is higher than that of LLA (95–99 °C), DLA (95–99 °C), and MLA (53–54 °C). Figure 8.2 shows the molecular structures of lactic acids and LAs [8, 54]. The industrial and laboratory method used for LA synthesis involves the thermal depolymerization of low-molecular-weight (LMW) PLAs followed by the distillation of LAs under reduced pressure. The obtained crude LAs are purified either by recrystallization in solvents such as ethylacetate or toluene, from the melt, or from the gas phase. The purification of LAs, that is, the removal of impurities such as solvents, free acids, catalysts, and moisture, is necessary to increase the molecular weight of the resulting PLA polymers. Detailed information regarding the procedures for the synthesis and purification of lactic acids and LAs is described in previous publications [1, 3, 15].

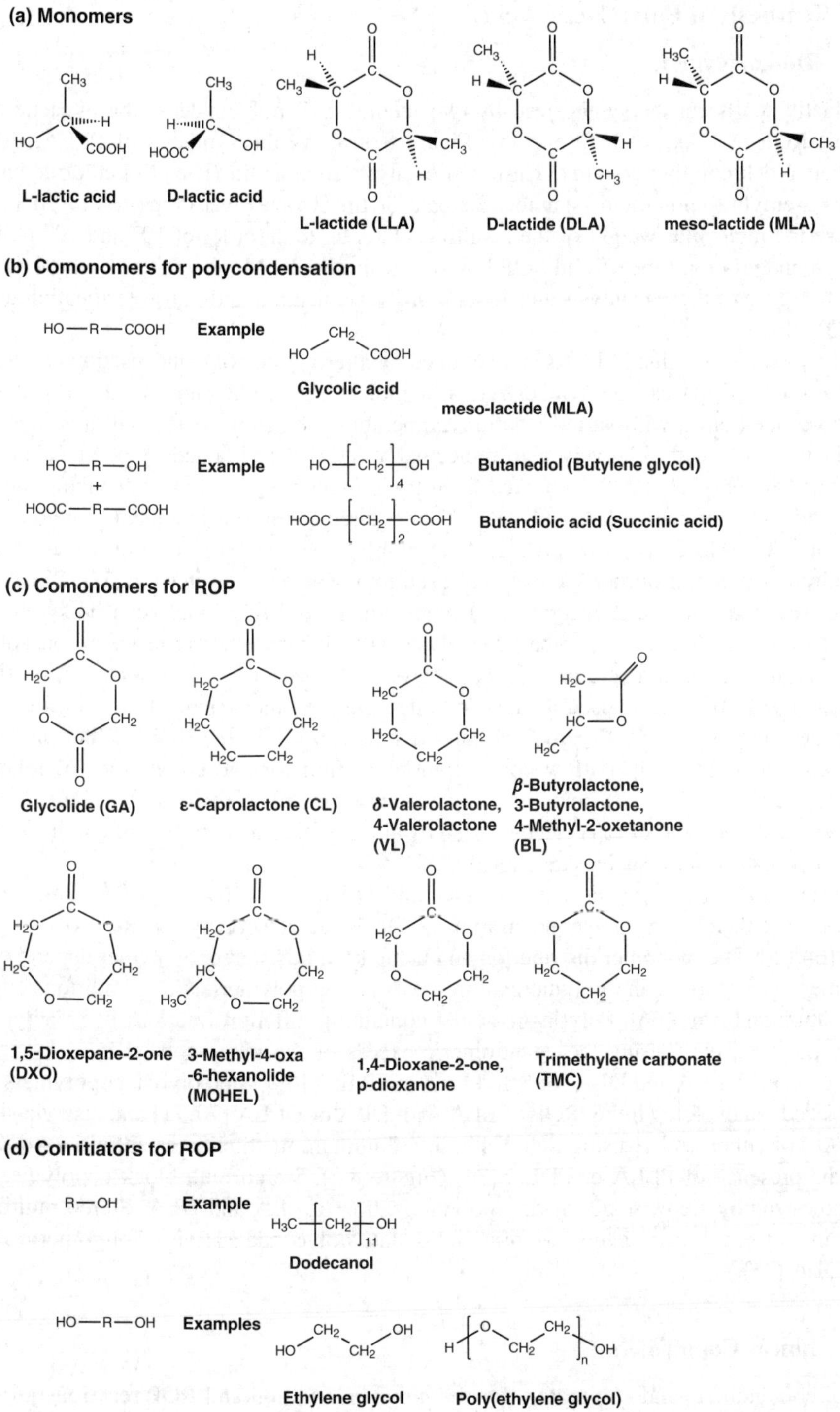

Figure 8.2 *Monomers, typical comonomers for polycondensation and ROP, and coinitiators [8, 54]. (Reproduced with permission from ref. [8] © 2002, Wiley-VCH Verlag GmbH & Co. KGaA and [54] © 1999, Research Trends Ltd.)*

8.4 Synthesis of Poly(Lactic Acid)

8.4.1 Homopolymers

Poly(lactic acid)s can be synthesized by two methods: (i) polycondensation of lactic acids and (ii) ROP of LAs [1, 3, 8, 54, 55]. Figure 8.3 shows the synthesis of PLLA, PDLA, and their block copolymers [11]. Tin-based catalysts such as tin (II or IV) chloride and tin (II) bis-2-ethylhexanoic acid (stannous octoate or tin octoate) were reported to effectively increase the molecular weight of the resulting PLAs up to an order of 10^5 and 10^6 g mol^{-1} for polycondensation [56, 57] and ROP [58–61], respectively. Moreover, bulk polymerization in the presence of these catalysts aids in avoiding racemization and transesterification during the ROP of LA.

The molecular weights of PLAs can be tuned by altering the polycondensation conditions. Poly(DL-lactic acid) can be synthesized in a molecular weight range of 2×10^3–2×10^4 via polycondensation with varied reaction temperatures, reaction times, and pressures [62]. Poly(lactic acid)s with relatively high molecular weights are obtained by polycondensation only when water is efficiently removed from the polymerization mixture, which comprises lactic acid and oligomers, PLA, and water. The required conditions include: (i) a temperature range of 180–200 °C, (ii) low pressure (<5 mmHg), (iii) a long reaction time, and (iv) the addition of an appropriate catalyst and azeotropic solvent of water [56, 57]. PLLA with a weight-average molecular weight (M_w) higher than 1×10^5 g mol^{-1} can be synthesized using Brønsted acid–activated tin-based catalysts with diphenylether as an azeotropic solvent. A very high reaction temperature reduces the optical purity of PLAs even when optically pure L- or D-lactic acid is used as the monomer. On the other hand, PLAs with very high molecular weights (>1×10^6 g mol^{-1}) can be synthesized via the ROP method under the following conditions [60]: (i) low concentrations of initiator and coinitiator, (ii) relatively low temperature (<120 °C), and (iii) long reaction time. The molecular weight of PLAs synthesized by the ROP of LAs can be varied by altering the concentration of coinitiators that feature hydroxyl groups such as alcohols [25, 54, 63].

Poly(lactic acid) comprising equimolar L- and D-lactyl (i.e. half of a LA) units can be synthesized either by the polycondensation of DL-lactic acid or by the ROP of DLLA or MLA [64, 65]. The monomer distribution and tacticity of this stereocopolymer depend on the monomer, the nature of the polymerization catalyst, and polymerization conditions such as temperature and time [66]. Poly(lactic acid)s containing different fractions of L-lactyl units can be prepared by altering the enantiomeric excess of the monomers, that is, L- and D-lactic acids [67], LLA and DLA [68], and LLA and MLA [69]. Stereoblock copolymers were synthesized via two-step living ROP of LLA with DLA or DLLA [70, 71] and also via single-step ROP of either DLLA using Schiff's base / aluminum methoxide as an initiator [72, 73] or in the presence of PLLA or PDLA [74] (Figure 8.3). Stereo multiblock copolymers can be synthesized by stepwise alternating polymerization of LLA and DLA. Stereo multiblock copolymers were directly synthesized by solid-state polycondensation of oligomeric PLLA and PDLA [75].

8.4.2 Linear Copolymers

Typical comonomers and coinitiators for the polycondensation and ROP reactions are summarized in Figure 8.2. For polycondensation, various hydroxycarboxylic acids or mixtures of diol and dicarboxylic acid are used as comonomers of lactic acid; of these, glycolic acid is most frequently used. On the other hand, for ROP, various heterocyclic compounds including

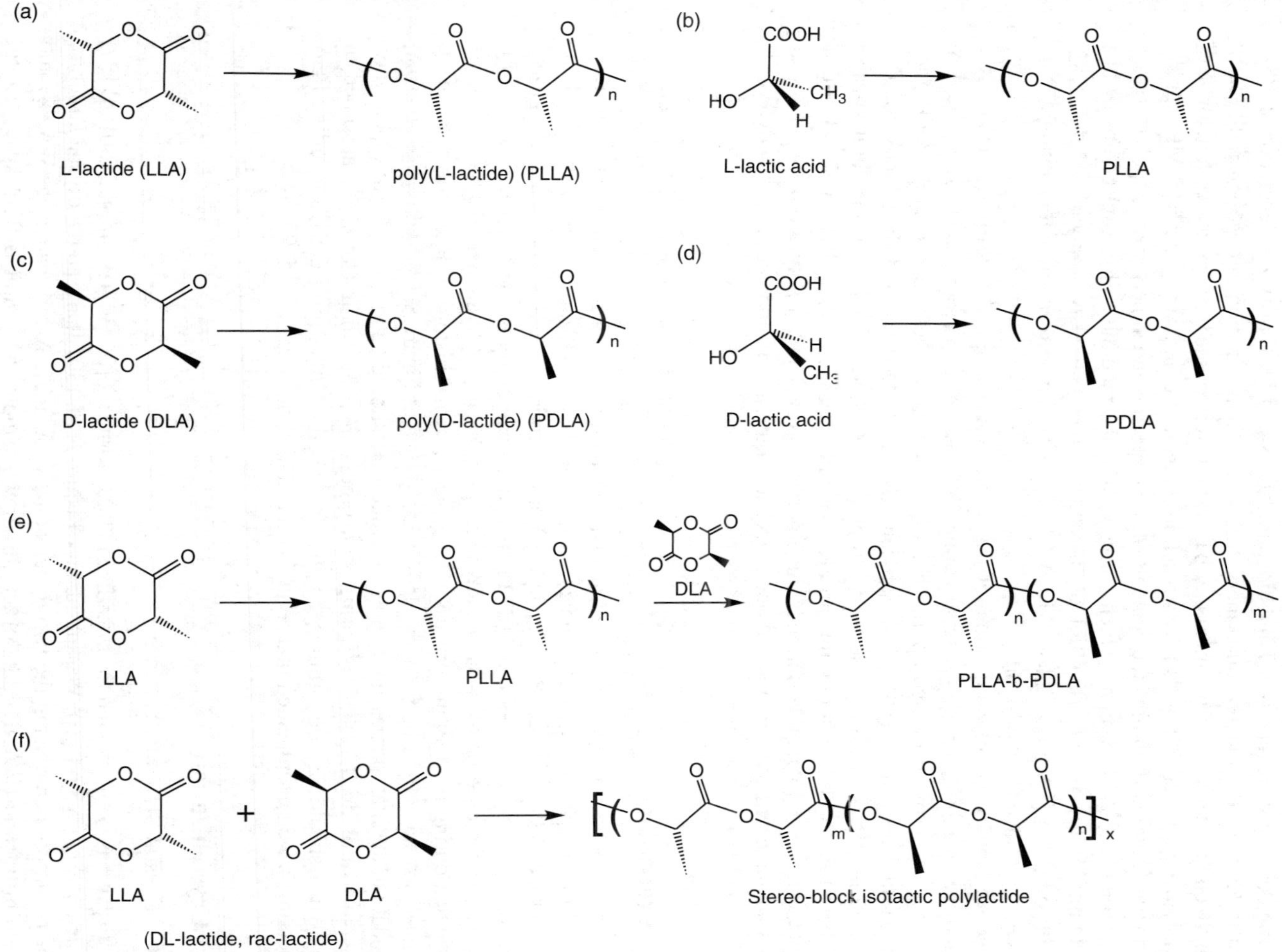

Figure 8.3 *Synthesis of PLLA, PDLA, and their block copolymers [11]. (Reproduced with permission from ref. [11]. Copyright © 2005 Wiley-VCH Verlag GmbH & Co. KGaA, Weinheim.)*

lactones and carbonates are utilized as comonomers of LA of which GA and CL are most often used [1, 2, 8, 36]. The copolymers of LA with glycolide (GA) or ε-caprolactone (CL) as well as LA homopolymers and stereocopolymers have been intensively studied as matrices for drug delivery systems [76, 77] and biodegradable scaffolds for tissue regeneration [1, 78]. As in the case of PLA stereocopolymers, the monomer distribution in copolymers depends strongly on the identities of the monomer pairs, the nature of the initiator and coinitiator, and the polymerization conditions such as reaction temperature and time. The ^{1}H NMR signals that are useful for determining the contents of the respective monomers in the copolymers are those around 5.2 and 1.6 ppm, which correspond to the methine and methyl groups of the lactyl unit [79, 80]. The incorporation of GA and CL imparts the LA copolymers with blocky structures P(LLA-GA) [81, 82] and P(LLA-CL) [83, 84]. Randomization of the monomer unit sequence is known to be difficult [85].

Two-step block copolymerization of LLA or DLLA with GA [86], CL [87], (R)-β-butyrolactone [88], or 1,5-dioxepan-2-one [89] was also performed. Hydrophobic LA sequences can grow from hydrophilic polymers such as poly(ethylene oxide) (PEO), that is, poly(ethylene glycol) [90], poly(propylene oxide) (PPO) [91], and their block copolymer (Pluronic™) [92] to increase the hydrophilicity, flexibility, and biodegradability of the PLA-based materials. Although this article does not discuss branched PLA in detail, it can be synthesized via polycondensation by incorporating polyol and/or polycarboxylic acid with more than three hydroxyl groups and/or carboxylic groups as comonomer(s) and via ROP with a polyol with more than three hydroxy groups as the coinitiator.

8.5 Processing

In order to process polymers, information regarding thermal properties such as glass transition and melting temperatures (i.e. T_g and T_m, respectively), crystallization behaviour, and rheological properties is required. Poly(lactic acid)-based products can be prepared using a normal processing procedure for petro-based polymers such as poly(ethylene terephthalate) (PET). However, in the case of PLA-based polymers, thermal and/or hydrolytic degradation must be avoided during the melting process. The typical thermal history of PLA polymer during one- and two-stage PLA bottle manufacturing is shown in Figure 8.4 [93]. To avoid thermal degradation, the melt processing temperature and time should be low and short, respectively. To avoid hydrolytic degradation during melt processing, the amount of water absorbed in the polymer should also be minimized before processing. Other information regarding PLA processing is well documented in a review article and book chapters [15, 93, 94].

8.6 Crystallization

8.6.1 Crystal Structures

Neat PLLA has been reported to crystallize in three forms: α [26–30] (Figure 8.5 [30]), β [27, 31], and γ (Figure 8.6 [32]). Wasanasuk and Tashiro *et al.* [30] analysed the crystal structure of the α-form in detail using the two-dimensional wide-angle X-ray scattering (WAXS) and neutron diffraction (WAND) data of an ultradrawn sample. Assuming a $P2_11$ space group, they found that the chain conformation took a distorted (10/3) helical form with 2_1 helical symmetry along the chain axis; the unit cell has monoclinic symmetry and an orthogonal shape (a = 10.86 Å, b = 6.16 Å, and c = 28.86 Å). The β-form of neat PLLA is reported to feature a rather frustrated structure with a trigonal unit cell (a = 10.52 Å, b = 10.52 Å, and

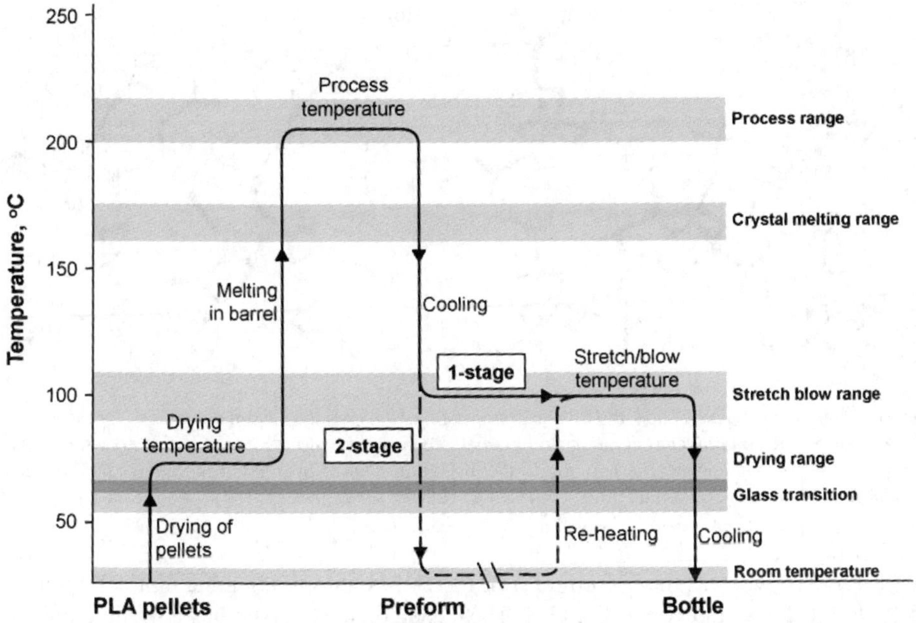

Figure 8.4 Thermal history of PLA polymer during one- and two-stage PLA bottle manufacturing. (Reprinted from Poly(lactic acid) – Synthesis, Structures, Properties, Processing, and Applications–, Chapter 14, L.-T. Lim., Processing of Poly(Lactic Acid), Copyright (2010), with permission from John Wiley & Sons, Inc. [94].)

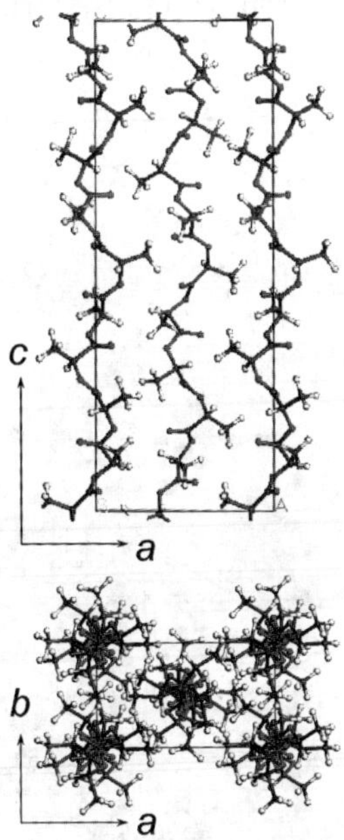

Figure 8.5 Crystal structure of α-form PLLA proposed by Wasanasuk and Tashiro. No symmetry exists along the chain axis under the space group of $P2_11$. The adjacent chains are packed upwards and downwards along the chain axis. (Reproduced with permission form [30]. Copyright © 2011, American Chemical Society.)

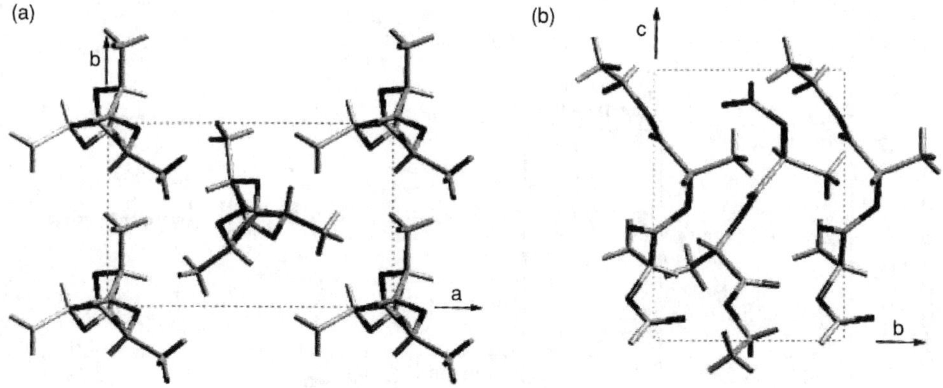

Figure 8.6 (a) Chain axis projection and (b) a-axis projection of the crystal structure determined for the γ-phase of PLLA produced by epitaxial crystallization on hexamethylbenzene at 140 °C. The structure corresponds to a minimum of the packing energy. The two antiparallel helices are linked by a 2_1 screw axis parallel to b [32]. (Reproduced with permission from ref. [32]. Copyright © Elsevier, 2000.)

$c = 8.8$ Å) [31], whereas the γ-form of neat PLLA has an orthorhombic unit cell with $a =$ 9.95 Å, $b = 6.25$ Å, and $c = 8.8$ Å [32]. The stereocomplex crystal has a triclinic unit cell ($a = 9.16$ Å, $b = 9.16$ Å, and $c = 8.70$ Å) in which the PLLA and PDLA chains assume a 3_1 helical conformation and are packed side-by-side in parallel [33] (Figure 8.7). A mechanism for the formation of the triangular stereocomplex crystal has been proposed by Brizzolara *et al.* [95]. The α'-form was thought to be a disordered α-form that is transformed to the α-form by increasing the temperature above 110–120 °C [96–100]. However, very recently, Wasanasuk

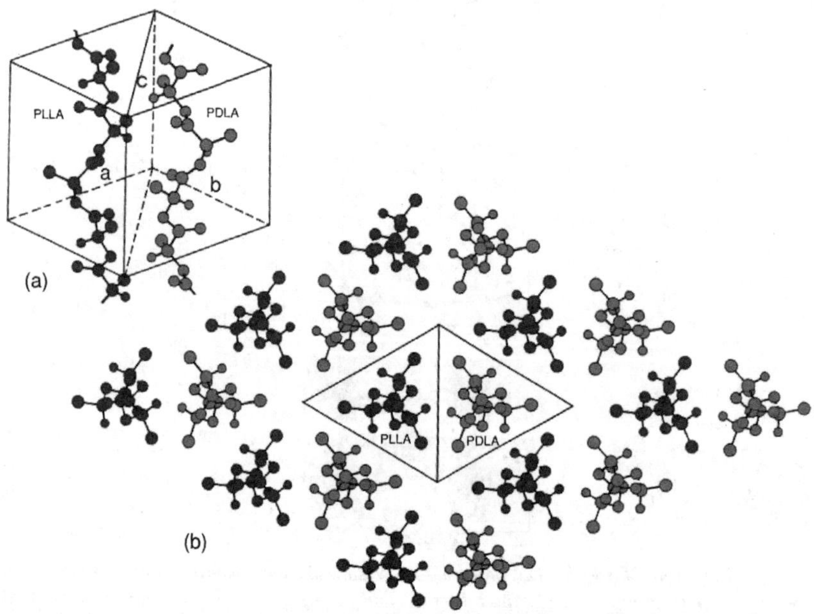

Figure 8.7 Crystal structure of PLA stereocomplex [33]. The lines between PLLA and PDLA chains were added to original figure. (Reproduced with permission from ref. [33]. Copyright © 1991, Taylor & Francis.)

and Tashiro *et al.* reported that the α'-form is a δ-form crystal rather than a disordered α-form crystal [101]. Polymorphism and isomorphism in biodegradable polyesters including PLA have been well documented in a review article [102]. Cold crystallization of melt-quenched PLLA is reported to result in the formation of α'-form (δ-form) crystallites probably due to the low crystallization temperature [103].

8.6.2 Crystalline Morphology

8.6.2.1 Spherulites

Spherulite morphology is useful to determine the degree of orientation and periodic rotation of lamellae, whereas the size of spherulites can be used to determine the number of spherulitic nuclei per unit area or volume. A spherulitic structure is observed for crystalline PLLA, PDLA, PLLA/PDLA blends, and PLA copolymers that have relatively long LLA or DLA unit sequences. The morphology and sizes of the spherulites depend on the molecular structures, blending, and crystallization conditions such as temperature and time in the case of isothermal crystallization and cooling rate in the case of nonisothermal crystallization. The spherulites of LA stereocopolymers retain well-defined structures even when the monomer unit sequence length approaches the critical value for crystallization [104]. Also, relatively random copolymerization of LLA with comonomer(s) did not change the morphology of the spherulites (Figure 8.8 [105]). Instead, disturbed spherulites or assemblies of crystallites appeared in the crystallizable PLLA/amorphous PDLLA blends with low PLLA contents [100, 106]. The morphology of the stereocomplex spherulites is similar to that of the normal spherulites of neat

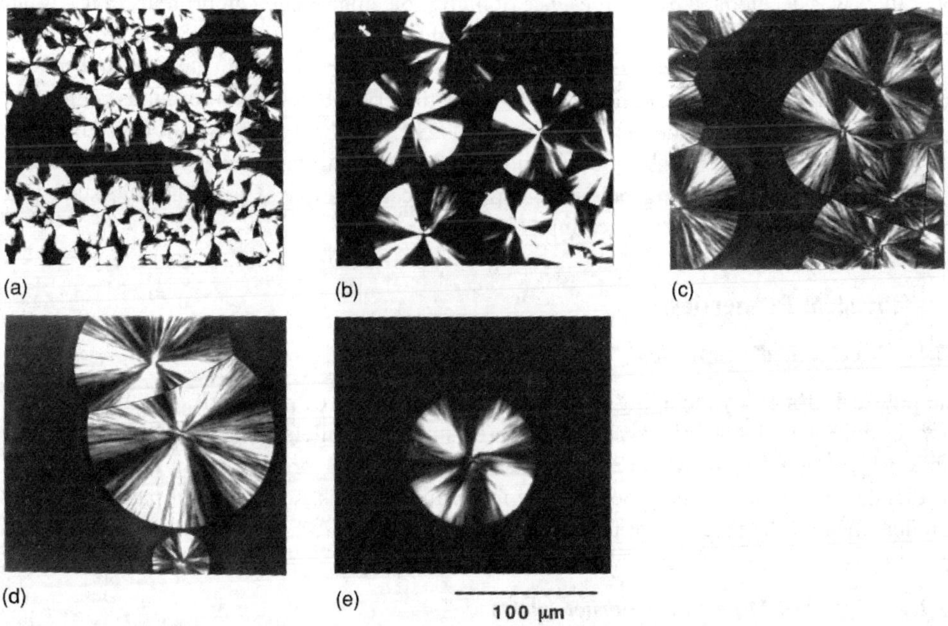

Figure 8.8 *Polarization optical photomicrographs of the films crystallized at $T_c = 120$ °C from the melt: (a) PLLA ($M_n = 1.0 \times 10^5$), $t_c = 5$ min; (b) PLLA ($M_n = 3.1 \times 10^4$), $t_c = 3$ min; (c) P(LLA-DLA) (95/5, $M_n = 9.6 \times 10^4$), $t_c = 50.5$ min; (d) P(LLA-GA) (95/5, $M_n = 3.8 \times 10^4$), t_c=22 min; (e) P(LLA-CL) (82/18, $M_n = 6.7 \times 10^4$), t_c = 20 min [105]. (Reproduced with permission from ref. [105]. Copyright © Elsevier, 2005.)*

PDLA or PLLA when solely stereocomplexation takes place [107]. However, spherulites with complicated morphologies are evident when stereocomplexation and homo-crystallization occur simultaneously [104, 107, 108]. Maillard and Prud'homme observed mirror-image stereocomplex crystalline assemblies for PLLA- and PDLA-rich PLLA/PDLA mixtures [109]. As observed for crystalline polymers, PLLA spherulites become smaller with decreasing temperature and time [110, 111].

8.6.2.2 *Single Crystals*

Poly(L-lactic acid) is reported to crystallize into lozenge-like [37, 112–114] and hexagonal-like [37, 114] single crystals in dilute solutions. In contrast, single crystals of the stereocomplex of PLLA and PDLA have a peculiar triangular shape when they are formed in p-xylene at solution concentrations as low as 0.04% [33, 115].

8.6.3 Crystallization Behaviour

The crystallization of PLLA in the presence of accelerating fillers is discussed in section 8.12.1. Spherulites form during the actual moulding process. The crystallization kinetics of PLLA was first reported by Vasanthakumari and Pennings [38]. They investigated the spherulite growth of PLLA with a wide viscosity-average molecular weight (M_v) of 10^3–10^5 g mol^{-1}and observed regime I–III kinetics. Normally, relatively low molecular weight PLLA shows bimodal peaks of radial growth rate of spherulites (G) as a function of the crystallization temperature (T_c) due to the transition from the from the $\alpha'(\delta)$-crystalline form to the α-form [96–100] or the regime transition from III to II at ~110–120 °C [63, 105, 116]. The radial growth of PLLA spherulites was decelerated by the presence of high-molecular-weight amorphous PDLLA but accelerated by the presence of low molecular weight PDLLA [117]. WAXS and small-angle X-ray scattering (SAXS) measurements of PLLA/PDLLA blends indicate that amorphous PDLLA is removed from the periodic stacking of the crystalline and amorphous regions of PLLA and does not affect the periodic structure of PLLA [100]. The overall crystallization rate is determined by the number of spherulite nuclei per unit mass and G and is maximized at ~110 °C [118, 119]. Aging below T_g induces the formation of a metastable state that accelerates crystallization during heating [120–124].

8.7 Physical Properties

8.7.1 Mechanical Properties

The physical properties including the mechanical properties of polymeric materials depend on their molecular characteristics, highly ordered structures, fillers, and material morphology. Table 8.1 [8] and Figure 8.9 [125] summarize the physical properties of PLA and other representative commercial polymers. In the following sections, some crucial parameters that significantly alter the mechanical properties are discussed.

8.7.1.1 *Effect of Molecular Characteristics*

Molecular weight is an important parameter to determine the mechanical properties of polymeric materials and is given by the following empirical equation:

$$P = P_0 - K/M_n \qquad (8.1)$$

Table 8.1 *Physical properties of some biodegradable aliphatic polyesters [8]. (Reproduced with permission from ref. [8] © 2002, Wiley-VCH Verlag GmbH & Co. KGaA.)*

	PLLA	PDLLA	syn-PLA	PLA stereocomplex	PCL	R-PHB	PGA
T_m (°C)	170–190	–	151	220–230	60	180	225–230
T_m^0 (°C)	205, 212, 215	–	–	279	71,79	188, 197	–
T_g (°C)	50–65	50–60	34	65–72	–60	5	40
ΔH_m^0 (X_c=100%) (J/g)	93,135, 142, 203	–	–	142, 146	142	146	180–207
Density (g/cm^3)	1.25–1.29	1.27	–	–	1.06–1.13	1.177–1.260	1.50–1.69
Solubility parameter (δ_p) (25 °C) [(J/cm^3)$^{0.5}$]	19–20.5, 22.7	21.1	–	–	20.8	20.6	–
$[\alpha]_{589}^{25}$ in chloroform (deg·dm^{-1}·g^{1}·cm^3)	–155 ± 1	0	–	–	0	+44[a]	–
WVTR[b] (g/m^2/day)	82–172	–	–	–	177	13[c]	–
σ_B[d] (GPa)	0.12–2.3[e]	0.04–0.05[f]	–	0.88[e]	0.1–0.8[e]	0.18–0.20[e]	0.08–1[e]
E[g] (GPa)	7–10[e]	1.5–1.9[f]	–	8.6[e]	–	5–6[e]	4–14[e]
ε_B[h] (%)	12–26[e]	5–10[f]	–	30[e]	20–120[e]	50–70[e]	30–40[e]

Notes: [a]300 nm, 23 °C; [b]water vapour transmission rate at 25 °C; [c]P(HB-HV) (94/6); [d]tensile strength; [e]oriented fibre; [f]nonoriented film; [g]Young's modulus; [h]elongation at break.

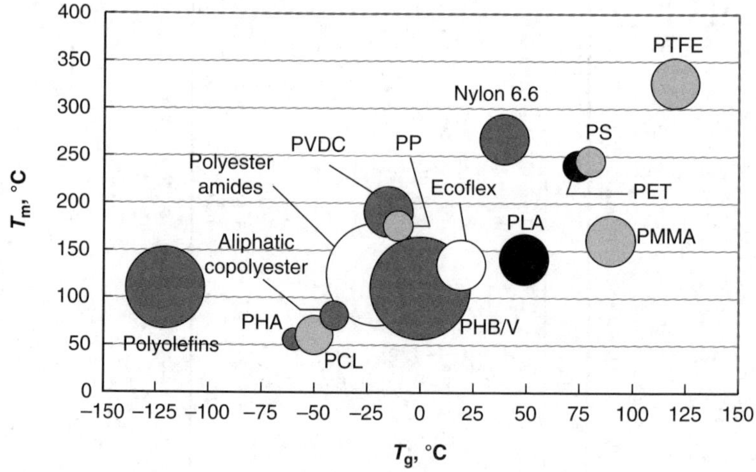

Figure 8.9 Comparison of glass transition and melting temperatures (T_g and T_m, respectively) of PLA with other thermoplastics [125]. (Reproduced with permission from [125]. Copyright © Elsevier, 2008.)

where P is the physical property of a polymeric material, P_0 is the P of a polymer with an infinite number-average molecular weight (M_n), and K is a constant. As-cast PLLA, PDLA, and PLLA/PDLA films have nonzero tensile strength (σ_B) below a $1/M_n$ value of 2.5×10^{-5} g^{-1} mol, which corresponds to an M_n of 4.0×10^4 g mol^{-1}; also, σ_B has been found to positively correlate with $1/M_n$ [4, 126] according to Equation 8.1 (Figure 8.10 [126]). Eling *et al.* [127] showed a similar σ_B dependence on $1/M_v$ for PLLA fibres.

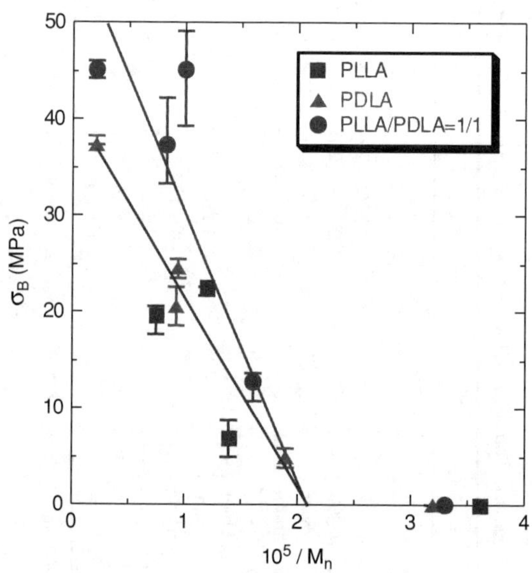

Figure 8.10 Tensile strength (σ_B) of as-cast neat PLLA and PDLA, and their blends as a function of M_n^{-1} [126]. (Reproduced with permission from ref. [126]. Copyright © Elsevier, 1999.)

Saturation of the flexural strength was observed for PDLLA and amorphous PLLA at viscosity-average molecular weight (M_v) values above 35 000 g mol^{-1} and 55 000 g mol^{-1}, respectively [128]. Optimal σ_B and Young's modulus (E) of fibres from LA copolymers P(LLA-CL) and P(LLA-DLA) are smaller than those of homopolymer PLLA fibres. The elongation-at-break (ε_B) of P(LLA-DLA) fibres becomes higher than that of PLLA fibres when they are melt-spun and thermally drawn [129]. Additionally, cross-linking effectively increases the impact strength of PLLA-based materials [130, 131].

8.7.1.2 *Effect of Highly Ordered Structures*

The mechanical properties of PLLA strongly depend on their highly ordered structure including orientation, crystallinity (X_c), and crystalline thickness (L_c) or T_m [111]. Of the highly ordered structure parameters, orientation has the most significant impact on the mechanical properties. The σ_B and E values of PLLA fibres increase and the ε_B value decreases with an increase in the degree of molecular orientation [127, 132–135]. A fibre with a σ_B value of 2.1 GPa and E value of 16 GPa was prepared from high-molecular-weight PLLA ($M_v = 9 \times 10^5$ g mol^{-1}) by dry-spinning and hot-drawing [39], whereas a fibre with a σ_B value of 275 MPa and E value of 9.1 GPa was prepared from low-molecular-weight PLLA (M_v = 13,100) using the zone-drawing method [136]. An increase in X_c increases the σ_B and E values of PLLA but decreases the ε_B value [111]. The decrease in the σ_B value of PLLA films prepared at high T_c may be ascribed to the formation of large spherulites and crystallites in addition to their high X_c. These results indicate that molecular orientation, X_c, and molecular weight are important parameters that influence the mechanical properties.

8.7.1.3 *Effect of Polymer Blending*

Blending is a commercially advantageous method of manipulating the mechanical properties of PLA-based materials [8]. The mechanical properties of miscible and phase-separated polymer blends respectively show continuous and discontinuous dependence on the polymer mixing ratio. Phase-separated blends have discontinuous mixing ratios where the inversion of the continuous and dispersed phases occurs. Even when phase separation occurs, the σ_B, yield stress (σ_Y), and E of blends of glassy PLLA (or PDLLA) with rubbery PEO [137] or PCL [138–140] can be widely varied by altering the polymer mixing ratio (Figure 8.11) [140]. In Figure 8.11, the weight fraction of PLLA (X_{PLLA}) is defined according to the following equation:

$$X_{PLLA} = W_{PLLA}/(W_{PCL} + W_{PLLA}) \qquad (8.2)$$

where W_{PLLA} and W_{PCL} are the weights of PLLA and PCL in the blend. The impact strength of PLLA is reported to increase via the addition of rubbery biodegradable polymers such as PCL [141]. In addition, compatibilizers of the two constituent polymers reduce the discontinuous behaviour and increase the ε_B [142]. Stereocomplexation between PLLA and PDLA enhances the tensile properties of blend films compared to those of neat PLLA or PDLA films [126].

8.7.1.4 *Effects of Material Morphology*

Porous PLA-based materials can be used as scaffolds for the regeneration of organs. These materials are prepared by the formation of a phase-separated structure of PLA-based materials

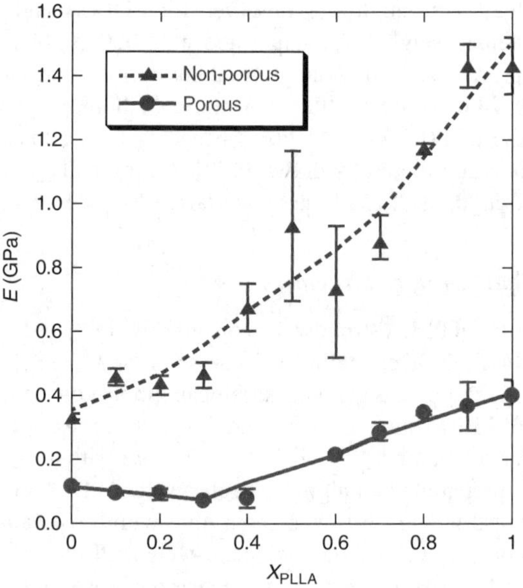

Figure 8.11 Young's modulus of nonporous and porous blends of PLLA and PCL as a function of weight fraction of PLLA (X_{PLLA}) in blends [140]. (Reproduced with permission from ref. [140]. Copyright © 2006 Society of Chemical Industry.)

and a second component such as a solid, liquid, or gas, and the removal of the second component [143, 144]. The formation of pores effectively lowers the E and elevates the ε_B of PLA-based materials [143]. The combination of polymer blending and pore formation can yield PLA-based materials with a wide range of E values [140]. For example, the E values of PLLA/PCL blends were tunable in the range of 0.1–1.4 GPa by altering the blending ratio and pore formation, as shown in Figure 8.11 [140].

Electrospinning of PLA and copolymers [12] and the PLA stereocomplex [145, 146] has been used to prepare nanofibres or nanomats. Similar to the effect of pores on the mechanical properties of PLA-based materials, the formation of nanomats effectively lowers the E and elevates the ε_B value. The morphology of the fibres and nanomats and the mechanical properties are affected by the applied voltage, the effluent rate and concentration of the solution, the type of solvent and so on (Figure 8.12 [147]).

8.7.2 Thermal Properties

Among the thermal properties, the T_m and glass transition temperature (T_g) are the most important to determine the thermal processing and drawing temperature and the usable temperature range. The T_g and T_m values of representative polymers are summarized in Figure 8.9. Values for the T_m, the equilibrium melting temperature (T_m^0), T_g, and the melting enthalpy of a crystal with infinite thickness (ΔH_m^0; ΔH_m at $X_c = 100\%$) of PLAs are given in Table 8.1. The Thomson–Gibbs expression that describes the relationship between T_m and L_c is given by the following equation [148]:

$$T_m = T_m^0(1 - 2\sigma/\Delta h^0 \rho_c L_c) \quad (8.3)$$

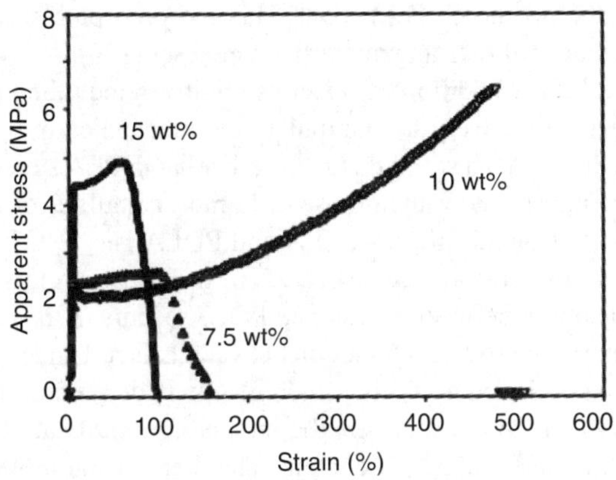

Figure 8.12 *Stress–strain curves for P(LLA-GA) (mol/mol=10/90) electrospun membranes prepared under the 2 kV/cm electric field, at a feed rate of 100 μl/min and different concentrations of 7.5, 10 and 15 wt%, respectively [147]. (Reproduced with permission from [147]. Copyright © Elsevier, 2003.)*

where σ, Δh^0, and ρ_c are the specific fold surface free energy, heat of fusion (per unit mass), and crystal density, respectively. Equation 8.3 implies that the T_m of polymeric materials, which increases with L_c, can be an indicator of L_c. Using the ΔH_m^0 values, the X_c values of the stereocomplex and homo-crystals in specimens can be evaluated via the following equation:

$$X_c(\%) = 100(\Delta H_{cc} + \Delta H_m)/\Delta H_m^0 \tag{8.4}$$

where ΔH_{cc} is the enthalpy of cold crystallization. By definition, ΔH_{cc} and ΔH_m are negative and positive values, respectively. T_m (L_c) and X_c decrease with decreasing crystallizable L- or D-lactyl unit sequence length in the LA homopolymers and copolymers including stereocopolymers; this can be caused by lower polymer molecular weights [126] and increased comonomer content [79, 104, 105]. However, the incorporation of polyfunctional comonomers such as 1,2,3,4-butanetetracarboxylic acid and 1,2,3,4,5,6-cyclohexanehexacarboxylic acid is reported to elevate the T_g values of the resultant copolymers to 80 and 86 °C, respectively [149]. On the other hand, the T_g of LA stereocopolymers is almost constant (~60 °C) regardless of LLA content [104].

T_m and X_c increase with crystallization in the temperature range of 70–160 °C [111, 150, 151] and the orientation of PLLA [132, 135, 152]. A crystallization temperature (T_c) that is too high or too low and is very close to T_m or T_g causes deviation from the appropriate super cooling ($\Delta T = T_m - T_c$), which results in no significant crystallization. The T_m and X_c values of PLLA can be altered within the range of 177–193 °C and 0–63%, respectively, by varying the T_c and crystallization time (t_c) and by adding a pretreatment such as melt quenching before crystallization [111, 150, 151]. Moreover, annealing at temperature ranges below T_g such as 25–50 °C increases the T_g and enthalpy of glass transition (ΔH_g) of PLLA and poly(L-lactide-co-D-lactide) [P(LLA-DLA)] (96/4) [121–124]. The T_g and ΔH_g values increase with increased annealing time (t_a) and annealing temperature (T_a) below T_g. This indicates that the inter-chain interactions of PLLA and P(LLA-DLA) (96/4) in the amorphous region increase and the polymer becomes more densely packed and stable by annealing at T_a values below the T_g. Due to this effect, crystallization during heating is accelerated by low temperature annealing below the T_g [122, 124]. This mesomorphic structure formation may be the result of conformational rearrangement from *gg* to *gt* [123].

The T_g is influenced by the addition of miscible additives including polymers [153–155]. This method can be used effectively to manipulate the mechanical properties of PLA-based materials and improve their impact strength. On the other hand, PLA stereocomplex crystallites have T_m and T_m^0 values much higher than those of homo-crystallites of neat PDLA or PLLA (Table 8.1). The T_g of 1 : 1 blend films of PLLA and PDLA is ~5 °C higher than that of the neat PLLA and PDLA films in the M_w range of 5×10^4 to 1×10^5 g mol^{-1}, which is when predominant stereocomplexation between PLLA and PDLA occurs in the blend films [126]. The increased T_g value is due to the strong interaction between L-lactyl and D-lactyl unit sequences in the amorphous region of the blend films, which results in dense and stable chain packing.

The thermal stability of PLAs depends on the molecular modifications, additives, residual initiators, catalysts, monomers, and water content. The thermal stability of PLLA synthesized using aluminum tri(isopropoxide) as an initiator is higher than that synthesized using tin(II) bis(2-ethylhexanoate) [156]. In addition, Jamshidi *et al.* found that the thermal stability of PLLA synthesized using tin(II) bis(2-ethylhexanoate) increases when the terminal carboxyl group is acetylated and that residual monomers enhance the thermal degradation of PLLA [157].

8.7.3 Permeability

First, the permeability of PLA-based materials was investigated with respect to their drug-release profile [158]. In addition, recently, the permeability of various gases through PLA-based materials has been studied since the permeability of PLLA towards water vapour, oxygen and carbon dioxide is important for packaging applications. The effects of the X_c value of PLLA on its water vapour permeability [159–161] have been investigated. Normally, the permeability of gases decreases with crystallinity and degree of orientation. The addition of fillers such as layered silicate is reported to reduce gas permeability [162–165] as is stereocomplexation between PLLA and PDLA [166] and polymer coating [167–169]. The effect of crystallinity depends on the type of gas [170]. The α crystal modification provides a better barrier to water vapour and higher Young's modulus compared to films containing the α'-(δ-)modification; however, it also results in a lower elongation-at-break value [171]. The water-barrier properties decrease in the presence of hydrophilic chitosan [167]. However, the tacticity and molecular weight ($M_n = 9 \times 10^4$–5×10^5 g mol^{-1}) had a very small effect of the water vapour temperature on gas permeability through PLA [10, 160, 172]. The effect of temperature was found to be similar across the glass transition temperature [170].

8.7.4 Surface Properties

Surface treatment effectively alters the nature of the surface without affecting the bulk material properties and can be applied *in situ* after the processing of the materials. Surface hydrophilicity is important when the materials are used for biomedical and pharmaceutical applications. Common methods for altering the surface hydrophilicity of polymers include plasma treatment, surface grafting, and coating [173–175]. However, a facile method involving alkaline treatment can be used to produce hydrolysable biodegradable polyesters including PLA, poly(glycolide) (PGA), and poly(ε-caprolactone) (PCL). This can be attained by the increase in the number of hydrophilic terminal groups, that is, hydroxyl and carboxyl groups, as a result of surface hydrolytic degradation in the presence of alkalis [176]. The hydrophilicity of the surface after alkaline treatment and polymer coating can be manipulated by varying the treatment temperature, treatment time, and the type and concentration of alkalis or secondary polymers [177–179]. Polymer surfaces with contact angles of around 70° exhibit the highest cell adhesion [180]. Therefore, adjusting the hydrophilicity of PLA-based materials by alkaline treatment leads to higher cell affinity for the material and adhesion.

8.7.5 Electrical Properties

Oriented PLLA materials have piezoelectric properties, and the piezoelectric constants of PLLA increase with the draw ratio and are maximized at a draw ratio of around 4–5 [181]. Oriented PLLA materials are reported to be useful for accelerating the regeneration of fractured bones in biomedical applications [181]. This effect is attributed to the piezoelectric current generated by strain caused by body movement. Poly(lactic acid) materials can also be used as electrical insulators. However, the addition of nanostructured carbons such as multiwall carbon nanotubes (MWCNTs) and single-wall carbon nanotubes (SWCNTs) can increase the conductivity of materials; the effectiveness depends on the shape and nature of the nanostructured carbon and the crystallinity of the polymers [182–184]. Of the nanostructured carbons, SWCNTs with the highest aspect ratio of 2000 most significantly decrease the resistivity (ρ) of PLLA materials especially at low additive concentrations (<5 wt%) [184] (Figure 8.13). In marked contrast to conventional carbon black (CB), which required an addition 3 wt% before having an effect, the resistivity decrease was significant even with the addition of merely 1 wt% of SWCNTs. This result is explained by the formation of a conductive network of SWCNTs even at 1 wt%.

8.7.6 Optical Properties

The visible and ultraviolet (UV) light transmittance of PLA (98% LLA) has been studied and compared with that of commercial polystyrene, PET, low density PE, and cellophane films in

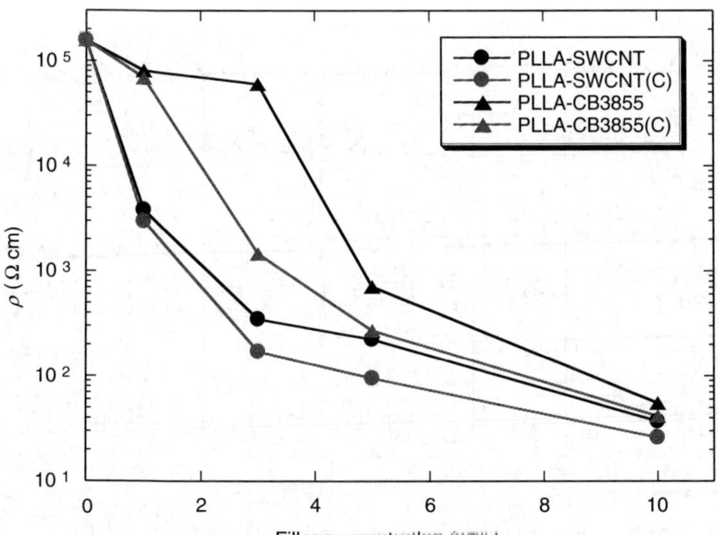

Figure 8.13 Resistivity (ρ) of PLLA-SWCNT and PLLA-CB3855 films at different carbon filler concentrations [184]. (Reproduced with permission from ref. [184]. Copyright © Elsevier, 2007.)

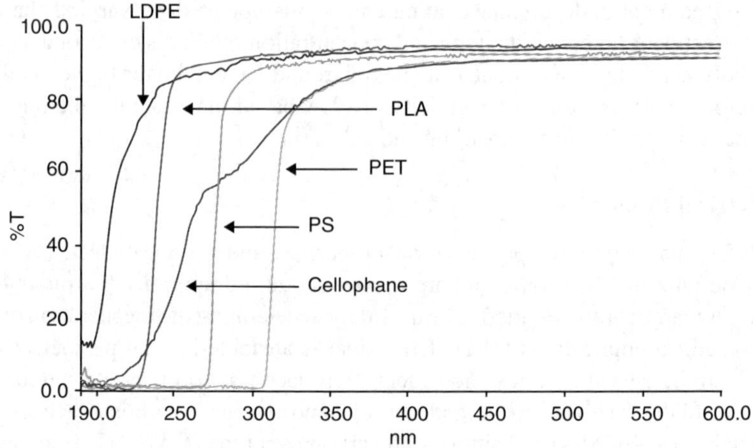

Figure 8.14 Percentage transmission versus wavelength for PLA (98% L-lactide), PS, LDPE, PET and cellophane films [10]. (Reproduced with permission from [10]. Copyright © 2004 Wiley-VCH, Verlag GmbH & Co. KGaA, Weinheim.)

the range of 190–600 nm (Figure 8.14 [10]). Although PLA does not transmit UV-C light, it is transparent to almost all UV-B and UV-A light. Therefore, the application of transparent PLA films to, for example, daily products may require additives to block UV light transmission [10]. The infrared, Raman, and solid-state ^{13}C NMR spectra of PLA and their peak ascriptions are shown in Figure 8.15 and Table 8.2 [185], respectively. Of the peaks, those in the 1250–1420 and 1650–1850 cm^{-1} regions in the Raman spectra are very sensitive to configurational, conformational, and crystallinity differences.

The monomers of L- and D-lactic acids, LLA, and DLA, and their homopolymers, PLLA and PDLA, are optically active. The specific optical rotation ($[\alpha]^{25}{}_{589}$) values of PLLA and PDLA

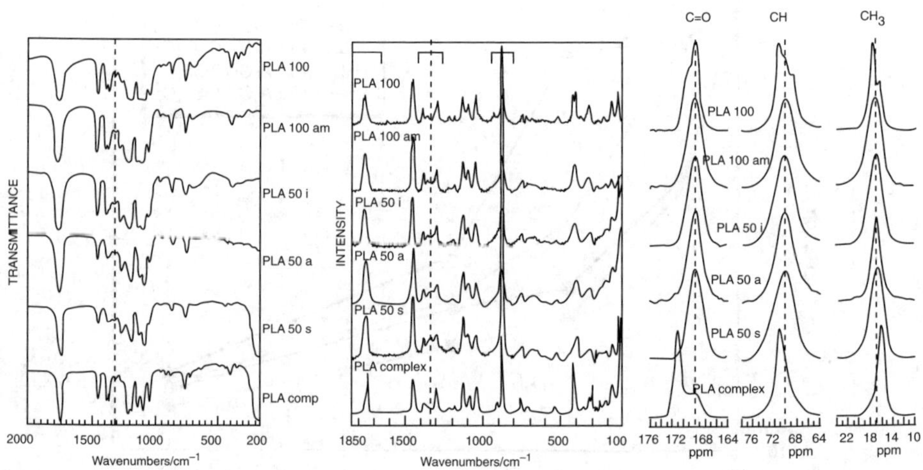

Figure 8.15 Infrared spectra (left), Raman spectra (centre), solid state CP/MAS ^{13}C NMR spectra (right) of poly(L-lactic acid)s: PLA100 (semi-crystalline), PLA100am (amorphous). PLA50i (isotactic), PLA50a (atactic), PLA50s (syndiotactic) and PLAcomplex (stereocomplex) [185]. (Reproduced with permission from [185]. Copyright © Elsevier, 1998.)

Table 8.2 *Wavenumbers (cm^{-1}) and vibrational assignments of semi-crystalline PLLA [185]. (Reproduced with permission from [185]. Copyright © Elsevier, 1998.)*

Raman ν/cm^{-1}	IR ν/cm^{-1}	Assignments
3014 M-3000 sh-2995 S	2997 M	$\nu_{as}CH_3$
2970 sh		$\nu_{as}CH_3$
2960 sh		$\nu_{as}CH_3$
2943 VS	2947 M	ν_sCH_3
2901 w-2877 M	2882 w	νCH
1773 S-1763 S-1769 sh-1749 S	1760 VS	$\nu C{=}O$
1452 S	1452 S	$\delta_{as}CH_3$
1394 sh-1388 M-1382 sh	1348–1388 S	δ_sCH_3
1371 M-1366 M-1353 M	1368 S-1360 S	$\delta_1CH + \delta_sCH_3$
1313 sh-1302 M-1293 M	1315–1300 M	δ_2CH
1270 w	1270 S	$\delta CH + \nu COC$
1216 w-1179 w	1215 VS-1185 VS	$\nu_{as}COC + r_{as}CH_3$
1128 S	1130 S	$\gamma_{as}CH_3$
1092 S-1088 sh	1100 VS-1090 sh	ν_sCOC
1050 sh-1042 S	1045 S	$\nu C - CH_3$
954 vw-920 M	960 w-925 w	$\gamma CH_3 + \nu CC$
873 VS-845 w	875 M-860 sh	νC-COO
760 sh-736 M	760 S-740 sh	$\delta C{=}O$
711 M-677 sh-650 sh	715 M-695 M	$\gamma C{=}O$
578 w-520 M	515 w	$\delta_1C\text{-}CH_3 + \delta CCO$
411 S-398 S	415 sh-400 M	δCCO
347 w	350 M	$\delta_2C\text{-}CH_3 + \delta COC$
325 sh-308 S-300 w	300 M-295 sh	$\delta COC + \delta_2C\text{-}CH_3$
251 w-234 w	240 M	τCC
218 w-206 w		τCC
160 S		skeletal torsion
117 S		skeletal torsion
77 S		skeletal torsion
60 M		skeletal torsion

Notes: VS = very strong; S = strong; M = medium; w = weak; sh = shoulder; s = symmetric; as = asymmetric.

in chloroform were reported to be around –160 and 160 deg·dm^{-1}·g^{-1}·cm^3, respectively [68, 104]. The birefringence (Δn) of PLLA fibres increases from 0.015 to 0.038 with the degree of molecular orientation [28, 132]. The intrinsic birefringence value that was estimated for highly drawn and annealed PLLA fibres containing solely α-form crystals using Stein's formula is 0.03–0.033 [186]. Kobayashi *et al.* found that the gyration tensor component, *g*33, of PLLA along the helical axis is extremely large [$(3.85 \pm 0.69) \times 10^{-2}$], which corresponds to a rotary power of $(9.2 \pm 1.7) \times 10^3$ °/mm and is about two orders of magnitude larger than those of ordinary crystals [28]. A similar rotary power of 7.2×10^3 °/mm was also reported by Tajitsu *et al.* [187].

8.8 Hydrolytic Degradation

Poly(lactic acid)s and their copolymers belong to the family of aliphatic polyesters; therefore, their ester groups are hydrolytically degraded in the presence of water according to the following reaction:

$$-COO- + H_2O \rightarrow -COOH + HO- .$$

Table 8.3 Material and media-related factors which affect degradation behaviour and degradation rate of PLA-based materials [189]. (Reproduced from NOVA.)

Material factors	Media-related factors
1. Molecular structures	1. Temperature
(1) Molecular weight and distribution	2. pH
(2) Tacticity (optical purity) and distribution	3. Solutes (kinds and concentrations)
(3) Comonomer structure, content, and distribution	4. Enzymes (kinds and concentrations)
(4) Terminal groups	5. Microbes (kinds, number, and culture conditions)
(5) Branching	
(6) Crosslinks	6. Stress or strain
2. Highly ordered structures	
(1) Crystallinity	
(2) Crystalline thickness	
(3) Spherulitic size and morphology	
(4) Orientation	
(5) Hybridization (blends and composites)	
3. Material morphology	
(1) Material shape and dimensions	
(2) Porosity and pore size	
4. Surface treatment	
(1) Coating	
(2) Alkaline treatment	

The hydrolytic degradation of PLAs and copolymers has been intensively studied, and, as a result, there are more than 250 papers related to this issue [188, 189]. The degradation mechanism, behaviour, and rate depend on material- and media-related factors, which are summarized in Table 8.3. In addition, the methods used to monitor the hydrolytic degradation of biodegradable polymers are summarized in Table 8.4. Detailed information regarding the hydrolytic degradation of PLA-based materials can be found in the related review articles [188–190].

8.8.1 Degradation Mechanism

8.8.1.1 Molecular Degradation Mechanism

The hydrolytic cleavage of ester groups occurs randomly regardless of the position of the ester group (i.e. chain-end or within the chain) as long as the polymers comprise only one type of

Table 8.4 Indexes (methods) for monitoring degradation [188]. (Reproduced with permission from ref. [188]. Copyright © 2006, John Wiley & Sons, Inc.)

Material-based indexes	Nonmaterial-based indexes
1. Weight remaining (gravimetry)	1. Dissolved organic carbon (DOC) or total organic carbon (TOC)
2. Molecular weight and distribution (gel permeation chromatography (GPC))	2. Biochemical oxygen demand (BOD)
3. Physical properties (differential scanning calorimetry (DSC), tensile testing)	3. Amount of released carbon dioxide
	4. Amount of released biogas
4. Material morphology (scanning electron microscopy (SEM), polarized optical microscopy (POM))	5. pH
	6. Absorbance (turbidity)

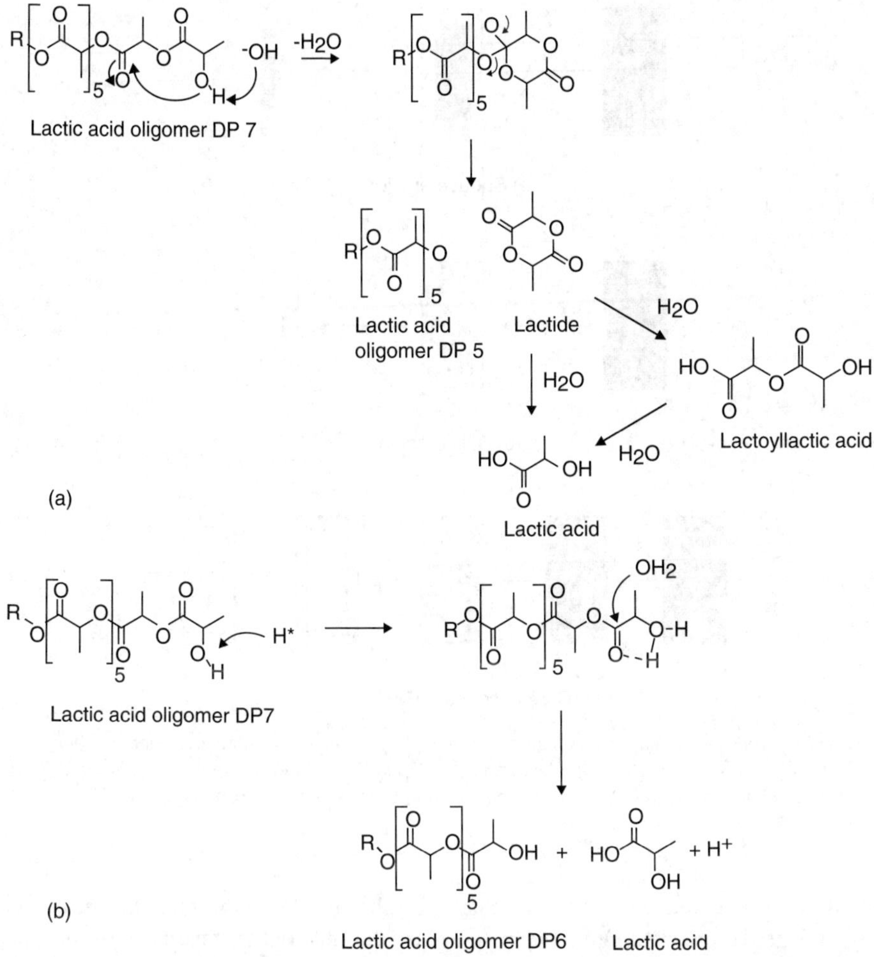

Figure 8.16 *Hydrolytic chain cleavage mechanism of PLLA in alkaline (a) and acidic (b) solutions [192]. (Reproduced with permission from [192]. Copyright © Elsevier, 2011.)*

monomer unit [191]. The specific chain cleavage mechanism of the terminal ester groups in acidic media [192–194], neutral media [195, 196], and alkaline media [192, 195, 197, 198] have been elucidated. De Jong *et al.* reported that the mechanism of the cleavage of PLA in solution depends on the media pH [192]. As shown in Figure 8.16 [192], in acidic solutions, hydrolytic degradation proceeds via chain-end scission of a lactyl monomer unit to form lactic acid, whereas in an alkaline solution, hydrolytic degradation occurs via backbiting to form an LA (lactyl dimer), which is further hydrolysed to give lactoyllactic acid.

8.8.1.2 *Material Degradation Mechanism*

Surface erosion occurs when the hydrolytic degradation rate of the material surface in contact with water containing catalytic substances such as alkali or enzymes is much higher than the

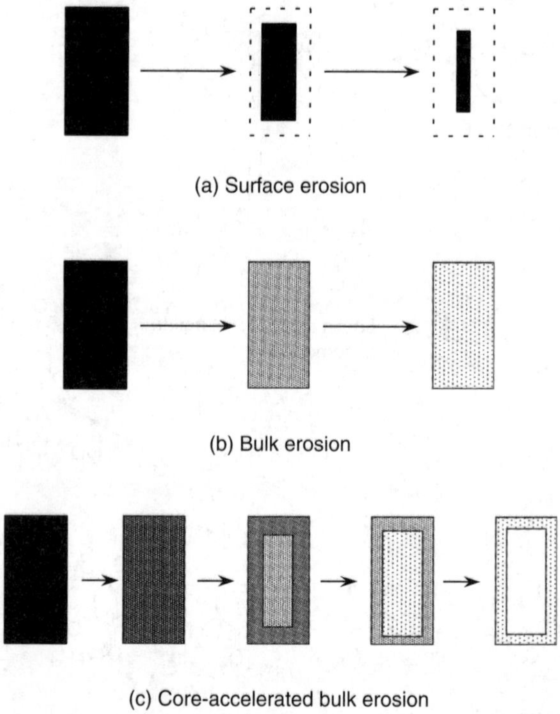

Figure 8.17 *Hydrolytic degradation mechanisms of bulky PLA materials: (a) surface erosion, (b) bulk erosion, (c) core-accelerated bulk erosion [8, 199]. (Reproduced with permission from ref. [8] © 2002, Wiley-VCH Verlag GmbH & Co. KGaA and from ref. [199], with permission from Corona Publishing Co., Ltd., Tokyo.)*

diffusion rate of the water molecules or catalytic substances into the material. In this case, hydrolytic degradation seemingly occurs solely on the material surface because the rate is much higher on the surface than in the core; therefore, the core remains intact (Figure 8.17(A)) [8, 199]. In contrast, a bulk erosion mechanism occurs when the hydrolytic degradation rate of the material surface is much lower than the diffusion rate. In this case, hydrolytic degradation occurs homogeneously and irrespectively of the depth from the material surface (Figure 8.17(B)). Material thickness is an important factor in determining the hydrolytic degradation mechanism. As can be expected, the hydrolytic degradation mechanism changes from the bulk erosion mechanism to the surface erosion mechanism when the material thickness is higher than the critical thickness ($L_{critical}$), as reported in the literature (Table 8.5) [200]. Core-accelerated erosion in neutral media was reported for PDLLA [201], PLLA [202,203], poly(L-lactide-*co*-glycolide) [i.e. poly(L-lactic acid-*co*-glycolic acid), which is abbreviated as P(LLA-*co*-GA)] [204], P(DLLA-*co*-GA) [204], PDLLA-*b*-PEO [205], and PLLA-*b*-PEO-*b*-PLLA [206] when the material is thicker than 0.5–2 mm. In such cases, oligomers and monomers that are formed via hydrolysis and are highly catalytic become trapped and accumulate in the core of the materials [207]. This results in accelerated hydrolytic degradation in the core, which is known as core-accelerated bulk erosion [Figure 8.17(C)]. Therefore, the hydrolytic degradation of PLA proceeds via bulk, core-accelerated, and surface-erosion mechanisms for materials with thicknesses <0.5–2 mm, between 0.5–2 mm and 7.4 cm, and higher than

Table 8.5 Critical thickness ($L_{critical}$) of biodegradable polyesters where hydrolytic degradation mechanism changes from bulk erosion to surface erosion [200]. (Reprinted from Biomaterials, vol. 23, von Burkersroda et al., Why degradation polymers undergo surface erosion or bulk erosion, pp. 4221–4231. Copyright © Elsevier, 2002.)

Polymer	Molecular structure / Example	$L_{critical}$
Poly(anhydride)	(–R–CO–O–CO–)$_n$	75 μm
Poly(ketal)	(–O–CR^{1}R^2–O–R^3–)$_n$	0.4 mm
Poly(ortho ester)	[-O-CR^1(OR^2)-O-R^3-]$_n$	0.6 mm
Poly(ε-hydroxycarboxylic acid)	[–O–(CH$_2$)$_5$–CO–]$_n$Poly(ε-caprolactone)	1.3 cm
Poly(acetal)	(–O–CHR^1–O–R^2–)$_n$	2.4 cm
Poly(α-hydroxycarboxylic acid)	[–O–CH(CH$_3$)–CO–]$_n$Poly(lactic acid), Poly(lactide)	7.4 cm
Poly(amide)	(–NH–R–CO–)$_n$	13.4 m

7.4 cm, respectively. In crystallized PLAs or copolymers, the chains in the crystalline regions are more resistant to hydrolysis than those in the amorphous regions because access of water molecules to the chains inside the rigid crystalline regions is impaired. This causes selective or predominant hydrolytic cleavage of chains in the amorphous regions and the removal of hydrolysis-formed water-soluble oligomers and monomers. As a result, the crystalline regions remain intact; such crystalline regions are called 'crystalline residues' [199, 208].

8.8.2 Effects of Surrounding Media

Of the parameters of the surrounding media shown in Table 8.3, pH and temperature are discussed here.

8.8.2.1 pH

The hydrolytic degradation of the ester groups of PLA and its copolymers is catalysed by the presence of hydronium and hydroxide ions. High concentrations of hydroxide ions in alkaline media significantly accelerate the hydrolytic degradation of PLA-based materials [197, 198, 209–215]. As in the case of alkaline media, acidic media accelerate the hydrolytic degradation of PLA-based materials [209, 210, 216]. However, as Mason *et al.* and Tsuji and Nakahara reported, hydronium ions have a negligible or very low catalytic effect on the hydrolytic degradation of PLA-based materials compared to that of alkalis when the molecular weight of PLA is higher than 1×10^5 g mol^{-1} [217–219].

8.8.2.2 Temperature

The hydrolytic degradation temperature (T_h) can be divided into three ranges: (1) $T_h < T_g$, (2) $T_g \leq T_h < T_m$, and (3) $T_m \leq T_h$. It has been reported that the neutral hydrolytic degradation (NHD) rate, as assessed by changes in the mechanical properties, dramatically increases when the degradation temperature is higher T_g [220]. The hydrolytic degradation mechanism of crystallizable PLA-based materials varies when the degradation temperature is higher than T_m, that is, when the materials are in the melt. At a T_h exceeding T_m, the crystalline regions melt and hydrolytic degradation takes place homogenously in the melt, as it does in noncrystallizable PDLLA. This causes rapid hydrolytic degradation above the T_m (Figure 8.18 [221]). Figure 8.18 shows the Arrhenius plot of the hydrolytic degradation rate

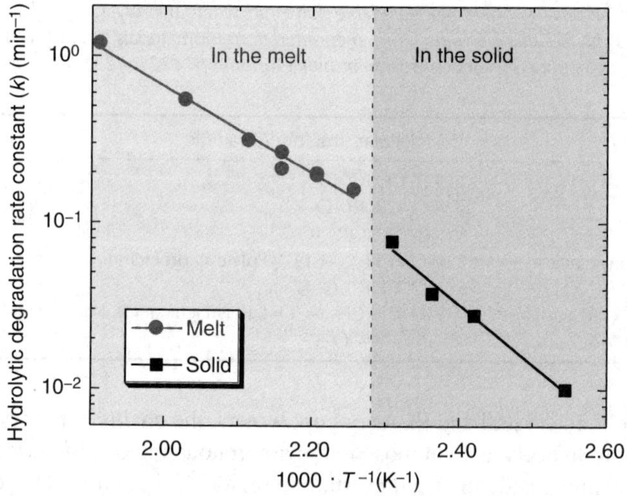

Figure 8.18 Arrhenius plot of k for PLLA [221]. (Reproduced with permission from ref. [221]. Copyright © Elsevier, 2008.)

constant (k) of PLLA in the melt and in the solid [221], as estimated from the following equation [222,223]:

$$\ln M_n(t_2) = \ln M_n(t_1) - k(t_2 - t_1). \tag{8.5}$$

The reported ΔE_h values and equations for k with respect to the degradation temperature for PLLA are summarized in Table 8.6 [224] along with those for other biodegradable aliphatic polyesters and aromatic PET [225]. The obtained ΔE_h value (50.9 kJ mol^{-1}) for PLLA in the melt in the temperature range of 180–250 °C [221] is lower than those of aliphatic polyesters, poly(butylene succinate) (PBS) and poly(butylene succinate / adipate) (PBSA) in the temperature range of 180–220 °C [224], PCL in the temperature range of 175–250 °C [226], poly[(R)-3-hydroxybutyrate], that is, poly[(R)-3-hydroxybutyric acid] (P(R-3HB)) in the temperature range of 180–220 °C [227], and PET in the temperature range of 250–280 °C [225]. This is a further indication that PLLA has the highest hydrolysability among the biodegradable aliphatic polyesters.

8.8.3 Effects of Material Parameters

Of the material parameters shown in Table 8.3, the molecular weight, copolymerization, crystallinity, and polymer blending/composites/additives are discussed here.

8.8.3.1 Molecular Weight

Figure 8.19 shows the changes in the M_n of PLLA with different molecular weights and comonomers including DLA, GA, and CL (i.e. P(LLA-*co*-DLA), P(LLA-*co*-GA), and P(LLA-*co*-CL) copolymers, respectively) [188, 228–230]. It is evident that the effect of copolymerization was dramatic, whereas that of M_n was insignificant or very small, in the range of 8×10^4 – 4×10^5 g mol^{-1} [228, 231]. However, at molecular weights below 4×10^4 g mol^{-1}, its effect

Table 8.6 *The activation energy of hydrolytic degradation (ΔE_h) in the melt (and in the solid only for PLLA (120–160 °C)) and hydrolytic degradation temperature dependence of hydrolytic degradation rate constant (k) values [224]. (Reproduced with permission from ref. [224]. Copyright © 2008 Wiley-VCH Verlag GmbH Co. KGaA, Weinheim.)*

Monomer type	Polyester	Method	Temperature range (°C)	ΔE_h (kJ mol^{-1})	ΔE_h (kcal mol^{-1})	k (min^{-1})
Aliphatic hydroxycarboxylic acid	PLLA [221]	M_n	120–160	69.6	16.6	$4.17 \times 10^8 \exp(-8.71 \times 10^3\, T^{-1})$
	PLLA [223]	M_n	180–250	50.9	12.2	$1.39 \times 10^5 \exp(-6.12 \times 10^3\, T^{-1})$
	P(R-3HB) [227]	M_n	180–220	126	30.0	$8.86 \times 10^{12} \exp(-1.51 \times 10^4\, T^{-1})$
	PCL [226]	M_n	175–250	72.2	17.3	$3.95 \times 10^6 \exp(-8.69 \times 10^3\, T^{-1})$
Aliphatic dicarboxylic acid / Aliphatic diol	PBS [224]	M_n	180–220	62.0	15.3	$1.41 \times 10^6 \exp(-7.70 \times 10^3\, T^{-1})$
	PBSA [224]	M_n	180–220	58.0	13.8	$2.79 \times 10^5 \exp(-6.97 \times 10^3\, T^{-1})$
Aromatic dicarboxylic acid / Aliphatic diol	PET [225]	Liquid water concentration	250–280	233	55.7	—

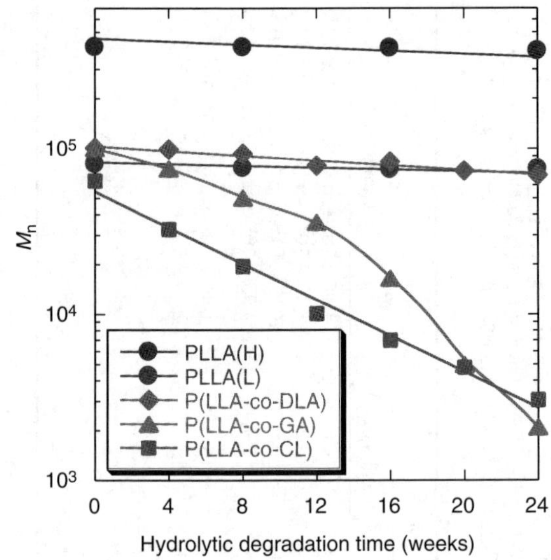

Figure 8.19 *M_n change of amorphous PLLA having high and low molecular weights [PLLA(H) and PLLA(L), respectively], poly(L-lactide-co-D-lactide) [P(LLA-DLA)] (77/23), poly(L-lactide-co-glycolide) [P(LLA-GA)] (81/19), and poly(L-lactide-co-ε-caprolactone) [P(LLA-CL)] (82/18) with respect to hydrolytic degradation in a phosphate-buffered solution [188, 228–230]. (Reproduced with permission from [188]. Copyright © 2006, John Wiley & Sons, Inc., and [229] Copyright © 2006, Copyright © 2006 WILEY-VCH Verlag GmbH Co. KGaA, Weinheim and [230] Copyright © 2006, Elsevier BV.)*

was reported to be significant for PDLLA [232, 233]. The latter result can be explained by four factors that are caused by a decrease in molecular weight: (i) elevated molecular mobility, (ii) increased density (number per unit mass) of hydrophilic terminal carboxyl and hydroxyl groups, (iii) increased density of catalytic terminal carboxyl groups, and (iv) higher probability of the formation of water-soluble oligomers and monomers upon hydrolytic degradation. Factors (i) and (ii) increase the diffusion rate and content of water, which enhances hydrolytic degradation. Factors (i)–(iv) should have a dramatic effect on PLA at molecular weight lower than 1×10^4 g mol^{-1}.

8.8.3.2 Copolymerization

Of the lactone comonomers, GA and CL units are frequently used to increase the hydrolytic degradation rate of PLA-based copolymers [188], as shown in Figure 8.19. Normally, the hydrolytic degradation rate increases with an increase in the hydrophilicity of the comonomer unit and the deviation of the concentration of the LA unit from 0 or 100% [188]. Essentially, the highest degradation rate will be attained at some comonomer concentration between 0 and 100% depending on the nature of the second monomer unit. In the case of P(LLA-CL), the maximum degradation rate, as determined by the molecular weight change, was observed at about 60% LA and 40% CL [234, 235]. In the case of block copolymers of PDLLA-*b*-PCL, the NHD rate, which was assessed by the viscosity and weight loss, increased with an increase in PDLLA content [236]. The incorporation of δ-valerolactone [237] and 3-methyl-4-oxa-6-hexanolide (MOHEL) [238] in PLLA have been reported to decrease the NHD rate.

Block copolymers are frequently synthesized by the polymerization of LLA, DLLA, LA/GA, and other combinations of LA with comonomers in the presence of hydrophilic polyethers and their copolymers such as poly(ethylene glycol) (i.e. poly(ethylene oxide) (PEO)), PPO, Pluronic, or poly(succinic anhydride-co-ethylene oxide) [8, 188]. Block copolymers including PEO can be used as matrices for drug release systems for hydrophilic drugs. During hydrolytic degradation, the high hydrophilicity of PEO accelerates the diffusion of water into the material, which increases its water content resulting in enhanced cleavage of the ester linkage. Concomitantly, the release of water-soluble PEO causes a rapid decrease in weight. Therefore, the incorporation of PEO increases the degradation rate of PLA-based homopolymers and copolymers, as determined by molecular weight changes and weight loss. Recently, homopolymers of various substituted DL-lactic acids were synthesized and their hydrolytic degradation behaviour was investigated [239]. It was determined that hydrolytic degradation had a complicated dependence on the carbon number of the side chains due to variations in steric hindrance, hydrophilicity, and interchain interactions.

8.8.3.3 Crystallinity

The crystallized PLA-based materials comprise two regions, that is, crystalline and amorphous regions, having different resistances to hydrolytic degradation, thus leading to different hydrolytic degradation rates. The chains in the crystalline regions are more resistant to hydrolysis than those in the amorphous regions. Therefore, alkaline and enzymatic hydrolytic degradation rates decrease with an increase in crystallinity (X_c) [211,212, 240–242]. However, for the NHD of PLLA [203, 243–246], PLLA/PDLA blends [247], and P(LLA-*co*-GA) [248] in neutral media and the *in vivo* degradation of PLLA in the human body [249–251], the hydrolytic degradation rate, which was assessed by changes in the weight loss, molecular weight, and mechanical properties, increases with an increase in X_c. A probable explanation is as follows: Upon crystallization of PLLA, the hydrophilic terminal groups (–OH and –COOH) and the catalytic terminal group (–COOH) condense in the amorphous region between the crystalline regions [199, 245]. The high densities of terminal groups cause looser chain packing in the amorphous region sandwiched between the crystalline regions than in completely amorphous film. The combination of loose chain packing and high densities of hydrophilic terminal groups enhances the diffusion of water molecules and increases the water content. The elevated water supply rate and content and the increase in the catalytic effect due to the high density of carboxylic groups synergize to significantly accelerate the hydrolytic degradation of crystallized PLLA films. Moreover, the terminal groups, which serve as sites for the consecutive formation of LA and lactic acid, condense in the amorphous region between the crystalline regions; therefore, the lactic acid formed should catalyse hydrolytic degradation in a restricted area of the amorphous region. This further enhances the synergistic effect mentioned above. However, according to a report by Li *et al.* [202], crystallized PLLA had a higher NHD rate than amorphous PLLA when degradation was monitored by L-lactic acid release, whereas the trend is reversed when it is monitored by weight-loss and tensile properties. In addition, for P(LLA-*co*-GA) (10/90), a higher X_c delayed the NHD, as assessed by weight loss [252].

8.8.3.4 Polymer Blending/Composites/Additives

Polymer blending is commercially advantageous to prepare PLA-based materials with a wide variety of physical and hydrolytic degradation properties. There are five primary factors of a second polymer or an additive that are proposed to influence on the rate, behaviour, and mechanism of the hydrolytic degradation of PLA-based materials in neutral media: (i) miscibility and dispersibility, (ii) hydrophilicity, (iii) acidity or basicity, (iv) molecular weight and distribution, (v) size and its distribution, and shape of the polymer domain (if the second polymer is immiscible with the first polymer) or additives. For example, the addition of hydrophilic polymers such as poly(vinyl alcohol) (PVA) [253, 254], PEO [255], PGA [256], or hydrophilic or catalytic compounds such as thioridazine [257], hydroxyapatite [258], β-tricalcium phosphate [259], lauric acid [260, 261], oligomeric PDLLA ($M_n = 2.4 \times 10^3$ g mol^{-1}) [262], and LA (either *in vitro* in a phosphate-buffered solution [233] or *in vivo* [249]) is known to enhance the NHD of PLLA. In contrast, the incorporation of enantiomeric PDLA decreases the hydrolytic degradation rate of PLLA [11, 247, 263–267]. In PLLA/PDLA blends, the interaction between PLLA and PDLA chains is much more significant than that between PLLA chains or PDLA chains, which results in the formation of a stereocomplex [11]. These strong interactions impair the diffusion of water into the material and lower the hydrolytic degradation rate regardless of whether the stereocomplex is formed in the blends or not [11, 247, 263–267]. Analysis via molecular modelling ascribed the higher hydrolytic resistance of PLLA/PDLA blends than that of P(LLA-DLA) stereocopolymers with the same L-lactyl unit content to the fact that blends form more stereocomplexes than stereocopolymers [268]. Initiators (or catalysts) such as zinc metal and derivatives or unreacted LA [269] that remain after polymerization can be regarded as additives and are reported to increase hydrolytic degradation [270]. Numerous reports of the effects of inorganic compounds have been reviewed in a published book chapter [188].

8.9 Thermal Degradation

As discussed below, the thermal degradation process at temperatures below 300 °C can be exploited to recycle PLA-based materials to LA. Volatile components, such as LAs, cyclic oligomers ranging from trimers to hexamers, CO, CO_2 acetaldehyde, and methylketene, are formed upon heating PLA above 250 °C [271]. Figure 8.20 [272] shows the probable nonradical reactions that occur during PLA thermal degradation [271]. Lactide and cyclic oligomers are formed by intramolecular transesterification reactions, (a) and (b). It is anticipated that intermolecular transesterification reaction (c) does not alter the overall molecular weight distribution of PLLA. In contrast, chain cleavage via hydrolytic degradation reaction (d) and cis-elimination reaction (e), which is similar to the Norrish II-type photodegradation reaction, and reduce the molecular weight and alters the terminal group structure.

The thermal stability and degradation rate of PLA and the formation rate, total yield and enantiomeric fractions of LA depend on a variety of factors: (i) the molecular structure, (ii) the type and concentration of initiator, co-initiator, catalyst, and additive, (iii) the concentration of LA and water in PLA, and (iv) the method and reaction conditions including the pressure of the surrounding gas, temperature, and time [157, 273–278].

Table 8.7 shows the fractions of LLA, DLA, and MLA formed by the thermal degradation (i.e. depolymerization) of PLLA at different temperatures and reaction times [272]. The yields of LA and LLA increase with an increase in degradation time within a degradation temperature range of 250–290 °C and become saturated at ~12 and 6%, respectively, when degradation proceeds for over 10 hours. The initial LA formation rate and the saturation

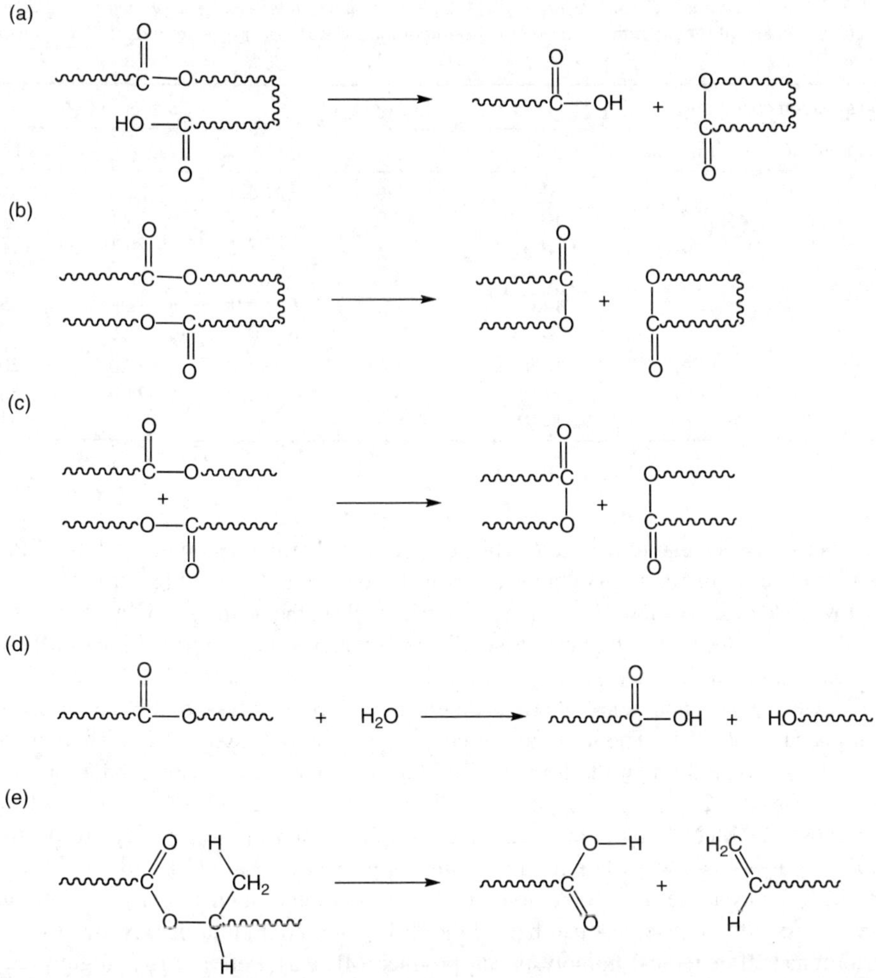

Figure 8.20 *Probable pathways for nonradical thermal degradation: formation of lactide and cyclic oligomers by intramolecular transesterification reactions (a, b), intermolecular transesterification reaction (c), hydrolytic degradation reaction (d), and cis-elimination reaction (e) [272]. (Reproduced with permission from ref. [272]. Copyright © Elsevier, 2003.)*

yield of LA increase with an increase in degradation temperature. The higher yield of LA at higher degradation temperatures implies that the formation of LA from PLLA is endothermic [279]. The highest yields of LA (14%) and LLA (8%) in Table 8.7 are much lower than those obtained by a small-scale distillation system (89% and 88%, respectively) [274] in which the LA formed was removed from the reactor by evaporation. In large-scale industrial plants for recycling PLLA to LA, most of the PLLA is exposed to closed-system conditions similar to this experiment; therefore, the concentrations of LA and LLA are much lower than those in a small-scale distillation system due to the highly viscous and bulky PLLA melt, which disturbs the diffusion of LA or LLA from the PLLA melt. Consequently, large-scale industrial plants

Table 8.7 Yields of LA and LLA, and fractions of LLA, DLA, and MLA formed as a result of thermal degradation of PLLA films at different temperatures and times [272]. (Reproduced with permission from ref. [272]. Copyright © Elsevier, 2003.)

Degradation conditions		Yield of LA	Yield of LLA	Fractions of LA		
Temperature (°C)	Time (h)	(%)	(%)	LLA (%)	DLA (%)	MLA (%)
250	5	3.9	3.6	90.6	1.0	8.4
	10	10.0	8.0	79.5	3.1	17.4
	15	9.4	5.4	57.2	10.1	32.7
270	5	8.9	5.9	66.4	6.9	26.7
	10	14.0	7.3	52.0	12.6	35.4
	15	10.6	4.3	40.7	19.8	39.5
290	5	10.8	5.6	52.1	27.0	20.9
	10	11.9	4.9	42.4	24.3	33.3
	15	12.9	6.1	47.2	20.9	31.9

have numerous issues that remain to be solved; however, thermal depolymerization of PLLA using a small-scale distillation system is effective for obtaining LLA at high yields.

The low yield of LLA shown in Table 8.7 is attributable to the formation of large amounts of MLA and DLA, especially at higher degradation temperatures. For the formation of MLA and DLA, configurational inversion at the asymmetric carbon atoms in the PLLA chains should occur. However, in a nonradical thermal degradation process, this configurational change does not occur [280, 281]. The inversion of configuration at the asymmetric carbon atoms is expected to transpire through the formation of carbon atoms with an sp^2 orbital via radical homolysis (Figure 8.21(a) and (b)) and by enolization (Figure 8.21(c)) [272]; these pathways were proposed by McNeill and Leiper [280] and Kopinke and Mackenzie [281], respectively. Radical homolysis pathways contribute significantly even at 230 °C [280]. As shown in Figure 8.21, the asymmetric carbon atoms with asterisks retain their initial L-configuration, whereas those without asterisks can have either a D- or L-configuration. Cyclization at the chain terminals after radical homolysis can produce MLA as well as LLA (Figure 8.21(a) and (b)]. It should be noted that terminal carbon atoms with sp^2 orbitals are reformed after LA production (Figure 8.21(a) and (b)) and can form another MLA via cyclization. After configurational inversion by enolization (Figure 8.21(C)), both MLA and DLA can form even if nonradical LA formation occurs. However, for the formation of DLA, the configuration of two sequential asymmetric carbon atoms should be inverted by either double enolization or by single enolization with subsequent radical homolysis at the neighbouring carbon atom and vice versa. The lower probability of configurational inversion at two sequential carbon atoms compared to that at only one carbon atom must be the cause of the lower fraction of DLA than MLA excluding the result at 290 °C and 5 min (Table 8.7). Moreover, the increased fractions of MLA and DLA in LA at higher temperatures strongly suggest that the contribution of the radical homolysis pathways and/or the enolization pathway increases with increasing degradation temperature.

As stated above, the hydrolytic degradation resistance of PLA-based materials was enhanced by stereocomplex formation. However, even in the molten state (i.e. above the T_m of the stereocomplex crystallites), the PLLA/PDLA blend has a higher thermal resistance than neat PLLA or PDLA [275, 276].

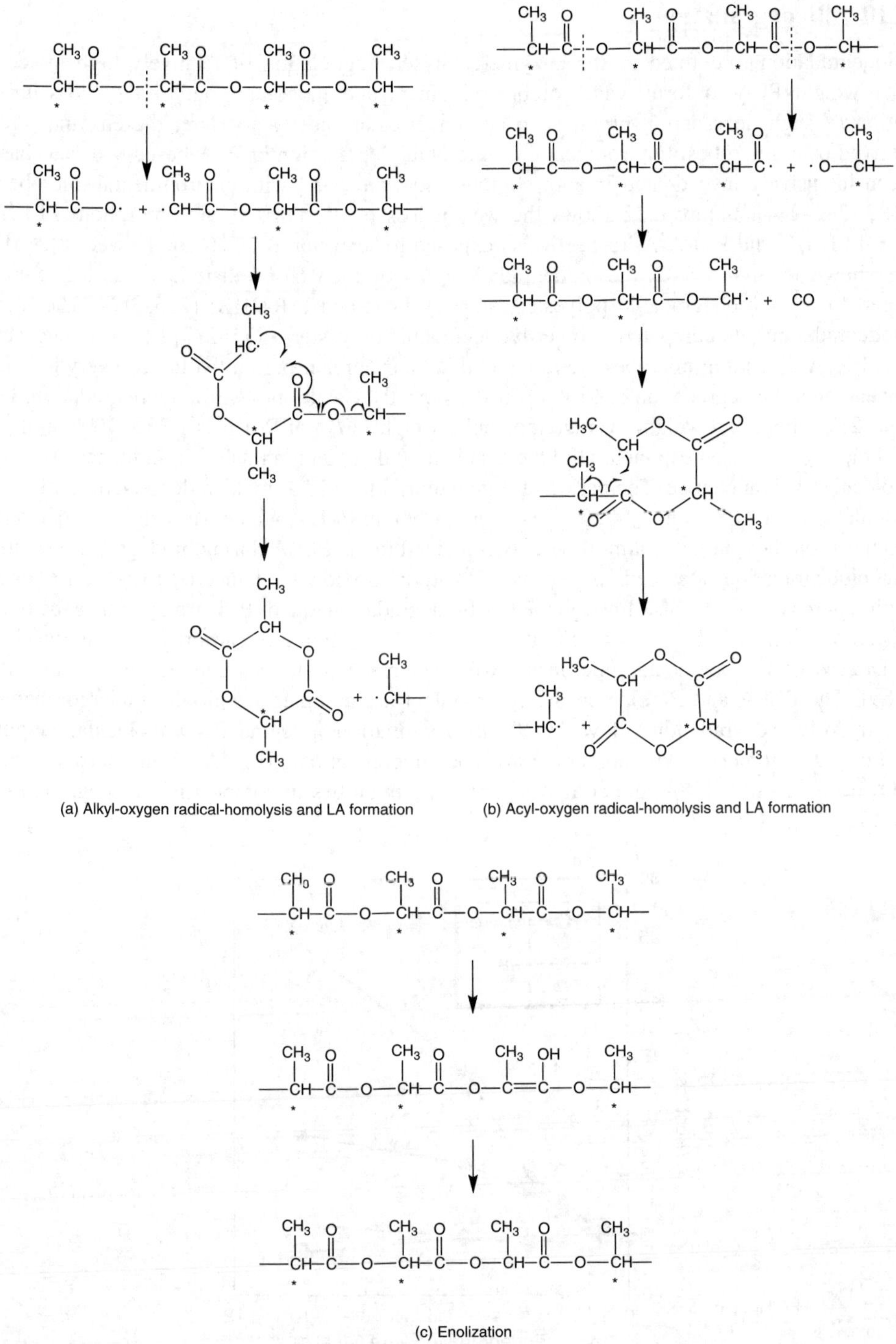

Figure 8.21 *Probable pathways for racemization during LA formation via radical pathways at asymmetric carbon atoms in PLLA: (a) Alkyl-oxygen radical-homolysis and LA formation; (b) Acyl-oxygen radical-homolysis and LA formation; (c) Enolization [272]. (Reproduced with permission from ref. [272]. Copyright © Elsevier, 2003.)*

8.10 Biodegradation

Biodegradation is defined as the enzyme-catalysed degradation of relatively high molecular weight PLA to form water-soluble oligomers and monomers and/or the bioassimilation of such degradation products to carbon dioxide and water. Here, the enzymes are derived from microbes. Environmental or microbial degradation of PLA-based materials has been intensively investigated in soil, compost, seawater, and with environmental microbes [189, 282–284]. Figure 8.22 shows the weight loss of crystallized and amorphous PLLA (i.e. PLLA-C and PLLA-A, respectively) exposed to seawater at 25 °C for 10 weeks [285]. As shown in Figure 8.22, the biodegradation rate of PLA-based materials was low compared to other biodegradable polyesters such as PCL and P(R-3HB) [285, 286]. The low biodegradation rate compared to other biodegradable polyesters [285–287] strongly suggests that PLLA-degrading microbes are rare and that their number per unit volume is very low. It appears that the degradation of PLLA proceeds mostly via abiotic hydrolytic degradation. In soil [287], the initial weight-average molecular weight (M_w) of PLLA is 1.33×10^6 g mol^{-1} and the M_w decreased exponentially; the exponential degradation rate is not influenced by the molecular weight change. The period of time required for PLLA to degrade to lactic acid was calculated to be 508 months (42.3 years). The carbon in PLLA, which was originally derived from carbon dioxide in the atmosphere, is contained in the PLLA during biodegradation. The low biodegradation rate of PLLA means that carbon dioxide is retained by PLLA for a long period in the environment. Although the low biodegradation rate of PLLA may reduce the rate of carbon cycling in the environment, it also reduces the amount of carbon dioxide in the air.

Dozens of environmental microbes have been reported to degrade PLLA or PDLLA [189]. The PLLA and PDLLA in the environment are assumed to initially undergo chemical hydrolytic degradation followed by the bioassimilation of degraded low molecular weight lactic acid oligomers and monomers [288]. The mineralization of PLLA by microbes to form CO_2 occurs in soils [289] and compost [290]. The microbes in compost induce degradation

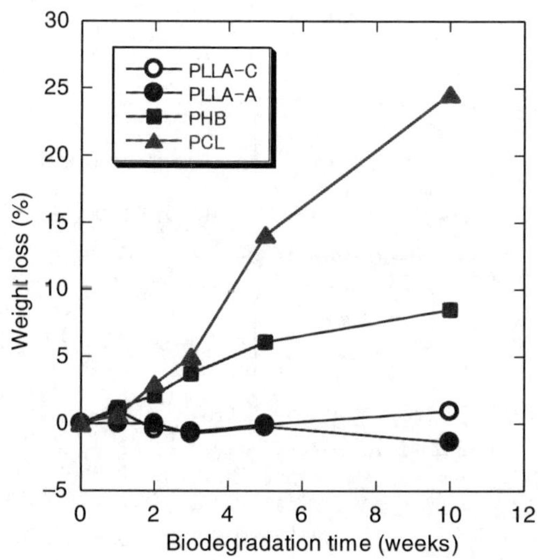

Figure 8.22 Weight loss of crystallized PLLA (PLLA-C), amorphous PLLA (PLLA-A), R-P(3HB), and PCL exposed to seawater at 25 °C [285]. (Reproduced with permission from ref. [285]. Copyright © Elsevier, 2002.)

near the chain ends of a PLLA material, which is preferred, and form lactic acid and lactoyl lactic acid [291]. Via this process, the ethyl ester of lactoyl lactic acid was formed. Jarerat *et al.* revealed that the presence of amino acids, peptides, and poly(L-amino acid) enhances microbial biodegradation of PLLA [292]. Poly(L-lactic acid) modified with peroxide has a lower degradation rate in soil [293]. A recent finding that PDLA was degraded by an environmental microbe [294] indicates that PLA stereocomplexed materials composed of PLLA and PDLA that are released into the environment will be degraded by environmental microbes. In seawater, the stress or strain from waves and currents causes the mechanical destruction of biodegradable aliphatic polyesters [286]. This effect was intense for glassy materials with high T_g values such as PLLA. The environmental or microbial degradation of biodegradable polyesters can be enhanced by pore formation, which increases the degradable surface area per unit mass [13, 295, 296] and alkaline surface treatment [295, 297], which induces the cleavage of ester groups on the film surface resulting in an increased number of carboxyl and hydroxyl groups. The increased surface hydrophilicity of biodegradable polyesters will attract microbes and elevate the rate of biodegradation.

8.11 Photodegradation

In the environment, PLA-based materials are exposed to biodegradation, hydrolytic degradation, and photodegradation by ultraviolet (UV) light. However, there have been few reports on the photodegradation of PLA-based materials. Ultraviolet light induced chain cleavage of PLA and PCL is reported to proceed via the Norrish II mechanism [298]. The Norrish II mechanism of PLLA and PCL is shown in Figure 8.23 [299]. From this figure, it is evident that PLLA has a higher number of ester groups per unit mass than PCL and is therefore expected to have a higher photodegradation rate. Figure 8.24 shows the change in M_n of crystallized PLLA (PLLA-C, $X_c = 34.6\%$), amorphous PLLA (PLLA-A, $X_c = 1.0\%$), and PCL ($X_c = 56.9\%$) films by photodegradation [299]. The photodegradation rate constant (k) can be calculated from Figure 8.24 according to Equation 8.5. The k values for PLLA-C, PLLA-A, and PCL films were 1.56×10^{-3}, 3.69×10^{-3}, and 3.50×10^{-3} h^{-1}, respectively. The k of the PLLA-A film was higher than that for the PLLA-C film. This indicates that the chains in the crystalline regions are more resistant to photodegradation than those in the amorphous region. Furthermore, despite the high number of ester groups per unit mass and the lower X_c value of the PLLA-C film compared to those of the PCL film, the k value of the PLLA-C film was lower than that for the PCL film. This reveals that the number of ester groups per unit mass is not the dominant factor for determining the photodegradation rate but the molecular structure adjacent to the ester group is crucial. It is also probable that a mechanism other than the Norrish II-type plays an important role in the photodegradation of biodegradable polyesters.

The accelerated photodegradation of PLLA in the presence of *N,N,N,N*-tetramethyl-1,4-phenylenediamine (TMPD) as a photosensitizer was exploited by Sakai *et al.* [300, 301]. The result for photodegradation of PLLA with TMPD indicated that the chains in the crystalline regions can be cleaved by UV irradiation in the early stage, which is in marked contrast to the results of the hydrolytic degradation investigations [302]. Ultraviolet irradiation is known to alter hydrolytic degradation behaviour and rate [303–306]. The decrease of the molecular weight of PLLA during hydrolytic degradation is reported to be accelerated by simultaneous irradiation with UV light [303]. Ho *et al.* studied the effects of sunlight and rain on the degradation of PLLA and PDLLA in banana fields [304]. During the early stage, hydrolytic degradation and proteinase K-catalysed enzymatic degradation were enhanced by

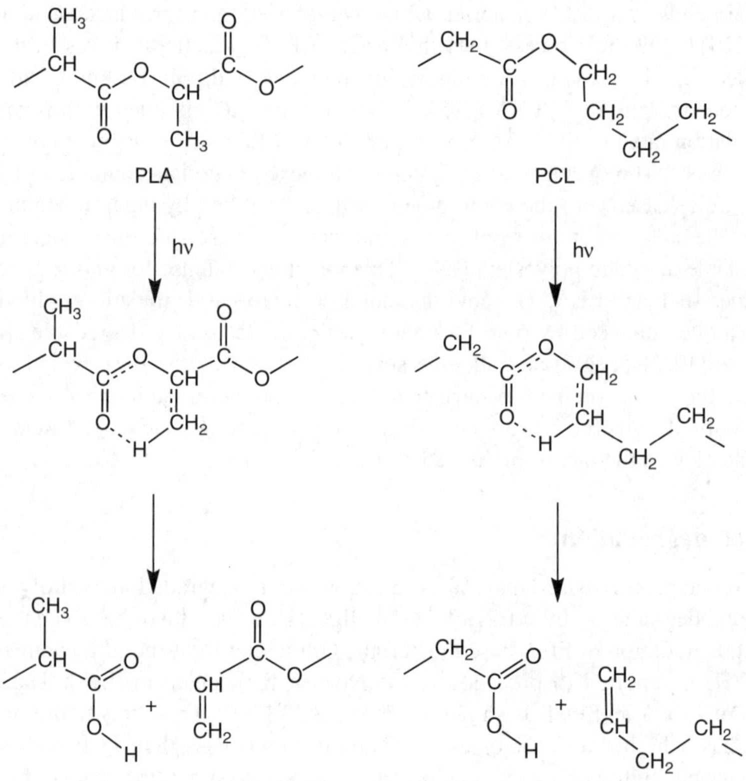

Figure 8.23 Photodegradation of PLA and PCL via Norrish II mechanism [299]. (Reproduced with permission from ref. [299]. Copyright © Elsevier, 2006.)

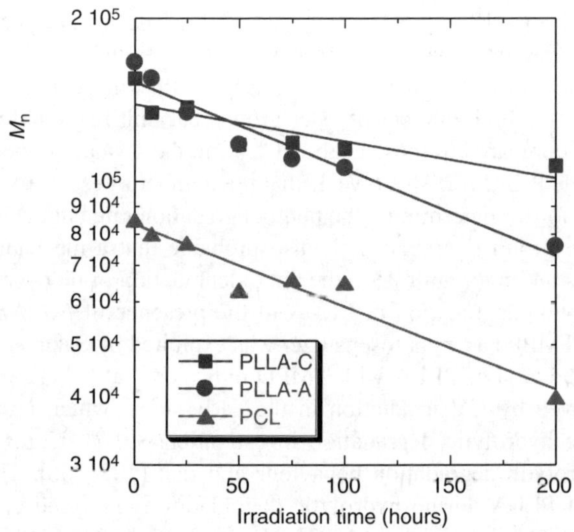

Figure 8.24 M_n of crystallized PLLA (PLLA-C), amorphous PLLA (PLLA-A), and PCL films as a function of UV-irradiation time [299]. (Reproduced with permission from ref. [299]. Copyright © Elsevier, 2006.)

UV irradiation, while they were disturbed at a later stage [305, 306]. The acceleration during the early stage is due to the lower molecular weight of PLLA caused by the chain cleavage of UV irradiation, whereas the disturbed degradation at a later stage is ascribed to the formation of double bonds and/or crosslinks by UV irradiation.

8.12 High-Performance Poly(Lactic Acid)-Based Materials

Effective methods to improve the thermal/hydrolytic degradation resistance of PLA-based materials include (i) increasing the crystallinity during processing by using nucleating or crystallization-accelerating agents, (ii) composite or nanocomposite formation with appropriate fillers, and (iii) stereocomplexation. In the following sections, these three methods are discussed. The additives or fillers are classified into subgroups: (i) on the basis of their origin (biobased and nonbiobased) and (ii) on the basis of biodegradability (biodegradable and nonbiodegradable). Biodegradable agents or fillers are required to prepare biodegradable PLA-based materials.

8.12.1 Nucleating or Crystallization-Accelerating Fillers

Most fillers (agents) either accelerate or decelerate the crystallization of polymers; numerous fillers fall into the former category. Representative fillers for PLA are discussed in this section. The effects that accelerate crystallization are divided into two groups: (i) the nucleating or nucleating-assisting effect, which increases the number of spherulite nuclei per unit mass and shortens the induction period for crystallization and (ii) the chain mobility enhancing effect, which increases the radial growth rate of spherulites. The overall crystallization effect of agents is determined by the combination of these two effects.

Nucleating or crystallization-accelerating fillers include talc, montmorillonite and nanostructured carbon [182, 184, 307–318], and low-molecular-weight organic compounds such as amide and hydrazide [318–320]. Recently, the PLLA/PDLA stereocomplex (SC) was applied as a biodegradable nucleating filler for PLLA [108, 309, 313, 321–323]. Narita *et al.* found that the crystallization of PLLA during cooling is significantly accelerated when SC crystallites are melted just above their endset melting temperature because of the decreased crystalline size of the SCs [324]. Two biopolymers, that is, PDLA-poly(menthide)-PDLA triblock copolymers, were used as crystal nucleating agents for PLLA [325]. In addition, PLLA crystallization is accelerated by the presence of biodegradable polyesters of PGA, P(R-3HB), and PCL [326, 327]. The acceleration of PLLA crystallization in the presence of the three biodegradable polyesters is due to the nucleation-assisting effect of PCL and PGA and the spherulite growth–accelerating effect of P(R-3HB); however, the incorporation of P(R-3HB) lowered the number of PLLA spherulites per unit area. More recently, we discovered that the homo-stereocomplex of P(L-2HB) and P(D-2HB) or hetero-stereocomplex of PLLA and P(D-2HB) or of PDLA and P(L-2HB) act as nucleating or crystallization accelerating agents [328]. Figure 8.25 shows the relative crystallinity (X_r) of PLLA films with no incorporated polymer, P(L-2HB), P(D-2HB), and P(L-2HB)/P(D-2HB), which were crystallized at crystallization temperature (T_c) of 135 °C as a function of crystallization time (t_c). At this T_c, the nucleating effects or crystallization accelerating effects of the homo-stereocomplex of P(L-2HB) and P(D-2HB) and hetero-stereocomplex of PLLA and P(D-2HB) are evident. The use of chemically modified thermoplastic starch as a biodegradable nucleating agent was investigated [329].

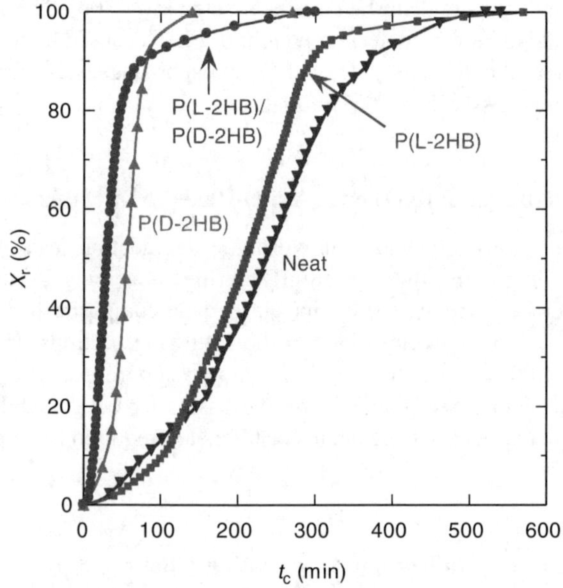

Figure 8.25 *Relative crystallinity (X_r) of PLLA films with no incorporated polymer (neat), P(L-2HB), P(D-2HB), and P(L-2HB)/P(D-2HB) crystallized at crystallization temperature (T_c) = 135 °C as a function of crystallization time (t_c) [328]. (Reproduced with permission from ref. [328]. Copyright © 2011 Wiley Periodicals, Inc.)*

The effects of the addition of two or more additives (i.e. talc, montmorillonite, C_{60}, PDLA, and various polysaccharides) have been studied by Tsuji *et al.* [313], while those of talc and/or triphenyl phosphate as a plasticizer were investigated by Xiao *et al.* [330]. Derivative *N,N,N*-tricyclohexyl-1,3,5-benzenetricarboxylamide was used as a model to tailor the crystalline superstructure of PLLA; three crystal morphologies including conelike, shish-kebab, and needlelike structures were obtained by melt crystallization [331].

8.12.2 Composites and Nanocomposites

Composite or nanocomposite formation is frequently used to modify the mechanical performance and barrier properties of PLLA [332–336]. As discussed in the section on crystallization-accelerating agents, these composite-forming fillers accelerate the PLLA crystallization effects; this further improves the heat resistance. The following parameters of fillers affect the mechanical and other physical properties: (i) physical properties, dimensions (aspect ratio), volume fraction, and dispersibility of the fillers, (ii) wettability and interfacial bonding, (iii) crystallinity and orientation of the PLA-based materials. Nonedible biobased fillers effectively reduce the edible fraction of raw materials for PLA-based materials

Nonfibrous PLLA fillers for industrial and biomedical applications are summarized in Tables 8.8 and 8.9, respectively [184, 308, 313, 314, 337–380]. Of the non-biobased inorganic fillers, nanostructured talc [$Mg_3Si_4O_{10}(OH)_2$] and montmorillonite [$(Na,Ca)_{0.33}(Al,Mg)_2(Si_4O_{10})(OH)_2\ nH_2O$] [308, 313, 332–339] are most frequently used for the preparation of nanocomposites. Conventional calcium carbonate ($CaCO_3$) is used to reduce the production cost and combustion energy released during disposal via incineration [340].

Table 8.8 *Fillers of PLA-based materials for industrial (commodity) applications.*

Type	Origin	Filler	Reference
Nonfibrous	Nonbiobased	Talc [$Mg_3Si_4O_{10}(OH)_2$]	Kolstad *et al.* [308]
		Montmorillonites (MMT) [$(Na,Ca)_{0.33}(Al,Mg)_2(Si_4O_{10})(OH)_2\ nH_2O$]	Ogata *et al.* [337]
		Organo-modified montmorillonites	Paul *et al.* [338]
		Organo-modified-clays (smectite, montmorillonite, and mica)	Maiti *et al.* [339]
		Calcium carbonate ($CaCO_3$)	Fukuda *et al.* [340]
		Titanium dioxide (TiO_2)	Fukuda and Tsuji [341]
		Aluminum hydroxide [$Al(OH)_3$]	Nishida *et al.* [342]
		Silicon dioxide (SiO_2)	Cao and Wu [343], Wang *et al.* [344]
		Zeolite	Yuzay *et al.* [345]
		Organomodified zinc aluminum layered double hydroxide	Wang *et al.* [346]
		Polyhedral oligomeric silsesquioxanes	Qiu *et al.* [347]
		Fullerene C_{60}	Tsuji *et al.* [313, 314]
		Carbon nanohorn (CNH), Carbon nanoballoon (CNB), and Ketjenblack (KB), Conductive grade and graphitized carbon black (CB)	Tsuji *et al.* [184]
		Graphene	Cao *et al.* [348]
		Nano-diamond	Zhao *et al.* [349]
	Biobased	Starch	Ke *et al.* [350], Yu *et al.* [351]
		Chitin	Tsuji *et al.* [313]
		Chitosan	Tsuji *et al.* [313]
		Glucosamine	Kondo *et al.* [352]
		Microcrystalline cellulose	Mathew *et al.* [353], Tsuji *et al.* [313]
		Cellulose nanocrystal	Xiang *et al.* [354]
		Soy protein	Liu *et al.* [355]
		Rice hull	Dimzoski *et al.* [356]
		Water bamboo husk	Wang *et al.* [357]
		Wood flour	Petinakis *et al.* [358]
		Spider silk fibroin	Zhou *et al.* [359]
Fibrous fillers	Nonbiobased	Multi wall carbon nanotube (MWCNT)	Moon *et al.* [182], Zhang *et al.* [183]
		Single wall carbon nanotube (SWCNT)	Tsuji *et al.* [184]
		POSS-MWNTs	Kim *et al.* [381]
		Net-like structure of carbon fibres	Nakamura *et al.* [382]
	Biobased	Cellulosic fibrous materials or fibre	Levit *et al.* [383], Funabashi *et al.* [384]
		Cellulose whiskers	Petersson *et al.* [385]
		Recycled cellulose fibre	Huda *et al.* [386]
		Cotton fibre	Graupner [387]
		Sugar beet pulp	Liu *et al.* [388]
		Flax	Wong *et al.* [389]
		Bamboo fibre	Lee *et al.* [390]
		Kenaf	Nishino *et al.* [391], Serizawa *et al.* [392], Pan *et al.* [393]

(continued)

Table 8.8 *(Continued)*

Type	Origin	Filler	Reference
		Papyrus	Nishino *et al.* [394]
		Hemp fibre	Masirek *et al.* [395]
		Cuphea, Lesquerella	Mohamed *et al.* [396]
		Ramie	Yu *et al.* [397]
		Rice straw fibre	Qin *et al.* [398]
		Red algae fibre	Sim *et al.* [399]
		Miscanthus fibre	Bourmaud *et al.* [400]
		Abaca fibre	Teramoto *et al.* [401]
		Milkweed	Mohamed *et al.* [402]
		Wood fibre	Huda *et al.* [403]
		Recycled newspaper fibre	Huda *et al.* [404]

Table 8.9 *Fillers of PLA-based materials for biomedical applications.*

Type	Filler	Reference
Nonfibrous	Hydroxyapatite [$Ca_5(PO_4)_3(OH)$]	Hyon *et al.* [361], Verhyne *et al.* [362]
	β-Tricalcium-Phosphate [$Ca_3(PO_4)_2$]	Kikuchi *et al.* [363]
	Biphasic calcium phosphate	Bleach *et al.* [364]
	Calcium sulfate ($CaSO_4$)	Mamidwar *et al.* [365]
	Wollastonite ($CaSiO_3$)	Li *et al.* [366]
	β-Dicalcium silicate(Ca_2SiO_4)	Chen *et al.* [367]
	Montmorillonites [$(Na,Ca)_{0.33}(Al,Mg)_2(Si_4O_{10})(OH)_2$ nH_2O]	Lee *et al.* [368]
	Talc	Kylmä and Seppälä [369]
	Bioglass	Leatherbury *et al.* [370], Blaker *et al.* [371]
	Resorbable ceramics	Ignatius *et al.* [372]
	Titania	Liu *et al.* [373]
	Polypyrrole nanoparticle	Wang *et al.* [374]
	Dextran	Pan *et al.* [375]
	Alginate	Milella *et al.* [376]
	Pectin	Liu *et al.* [377]
	Gelatin	Moon *et al.* [378]
	Collagen	Lee *et al.* [379]
	Silk fibroin	Kesenci *et al.* [380]
Fiborous	Carbon and inorganic fibres	Daniels *et al.* [406]
	PLLA (self-reinforcement)	Majola *et al.* [407], Rokkanen *et al.* [408]
	Poly(p-dioxane) fibre	Park *et al.* [409]
	Chitin	Li *et al.* [410]
	Bioactive glass, Chitosan fibre, Polyester amides	Park *et al.* [411]
	Hydroxyapatite fibre	Thomson *et al.* [412]
	Hydroxyapatite whiskers	Mizutani *et al.* [413]
	Halloysite ($Al_2Si_2O_5(OH)_4$) nanotubes	Qi *et al.* [414]
	Fibre from different tissue types of *Picea sitchensis*	Gregorova *et al.* [415]

Ultrafine anatase-type titanium dioxide (TiO_2) exhibits a higher photocatalytic activity than rutile-type TiO_2, which is more suitable for white pigments and coatings because of its high refractivity and notable chemical inertia [341]. The diverse characteristics of TiO_2 are anticipated to provide PLLA/TiO_2 composites with a wide variety of properties. Other fillers that have been reported so far are mica [339], aluminum hydroxide [$Al(OH)_3$] [342], silicon dioxide (SiO_2) [343,344], zeolite [345], organomodified zinc aluminum–layered double hydroxide [346], polyhedral oligomeric silsesquioxanes [347]. For more information, consult the related review articles and book chapters [332–336].

Recently, nanostructured carbon-based fillers such as C_{60} [313, 314], single-wall carbon nanotubes, carbon nanohorns (CNHs), carbon nanoballoons (CNBs), ketjenblack (KB), conductive grade and graphitized carbon black (CB) [184], graphene [348], and nanodiamonds [349] have been used to prepare PLA-based composites. These fillers enhance the crystallization of PLLA [184, 313, 314]. Nanocomposites incorporating fibrous MWCNTs and SWCNTs are discussed in the section on fibre-reinforced plastics (section 8.12.3).

Biobased fillers have recently been intensively studied in order to reduce negative environmental impacts. Representative biobased fillers are polysaccharides, which include starches [313, 350, 351], chitin and chitosan [313], glucosamine [352], microcrystalline cellulose [313, 353] and cellulose nanocrystals [354]. In contrast to animal-derived chitin and chitosan, plant-derived starches are frequently used due to their increased cost effectiveness. Other plant-derived fillers that have been reported to date include soy proteins [355], rice hulls [356], water bamboo husks [357], and wood flour [358]. Spider silk fibroin, which is animal based, has also been reported as filler for PLLA [359]. Biobased fillers that are used for biomedical applications are discussed below. Very recently, interesting polymer composites with pollen as biobased filler were reported although PLLA polymers were not investigated [360]. Pollen has an exine shell with remarkable chemical stability, high strength and unique microstructures. Under certain solvent and annealing conditions, the mechanical properties of polystyrene increased upon the addition of pollen, while those of PCL decreased, which is in good agreement with the wetting behaviour of the polymer-pollen interface. One of the issues of using pollen as filler is the cost-effective collection of sufficient amounts of pollen for polymer composites.

For biomedical applications, calcium-based fillers such as hydroxyapatite [$Ca_5(PO_4)_3(OH)$] [361, 362] and β-tricalcium-phosphate [$Ca_3(PO_4)_2$] [363] are frequently used to enhance bone regeneration and the affinity of bone to these materials. Additionally, various types of inorganic fillers are used for biomedical applications. Other calcium-based or calcium-containing fillers include biphasic calcium phosphate [364], biphasic calcium sulfate ($CaSO_4$) [365], wollastonite ($CaSiO_3$) [366], β-dicalcium silicate (Ca_2SiO_4) [367], and montmorillonites [$(Na,Ca)_{0.33}(Al,Mg)_2(Si_4O_{10})(OH)_2\ nH_2O$] [368]. Talc [369], bioglass [370, 371], bioresorbable ceramics [372], and titanium dioxide (TiO_2, brookite nanosize ceramic grains) [373] are examples of noncalcium-based inorganic fillers. Organic fillers used include polypyrrole nanoparticles [374], dextran [375], alginate [376], pectin [377], gelatin [378], collagen [379], and silk fibroin [380].

8.12.3 Fibre-Reinforced Plastics (FRPs)

Fibrous fillers of PLA-based materials for industrial and biomedical applications are summarized in Tables 8.8 and 8.9, respectively [182–184, 382–416]. Nonbiobased fibres that are used for the preparation of fibre-reinforced PLA-based plastics are multiwall carbon nanotubes (MWCNTs) [182, 183], single wall carbon nanotubes (SWCNTs) [184], polyhedral oligosilsesquioxane (POSS)-modified MWCNTs [381], and carbon fibre [382].

Since the early 2000s, plant-derived biobased fibrous fillers have been frequently used for the reinforcement of PLA-based materials. A well known and frequently used plant-derived fibrous filler is kenaf. Like the case of pollen as an additive, the wettability between PLLA and kenaf should be improved by the addition of a compatibilizer. Other fibrous materials are cellulose fibrous materials or fibres [383, 384], cellulose whiskers [385], recycled cellulose fibre [386], cotton fibre [387], sugar beet pulp [388], flax [389], bamboo fibre [390], kenaf [391–393], papyrus [394], hemp fibre [395], cuphea and lesquerella [396], ramie [397], rice straw fibre [398], red algae fibre [399], miscanthus fibre [400], abaca fibre [401], milkweed [402], wood fibre [403] and recycled newspaper fibre [404]. Poly(L-lactic acid) fibre can also be used to reinforce soft plastics such as PCL [405].

Fibrous fillers for biomedical PLA-based FRPs include carbon and inorganic fibres [406], PLLA (i.e. self-reinforcement) [407,408], poly(p-dioxane) fibre [409], chitin [410], biodegradable fibre (e.g. bioactive glass, chitosan fibre, polyester amides) [411], hydroxyapatite fibre [412], hydroxyapatite whiskers [413], halloysite ($Al_2Si_2O_5(OH)_4$) nanotubes [414] and the fibre from different tissue types of *Picea sitchensis* [415].

8.12.4 Stereocomplexation

Homostereocomplexes are formed from PLLA and PDLA and enhance the mechanical properties, hydrolytic and thermal degradation resistance [11, 43–45, 416–418], and gas-barrier properties [166] of PLA-based materials. Recently, it was reported that P(L-2HB) and P(D-2HB), that is, substituted PLLA and PDLA, form homo-stereocomplexes [49, 418], and P(L-2HB) and PDLA or P(D-2HB) and PLLA form hetero-stereocomplexes [50, 51]. Similar to that between PLLA and PDLA, the stereocomplexation between P(L-2HB) and P(D-2HB) accelerates crystallization and enhances the hydrolytic and thermal degradation resistance [418]. Assuming that the lattice types of these stereocomplexes are the same as that of PLLA/PDLA stereocomplex, the Interplane distance (*d*) values (Table 8.10 [49, 50]) decrease in the following order: P(L-2HB)/(D-2HB) homo-stereocomplex > P(L-2HB)/PDLA or P(D-2HB)/PLLA hetero-stereocomplex > PLLA/PDLA homo-stereocomplex. This is due to the larger side chains of P(L-2HB) and P(D-2HB) compared to those of PLLA and PDLA. In the PDLA and P(L-2HB) polymer blend, hetero-stereocomplexation occurs at a restricted crystallization temperature, while block copolymerization of PDLA and P(L-2HB) facilitates hetero-stereocomplexation over a wide crystallizable temperature range [51]. Zhang *et al.* reported that PLLA/PDLA blends with relatively high molecular weights (M_w = 8.4 and

Table 8.10 *Interplane distance (d) values of P(L-2HB)/PDLA hetero-stereocomplex, PLLA/PDLA and P(L-2HB)/P(D-2HB) homo-stereocomplexes [49, 50]. (Reproduced with permission from ref. [49]. Copyright © 2009, American Chemical society and from ref. [50]. Copyright © 2010, American Chemical Society.)*

	d (nm) (2*θ* (degrees))		
Stereocomplex	(100)	(110)	(210)
PLLA/PDLA homo-stereocomplex[a]	0.744 (11.9)	0.429 (20.7)	0.372 (23.9)
P(L-2HB)/PDLA hetero-stereocomplex	0.783 (11.3)	0.453 (19.6)	0.395 (22.5)
P(L-2HB)/P(D-2HB) homo-stereocomplex[a]	0.827 (10.7)	0.480 (18.5)	0.415 (21.4)

Note: [a]Data from supporting information of ref. [49].

7.7×10^5 g mol^{-1}) crystallize in the α'-form during solvent evaporation; this reorganizes into the α-form at 120 °C, melts at 180 °C and further recrystallizes into stereocomplex crystallites [99].

The stereocomplex formation of PLLA and PDLA is enhanced in the presence of stereoblock poly(lactic acid) [419]. Stereocomplex formation was proposed to be a novel method of controlling the protein- and cell-adhesive properties of biodegradable matrices composed of PEG-PLA copolymers [420]. A triangular growth mechanism was proposed for the PLLA/PDLA stereocomplex [95], and the nucleating mechanism is not altered by the PLLA/PDLA blending ratio [421].

8.13 Applications

8.13.1 Alternatives to Petro-Based Polymers

Poly(L-lactic acid) is an effective alternative to petro-based polymers in applications such as packaging material, automotive materials, including floor mats and spare tire covers, and the chassis of electrical appliances such as computers, mobile phones, remote controls, and optical disks [422, 423]. Figure 8.26 shows the commodity applications of PLLA. For

Figure 8.26 Commodity applications of PLLA. (Reproduced courtesy of Mitsui Chemicals Tohcello Co. Ltd, Tokyo, Japan.)

PLA-based materials to be effective alternatives to petro-based polymers such as PP or PET. The hydrolytic and thermal degradation resistance, mechanical performance and gas barrier properties of PLLA-based materials should be improved. This can be achieved through various techniques including the addition of nucleating agents to improve crystallinity and dimensional stability at high temperatures, composite and nanocomposite formation to improve dimensional stability, mechanical performance and gas barrier properties, fibre reinforcement to improve dimensional stability and mechanical performance, and stereocomplexation to improve hydrolytic and thermal degradation resistance, mechanical performance, dimensional stability, and gas barrier properties. These techniques were discussed in previous sections. For optical applications such as optical disks, blending with polycarbonate is considered to increase the transmittance of shorter wavelength light and the dimensional stability as optical disks may be used in high temperature disk drives. The application of these techniques may have a negative environmental impact; therefore, the balance between performance and environmental impact should be considered before implementation.

8.13.2 Biomedical

Biodegradable or bioabsorbable polyesters including PLAs can be applied as biomedical materials [41]. Table 8.11 [4] shows the functions and purposes of bioabsorbable polymers. Bioabsorbable biomedical materials should be nontoxic, effective, sterilizable and biocompatible. In addition, bioabsorbable biomedical materials should have the appropriate degradation rate required for each application (Figure 8.27 [13]).

The *in vivo* degradation rate of PLA-based materials is reported to be very similar to that of *in vitro* degradation [424, 425]. Because of its relatively slow *in vivo* degradation, PLA-based materials are not used for normal suture materials such as PGA or its copolymers with LLA (90 : 10) but are used for applications that require long-term retention of mechanical strength. Preclinical and clinical applications of PLA-based materials include fibres for ligament and tendon reconstruction, stents for vascular and urological surgery, meshes for guided bone regeneration, and plates, pins, screws and wires for bone fixation (Figure 8.28 [13]). For bone-regeneration devices, oriented PLLA plates are reported to have piezoelectric properties that are sensitive to mechanical stress imposed on the bone; therefore, connected oriented

Table 8.11 *Medical applications of bioabsorbable polymers [4]. (Reproduced with permission from ref. [4]. Copyright © 2000 Wiley-VCH Verlag GmbH, Weinheim.)*

Function	Purpose	Examples
Bonding	Suturing Fixation Adhesion	Vascular and intestinal anastomosis Fractured bone fixation Surgical adhesion
Closure	Covering Occlusion	Wound cover, local hemostasis Vascular embolization
Separation	Isolation Contact inhibition	Organ protection Adhesion prevention
Scaffold	Cellular proliferation Tissue guide	Skin reconstruction, blood vessel reconstruction Nerve reunion
Capsulation	Controlled drug delivery	Sustained drug release

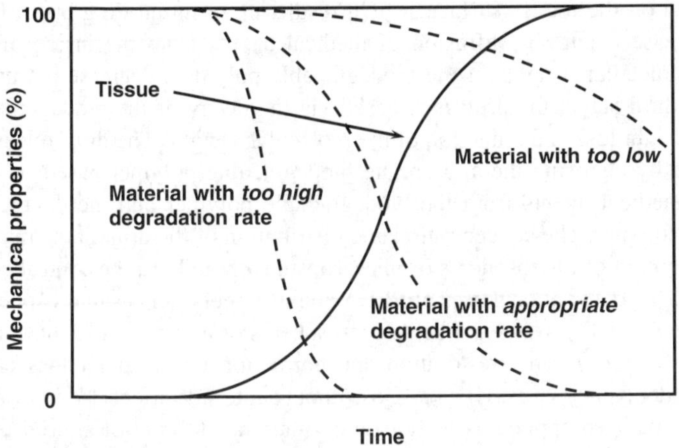

Figure 8.27 *Mechanical properties of tissue and bioabsorbable material during medical treatment [13]. (Reproduced with permission from Yoneda Shuppan, Japan.)*

PLLA plates give rise to voltage, which enhances bone regeneration [181]. Meanwhile, self-reinforced (SR)-PLLA and PDLLA are used as stent [426] and screw [427] materials.

Biodegradable polymers including PLLA and its copolymers have been widely investigated for drug-delivery systems for various medicines including contraceptives, narcotic antagonists, local anaesthetics, and vaccines. Drug delivery or release from biodegradable polymer matrices proceeds via three stages. In the first stage, drug release results from desorption of the medical

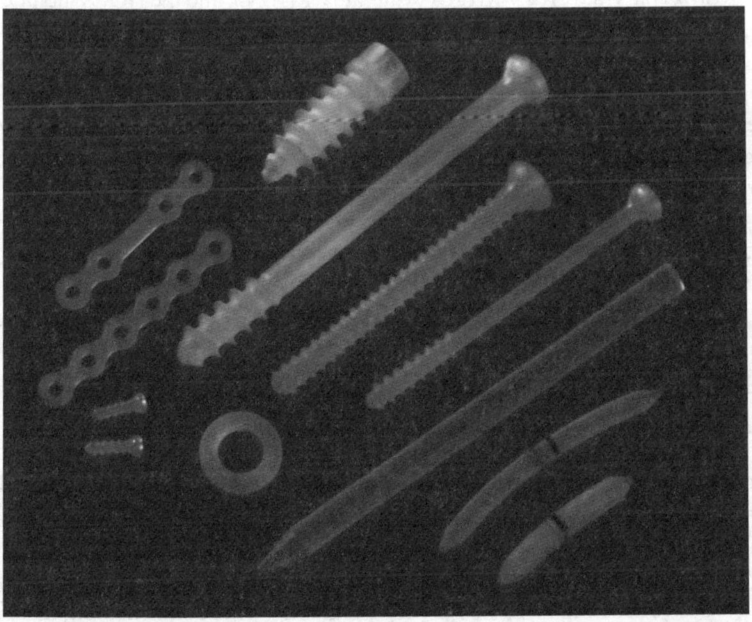

Figure 8.28 *PLA based bone fixation devices with courtesy of Gunze Ltd [13]. (Reproduced with permission from Gunze Ltd, Japan.)*

agents adsorbed on the matrix surface, which results in an initial drug burst. In the second stage, drug release occurs by diffusion of medical agents from the inside of matrices; in this step, the molecular weights of the biodegradable polymers decrease but no weight loss occurs. In the third stage, the drug is released via the decay of the biodegradable matrices causing both weight loss and reduction of the molecular weight. The drug-release profile can be manipulated by choosing the appropriate biodegradable polymer based on the diffusion coefficient of medical agents from the biodegradable polymer and the degradation rate of biodegradable polymer. The concentration and distribution of the drug in the matrices and the sizes and distribution of microspheres or microcapsules should also be controlled.

In addition to the conventional uses of PLLA and its copolymers as microspheres or microcapsules for drug delivery systems, PLLA microspheres or microcapsules have recently been approved by the FDA for the restoration and correction of facial fat loss caused by the human immunodeficiency virus (HIV)/AIDS without acute side effects [41]. Moreover, PLLA microspheres have been applied as embolization materials to control arterial malformation, haemorrhage and tumours without complications [41].

8.13.3 Environmental Applications

8.13.3.1 Absorbents

Poly(L-lactic acid) and its copolymers have attracted attention for applications as materials to absorb harmful and persistent contaminants, as denitrification-assisting materials, and as bioremediation-assisting materials [428]. The removal of hydrophobic organic contaminants (HOCs) from aqueous solution using biodegradable polyesters (BDPs) was investigated [429]. In this study, the BDPs include PLLA, PCL, PBS, PBSA, P(R-3HB), and poly((R)-3-hydroxybutyrate-(R)-3-hydroxyvalerate) (PHBV), and the HOCs include biphenyl, bisphenol A, dibenzofuran, diethylstilbestrol, nonylphenol, and its chlorinated derivatives, 2-monochlorophenol and 3,5-dichlorophenol. All BDPs with T_g values below room temperature strongly absorbed HOCs. However, PLLA with a T_g value of 60 °C, higher than room temperature, showed weak absorption of HOCs. This result suggests that segmental mobility is crucial to determine the absorption properties of HOCs; an increase of the absorption temperature above T_g will elevate the absorption of HOCs by PLLA.

8.13.3.2 Denitrification-Assisting Materials

During the denitrification process, biodegradable polymers such as P(R-3HB) can be used as electron-donating substrates (Figure 8.29) [428]. In contrast with P(R-3HB), which is relatively easily biodegradable, PLLA is not considered to be applicable for denitrification-assisting materials because of its low biodegradability despite its cost-effectiveness. However, the biodegradability of PLLA can be improved by decreasing its molecular weight via high-temperature hydrolytic degradation. Consequently, degraded PLLA with a molecular weight of 9.9×10^3 g mol^{-1} showed a nitrogen removal rate of 50 mg NO_3^--N/[g (dry weight) h]. This value is higher than that reported for PHBV (20 mg NO_3^--N/[g (dry weight) h] [430,431]. However, PLLA pellets with molecular weights higher than 1.2×10^4 g mol^{-1} had a nitrogen removal rate of less than 1 mg NO_3^--N/[g (dry weight) h]. This confirms that the slow release of lactic acid is the most important issue for high molecular weight PLLA, although it can be solved by predegradation of PLLA.

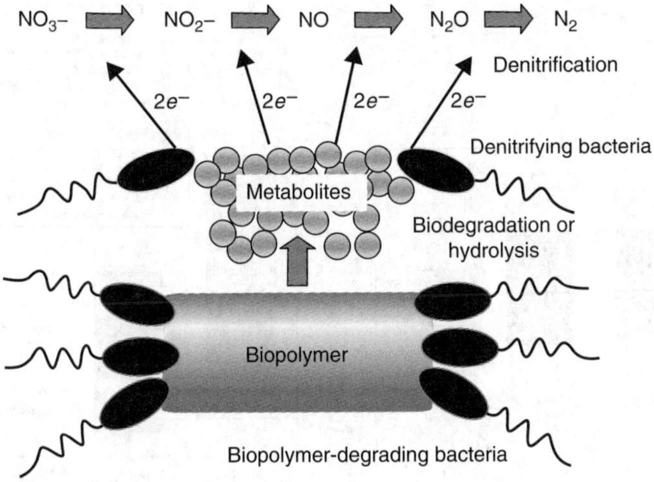

Figure 8.29 *Outline of the solid-phase denitrification process using biodegradable polymers [428]. (Reproduced with permission from ref. [428]. Copyright © John Wiley & Sons, Inc., 2010.)*

8.13.3.3 Bioremediation-Assisting Materials

Bioremediation is defined as the removal of harmful chemical contaminants from the environment by micro-organisms. In engineered bioremediation of anaerobically biodegradable contaminants, anaerobic microbial redox processes work efficiently when electron-donating substrates are supplied over a long period of time [428]. A variety of halo-organic compounds including chlorinated ethenes such as tetrachloroethene (PCE) and trichloroethene are produced for solvents. Environmental contamination by halo-organic compounds threatens human health and other living things. With the incorporation of lactate as an electron donor, the degradation rate of PCE to ethene was the highest [432], thus PLA, which releases lactic acid over a long period of time, is expected to act as an effective electron donor for PCE degradation.

8.14 Recycling

Use of PLA-based materials for several years significantly lowers the molecular weights. Moreover, PLA-based materials have a relatively low resistance to heat during melt processing. Therefore, the methods of recycling used for PET are inappropriate for PLA-based materials. Instead, the recycling of PLA-based materials to its lactic acid or LA monomers is preferable as it prolongs the period that carbon is held in the materials; upon disposal via incineration or biodegradation, carbon dioxide will be released. In addition, it is reported that this type of chemical recycling will reduce the impact on the environment by reducing the consumption of fossil resources, energy, and released carbon dioxide for the synthesis of lactic acids or LAs from plant resources such as corn starch (Figure 8.30 [433]). However, with respect to eutrophication, ecotoxicity, and ozone depletion and so on, PLLA can have a higher impact on the environment compared to petro-based polymers [434]. The impact on eutrophication resulting from intensive agriculture can be mitigated during corn production or reduced by using alternative, less agricultural-chemical intensive feedstock.

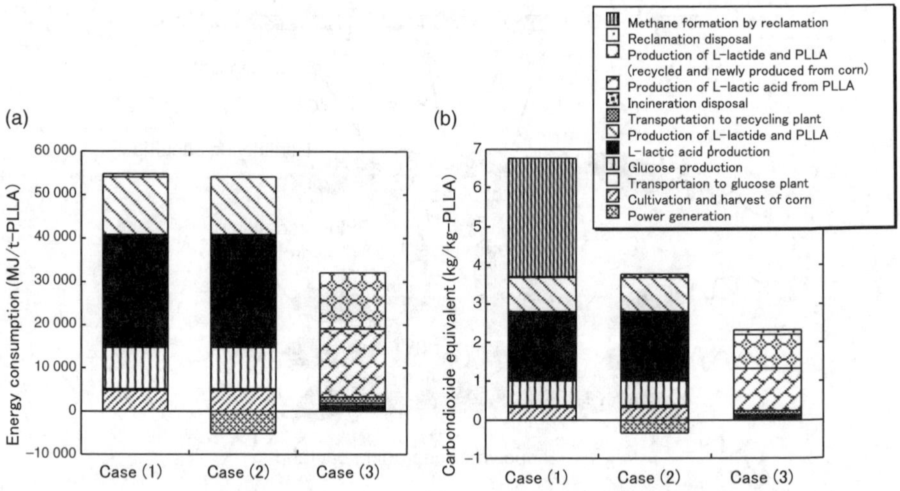

Figure 8.30 *Energy consumption (a) and carbon dioxide equivalent amount (b) for PLLA production. In Case (1), used PLLA is disposed by reclamation and equivalent amount of PLLA is produced from corn. In Case (2), used PLLA is disposed by incineration, the energy is utilized for power generation, and equivalent amount of PLLA is produced from corn. In Case (3), used PLLA is recycled by hydrolytic degradation, lactide production and polymerization, and PLLA with equivalent amount of which is lost during recycling is produced from corn [433]. (Reproduced from [433] with permission from Society of Environmental Science, Japan © 2009.)*

The recycling of PLA to lactic acid is readily performed using high-temperature water without catalytic compounds, which, if included, must be removed after the reaction. Figure 8.31 [223] shows the yield of L-lactic acid upon the hydrolytic degradation of PLLA with high-temperature water and without added catalyst. LA formation is known to be effective for recycling (see the thermal degradation section) although there is the issue of low recovery of the LA formed in large-scale plants due to the high viscosity, which impairs diffusion of LA

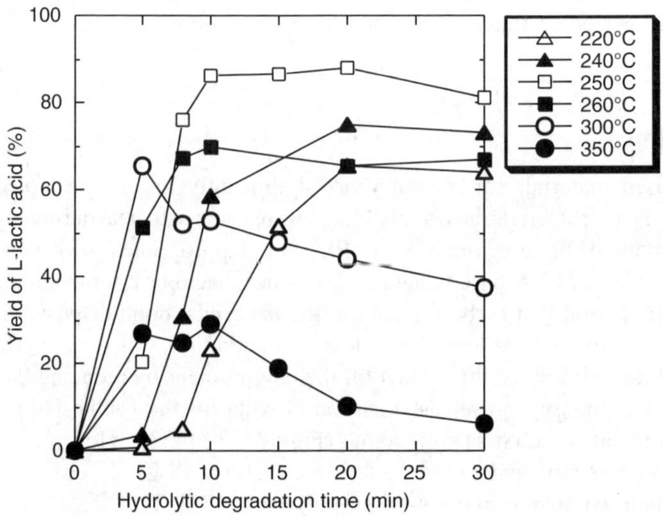

Figure 8.31 *Yield of L-lactic acid formed by hydrolytic degradation of PLLA at different temperatures as a function of hydrolytic degradation time [223]. (Reproduced with permission from ref. [223]. Copyright © 2003, American Chemical Society.)*

from the PLLA melt. Although most of the technical issues have been solved, crucial issues such as imperfect legislation and social system for recycling still remain.

8.15 Conclusions

This article presented an overview of the basic aspects of synthesis, processing, structures, physical properties, and degradation of PLAs. The crucial issues for the use of PLA-based materials as alternatives to petro-based polymeric materials include reducing the production cost and enhancing the mechanical performance and resistance to hydrolytic/thermal degradation. Other important goals are increasing the production of PLA from nonedible raw materials and reducing the fraction of edible raw materials used for PLA-based materials, which can be attained by using nonedible fillers. The production of PLA from corn is not preferable due to the limited production mass of provisions. To produce high-performance PLA-based materials, composite and fibre reinforced plastics formation are effective. However, the balance between high performance and environmental impact (and production cost) should be considered. With respect to their performance, hydrolytic degradation, environmental impact, and production cost, greatly modified PLA-based materials are futile due to the high intrinsic performance of polymeric materials such as aromatic polyamide- or polyimide-based materials. For medical and environmental applications, PLA should be used more frequently when the functionality of hydrolysability is required.

References

1. Kharas, G.B., Sanchez-Riera, F. and Severson, D.K. (1994) Polymer of lactic acids, in *Plastics from Microbes* (ed. D.P. Mobley), Hanser Publishers, New York, pp. 93–137.
2. Coombes, A.G.A. and Meikle, M.C. (1994) Resorbable synthetic polymers as replacements for bone graft. *Clinical Materials*, **17**, 35–67.
3. Hartmann, M.H. (1998) *High Molecular Weight Polylactic Acid Polymers, Biopolymers from Renewable Resources* (ed. D.L. Kaplan), Springer, Berlin, pp. 367–411.
4. Ikada Y. and Tsuji, H. (2000) Biodegradable polyesters for medical and ecological applications. *Macromolecular Rapid Communications*, **21**, 117–132.
5. Södergård, A. and Stolt, M. (2002) Properties of lactic acid based polymers and their correlation with composition. *Progress in Polymer Science*, **27**, 1123–1163.
6. Garlotta, D. (2001) A literature review of poly(lactic acid). *Journal of Polymers and the Environment*, **9**, 63–84.
7. Albertsson, A.-C. (ed.) (2002) *Degradable Aliphatic Polyesters* (Advances in Polymer Science, vol. **157**), Springer, Berlin.
8. Tsuji, H. (2002) Polylactide, in *Polyesters 3* (Biopolymers, vol. **4**), (eds Y. Doi and A. Steinbüchel), Wiley-VCH, Weinheim (Germany), pp. 129–177.
9. Auras, R., Lim, L.-T., Selke, S.E.M. and Tsuji, H. (eds) (2010) *Poly(Lactic Acid): Synthesis, Structures, Properties, Processing, and Applications* (Wiley Series on Polymer Engineering and Technology), John Wiley & Sons, Inc., NJ.
10. Auras, R., Harte, B. and Selke, S. (2004) An overview of polylactides as packaging materials. *Macromolecular Bioscience*, **4**, 835–864.
11. Tsuji, H. (2005) Poly(lactide) stereocomplexes: Formation, structure, properties, degradation, and applications. *Macromolecular Bioscience*, **5**, 569–597.
12. Gupta, B., Revagade, N. and Hilborn, J. (2007) Poly(lactic acid) fiber: An overview. *Progress in Polymer Science*, **32**, 455–448.
13. Tsuji, H. (2008) *Poly(Lactic Acid) –Fundamentals and Applications of Plant-Derived Plastics-* (in Japanese), Yoneda Shuppan, Tokyo (Japan).

14. Jamshidian, M., Tehrany, E.A., Imran, M. *et al.* (2010) Poly-lactic acid: Production, applications, nanocomposites, and release studies. *Comprehensive Review in Food Science Food Safety*, **9**, 552–571.
15. Inkinen, S., Hakkarainen, M., Albertsson, A.-C. and Södergård, A. (2011) From lactic acid to poly(lactic acid) (PLA): Characterization and analysis of PLA and Its precursors. *Biomacromolecules*, **12**, 523–532.
16. Madhavan Nampoothiri, K., Nair, N.R. and John, R.P. (2010) An overview of the recent developments in polylactide (PLA) research. *Bioresource Technology*, **101**, 8493–8501.
17. Pang, X., Zhuang, X., Tang, Z. and Chen, X. (2010) Polylactic acid (PLA): Research, development and industrialization. *Biotechnology Journal*, **5**, 1125–1136.
18. Ren, J. (ed.) (2010) *Biodegradable Poly(Lactic Acid), Synthesis, Modification, Processing and Applications*, Tsinghua University Press, Beijing (Springer, Heidelberg).
19. Pelouze, P.M.J. (1845) Mémoire sur l'acide lactique. *Annales de Chimie et de Physique, Series 3*, **13**, 257–268.
20. Nef, J.U. (1914) Dissoziationsvorgänge in der zuckergruppe. *Liebigs Annalen der Chemie*, **403**, 204–383.
21. Carothers, W.H., Dorough, G.L. and Van Natta, F.J. (1932) Studies of polymerization and ring formation. X. The reversible polymerization of six-membered cyclic esters. *Journal of the American Chemical Society*, **54**, 761–772.
22. Lowe, C.E. (1954) Preparation of high molecular weight polyhydroxyacetic ester. US Patent 2 668 162.
23. Kulkarni, R.K., Pani, K.C., Neuman, C. and Leonard, F. (1966) Polylactic acid for surgical implant. *Archives of Surgery*, **93**, 839–843.
24. Wise, D.L., Fellmann, T.D., Sanderson, J.E. and Wentworth, R.L. (1979) Lactic/glycolic acid polymers, in *Drug Carriers in Biology and Medicine* (ed. G. Gregoriadis), Academic Press, pp. 237–270.
25. Shneider, A.K. (1972) Polyalctide sutures. US Patent 3 636 956.
26. De Santis P. and Kovacs, A.J. (1968) Molecular conformation of poly(S-lactic acid). *Biopolymer*, **6**, 299–306.
27. Hoogsteen, W., Postema, A.R., Pennings, A.J. *et al.* (1990) Crystal structure, conformation, and morphology of solution-spun poly(L-lactide) fibers. *Macromolecules*, **23**, 634–642.
28. Kobayashi, J., Asahi, T., Ichiki, M. *et al.* (1995) Structural and optical properties of poly lactic acids. *Journal of Applied Physics*, **77**, 2957–2973.
29. Sasaki S. and Asakura, T. (2003) Helix distortion and crystal structure of the α-form of poly(L-lactide). *Macromolecules*, **36**, 8385–8390.
30. Wasanasuk, K., Tashiro, K., Hanesaka, M. *et al.* (2011) Crystal structure analysis of poly(L-lactic acid) α form based on the dimensional wide-angle synchrotoron X-ray and neutron diffraction measurements. *Macromolecules*, **44**, 6441–6452.
31. Puiggali, J., Ikada, Y., Tsuji, H. *et al.* (2000) The frustrated structure of poly(L-lactide). *Polymer*, **41**, 8921–8930.
32. Cartier, L., Okihara, T., Ikada, Y. *et al.* (2000) Epitaxial crystallization and crystalline polymorphism of polylactides. *Polymer*, **41**, 8909–8919.
33. Okihara, T., Tsuji, M., Kawaguchi, A. *et al.* (1991) Crystal structure of stereocomplex of poly(L-lactide) and poly(D-lactide). *Journal of Macromolecular Science: Physics*, **B30**, 119–140.
34. Kricheldorf, H.R. and Kreiser-Sauders, I. (1996) Polylactides-synthesis, characterization and medical application. *Macromolecular Symposia*, **103**, 85–102.
35. Dubois, P., Degée, P., Ropson, N. and Jérome, R. (1997) Macromolecular engineering of polylactones and polylactides by ring-opening polymerization, in *Macromolecular Design of Polymeric Materials* (eds K. Hatada, T. Kitayama and O. Vogl), Marcel Dekker, Inc., New York, pp. 247–272.
36. Albertsson, A.-C., Varma, I.K., Lochab, B. *et al.* (2010) Design and synthesis of different types of poly(lactic acid), in *Poly(Lactic Acid): Synthesis, Structures, Properties, Processing, and Applications* (Wiley Series on Polymer Engineering and Technology) (eds R. Auras, L.-T. Lim, S.E.M. Selke and H. Tsuji), John Wiley & Sons, Inc., NJ, pp. 43–58.
37. Kalb, B. and Pennings, A.J. (1980) General crystallization behaviour of poly(L-lactic acid). *Polymer*, **21**, 607–612.

38. Vasanthakumari, R. and Pennings, A.J. (1983) Crystallization kinetics of poly(L-lactic acid). *Polymer*, **24**, 175–178.
39. Leenslag, J.W. and Pennings, A.J. (1987) High-strength poly(L-lactide) fibres by a dry-spinning/hot-drawing process. *Polymer*, **28**, 1695–1702.
40. Li, S. and Vert, M. (1995) Biodedegradation of aliphatic polyesters, Chapter 4, in *Degradable Polymers-Principles and Appliactions* (eds G. Scott and D. Gilead), Chapman & Hall, London, pp. 43–87.
41. Suzuki S. and Ikada, Y. (2010) Medical applications, in *Poly(Lactic Acid): Synthesis, Structures, Properties, Processing, and Applications* (Wiley Series on Polymer Engineering and Technology) (eds R. Auras, L.-T. Lim, S.E.M. Selke and H. Tsuji), John Wiley & Sons, Inc., NJ, pp. 445–456.
42. Ikada, Y., Jamshidi, K., Tsuji, H. and Hyon, S.-H. (1987) Stereocomplex formation between enantiomeric poly(lactides). *Macromolecules*, **20**, 904–906.
43. Slager, J. and Domb, A.J. (2003) Biopolymer stereocomplexes. *Advanced Drug Delivery Reviews*, **55**, 549–583.
44. Fukushima, K. and Kimura, Y. (2006) Stereocomplexed polylactides (Neo-PLA) as high-performance bio-based polymers: their formation, properties, and application. *Polymer International*, **55**, 626–642.
45. Hirata, M. and Kimura, T. (2010) Structure and properties of stereocomplex-type poly(lactic acid), in *Poly(Lactic Acid): Synthesis, Structures, Properties, Processing, and Applications* (eds R. Auras, L.-T. Lim, S.E.M. Selke and H. Tsuji), John Wiley & Sons, Inc., NJ, pp. 59–65.
46. Yin, M. and Baker, G.L. (1999) Preparation and characterization of substituted polylactides. *Macromolecules*, **32**, 7711–7718.
47. Jing, F., Smith, M.R. III and Baker, G.L. (2007) Cyclohexyl-substituted polyglycolides with high glass transition temperatures. *Macromolecules*, **40**, 9304–9312.
48. Liu, T., Simmons, L., Bohnsack, D.A. *et al.* (2007) Synthesis of polymandelide: a degradable polylactide derivative with polystyrene-like properties. *Macromolecules*, **40**, 6040–6047.
49. Tsuji, H. and Okumura, A. (2009) Stereocomplex formation between enantiomeric substituted poly(lactide)s: blends of poly[(S)-2-hydroxybutyrate] and poly[(R)-2-hydroxybutyrate]. *Macromolecules*, **42**, 7263–7266.
50. Tsuji, H., Yamamoto, S., Okumura, A. and Sugiura, Y. (2010) Hetero-stereocomplexation between biodegradable and optically active polyesters as a versatile preparation method for biodegradable materials. *Biomacromolecules*, **11**, 252–258.
51. Tsuji, H., Shimizu, K., Sakamoto, Y. and Okumura, A. (2011) Hetero-stereocomplex formation of stereoblock copolymer of substituted and non-substituted poly(lactide)s. *Polymer*, **52**, 1318–1325.
52. Jung, Y.K., Kim, T.Y., Park, S.J. and Lee, S.Y. (2010) Metabolic engineering of *Escherichia coli* for the production of polylactic acid and its copolymers. *Biotechnology and Bioengineering*, **105**, 161–171.
53. Han, X., Satoh, Y., Satoh, T. *et al.* (2011) Chemo-enzymatic synthesis of polyhydroxyalkanoate (PHA)incorporating 2-hydroxybutyrate by wild-type class I PHA synthase from *Ralstonia eutropha*. *Applied Microbiology and Biotechnology*, **92**, 509–517.
54. Tsuji, H. and Ikada, Y. (1999) Physical properties of polylactides, in *Current Trends in Polymer Science*, vol. **4** (eds K.L. DeVries *et al.*), Editorial advisory board, Research Trends, Trivandrum (India), pp. 27–46.
55. Mehta, R., Kumar, V., Bhunia, H. and Upadhyay, S.N. (2005) Synthesis of poly(lactic acid): A review. *Journal of Macromolecular Science-Polymer Reviews*, **45**, 325–349.
56. Ajioka, M., Enomoto, E., Suzuki, K. and Yamaguchi, A. (1995) Basic properties of polylactic acid produced by the direct condensation polymerization of lactic acid. *Bulletin of the Chemical Society of Japan*, **68**, 2125–2131.
57. Moon, S.I., Lee, C.W., Miyamoto, M. and Kimura, Y. (2000) Melt polycondensation of L-lactic acid with Sn(II) catalysts activated by various proton acids: A direct manufacturing rout to high molecular weight poly(L-lactic acid). *Journal of Polymer Science Part A: Polymer Chemistry*, **38**, 1673–1679.
58. Lillie, E. and Schulz, R.C. (1975) ^{1}H- and ^{13}C-{^{1}H}-NMR spectra of stereocopolymers of lactide. *Makromolekulare Chemie*, **176**, 1901–1906.

59. Kohn, F.E., Van Den Berg, J.W.A., Van De Ridder, G. and Feijen, J. (1984) The ring-opening polymerization of D,L-lactide in the melt initiated with tetraphenyltin. *Journal of Applied Polymer Science*, **29**, 4265–4277.
60. Leenslag, J.W. and Pennings, A.J. (1987) Synthesis of high-molecular-weight poly(L-lactide) initiated with tin 2-ethylhexanoate. *Makromolekulare Chemie*, **188**, 1809–1814.
61. Kricheldorf, H.R. and Sumbél, M. (1989) Polylactones-18. Polymerization of L.L-lactide with Sn(II) and Sn(IV) halogenides. *European Polymer Journal*, **25**, 585–591.
62. Hyon, S.-H., Jamshidi, K. and Ikada, Y. (1997) Synthesis of polylactides with different molecular weights. *Biomaterials*, **18**, 1503–1508.
63. Tsuji, H., Miyase, T., Tezuka, Y. and Saha, S.K. (2005) Physical properties and crystallization, and spherulite growth of linear and 3-arm poly(L-lactide)s. *Biomacromolecules*, **6**, 244–254.
64. Schindler, A. and Gaetano, K.D. (1988) Poly(lactate) III. Stereoselective polymerization of meso dilactide. *Journal of Polymer Science Part C: Polymer Letters*, **26**, 47–48.
65. Kricheldorf, H.R. and Boettcher, C. (1993) Polylactones. XXV. Polymerizations of racemic- and meso-D,L-lactide with Zn, Pb, Sb, and Bi salts – Stereochemical aspects. *Journal of Macromolecular Science: Pure and Applied Chemistry*, **A30**, 441–448.
66. Kasperczyk, J.E. (1995) Microstructure analysis of poly(lactic acid) obtained by lithium tert-butoxide as initiator. *Macromolecules*, **28**, 3937–3939.
67. Fukuzaki, H., Yoshida, M., Asano, M. and Kumakura, M. (1989) Synthesis of copoly(D,L-lactic acid) with relatively low molecular weight and in vitro degradation. *European Polymer Journal*, **25**, 1019–1026.
68. Tsuji, H. and Ikada, Y. (1992) Stereocomplex formation between enantiomeric poly(lactic acid)s. 6. Binary blends from copolymers. *Macromolecules*, **25**, 5719–5723.
69. MacDonald, R.T., McCarthy, S.P. and Gross, R.A. (1996) Enzymatic degradability of poly(lactide): Effects of chain stereochemistry and material crystallinity. *Macromolecules*, **29**, 7356–7361.
70. Yui, N., Dijkstra, P.J. and Feijen, J. (1990) Stereo block copolymers of L- and D-lactides. *Makromolekulare Chemie*, **191**, 481–488.
71. Tsuji, H., Wada, T., Sakamoto, Y. and Sugiura, Y. (2010) Stereocomplex crystallization and spherulite growth behavior of poly(L-lactide)-*b*-poly(D-lactide) stereodiblock copolymers. *Polymer*, **51**, 4937–4947.
72. Spassky, N., Pluta, C., Simic, V. *et al.* (1998) Stereochemical aspects of the controlled ring-opening polymerization of chiral cyclic esters. *Macromolecular Symposia*, **128**, 39–51.
73. Sarasua, J.-R., Prud'homme, R.E., Wisniewski, M. *et al.* (1998) Crystallization and melting behavior of polylactides. *Macromolecules*, **31**, 3895–3905.
74. Spinu, M., Jackson, C., Keating, M.Y. and Gardner, K.H. (1996) Material design in poly(lactic acid) system: Block copolymers, star homo- and copolymers, and sereocomplexes. *Journal of Macromolecular Science: Pure and Applied Chemistry*, **A33**, 1497–1530.
75. Fukushima, K., Hirata, M. and Kimura, Y. (2007) Synthesis and characterization of stereoblock poly(lactic acid)s with nonequivalent D/L sequence ratios. *Macromolecules*, **40**, 3049–3055.
76. Lewis, D.H. (1990) Controlled release of bioactive agents from lactide/glycolide polymers, Chapter 1, in *Biodegradble Polymers as Drug Delivery Systems* (eds M. Chasin and R. Langer), Marcel Dekker, Inc., New York, pp. 1–42.
77. Pitt, C.G. (1990) Poly(ε-caprolactone) and its copolymers, Chapter 3, in *Biodegradble Polymers as Drug Delivery Systems* (M. Chasin and R. Langer), Marcel Dekker, Inc., New York, pp. 71–120.
78. Agrawal, C.M., Athanasiou, K.A. and Heckman, J.D. (1997) Biodegradable PLA-PGA polymers for tissue engineering in orthopedics. *Materials Science Forum*, **250**, 115–128.
79. Tsuji, H. and Ikada, Y. (1994) Stereocomplex formation between enantiomeric poly(lactic acid)s. X. Binary blends from poly(D-lactide-co-glycolide) and poly(L-lactide-co-glycolide). *Journal of Applied Polymer Science*, **53**, 1061–1071.
80. Tsuji, H. and Ikada, Y. (2000) Enhanced crystallization of poly(L-lactide-co-ε-caprolactone) during storage at room temperature. *Journal of Applied Polymer Science*, **76**, 947–953.

81. Gilding, D.K. and Reed, A.M. (1979) Biodegradable polymers for use in surgery-polyglcolic/poly(lactic acid) homo- and copolymers: 1. *Polymer*, **20**, 1459–1464.
82. Kricheldorf, H.R. and Kreiser, I. (1987) Polylactones. 11. Cationic copolymerization of glycolide with L,L-dilactide. *Makromolekulare Chemie*, **188**, 1861–1873.
83. Kricheldorf, H.R. and Kreiser, I. (1987) Polylactones. 13. Transesterification of poly(L-lactide) with poly(glycolide), poly(β-propiolactone), and poly(ε-caprolactone). *Journal of Macromolecular Science: Part A - Chemistry*, **A24**, 1345–1356.
84. Choi, E.-J., Park, J.-K. and Chang, H.-N. (1994) Effect of polymerization catalysts on the microstructure of P(LLA-co-εCL). *Journal of Polymer Science Part B: Polymer Physics*, **32**, 2481–2489.
85. Shen, Y., Zhu, K.J., Shen, Z. and Yao, K.-M. (1996) Synthesis and characterization of highly random copolymer of ε-caprolactone and D,L-lactide using rare earth catalyst. *Journal of Polymer Science Part A: Polymer Chemistry*, **34**, 1799–1805.
86. Rafler, G. and Jobmann, M. (1994) Controlled release systems of biodegradable polymers. 3rd communication: Degradation and release properties of poly(glycolide(50)-co-lactide(50))s with random and non-random monomer distribution. *Drug Made in Germany*, **38**, 20–22.
87. Song, C.X. and Feng, X.D. (1984) Synthesis of ABA triblock copolymers of ε-caprolactone and DL-lactide. *Macromolecules*, **17**, 2764–2767.
88. Hiki, S., Miyamoto, M. and Kimura, Y. (2000) Synthesis and characterization of hydroxy-terminated (RS)-poly(3-hydroxybutyrate) and its utilization to block copolymerization with l-lactide to obtain a biodegradable themoplastic elastomer. *Polymer*, **41**, 7369–7379.
89. Stridsberg, K. and Albertsson, A.-C. (2000) Changes in chemical and thermal properties of the tri-block copolymer poly(L-lactide-b-1,5-dioxepan-2-one-b-L-lactide) during hydrolytic degradation. *Polymer*, **41**, 7321–7330.
90. Younes, H. and Cohn, D. (1987) Morphological study of biodegradable PEO/PLA block copolymers. *Journal of Biomedical Materials Research*, **21**, 1301–1316.
91. Kimura, Y., Matsuzaki, Y., Yamane, H. and Kitao, T. (1989) Preparation of block copoly(ester-ether) comprising poly(L-lactide) and poly(oxypropylene) and degradation of its fibre in vitro and in vivo. *Polymer*, **30**, 1342–1349.
92. Yamaoka, T., Takahashi, Y., Ohta, T. *et al.* (1999) Synthesis and properties of multiblock copolymers consisting of poly(L-lactic acid) and poly(oxypropylene-co-oxyethylene) prepared by direct polycondensation. *Journal of Polymer Science Part A: Polymer Chemistry*, **37**, 1513–1522.
93. Lim, L.-T., Auras, R. and Rubino, M. (2008) Processing technologies for poly(lactic acid). *Progress in Polymer Science*, **33**, 820–852.
94. Lim, L.-T. and Vanyo, C. (2010) Processing of poly(lactic acid), in *Poly(Lactic Acid): Synthesis, Structures, Properties, Processing, and Applications* (Wiley Series on Polymer Engineering and Technology) (eds R. Auras, L.-T. Lim, S.E.M. Selke and H. Tsuji), John Wiley & Sons, Inc., NJ, pp. 191–215.
95. Brizzolara, D., Cantow, H.-J., Diederichs, K. *et al.* (1996) Mechanism of stereocomplex formation between enantiomeric poly(lactide)s. *Macromolecules*, **29**, 191–197.
96. Pan, P., Kai, W., Zhu, B. *et al.* (2007) Polymorphous crystallization and multiple melting behavior of poly(L-lactide): molecular weight dependence. *Macromolecules*, **40**, 6898–6905.
97. Pan, P., Zhu, B., Kai, W. *et al.* (2008) Effect of crystallization temperature on crystal modifications and crystallization kinetics of poly(L-lactide). *Journal of Applied Polymer Science*, **107**, 54–62.
98. Kawai, T., Rahma, N., Matsuba, G. *et al.* (2007) Crystallization and melting behavior of poly (L-lactic acid). *Macromolecules*, **40**, 9463–9469.
99. Zhang, J., Tashiro, K., Tsuji, H. and Domb, A.J. (2007) Investigation of phase transitional behavior of poly(L-lactide)/poly(D-lactide) blend used to prepare the highly-oriented stereocomple. *Macromolecules*, **40**,1049–1054.
100. Bouapao, L., Tsuji, H., Tashiro, K. *et al.* (2009) Crystallization, spherulite growth, and structure of blends of crystalline and amorphous poly(lactide)s. *Polymer*, **50**, 4007–4017.
101. Wasanasuk, K. and Tashiro, K. (2011) Crystal structure and disorder in Poly(L-lactic acid) δ form (α′ form) and phase transition mechanism to the ordered α form. *Polymer*, **52**, 6097–6109.

102. Pan, P. and Inoue, Y. (2009) Polymorphism and isomorphism in biodegradable polyesters. *Progress in Polymer Science*, **34**, 605–640.
103. Zhang, J., Duan, Y., Sato, H. *et al.* (2005) Crystal modifications and thermal behavior of poly(L-lactic acid) revealed by infrared spectroscopy. *Macromolecules*, **38**, 8012–8021.
104. Tsuji, H. and Ikada, Y. (1996) Crystallization from the melt of poly(lactide)s with different optical purities and their blends. *Macromolecular Chemistry and Physics*, **197**, 3483–3499.
105. Tsuji, H., Tezuka, Y., Saha, S.K. *et al.* (2005) Spherulite growth of L-lactide copolymers: effects of tacticity and comonomers. *Polymer*, **46**, 4917–4927.
106. Tsuji, H. and Ikada, Y. (1995) Blends of isotactic and atactic poly(lactide)s. I. Effects of mixing ratio of isomers on crystallization of blends from melt. *Journal of Applied Polymer Science*, **58**, 1793–1802.
107. Tsuji, H. and Ikada, Y. (1993) Stereocomplex formation between enantiomeric poly(lactic acid)s. 9. Stereocomplexation from the melt. *Macromolecules*, **26**, 6918–6926.
108. Brochu, S., Prud'homme, R.E., Barakat, I. and Jérome, R. (1995) Stereocomplexation and morphology of polylactides. *Macromolecules*, **28**, 5230–5239.
109. Maillard, D. and Prud'homme, R.E. (2010) Difference between crystals obtained in PLLA-rich or PDLA-rich stereocomplex mixture. *Macromolecules*, **43**, 4006–4010.
110. Marega, C., Marigo, A., Di Noto, V. *et al.* (1992) Structure and crystallization kinetics of poly(L-lactic acid). *Makromolekulare Chemie*, **193**, 1599–1606.
111. Tsuji, H. and Ikada, Y. (1995) Properties and morphologies of poly(L-lactide): 1. Annealing condition effects on properties and morphologies of poly(L-lactide). *Polymer*, **36**, 2709–2716.
112. Fischer, E.W., Sterzel, H.J. and Wegner, G. (1973) Investigation of the structure of solution grown crystals of lactide copolymers by means of chemical reaction. *Kolloid-Zeitschrift und Zeitschrift für Polymere*, **251**, 980–990.
113. Miyata, T. and Masuko, T. (1998) Crystallization behavior of poly(L-lactide). *Polymer*, **39**, 5515–5521.
114. Iwata, T. and Doi, Y. (1998) Morphology and enzymatic degradation of poly(L-lactic acid) single crystals. *Macromolecules*, **31**, 2461–2467.
115. Tsuji, H., Hyon, S.-H. and Ikada, Y. (1992) Stereocomplex formation between enantiomeric poly(lactic acid)s. 5. Calorimetric and morphological studies on the stereocomplex formed in acetonitrile solution. *Macromolecules*, **25**, 2940–2946.
116. Di Lorenzo, M.L. (2001) Determination of spherulite growth rates of poly(L-lactic acid) using combined isothermal and non-isothermal procedures. *Polymer*, **42**, 9441–9446.
117. Tsuji, H. and Ikada, Y. (1996) Blends of isotactic and atactic poly(lactide): 2. Molecular-weight effects of atactic component on crystallization and morphology of equimolar blends from the melt. *Polymer*, **37**, 595–602.
118. Di Lorenzo, M.L. (2005) Crystallization behavior of poly(L-lactic acid). *European Polymer Journal* **41**, 569–575.
119. Yasuniwa, M., Tsubakihara, S., Iura, K. *et al.* (2006) Crystallization behavior of poly(L-lactic acid). *Polymer*, **47**, 7554–7563.
120. Matsuba, G., Kaji, K., Nishida, K. *et al.* (1999) Conformational change and orientation fluctuations prior to the crystallization of syndiotactic polystyrene. *Macromolecules*, **32**, 8932–8937.
121. Cai, H., Dave, V., Gross, R.A. and McCarthy, S.P. (1996) Effects of physical aging, crystallinity, and orientation on the enzymatic degradation of poly(lactic acid). *Journal of Polymer Science Part B: Polymer Physics*, **34**, 2701–2708.
122. Pan, P., Zhu, B. and Inoue, Y. (2007) Enthalpy relaxation and embrittlement of poly(L-lactide) during physical aging. *Macromolecules*, **40**, 9664–9671.
123. Pan, P., Liang, Z., Zhu, B. *et al.* (2008) Roles of physical aging on crystallization kinetics and induction period of poly(L-lactide). *Macromolecules*, **41**, 8011–8019.
124. Tsuji, H. and Sawada, M. (2010) Accelerated crystallization of poly(L-lactide) by physical aging. *Journal of Applied Polymer Science*, **116**, 1190–1196.
125. Lim, L.-T., Auras, R. and Rubino, M. (2008) Processing technologies for poly(lactic acid). *Progress in Polymer Science*, **33**, 820–852.

126. Tsuji, H. and Ikada, Y. (1999) Stereocomplex formation between enantiomeric poly(lactic acid)s. 11. Mechanical properties and morphology of solution-cast films. *Polymer*, **40**, 6699–6708.
127. Eling, B., Gogolewski, S. and Pennings, A.J. (1982) Biodegradable materials of poly(L-lactic acid): 1. Melt-spun and solution spun fibres. *Polymer*, **23**, 1587–1593.
128. Perego, G., Cella, G.D. and Bastioli, C. (1996) Effects of molecular weight and crystallinity on poly(lactic acid) mechanical properties. *Journal of Applied Polymer Science*, **59**, 37–43.
129. Penning, J.P., Dijkstra, H. and Pennings, A.J. (1993) Preparation and properties of absorbable fibers from L-lactide copolymers. *Polymer*, **34**, 942–951.
130. Grijpma, D.W., Kroeze, E., Nijenhuis, A.J. and Pennings, A.J. (1993) Poly(L-lactide) crosslinked with spiro-bis-dimethylene-carbonate. *Polymer*, **34**, 1496–1503.
131. Takayama, T., Todo, M. and Tsuji, H. (2011) Effect of annealing on the mechanical properties of PLA/PCL and PLA/PCL/LTI polymer blends. *Journal of the Mechanical Behavior of Biomedical Materials*, **4**, 255–260.
132. Hyon, S.-H., Jamshidi, K. and Ikada, Y. (1984) Melt spinning of poly-L-lactide and hydrolysis of the fiber in vitro, in *Polymers as Biomaterials* (eds S.W. Shalaby, A.S. Hoffman, B.D. Ratner and T.A. Horbett), Prenum Press, New York, pp. 51–65.
133. Horacek, I. and Kalísek, V. (1994) Polylactide. I. Continuous dry spinning-hot drawing preparation of fibers. *Journal of Applied Polymer Science*, **54**, 1751–1757.
134. Horacek, I. and Kalísek, V. (1994) Polylactide. II. Discontinuous dry spinning-hot drawing preparation of fibers. *Journal of Applied Polymer Science*, **54**, 1759–1765.
135. Fambri, L., Pegoretti, A., Fenner, R. *et al.* (1997) Biodegradable fibres of poly(L-lactic acid) produced by melt-spinning. *Polymer*, **38**, 79–85.
136. Okuzaki, H., Kubota, I. and Kunugi, T. (1999) Mechanical properties and structure of the zone -drawn poly(L-lactic acid) fibers. *Journal of Polymer Science Part B: Polymer Physics*, **37**, 991–996.
137. Sheth, M., Kumar, R.A., Davé, V. *et al.* (1997) Biodegradable polymer blends from poly(lactic acid) and poly(ethylene glycol). *Journal of Applied Polymer Science*, **66**, 1495–1505.
138. Tsuji, H. and Ikada, Y. (1996) Blends of aliphatic polyesters. I. Physical properties and morphologies of solution-cast blends from poly(DL-lactide) and poly(ε-caprolactone). *Journal of Applied Polymer Science*, **60**, 2367–2375.
139. Tsuji, H. and Ikada, Y. (1998) Blends of aliphatic polyesters. II. Hydrolysis of solution-cast blends from poly(L-lactide) and poly(ε-caprolactone) in phosphate-buffered solution. *Journal of Applied Polymer Science*, **67**, 405–415.
140. Tsuji, H. and Horikawa, G. (2007) Porous biodegradable polyester blends of poly(L-lactic acid) and poly(ε-caprolactone): physical properties, morphology, and biodegradation. *Polymer International*, **56**, 258–266.
141. Grijpma, D.W., van Hofslot, R.D.A., Supèr, H. *et al.* (1994) Rubber toughing of poly(lactide) by blending and block copolymerization. *Polymer Engineering and Science*, **34**, 1674–1684.
142. Tsuji, H., Yamada, T., Suzuki, M. and Itsuno, S. (2003) Blends of aliphatic polyesters: VII. Effects of poly(L-lactide-co-ε-caprolactone) on morphology, structure, crystallization, and physical properties of blends of poly(L-lactide) and poly(ε-caprolactone). *Polymer International*, **52**, 269–275.
143. Thomson, R.C., Wake, M.C., Yaszemski, M.J. and Mikos, A.G. (1995) Biodegradable polymer scaffolds to regenerate organs, in *Advances in Polymer Science, Biopolymer II*, vol. **122** (eds N.A. Peppas and R.S. Langer), Springer-Verlag, Berlin, pp. 245–274.
144. Tsuji, H., Smith, R., Bonfield, W. and Ikada, Y. (2000) Porous biodegradable polyesters. 1. Preparation of porous poly(L-lactide) films by extraction of poly(ethylene oxide) from their blends. *Journal of Applied Polymer Science*, **75**, 629–637.
145. Tsuji, H., Nakano, M., Hashimoto, M. *et al.* (2006) Electrospinning of poly(lactic acid) stereocomplex nanofibers. *Biomacromolecules*, **7**, 3316–3320.
146. Ishii, D., Ying, T.H., Mahara, A. *et al.* (2009) In vivo tissue response and degradation behavior of PLLA and stereocomplexed PLA nanofibers. *Biomacromolecules*, **10**, 237–242.

147. Zong, X., Ran, S., Fang, D. *et al.* (2003) Control of structure, morphology and property in electrospun poly(glycolide-*co*-lactide) non-woven membranes via post-draw treatments. *Polymer*, **44**, 4959–4967.
148. Gedde, U.W. (1995) *Polymer Physics*, Chapman & Hall, London, Chapters 7 and 8, pp. 131–198.
149. Inkinen, S., Stolt, M. and Södergård, A. (2010) Readily controllable step-growth polymerization method for poly(lactic acid) copolymers having a high glass transition temperature. *Biomacromolecules*, **11**, 1196–1201.
150. Migliaresi, C., Cohn, D., De Lollis, A. and Fambri, L. (1991) Dynamic mechanical and calorimetric analysis of compression-molded PLLA of different molecular weights: Effect of thermal treatments. *Journal of Applied Polymer Science*, **43**, 83–95.
151. Migliaresi, C., De Lollis, A., Fambri, L. and Cohn, D. (1991) The effect of thermal history on the crystallinity of different molecular weight PLLA biodegradable polymers. *Clinical Materials*, **8**, 111–118.
152. Jamshidi, K., Hyon, S.-H., Nakamura, T. *et al.* (1986) In vitro and in vivo degradation of poly-L-lactide fibers, in *Biological and Biomechanical Performance of Biomaterials* (eds P. Christel, A. Meunier and A.J.C. Lee), Elsevier Science Publisher B.V., Amsterdam (The Netherlands), pp. 227–232.
153. Koyama, N. and Doi, Y. (1997) Miscibility of binary blends of poly((R)-3-hydroxybutyric acid) and poly((S)-lactic acid). *Polymer*, **38**, 1589–1594.
154. Ohkoshi, I., Abe, H. and Doi, Y. (2000) Miscibility and solid-state structures for blends of poly[(S)-lactide] with atactic poly[(R,S)-3-hydroxybutyrate]. *Polymer*, **41**, 5985–5992.
155. Edlund, U. and Albertsson, A.-C. (2000) Microspheres from poly(D,L-lactide)/poly(1,5-dioxepan-2-one) miscible blends for controlled drug delivery. *Journal of Bioactive and Compatible Polymers*, **15**, 214–229.
156. Degée, P., Dubois, P. and Jérome, R. (1997) Bulk polymerization of lactides initiated by aluminium isopropoxide, 3 Thermal stability and viscoelastic properties. *Macromolecular Chemistry and Physics*, **198**, 1985–1995.
157. Jamshidi, K., Hyon, S.-H. and Ikada, Y. (1988) Thermal characterization of polylactides. *Polymer*, **29**, 2229–2234.
158. Pitt, C.G., Jeffcoat, R., Zweidinger, R.A. and Schindler, A. (1979) Sustained drug delivery systems. I. The permeability of poly(ε-caprolactone), poly(DL-lactic acid), and their copolymers. *Journal of Biomedical Materials Research*, **13**, 497–507.
159. Shogren, R. (1997) Water vapor permeability of biodegradable polymers. *Journal of Environmental Polymer Degradation*, **5**, 91–95.
160. Tsuji, H., Okino, R., Daimon, H. and Fujie, K. (2006) Water vapor permeability of poly(lactide)s: Effects of molecular characteristics and crystallinity. *Journal of Applied Polymer Science*, **99**, 2245–2252.
161. Drieskens, M., Peeters, R., Mullens, J. *et al.* (2009) Structure versus properties relationship of poly(lactic acid). I. Effect of crystallinity on barrier properties. *Journal of Polymer Science Part B: Polymer Physics*, **47**, 2247–2258.
162. Chang, J.-H., An, Y.U. and Sur, G.S. (2002) Poly(lactic acid) nanocomposites with various organoclays. I. Thermomechanical properties, morphology, and gas permeability. *Journal of Polymer Science Part B: Polymer Physics*, **41**, 94–103.
163. Bhatia, A., Gupta, R.K., Bhattacharya, S.N. and Choi, H.J. (2009) Effect of clay on thermal, mechanical and gas barrier properties of biodegradable poly(lactic acid)/poly(butylene succinate) (PLA/PBS) nanocomposites. *International Polymer Processing*, **25**, 5–14.
164. Shafiei Sabet, S. and Katbab, A.A. (2009) Interfacially compatibilized poly(lactic acid) and poly(lactic acid)/polycaprolactone/organoclay nanocomposites with improved biodegradability and barrier properties: effects of the compatibilizer structural parameters and feeding route. *Journal of Applied Polymer Science*, **111**, 1954–1963.
165. Singh, S., Gupta, R.K., Ghosh, A.K. *et al.* (2010) Poly (L-lactic acid)/layered silicate nanocomposite blown film for packaging application: thermal, mechanical and barrier properties. *Journal of Polymer Engineering*, **30**, 361–375.

166. Tsuji, H. and Tsuruno, T. (2010) Water vapor permeability of poly(L-lactide)/poly(D-lactide) stereocomplexes. *Macromolecular Materials and Engineering*, **295**, 709–715.
167. Suyatma, N.E., Copinet, A., Tighzert, L. and Coma, V. (2004) Mechanical and barrier properties of biodegradable films made from chitosan and poly(lactic acid) blends. *Journal of Polymers and the Environment*, **12**, 1–6.
168. Uemura, Y., Maetsuru, Y.-S., Fujita, T. *et al.* (2006) The effect of coatings formed by low temperature tetramethoxysilane plasma treatment on water-vapor permeability of poly(L-lactic acid) film. *Korean Journal of Chemical Engineering*, **23**, 144–147.
169. Iotti, M., Fabbri, P., Messori, M. *et al.* (2009) Organic-inorganic hybrid coatings for the modification of barrier properties of poly(lactic acid) films for food packaging applications. *Journal of Polymers and the Environment*, **17**, 10–19.
170. Komatsuka, T. and Nagai, K. (2009) Temperature dependence on gas permeability and permselectivity of poly(lactic acid) blend membranes. *Polymer Journal*, **41**, 455–458.
171. Cocca, M., Lorenzo, M.L.D., Malinconico, M. and Frezza, V. (2011) Influence of crystal polymorphism on mechanical and barrier properties of poly(L-lactic acid). *European Polymer Journal*, **47**, 1073–1080.
172. Auras, R.A., Harte, B., Selke, S. and Hernandez, R. (2003) Mechanical, physical, and barrier properties of poly(lactide) films. *Journal of Plastic Film and Sheeting*, **19**, 123–135.
173. Ikada, Y. (2006) *Tissue Engineering: Fundamentals and Applications (Interface Science and Technology, vol. 8)* (series ed. A. Hubbard), Elsevier B.V., Amsterdam (The Netherlands), Chapter 3, pp. 235–421.
174. Desmet, T., Morent, R., De Geyter, N. *et al.* (2009) Nonthermal plasma technology as a versatile strategy for polymeric biomaterials surface modification: a review. *Biomacromolecules*, **10**, 2351–2378.
175. Wang, S., Cui, W. and Bei, J. (2005) Bulk and surface modifications of polylactide. *Analytical and Bioanalytical Chemistry*, **381**, 547–556.
176. Tsuji, H., Nishikawa, M., Osanai, Y. and Matsumura, S. (2007) New strategy for controlling biodegradability of biodegradable polyesters by enzyme-catalyzed surface grafting. *Macromolecular Rapid Communications*, **28**, 1651–1656.
177. Tsuji, H. and Ishida, T. (2003) Surface hydrophilicity and enzymatic hydrolyzability of biodegradable polyesters: 2. Effects of hydrophilic polymer coating. *Macromolecular Bioscience*, **3**, 51–59.
178. Tsuji, H. and Ishida, T. (2003) Poly(L-lactide): X. Enhanced surface hydrophilicity and chain-scission mechanisms of poly(L-lactide) film in enzymatic, alkaline, and phosphate-buffered solutions. *Journal of Applied Polymer Science*, **87**, 1628–1633.
179. Tsuji, H., Ishida, T. and Fukuda, N. (2003) Surface hydrophilicity and enzymatic hydrolyzability of biodegradable polyesters: 1. Effects of alkaline treatment. *Polymer International*, **52**, 843–852.
180. Ikada, Y. (2006) *Tissue Engineering: Fundamentals and Applications (Interface Science and Technology, vol. 8)* (series ed. A. Hubbard), Elsevier B.V., Amsterdam (The Netherlands), Chapter 4, pp. 424–462.
181. Ikada, Y., Shikinami, Y., Hara, Y. *et al.* (1996) Enhancement of bone formation by drawn poly(L-lactide). *Journal of Biomedical Materials Research*, **30**, 553–558.
182. Moon, S.-I., Jin, F., Lee, C. *et al.* (2005) Novel carbon nanotube/poly(L-lactic acid) nanocomposites; their modulus, thermal stability, and electrical conductivity. *Macromolecular Symposia*, **224**, 287–295.
183. Zhang, D., Kandadai, M.A., Cech, J. *et al.* (2006) Poly(L-lactide) (PLLA)/multiwalled carbon nanotube (MWCNT) composite: Characterization and biocompatibility evaluation. *Journal of Physical Chemistry*, **B110**(26), 12910–12915.
184. Tsuji, H., Kawashima, Y., Takikawa, H. and Tanaka, S. (2007) Poly(L-lactide)/nano-structured carbon composites: Conductivity, thermal properties, crystallization, and biodegradation. *Polymer*, **48**, 4213–4225.
185. Kister, G., Cassanas, G. and Vert, M. (1989) Effects of morphology, conformation and configuration on the IR and Raman spectra of various poly(lactic acid)s. *Polymer*, **39**, 267–273.
186. Ohkoshi, Y., Shirai, H., Gotoh, Y. and Nagura, M. (1999) Intrinsic birefringence of poly(L-lactide). *Sen'i Gakkaishi*, **55**, 62–68.
187. Tajitsu, Y., Hosoya, R., Murayama, T. *et al.* (1999) Huge optical rotatory power of uniaxially oriented film of poly-L-lactic acid. *Journal of Materials Science Letters*, **18**, 1785–1787.

188. Tsuji, H. (2010) *Hydrolytic degradation, in Poly(Lactic Acid): Synthesis, Structures, Properties Processing, and Applications* (Wiley Series on Polymer Engineering and Technology) (eds R. Auras, L.-T. Lim, S.E.M. Selke and H. Tsuji), John Wiley & Sons, Inc., NJ, pp. 345–381.
189. Tsuji, H. (2007) *Degradation of Poly(lactide)-Based Biodegradable Materials*, Nova Science Publishers, NY.
190. Li, S. (1999) Hydrolytic degradation characteristics of aliphatic polyesters derived from lactic and glycolic acids. *Journal of Biomedical Materials Research*, **48**, 342–353.
191. Hakarainen, M., Karlsson, S. and Albertsson, A.C. (2000) Rapid (bio)degradation of polylactide by mixed culture of compost microorganisms-low molecular weight products and matrix changes. *Polymer*, **41**, 2331–2338.
192. de Jong, S.J., Arias, E.R., Rijkers, D.T.S. *et al.* (2001) New insights into the hydrolytic degradation of poly(lactic acid): participation of the alcohol terminus. *Polymer*, **42**, 2795–2802.
193. Shih, C. (1995) A graphical method for the determination of the mode of hydrolysis of biodegradable polymers. *Pharmaceutical Research*, **12**, 2036–2040.
194. Shih, C. (1995) Chain-end scission in acid catalyzed hydrolysis of poly(D,L-lactide) in solution. *Journal of Controlled Release*, **34**, 9–15.
195. Braud, C., Devarieux, R., Garreau, H. and Vert, M. (1996) Capillary electrophoresis to analyze water-soluble oligo(hydroxyacids) issued from degraded or biodegraded aliphatic polyesters. *Journal of Environmental Polymer Degradation*, **4**, 135–148.
196. van Nostrum, C.F., Veldhuis, T.F.J., Bos, G.W. and Hennink, W.E. (2004) Hydrolytic degradation of oligo(lactic acid): a kinetic and mechanistic stud. *Polymer*, **45**, 6779–6787.
197. Kulkarni, A., Reiche, J. and Lendlein, A. (2007) Hydrolytic degradation of poly(rac-lactide) and poly[(rac-lactide)-co-glycolide] at the air–water interface. *Surface and Interface Analysis*, **39**, 740–746.
198. Thanki, P.N., Dellacherie, E. and Six, J.L. (2005) Prevailing mechanisms of the hydrolytic degradation of oligo(d,l-lactide)-grafted dextran. *European Polymer Journal*, **41**, 1546–1553.
199. Tsuji, H. (2002) *Biodegradable Polymers* (in Japanese), Corona Publishing Co., Ltd., Tokyo.
200. von Burkersroda, F., Schedl, L. and Göpferich, A. (2002) Why degradable polymers undergo surface erosion or bulk erosion. *Biomaterials*, **23**, 4221–4231.
201. Li, S.M., Garreau, H. and Vert, M. (1990) Structure-property relationships in the case of the degradation of massive aliphatic poly-(α-hydroxy acids) in aqueous media, Part 1: Poly(DL-lactic acid). *Journal of Materials Science: Materials in Medicine*, **1**, 123–130.
202. Li, S.M., Garreau, H., Vert, M. *et al.* (1990) Structure-property relationships in the case of the degradation of massive poly(α-hydroxy acids) in aqueous media, Part 3 Influence of the morphology of poly(L-lactic acid). *Journal of Materials Science: Materials in Medicine*, **1**, 198–206.
203. Södergård, A., Selin, J.F. and Näsman, J.H. (1996) Hydrolytic degradation of peroxide modified poly(L-lactide). *Polymer Degradation and Stability*, **51**, 351–359.
204. Li, S.M., Garreau, H. and Vert, M. (1990) Structure-property relationships in the case of the degradation of massive poly(α-hydroxy acids) in aqueous media, Part 2 Degradation of lactide-glycolide copolymers:PLA37.5GA25 and PLA75GA25. *Journal of Materials Science: Materials in Medicine*, **1**, 131–139.
205. Stefani, M., Coudane, J. and Vert, M. (2006) Effects of polymerization conditions on the in vitro hydrolytic degradation of plaques of poly(dl-lactic acid-block-ethylene glycol) diblock copolymers. *Polymer Degradation and Stability*, **91**, 2853–2859.
206. Li, S.M., Anjard, S., Rashkov, I. and Vert, M. (1998) Hydrolytic degradation of PLA/PEO/PLA triblock copolymers prepared in the presence of Zn metal or CaH_2. *Polymer*, **39**, 5421–5430.
207. Li, S.M. and Vert, M. (1995) Biodegradation of aliphatic polyesters, in *Biodegradable Polymers. Principles and Applications* (eds G. Scott and D. Gilead), Chapman & Hall, Cambridge, pp. 43–87.
208. Tsuji, H. and Ikarash, K. (2004) In vitro hydrolysis of poly(L-lactide) crystalline residues as extended-chain crystallites: II. Effects of hydrolysis temperature. *Biomacromolecules*, **5**, 1021–1028.
209. Makino, K., Arakawa, M. and Kondo, T. (1985) Preparation and in vitro degradation properties of polylactide microcapsules. *Chemical and Pharmaceutical Bulletin*, **33**, 1195–1201.

210. Makino, K., Ohshima, H. and Kondo, T. (1986) Mechanism of hydrolytic degradation of poly(L-lactide) microcapsules: effects of pH, ionic strength and buffer concentration. *Journal of Microencapsulation*, **3**, 203–212.
211. Cam, D., Hyon, S.H. and Ikada, Y. (1995) Degradation of high molecular weight poly(-lactide) in alkaline medium. *Biomaterials*, **16**, 833–843.
212. Tsuji, H. and Ikada, Y. (1998) Properties and morphology of poly(L-lactide). II. Hydrolysis in alkaline solution. *Journal of Polymer Science Part A: Polymer Chemistry*, **36**, 59–66.
213. Yuan, X., Mak, A.F.T. and Yao, K. (2003) Surface degradation of poly(-lactic acid) fibres in a concentrated alkaline solution. *Polymer Degradation and Stability*, **79**, 45–52.
214. Taddei, P., Monti, P. and Simoni, R. (2002) Vibrational and thermal study on the in vitro and in vivo degradation of a poly(lactic acid)-based bioabsorbable periodontal membrane. *Journal of Materials Science: Materials in Medicine*, **13**, 469–475.
215. Chu, C.C. (1982) A Comparison of the effect of pH on the biodegradation of two synthetic absorbable sutures. *Annals of Surgery*, **195**, 55–59.
216. Belbella, A., Vauthier, C., Fessi, H. *et al.* (1996) In vitro degradation of nanospheres from poly(D,L-lactides) of different molecular weights and polydispersities. *International Journal of Pharmaceutics*, **129**, 95–102.
217. Tsuji, H. and Nakahara, K. (2002) Poly(L-lactide): IX. Hydrolysis in acid media. *Journal of Applied Polymer Science*, **86**, 186–194.
218. Mason, N.S., Miles, C.S. and Sparks, R.E. (1981) Hydrolytic degradation of poly DL-(lactide). *Polymer Science and Technology*, **14**, 279–291.
219. Jung, J.H., Ree, M. and Kim, H. (2006) Acid- and base-catalyzed hydrolyses of aliphatic polycarbonates and polyesters. *Catalysis Today*, **115**, 283–287.
220. Reed, A.M. and Gilding, D.K. (1981) Biodegradable polymers for use in surgery- poly(glycolic)/poly(lactic acid) homo and copolymers: 2. In vitro degradation-. *Polymer*, **22**, 494–498.
221. Tsuji, H., Saeki, T., Tsukegi, T. *et al.* (2008) Comparative study on hydrolytic degradation and monomer recovery of poly(L-lactic acid) in the solid and in the melt. *Polymer Degradation and Stability*, **93**, 1956–1963.
222. Cha, Y. and Pitt, C.G. (1990) The biodegradability of polyester blends. *Biomaterials*, **11**, 108–112.
223. Tsuji, H., Daimon, H. and Fujie, K. (2003) A new strategy for recycling and preparation of poly(L-lactic acid): hydrolysis in the melt. *Biomacromolecules*, **4**, 835–840.
224. Tsuji, H., Yamamura, Y., Ono, T. *et al.* (2008) Hydrolytic degradation and monomer recovery of poly(butylene succinate) and poly(butylene succinate/adipate) in the melt. *Macromolecular Reaction Engineering*, **2**, 522–528.
225. Campanelli, J.R., Kamal, M.R. and Cooper, D.G. (1993) A kinetic study of the hydrolytic degradation of polyethylene terephthalate at high temperatures. *Journal of Applied Polymer Science*, **48**, 443–451.
226. Tsuji, H., Ono, T., Saeki, T. *et al.* (2005) Hydrolytic degradation of poly(ε-caprolactone) in the melt. *Polymer Degradation and Stability*, **89**, 336–343.
227. Saeki, T., Tsukegi, T., Tsuji, H. *et al.* (2005) Hydrolytic degradation of poly[(R)-3-hydroxybutyric acid] in the melt. *Polymer*, **46**, 2157–2162.
228. Saha, S.K. and Tsuji, H. (2006) Effects of molecular weight and small amounts of D-lactide units on hydrolytic degradation of poly(L-lactic acid)s. *Polymer Degradation and Stability*, **91**, 1665–1673.
229. Saha, S.K. and Tsuji, H. (2006) Hydrolytic degradation of amorphous-made films of L-lactide copolymers with glycolide and D-lactide. *Macromolecular Materials and Engineering*, **291**, 357–368.
230. Saha, S.K. and Tsuji, H. (2006) Effects of rapid crystallization on hydrolytic degradation and mechanical properties of poly(L-lactide-co-ε-caprolactone). *Reactive and Functional Polymers*, **66**, 1362–1372.
231. Migliaresi, G., Fambri, L. and Cohn, D. (1994) A study on the in vitro degradation of poly(lactic acid). *Journal of Biomaterials Science, Polymer Edition*, **5**, 591–606.
232. Çellikkaya, E., Denkbas, E.B. and Piskin, E. (1996) Poly(DL-lactide)/poly(ethylene glycol) copolymer particles. I. Preparation and characterization. *Journal of Applied Polymer Science*, **61**, 1439–1446.

233. Hyon, H., Jamshidi, K. and Ikada, Y. (1998) Effects of residual monomer on the degradation of DL-lactide polymer. *Polymer International*, **46**, 196–202.
234. Nakamura, T., Hitomi, S., Shimamoto, T. *et al.* (1987) Surgical application of biodegradable films prepared from lactide and ε-caprolactone, in *Biomaterials and Clinical Applications* (eds A. Pizzoferrato, P.G. Marchetti, A. Ravaglioli and A.J.C. Lee), Elsevier Science Publishers B.V., Amsterdam (The Netherlands), pp. 759–764.
235. Malin, M., Hiljanen-Vainio, M., Karjalainen, T. and Seppälä, J. (1996) Biodegradable lactone copolymers. II. Hydrolytic study of ε-caprolactone and lactide copolymers. *Journal of Applied Polymer Science*, **59**, 1289–1298.
236. Ye, W.P., Du, F.S., Jin, W.H. *et al.* (1997) In vitro degradation of poly(caprolactone), poly(lactide) and their block copolymers: influence of composition, temperature and morphology. *Reactive & Functional Polymers*, **32**, 161–168.
237. Nakayama, A., Kawasaki, N., Maeda, Y. *et al.* (1997) Study of biodegradability of poly(δ-valerolactone-co-L-lactide). *Journal of Applied Polymer Science*, **66**, 741–748.
238. Shirahama, H. and Yasuda, H. (2001) Synthesis of poly(lactide-ran-MOHEL) and its biodegradation with proteinase K. *Journal of Polymer Science Part A: Polymer Chemistry*, **39**, 1374–1381.
239. Tsuji, H., Eto, T. and Sakamoto, Y. (2011) Synthesis and hydrolytic degradation of substituted poly(DL-lactic acid)s. *Materials*, **4**, 1384–1398.
240. Reeve, M.S., McCarthy, S.P., Downey, M.J. and Gross, R. (1994) Polylactide stereochemistry: effect on enzymic degradability. *Macromolecules*, **27**, 825–831.
241. Tsuji, H. and Miyauchi, S. (2001) Poly(L-lactide). 6. Effects of crystallinity on enzymatic hydrolysis of poly(L-lactide) without free amorphous region. *Polymer Degradation and Stability*, **71**, 415–424.
242. Tsuji, H. and Miyauchi, S. (2001) Poly(L-lactide): 7. Enzymatic hydrolysis of free and restricted amorphous regions in poly(L-lactide) films with different crystallinities and a fixed crystalline thickness. *Polymer*, **42**, 4463–4467.
243. Duek, E.A.R., Zavaglia, C.A.C. and Belangero, W.D. (1999) In vitro study of poly (lactic acid) pin degradation. *Polymer*, **40**, 6465–6473.
244. Tsuji, H., Mizuno, A. and Ikada, Y. (2000) Properties and morphology of poly(L-lactide). 3. Effects of crystallinity on long-term in vitro hydrolysis of poly(L-lactide) film in phosphate-buffered solution. *Journal of Applied Polymer Science*, **77**, 1452–1464.
245. Tsuji, H. and Ikada, Y. (2000) Properties and morphology of poly(L-lactide). 4. Effects of structural parameters on long-term in vitro hydrolysis of poly(L-lactide) in phosphate-buffered solution. *Polymer Degradation and Stability*, **67**, 179–189.
246. Tsuji, H., Nakahara, K. and Ikarashi, K. (2001) Poly(L-lactide): 8. High-temperature hydrolysis of poly(L-lactide) films with different crystallinities and crystalline thicknesses in phosphate-buffered solution. *Macromolecular Materials and Engineering*, **286**, 398–406.
247. Tsuji, H. and Del Carpio, C.A. (2003) In vitro hydrolysis of blends from enantiomeric poly(lactide)s. 3. Homo-crystallized and amorphous blend films. *Biomacromolecules*, **4**, 7–11.
248. Loo, S.C.J., Ooi, C.P., Wee, S.H.E. and Boey, Y.C.F. (2005) Effect of isothermal annealing on the hydrolytic degradation rate of poly(lactide-co-glycolide) (PLGA). *Biomaterials*, **26**, 2827–2833.
249. Nakamura, T., Hitomi, S., Watanabe, S. *et al.* (1989) Bioabsorption of polylactides with different molecular properties. *Journal of Biomedical Materials Research*, **23**, 1115–1130.
250. Pistner, H., Bendix, D.R., Mühling, J. and Reuther, J.F. (1993) Poly(L-lactide): a long-term degradation study in vivo: Part III. Analytical characterization. *Biomaterials*, **14**, 291–298.
251. Pistner, H., Stallforth, H., Gutwald, R. *et al.* (1994) Poly(L-lactide): a long-term degradation study in vivo: Part II: physico-mechanical behaviour of implants. *Biomaterials*, **15**, 439–450.
252. Zong, X.H., Wang, Z.G., Hsiao, B.S. *et al.* (1999) Structure and morphology changes in absorbable poly(glycolide) and poly(glycolide-co-lactide) during in vitro degradation. *Macromolecules*, **32**, 8107–8114.
253. Pitt, C.G., Cha, Y., Shah, S.S. and Zhu, K.J. (1992) Blends of PVA and PGLA: control of the permeability and degradability of hydrogels by blending. *Journal of Controlled Release*, **19**, 189–200.

254. Tsuji, H. and Muramatsu, H. (2001) Blends of aliphatic polyesters. 5. Non-enzymatic and enzymatic hydrolysis of blends from hydrophobic poly(L-lactide) and hydrophilic poly(vinyl alcohol). *Polymer Degradation and Stability*, **71**, 403–415.
255. Nijenhuis, A.J., Colstee, E., Grijpma, D.W. and Pennings, A.J. (1996) High molecular weight poly(L-lactide) and poly(ethylene oxide) blends: thermal characterization and physical properties. *Polymer*, **37**, 5849–5857.
256. You, Y., Lee, S.W., Youk, J.H. *et al.* (2005) In vitro degradation behaviour of non-porous ultra-fine poly(glycolic acid)/poly(L-lactic acid) fibres and porous ultra-fine poly(glycolic acid) fibres. *Polymer Degradation and Stability*, **90**, 441–448.
257. Kishida, A., Yoshioka, S., Takeda, Y. and Uchiyama, M. (1989) Formulation-assisted biodegradable polymer matrices. *Chemical and Pharmaceutical Bulletin*, **37**, 1954–1956.
258. Shikinami, Y. and Okuno, M. (1999) Bioresorbable devices made of forged composites of hydroxyapatite (HA) particles and poly-L-lactide (PLLA): Part I. Basic characteristics. *Biomaterials*, **20**, 859–877.
259. Ignatius, A.A., Augat, P. and Claes, L.E. (2001) Degradation behavior of composite pins made of tricalcium phosphate and poly(L,DL-lactide). *Journal of Biomaterials Science, Polymer Edition*, **12**, 185–194.
260. Renouf-Glauser, A.C., Rose, J., Farrar, D.F. and Cameron, R.E. (2005) A degradation study of PLLA containing lauric acid. *Biomaterials*, **26**, 2415–2422.
261. Renouf-Glauser, A.C., Rose, J., Farrar, D.F. and Cameron, R.E. (2006) Comparison of the hydrolytic degradation and deformation properties of a PLLA-lauric acid based family of biomaterials. *Biomacromolecules*, **7**, 612–617.
262. Mauduit, J., Pérouse, E. and Vert, M. (1996) Hydrolytic degradation of films prepared from blends of high and low molecular weight poly(DL-lactic acid)s. *Journal of Biomedical Materials Research*, **30**, 201–207.
263. Tsuji, H. (2000) In vitro hydrolysis of blends from enantiomeric poly(lactide)s. 1. Well-stereocomplexed blend and non-blended films. *Polymer*, **41**, 3621–3630.
264. Tsuji, H. and Suzuki, M. (2001) In vitro hydrolysis of blends from enantiomeric poly(lactide)s. 2. Well-stereocomplexed fiber and films. *Sen'i Gakkaishi*, **57**, 198–202.
265. Tsuji, H. (2002) Autocatalytic hydrolysis of amorphous-made polylactides: effects of L-lactide content, tacticity, and enantiomeric polymer blending. *Polymer*, **43**, 1789–1796.
266. Tsuji, H. (2003) In vitro hydrolysis of blends from enantiomeric poly(lactide)s. 4. Well-homo-crystallized blend and nonblended films. *Biomaterials*, **24**, 537–547.
267. Lee, W.K., Iwata, T. and Gardella, J.A., Jr. (2005) Hydrolytic behavior of enantiomeric poly(lactide) mixed monolayer films at the air/water interface: Stereocomplexation effects. *Langmuir*, **21**, 11180–11184.
268. Karst, D. and Yang, Y. (2008) Effect of arrangement of L-lactide and D-lactide in poly[(L-lactide)-co-(D-lactide)] on its resistance to hydrolysis studied by molecular modeling. *Macromolecular Chemistry and Physics*, **209**, 168–174.
269. Leenslag, J.W., Pennings, A.J., Bos, R.R.M. *et al.* (1987) Resorbable materials of poly(L-lactide): VII. In vivo and in vitro degradation. *Biomaterials*, **8**, 311–314.
270. Schwach, G., Coudane, J., Engel, R. and Vert, M. (2002) Influence of polymerization conditions on the hydrolytic degradation of poly(DL-lactide) polymerized in the presence of stannous octoate or zinc-metal. *Biomaterials*, **23**, 993–1002.
271. Kopinke, F.-D., Remmler, M., Mackenzie, K. *et al.* (1996) Thermal decomposition of biodegradable polyester-II. Poly(lactic acid). *Polymer Degradation and Stability*, **53**, 329–342.
272. Tsuji, H., Fukui, I., Daimon, H. and Fujie, K. (2003) Poly(L-lactide): XI. Lactide formation by thermal depolymerization of poly(L-lactide) in a closed system. *Polymer Degradation and Stability*, **81**, 501–509.
273. Cam, D. and Marucci, M. (1997) Influence of residual monomers and metals on poly (L-lactide) thermal stability. *Polymer*, **38**, 1879–1884.

274. Noda, M. and Okuyama, H. (1999) Thermal catalytic depolymerization of poly(L-lactic acid) oligomer into LL-lactide: Effects of Al, Ti, Zn and Zr compounds as catalysts. *Chemical & Pharmaceutical Bulletin*, **47**, 467–471.
275. Tsuji, H. and Fukui, I. (2003) Enhanced thermal stability of poly(lactide)s in the melt by enantiomeric polymer blending. *Polymer*, **44**, 2891–2896.
276. Fan, Y., Nishida, H., Shirai, Y. *et al.* (2004) Thermal degradation behaviour of poly(lactic acid) stereocomplex. *Polymer Degradation and Stability*, **86**, 197–208.
277. Mori, T., Nishida, H., Shirai, Y. and Endo, T. (2004) Effects of chain end structures on pyrolysis of poly(L-lactic acid) containing tin atoms. *Polymer Degradation and Stability*, **84**. 243–251.
278. Fan, Y., Nishida, H., Mori, T. *et al.* (2004) Thermal degradation of poly(L-lactide): effect of alkali earth metal oxides for selective L,L-lactide formation. *Polymer*, **45**. 1197–1205.
279. Atkins, P.W. (1998) *Physical Chemistry*, 6th edn, Oxford University Press, Oxford, pp. 224–227.
280. McNeill, I.C. and Leiper, H.A. (1985) Degradation studies of some polyesters and polycarbonates: 2. Polylactide: Degradation under isothermal conditions, thermal degradation mechanism and photolysis of the polymer. *Polymer Degradation and Stability*, **11**, 309–326.
281. Kopinke, F.-D. and Mackenzie, K.J. (1997) Mechanistic aspects of the thermal degradation of poly(lactic acid) and poly(β-hydroxybutyric acid). *Anal. Appl. Pyrolysis*, **40–41**, 43–53.
282. Tokiwa, Y. and Jarerat, A. (2004) Biodegradation of poly(L-lactide). *Biotechnology Letters*, **26**, 771–777.
283. Tokiwa, Y. and Calabia, P. (2006) Biodegradability and biodegradation of poly(lactide). *Applied Microbiology and Biotechnology*, **72**, 244–251.
284. Calabia, B.P., Tokiwa, Y., Ugwu, C.U. and Aiba, S. (2010) Biodegradation, in *Poly(Lactic Acid): Synthesis, Structures, Properties, Processing, and Applications* (Wiley Series on Polymer Engineering and Technology) (eds R. Auras, L.-T. Lim, S.E.M. Selke and H. Tsuji), John Wiley & Sons, Inc., NJ, pp. 423–430.
285. Tsuji, H. and Suzuyoshi, K. (2002) Environmental degradation of biodegradable polyesters: 1. Poly(ε-caprolactone), poly[(R)-3-hydroxybutyrate], and poly(L-lactide) films in controlled static seawater. *Polymer Degradation and Stability*, **75**, 347–355.
286. Tsuji, H. and Suzuyoshi, K. (2002) Environmental degradation of biodegradable polyesters: 2. Poly(ε-caprolactone), poly[(R)-3-hydroxybutyrate], and poly(L-lactide) films in natural dynamic seawater. *Polymer Degradation and Stability*, **75**, 357–363.
287. Tsuji, H., Mizuno, A. and Ikada, Y. (1998) Blends of aliphatic polyesters: III. biodegradation of solution-cast blends from poly(L-lactide) and poly(ε-caprolactone). *Journal of Applied Polymer Science*, **70**, 2259–2268.
288. Torres, A., Li, S.M., Roussos, S. and Vert, M. (1996) Poly(lactic acid) degradation in soil or under controlled conditions. *Journal of Applied Polymer Science*, **62**, 2295–2302.
289. Ho, K.-L.G. and Pometto, A.L. III (1999) Temperature effects on soil mineralization of polylactic acid plastic in laboratory respirometers. *Journal of Environmental Polymer Degradation*, **7**, 101–108.
290. Buchanan, C.M., Dorschel, D.D., Gardner, R.M. *et al.* (1995) Biodegradation of cellulose esters: Composting of cellulose ester-diluent mixture. *Journal of Macromolecular Science, Part A: Pure and Applied Chemistry*, **A32**, 683–697.
291. Hakkarainen, M., Karlsson, S. and Albertsson, A.-C. (2000) Rapid (bio)degradation of polylactide by mixed culture of compost microorganisms-low molecular weight products and matrix changes. *Polymer*, **41**, 2331–2338.
292. Jarerat, A., Tokiwa, Y. and Tanaka, H. (2004) Microbial poly(L-lactide)-degrading enzyme induced by amino acids, peptides and poly(L-amino acids). *Journal of Polymers and the Environment*, **12**, 139–146.
293. Södergård, A., Selin, J.-F. and Pantke, M. (1996) Environmental degradation of peroxide modified poly(L-lactide). *International Biodeterioration and Biodegradation*, **38**, 101–106.
294. Tomita, K., Tsuji, H., Nakajima, T. *et al.* (2003) Degradation of poly(D-lactic acid) by a thermophile. *Polymer Degradation and Stability*, **81**, 167–171.

295. Tsuji, H. and Suzuyoshi, K. (2003) Environmental degradation of biodegradable polyesters: 4. Effects of pores and surface hydrophilicity on biodegradation of poly(ε-caprolactone) and poly[(R)-3-hydroxybutyrate] films in controlled seawater. *Journal of Applied Polymer Science*, **90**, 587–593.
296. Tsuji, H. and Ishizaka, T. (2001) Porous biodegradable polyesters. 2. Physical properties, morphology, and enzymatic and alkaline hydrolysis of porous poly(ε-caprolactone) Films. *Journal of Applied Polymer Science*, **80**, 2281–2291.
297. Tsuji, H., Suzuyoshi, K., Tezuka, Y. and Ishida, T. (2003) Environmental degradation of biodegradable polyesters: 3. Effects of alkali-treatment on biodegradation of poly(ε-caprolactone) and poly[(R)-3-hydroxybutyrate] films in controlled soil. *Journal of Polymers and the Environment*, **11**, 57–65.
298. Ikada, E. (1997) Relationship between photodegradability and biodegradability of some aliphatic polyesters. *Journal of PhotoPolymer Science and Technology*, **10**, 265–270.
299. Tsuji, H., Echizen, Y. and Nishimura, Y. (2006) Photodegradation of biodegradable polyesters: A comprehensive study on poly(L-lactide) and poly(ε-caprolactone). *Polymer Degradation and Stability*, **91**, 1128–1137.
300. Sakai, W., Kinoshita, M., Nagata, M. and Tsutsumi, N. (2001) ESR studies of photosensitized degradation of poly(L-lactic acid) via photoionization of dopant. *Journal of Polymer Science Part A: Polymer Chemistry*, **39**, 706–714.
301. Sakai, W., Sadakane, T., Nishimoto, W. *et al.* (2002) Photosensitized degradation and crosslinking of linear aliphatic polyesters studied by GPC and ESR. *Polymer*, **43**, 6231–6238.
302. Tsuji, H., Echizen, Y., Saha, S.K. and Nishimura, Y. (2005) Photodegradation of poly(L-lactic acid): Effects of photosensitizer. *Macromolecular Materials and Engineering*, **290**, 1192–1203.
303. Copinet, A., Bertrand, C., Govindin, S. *et al.* (2004) Effects of ultraviolet light (315 nm), temperature and relative humidity on the degradation of polylactic acid plastic films. *Chemosphere*, **55**, 763–773.
304. Ho, K.-L.G., Pometto, A.L. III, Hinz, P.N. *et al.* (1999) Field exposure study of polylactic acid (PLA) plastic films in the banana fields of Costa Rica. *Journal of Polymers and the Environment*, **7**, 167–172.
305. Tsuji, H., Echizen, Y. and Nishimura, Y., Y. (2006) Enzymatic degradation of poly(L-lactic acid): Effects of UV irradiation. *Journal of Polymers and the Environment*, **14**, 239–248.
306. Tsuji, H., Shimizu, K. and Sato, Y. (2012) Hydrolytic degradation of poly(L-lactic acid): combined effects of UV-treatment and crystallization. submitted to *Journal of Applied Polymer Science*, **125**, 2394–2406.
307. Thakur, K.A.M., Kean, R.T., Zupfer, J.M. *et al.* (1996) Solid state ^{13}C CP-MAS NMR studies of the crystallinity and morphology of poly(L-lactide). *Macromolecules*, **29**, 8844–8851.
308. Kolstad, J.J. (1996) Crystallization kinetics of poly(L-lactide-co-meso-lactide). *Journal of Applied Polymer Science*, **62**, 1079–1091.
309. Schmidt, S.C. and Hillmyer, M.A. (2001) Polylactide stereocomplex crystallites as nucleating agents for isotactic polylactide. *Journal of Polymer Science Part B: Polymer Physics*, **39**, 300–313.
310. Nam, J.Y., Ray, S.S. and Okamoto, M. (2003) Crystallization behavior and morphology of biodegradable polylactide/layered silicate nanocomposite. *Macromolecules*, **36**, 7126–7131.
311. Pluta, M. (2004) Morphology and properties of polylactide modified by thermal treatment, filling with layered silicates and plasticization. *Polymer*, **45**, 8239–8251.
312. Krikorian, V. and Pochan, D.J. (2005) Crystallization behavior of poly(L-lactic acid) nanocomposites: nucleation and growth probed by infrared spectroscopy. *Macromolecules*, **38**, 6520–6527.
313. Tsuji, H., Takai, H., Fukuda, N. and Takikawa, H. (2006) Non-isothermal crystallization behavior of poly(L-lactic acid) in the presence of various additives. *Macromolecular Materials and Engineering*, **291**, 325–335.
314. Tsuji, H., Kawashima, Y. and Takikawa, H. (2007) Poly(L-lactide) / C_{60} nanocomposites: effects of C_{60} on crystallization of poly(L-lactide). *Journal of Polymer Science Part B: Polymer Physics*, **45**, 2167–2176.
315. Liao, R., Yang, B., Yu, W. and Zhou, C. (2007) Isothermal cold crystallization kinetics of polylactide/nucleating agents. *Journal of Applied Polymer Science*, **104**, 310–317.
316. Pluta, M., Murariu, M., Ferreira, A.D.S. *et al.* (2007) Polylactide compositions. II. Correlation between morphology and main properties of PLA/calcium sulfate composites. *Journal of Polymer Science Part B: Polymer Physics*, **45**, 2770–2780.

317. Pan, P., Zhu, B., Dong, T. and Inoue, Y. (2008) Poly(L-lactide)/layered double hydroxides nanocomposites: Preparation and crystallization behavior. *Journal of Polymer Science Part B: Polymer Physics*, **46**, 2222–2233.
318. Pan, P., Liang, Z., Cao, A. and Inoue, Y. (2009) Layered metal phosphonate reinforced poly(L-lactide) composites with a highly enhanced crystallization rate. *Applied Materials and Interfaces*, **1**, 402–411.
319. Nam, J.Y., Okamoto, M., Okamoto, H. *et al.* (2006) Morphology and crystallization kinetics in a mixture of low-molecular weight aliphatic amide and polylactide. *Polymer*. **47**, 1340–1347.
320. Kawamoto, N., Sakai, A., Horikoshi, T. *et al.* (2007) Nucleating agent for poly(L-lactic acid)-An optimization of chemical structure of hydrazide compound for advanced nucleation ability. *Journal of Applied Polymer Science*, **103**, 198–203.
321. Yamane, H. and Sasai, K. (2003) Effect of the addition of poly(D-lactic acid) on the thermal property of poly(L-lactic acid). *Polymer*, **44**, 2569–2575.
322. Anderson, K.S. and Hillmyer, M.A. (2006) Melt preparation and nucleation efficiency of polylactide stereocomplex crystallites. *Polymer*, **47**, 2030–2035.
323. Tsuji, H., Takai, H. and Saha, S.K. (2006) Isothermal and non-isothermal crystallization behavior of poly(L-lactic acid): Effects of stereocomplex as nucleating agent. *Polymer*, **47**, 3826–3837.
324. Narita, J., Katagiri, M. and Tsuji, H. (2011) Highly enhanced nucleating effect of melt-recrystallized stereocomplex crystallites on poly(L-lactic acid) crystallization. *Macromolecular Materials and Engineering*, **296**, 887–893.
325. Wanamaker, C.L., Tolman, W.B. and Hillmyer, M.A. (2009) 2-Biopolymers: poly(D-lactide)-poly(menthide)-poly(D-lactide) triblock copolymers as crystal nucleating agents for poly(L-lactide). *Macromolecular Symposia*, **283–284**, 130–138.
326. Tsuji, H., Tashiro, K., Bouapao, L. and Narita, J. (2008) Poly(glycolide) as biodegradable nucleating agent of poly(L-lactide). *Macromolecular Materials and Engineering*, **293**, 947–951.
327. Tsuji, H., Sawada, M. and Bouapao, L. (2009) Biodegradable polyesters as crystallization-accelerating agents of poly(L-lactide). *ACS Applied Materials and Interfaces*, **1**, 1719–1730.
328. Tsuji, H., Yamamoto, S. and Okumura, A. (2011) Homo- and hetero-stereocomplexes of substituted poly(lactide)s as promising biodegradable crystallization-accelerating agents of poly(L-lactide). *Journal of Applied Polymer Science*, **122**, 321–333.
329. Kang, K.S., Lee, S.I., Lee, T.J. *et al.* (2008) Effect of biobased and biodegradable nucleating agent on the isothermal crystallization of poly(lactic acid). *Korean Journal of Chemical Engineering*, **25**, 599–608.
330. Xiao, H., Yang, L., Ren, X. *et al.* (2010) Kinetics and crystal structure of poly(lactic acid) crystallized nonisothermally: Effect of plasticizer and nucleating agent. *Polymer Composites*, **31**, 2057–2068.
331. Bai, H., Zhang, W., Deng, H. *et al.* (2011) Control of crystal morphology in poly(L-lactide) by adding nucleating agent. *Macromolecules*, **44**, 1233–1237.
332. Lebaron, P.C., Wang, Z. and Pinnavaia, T.J. (1999) Polymer-layered silicate nanocomposites: An overview. *Applied Clay Science*, **15**, 11–29.
333. Ghosh, S.B., Bandyopadhyay-Ghosh, S. and Sain, M. (2010) Composites, in *Poly(Lactic Acid): Synthesis, Structures, Properties, Processing, and Applications* (Wiley Series on Polymer Engineering and Technology) (eds R. Auras, L.-T. Lim, S.E.M. Selke and H. Tsuji), John Wiley & Sons, Inc., NJ, pp. 293–310.
334. Ray, S.S. (2010) Nanocomposites, in *Poly(Lactic Acid): Synthesis, Structures, Properties, Processing, and Applications* (Wiley Series on Polymer Engineering and Technology), (eds R. Auras, L.-T. Lim, S.E.M. Selke and H. Tsuji), John Wiley & Sons, Inc., NJ, pp. 311–321.
335. Ray, S.S. and Okamoto, M. (2003) Polymer/layered silicate nanocomposites: a review from preparation to processing. *Progress in Polymer Science*, **28**, 1539–1641.
336. Ray, S.S. and Okamoto, M. (2003) Biodegradable polylactide and its nanocomposites: opening a new dimension for plastics and composites. *Macromolecular Rapid Communications*, **24**, 815–840.
337. Ogata, N., Jimenez, G., Kawai, H. and Ogihara, T. (1997) Structure and thermal/mechanical properties of poly(L-lactide)-clay blend. *Journal of Polymer Science Part B: Polymer Physics*, **35**, 389–396.

338. Paul, M.-A., Alexandre, M., Degée, P. *et al.* (2002) New nanocomposite materials based on plasticized poly(L-lactide) and organo-modified montmorillonites: Thermal and morphological study. *Polymer*, **44**, 443–450.
339. Maiti, P., Yamada, K., Okamoto, M. *et al.* (2002) New polylactide/layered silicate nanocomposites: Role of organoclays. *Chemistry of Materials*, **14**, 4654–4661.
340. Fukuda, N., Tsuji, H. and Ohnishi, Y. (2002) Physical properties and enzymatic hydrolysis of poly(L-lactide)-$CaCO_3$ composites. *Polymer Degradation and Stability*, **78**, 119–127.
341. Fukuda, N. and Tsuji, H. (2005) Physical properties and enzymatic hydrolysis of poly(L-lactide)–TiO_2 composites. *Journal of Applied Polymer Science* **96**, 190–199.
342. Nishida, H., Fan, Y., Mori, T. *et al.* (2005) Feedstock recycling of flame-resisting poly(lactic acid)/aluminum hydroxide composite to L,L-lactide. *Industrial and Engineering Chemistry Research*, **44**, 1433–143.
343. Cao, D. and Wu, L. (2009) Poly(L-lactic acid)/silicon dioxide nanocomposite prepared via the in situ melt polycondensation of L-lactic acid in the presence of acidic silica sol: Isothermal crystallization and melting behaviors. *Journal of Applied Polymer Science*, **111**, 1045–1050.
344. Wang, R., Wang, S. and Zhang, Y. (2009) Morphology, mechanical properties, and thermal stability of poly(L-lactic acid)/poly(butylene succinate-co-adipate)/ silicon dioxide composites. *Journal of Applied Polymer Science*, **113**, 3630–3637.
345. Yuzay, I.E., Auras, R. and Selke, S. (2010) Poly(lactic acid) and zeolite composites prepared by melt processing: Morphological and physical-mechanical properties. *Journal of Applied Polymer Science*, **115**, 2262–2270.
346. Wang, D.-Y., Leuteritz, A., Wang, Y.-Z. *et al.* (2010) Preparation and burning behaviors of flame retarding biodegradable poly(lactic acid) nanocomposite based on zinc aluminum layered double hydroxide. *Polymer Degradation and Stability*, **95**, 2474–2480.
347. Qiu, Z. and Pan, H. (2010) Preparation, crystallization and hydrolytic degradation of biodegradable poly(L-lactide)/polyhedral oligomeric silsesquioxanes nanocomposite. *Composites Science and Technology*, **70**, 1089–1094.
348. Cao, Y., Feng, J. and Wu, P. (2010) Preparation of organically dispersible graphene nanosheet powders through a lyophilization method and their poly(lactic acid) composites. *Carbon*, **48**, 3834–3839.
349. Zhao, Y.-Q., Lau, K.-T. and Li, H.-L. (2008) Manufacture of a homogenous nano-diamond/poly (lactic acid) bio-engineered composite. *Advanced Materials Research*, **47–50**, 1221–1224.
350. Ke, T. and Sun, X. (2003) Melting behavior and crystallization kinetics of starch and poly(lactic acid) composites. *Journal of Applied Polymer Science*, **89**, 1203–1210.
351. Yu, L., Petinakis, E., Dean, K. and Lie, H. (2010) Poly(lactic acid)/starch blends, in *Poly(Lactic Acid): Synthesis, Structures, Properties, Processing, and Applications* (Wiley Series on Polymer Engineering and Technology) (eds R. Auras, L.-T. Lim, S.E.M. Selke and H. Tsuji), John Wiley & Sons, Inc., NJ, pp. 217–226.
352. Kondo, Y., Harima, Y., Hyakutake, K. *et al.* (2007) Synthesis and characterization of poly(lactic acid) and glucosamine-formaldehyde / dodecylbenzenesulfonate composite films. *Colloid and Polymer Science*, **285**, 1019–1027.
353. Mathew, A.P., Oksman, K. and Sain, M. (2005) Mechanical properties of biodegradable composites from poly lactic acid (PLA) and microcrystalline cellulose (MCC). *Journal of Applied Polymer Science*, **97**, 2014–2025.
354. Xiang, C., Joo, Y.L., Frey, M.W. (2009) Nanocomposite fibers electrospun from poly(Lactic Acid)/cellulose nanocrystals. *Journal of Biobased Materials and Bioenergy*, **3**, 147–155.
355. Liu, B., Jiang, L., Liu, H. and Zhang, J. (2010) Synergetic effect of dual compatibilizers on in situ formed poly(lactic acid)/ soy protein composites. *Industrial and Engineering Chemistry Research*, **49**, 6399–6406.
356. Dimzoski, B., Bogoeva-Gaceva, G., Gentile, G. *et al.* (2008) Preparation and characterization of poly(lactic acid)/rice hulls based biodegradable composite. *Journal of Polymer Engineering*, **28**, 369–384.

357. Wang, K.-H., Wu, T.-M., Shih, Y.-F. and Huang, C.-M. (2008) Water bamboo husk reinforced poly(lactic acid) green composite. *Polymer Engineering and Science*, **48**, 1833–1839.
358. Petinakis, E., Yu, L., Edward, G. *et al.* (2009) Effect of matrix-particle interfacial adhesion on the mechanical properties of poly(lactic acid)/wood-flour micro-composites. *Journal of Polymers and the Environment*, **17**, 83–94.
359. Zhou, S., Peng, H., Yu, X. *et al.* (2008) Preparation and characterization of a novel electrospun spider silk fibroin/poly(D,L-lactide) composite fiber. *Journal of Physical Chemistry B*, **112**, 11209–11216.
360. Lee, J.H., Suttle, B.M., Kim, H.J. and Meredith, J.C. Pollen: a novel, biorenewable filler for polymer composites. *Macromolecular Materials and Engineering*, **296**, 1055–1062.
361. Hyon, S.H., Jamshidi, K., Ikada, Y. *et al.* (1985) Bone Filler from Poly(lactic acid)-Hydroxyapatite Composites. *Kobunshi Ronbunshu*, **42**, 771–776.
362. Verheyen, C.C.P.M., De Wijn, J.R., Van Blitterswijk, C.A. and De Groot, K. (1992) Evaluation of hydroxylapatite/poly(L-lactide) composites: Mechanical behavior. *Journal of Biomedical Materials Research*, **26**, 1277–1296.
363. Kikuchi, M., Cho, S.-B., Tanaka, J. *et al.* (1998) Preparation and biocompatibility of β-tricalcium-phosphate/copolymerized-poly-L-lactide composite. *Journal of the Japan Society of Powder and Powder Metallurgy*, **45**, 36–40.
364. Bleach, N.C., Nazhat, S.N., Tanner, K.E. *et al.* (2002) Effect of filler content on mechanical and dynamic mechanical properties of particulate biphasic calcium phosphate - Polylactide composites. *Biomaterials*, **23**, 1579–1585.
365. Mamidwar, S., Weiner, M., Alexander, H. and Ricci, J. (2008) In vivo bone response to calcium sulfate/poly L-lactic acid composite. *Implant Dentistry*, **17**, 208–216.
366. Li, H. and Chang, J. (2004) Preparation and characterization of bioactive and biodegradable Wollastonite/poly(D,L-lactic acid) composite scaffolds. *Journal of Materials Science: Materials in Medicine*, **15**, 1089–1095.
367. Cheng, W., Li, H. and Chang, J. (2005) Fabrication and characterization of β-dicalcium silicate/poly(D,L-lactic acid) composite scaffolds. *Materials Letters*, **59**, 2214–2218.
368. Lee, J.H., Park, T.G., Park, H.S. *et al.* (2003) Thermal and mechanical characteristics of poly(L-lactic acid) nanocomposite scaffold. *Biomaterials*, **24**, 2773–2778.
369. Kylmä, J. and Seppälä, J. (2001) Ternary-phase poly(ester-urethane)/elastomer/filler composites. *Journal of Applied Polymer Science*, **79**, 1531–1539.
370. Leatherbury, N.C., Niederauer, G.G., Kieswetter, K. and Slivka, M.A. (1998) Bioglass increases the compressive stiffness of composites of poly(DL-lactide-co-glycolide). Proceedings of the Southern Biomedical Engineering Conference, p. 68.
371. Blaker, J.J., Maquet, V., Jérôme, R. *et al.* (2005) Mechanical properties of highly porous PDLLA/Bioglass® composite foams as scaffolds for bone tissue engineering. *Acta Biomaterialia*, **1**, 643–652.
372. Ignatius, A.A., Wolf, S., Augat, P. and Claes, L.E. (2001) Composites made of rapidly resorbable ceramics and poly(lactide) show adequate mechanical properties for use as bone substitute materials. *Journal of Biomedical Materials Research*, **57**, 126–131.
373. Liu, H., Slamovich, E.B. and Webster, T.J. (2005) Increased osteoblast functions on nanophase titania dispersed in poly-lactic-co-glycolic acid composites. *Nanotechnology*, **16**, S601–S608.
374. Wang, Z., Roberge, C., Wan, Y. *et al.* (2003) Biodegradable electrical bioconductor made of polypyrrole nanoparticle/poly(D,L-lactide) composite: A preliminary in vitro biostability study. *Journal of Biomedical Materials Research – Part A*, **66**, 738–746.
375. Pan, H., Jiang, H. and Chen, W. (2006) Interaction of dermal fibroblasts with electrospun composite polymer scaffolds prepared from dextran and poly lactide-co-glycolide. *Biomaterials*, **27**, 3209–3220.
376. Milella, E., Barra, G., Ramires, P.A. *et al.* (2001) Poly(L-lactide)acid/alginate composite membranes for guided tissue regeneration. *Journal of Biomedical Materials Research*, **57**, 248–257.

377. Liu, L., Won, Y.J., Cooke, P.H. *et al.* (2004) Pectin/poly(lactide-co-glycolide) composite matrices for biomedical applications. *Biomaterials*, **25**, 3201–3210.
378. Moon, Y.S., Uyama, H., Inoue, S. and Tabata, Y. (2006) Fabrication of non-woven mats of gelatin/poly(L-lactic acid) composites by electrospinning and their application for scaffold of cell proliferation. *Chemistry Letters*, **35**, 564–565.
379. Lee, S.J., Lim, G.J., Lee, J.-W. *et al.* (2006) In vitro evaluation of a poly(lactide-co-glycolide)-collagen composite scaffold for bone regeneration. *Biomaterials*, **27**, 3466–3472.
380. Kesenci, K., Motta, A., Fambri, L. and Migliaresi, C. (2001) Poly(ε-caprolactone-co-D,L-lactide)/silk fibroin composite materials: Preparation and characterization. *Journal of Biomaterials Science, Polymer Edition*, **12**, 337–351.
381. Nakamura, A. and Iji, M. (2009) Enhancement of thermal diffusivity of poly(L-lactic acid) composites with a net-like structure of carbon fibers. *Journal of Materials Science*, **44**, 4572–4576.
382. Kim, B.S., Kim, K.O. and Kim, I.S. (2011) Poly(L-lactide) composite nanofibers incorporating POSS-MWNTs. *Advanced Materials Research*, **175–176**, 341–344.
383. Levit, M.R., Farrel, R.E., Gross, R.A. and McCarthy, S.P. (1996) Composites based on poly(lactic acid) and cellulosic fibrous materials: Mechanical properties and biodegradability. *Journal of Engineering and Applied Science*, **2**, 1387–1391.
384. Funabashi, M. and Kunioka, M. (2005) Biodegradable composites of poly(lactic acid) with cellulose fibers polymerized by aluminum triflate. *Macromolecular Symposia*, **224**, 309–321.
385. Petersson, L., Kvien, I. and Oksman, K. (2007) Structure and thermal properties of poly(lactic acid)/cellulose whiskers nanocomposite materials. *Composites Science and Technology*, **67**, 2535–254.
386. Huda, M.S., Mohanty, A.K., Drzal, L.T. *et al.* (2005) 'Green' composites from recycled cellulose and poly(lactic acid): Physico-mechanical and morphological properties evaluation. *Journal of Materials Science*, **40**, 4221–4229.
387. Graupner, N. (2008) Application of lignin as natural adhesion promoter in cotton fibre-reinforced poly(lactic acid) (PLA) composites. *Journal of Materials Science*, **43**, 5222–5229.
388. Liu, L., Fishman, M.L., Hicks, K.B. and Liu, C.-K. (2005) Biodegradable composites from sugar beet pulp and poly(lactic acid). *Journal of Agricultural and Food Chemistry*, **53**, 9017–9022.
389. Wong, S., Shanks, R.A. and Hodzic, A. (2003) Poly(L-lactic acid) composites with flax fibers modified by plasticizer absorption. *Polymer Engineering & Science*, **43**, 1566–1575.
390. Lee, S.-H., Ohkita, T. and Kitagawa, K. (2004) Eco-composite from poly(lactic acid) and bamboo fiber. *Holzforschung*, **58**, 529–536.
391. Nishino, T., Hirao, K. and Kotera, M. (2006) X-ray diffraction studies on stress transfer of kenaf reinforced poly(l-lactic acid) composite. *Composites Part A: Applied Science and Manufacturing*, **37**, 2269–2273.
392. Serizawa, S., Inoue, K. and Iji, M. (2006) Kenaf-fiber-reinforced poly(lactic acid) used for electronic products. *Journal of Applied Polymer Science*, **100**, 618–624.
393. Pan, P., Zhu, B., Kai, W. *et al.* (2007) Crystallization behavior and mechanical properties of bio-based green composites based on poly(L-lactide) and kenaf fiber. *Journal of Applied Polymer Science*, **105**, 1511–1520.
394. Nishino, T., Hirao, K. and Kotera, M. (2007) Papyrus reinforced poly(L-lactic acid) composite, Advanced Composite Materials. *The Official Journal of the Japan Society of Composite Materials*, **16**, 259–267.
395. Masirek, R., Kulinski, Z., Chionna, D. *et al.* (2007) Composites of poly(L-lactide) with hemp fibers: Morphology and thermal and mechanical properties. *Journal of Applied Polymer Science*, **105**, 255–268.
396. Mohamed, A., Finkenstadt, V.L., Rayas-Duarte, P. *et al.* (2009) Thermal properties of extruded and injection-molded poly(lactic acid)-based cuphea and lesquerella bio-composites. *Journal of Applied Polymer Science*, **111**, 114–124.
397. Yu, T., Ren, J., Li, S. *et al.* (2010) Effect of fiber surface-treatments on the properties of poly(lactic acid)/ramie composites. *Composites Part A: Applied Science and Manufacturing*, **41**, 499–505.
398. Qin, L., Qiu, J., Liu, M. *et al.* (2010) Mechanical and thermal properties of poly(lactic acid) composites with rice straw fiber modified by poly(butyl acrylate). *Chemical Engineering Journal*, **166**, 772–778.

399. Sim, K.J., Han, S.O. and Seo, Y.B. (2010) Dynamic mechanical and thermal properties of red algae fiber reinforced poly(lactic acid) biocomposites. *Macromolecular Research*, **18**, 489–495.
400. Bourmaud, A. and Pimbert, S. (2008) Investigations on mechanical properties of poly(propylene) and poly(lactic acid) reinforced by miscanthus fibers. *Composites Part A: Applied Science and Manufacturing*, **39**, 1444–1454.
401. Teramoto, N., Urata, K., Ozawa, K. and Shibata, M. (2004) Biodegradation of aliphatic polyester composites reinforced by abaca fiber. *Polymer Degradation and Stability*, **86**, 401–409.
402. Mohamed, A.A., Finkenstadt, V.L., Palmquist, D.E. and Rayas-Duarte, P. (2009) Thermal properties of extruded injection-molded poly(lactic acid) and milkweed composites: Degradation kinetics and enthalpic relaxation. *Journal of Applied Polymer Science*, **111**, 175–184.
403. Huda, M.S., Drzal, L.T., Misra, M. and Mohanty, A.K. (2006) Wood-fiber-reinforced poly(lactic acid) composites: Evaluation of the physicomechanical and morphological properties. *Journal of Applied Polymer Science*, **102**, 4856–4869.
404. Huda, M.S., Drzal, L.T., Misra, M. *et al.* (2005) A study on biocomposites from recycled newspaper fiber and poly(lactic acid). *Industrial and Engineering Chemistry Research*, **44**, 5593–5601.
405. Tsuji, H., Kidokoro, Y., Mochizuki, M. (2006) Enzymatic degradation of biodegradable polyester composites of poly(L-lactic acid) and poly(ε-caprolactone). *Macromolecular Materials and Engineering*, **291**, 1245–1254.
406. Daniels, A.U., Chang, M.K. and Andriano, K.P. (1990) Mechanical properties of biodegradable polymers and composites proposed for internal fixation of bone. *Journal of Applied Biomaterials*, **1**, 57–78.
407. Majola, A., Vainionpää, S., Vihtonen, K. *et al.* (1991) Absorption, biocompatibility, and fixation properties of polylactic acid in bone tissue: An experimental study in rats. *Clinical Orthopaedics and Related Research* **268**, 260–269.
408. Rokkanen, P.U., Böstman, O., Hirvensalo, E. *et al.* (2000) Bioabsorbable fixation in orthopaedic surgery and traumatology. *Biomaterials*, **21**, 2607–2613.
409. Park, J.-M., Kim, D.-S. and Kim, S.-R. (2004) Nondestructive evaluation of interfacial damage properties for plasma-treated biodegradable poly(p-dioxanone) fiber/poly(L-lactide) composites by micromechanical test and surface wettability. *Composites Science and Technology*, **64**, 847–860.
410. Li, X. and Feng, Q. (2005) Porous poly-L-lactic acid scaffold reinforced by chitin fibers. *Polymer Bulletin*, **54**, 47–55.
411. Park, J.-M., Kim, D.-S. and Kim, S.-R. (2003) Interfacial properties and microfailure degradation mechanisms of bioabsorbable fibers/poly-L-lactide composites using micromechanical test and nondestructive acoustic emission. *Composites Science and Technology*, **63**, 403–419.
412. Thomson, R.C., Yaszemski, M.J., Powers, J.M. and Mikos, A.G. (1998) Hydroxyapatite fiber reinforced poly(α-hydroxy ester) foams for bone regeneration. *Biomaterials*, **19**, 1935–1943.
413. Mizutani, Y., Hattori, M., Okuyama, M. *et al.* (2005) Preparation of porous poly(L-lactic acid) composite containing hydroxyapatite whiskers. *Chemistry Letters*, **34**, 1110–1111.
414. Qi, R., Guo, R., Shen, M. *et al.* (2010) Electrospun poly(lactic-co-glycolic acid)/halloysite nanotube composite nanofibers for drug encapsulation and sustained release. *Journal of Materials Chemistry*, **20**, 10622–10629.
415. Gregorova, A., Hrabalova, M., Wimmer, R. *et al.* (2009) Poly(lactide acid) composites reinforced with fibers obtained from different tissue types of Picea sitchensis. *Journal of Applied Polymer Science*, **114**, 2616–2623.
416. Tsuji, H. and Ikada, Y. (2008) Stereocomplex between Enantiomeric Poly(lactide)s, in *Biodegradable Polymer Blends and Composites from Renewable Resources* (ed. L. Yu), Wiley, NJ, pp. 165–190.
417. Tsuji, H. and Tsuruno, T. (2010) Accelerated hydrolytic degradation of poly(L-lactide)/poly(D-lactide) stereocomplex up to late stage. *Polymer Degradation and Stability*, **95**, 477–484.
418. Tsuji, H. and Okumura, A. (2011) Crystallization and hydrolytic/thermal degradation of novel stereocomplexationable blend of poly(L-2-hydroxybutyrate) and poly(D-2-hydroxybutyrate). *Polymer Journal*, **43**, 317–324.

419. Fukushima, K., Chang, Y.-H. and Kimura, Y. (2007) Enhanced stereocomplex formation of poly(L-lactic acid) and poly(D-lactic acid) in the presence of stereoblock poly(lactic acid). *Macromolecular Bioscience*, **7**, 829–835.
420. Nagahama, K., Nishimura, Y., Ohya, Y. and Ouchi, T. (2007) Impacts of stereoregularity and stereocomplex formation on physicochemical, protein adsorption and cell adhesion behaviors of star-shaped 8-arms poly(ethylene glycol)-poly(lactide) block copolymer films. *Polymer*, **48**, 2649–2658.
421. Bouapao, L. and Tsuji, H. (2009) Stereocomplex crystallization and spherulite growth of low molecular weight poly(L-lactide) and poly(D-lactide) from the melt. *Macromolecular Chemistry and Physics*, **210**, 993–1002.
422. Obuchi, S. and Ogawa, S. (2010) Packing and other commercial applications, in *Poly(Lactic Acid): Synthesis, Structures, Properties, Processing, and Applications* (Wiley Series on Polymer Engineering and Technology) (eds R. Auras, L.-T. Lim, S.E.M. Selke and H. Tsuji), John Wiley & Sons, Inc., Hoboken, NJ, pp. 457–467.
423. Mochizuki, M. (2010) Textile applications, in *Poly(Lactic Acid): Synthesis, Structures, Properties, Processing, and Applications* (Wiley Series on Polymer Engineering and Technology) (eds R. Auras, L.-T. Lim, S.E.M. Selke and H. Tsuji), John Wiley & Sons, Inc., NJ, pp. 469–476.
424. Therin, M., Christel, P. and Li, S.M. *et al.* (1992) *In vivo* degradation of massive poly(α-hydroxy acids): Validation of In vitro findings. *Biomaterials*, **13**, 594–600.
425. Matsusue, Y., Yamamuro, T., Oka, M. *et al.* (1992) *In vitro* and *in vivo* studies on bioabsorbable ultra-high-strength poly(L-lactide) rods *Journal Biomaterial Resource*, **26**, 1553–1567.
426. Isotalo, T.M., Nuutine, J.-P., Vaajanent, A. *et al.* (2006) Biocompatibility properties of a new braided biodegradable urethral stent: a comparison with a biodegradable spiral and a braided metallic stent in the rabbit urethra. *BJU International*, **97**, 856–859.
427. Laine, P., Kontio, R., Lindqvist, C. and Suuronen, R. (2004) Are there any complications with bioabsorbable fixation devices? A 10 year review in orthognathic surgery. *International Journal of Oral and Maxillofacial Surgery*, **33**, 240–244.
428. Hiraishi, A. (2010) Environmental applications, in *Poly(Lactic Acid): Synthesis, Structures, Properties, Processing, and Applications* (Wiley Series on Polymer Engineering and Technology) (eds R. Auras, L.-T. Lim, S.E.M. Selke and H. Tsuji), John Wiley & Sons, Inc., NJ, pp. 477–486.
429. Matsuzawa, Y., Kimura, Z., Nishimura, T. *et al.* (2010) Removal of hydrophobic organic contaminants from aqueous solutions by sorption onto biodegradable polyesters. *Journal of Water Resource and Protection*, **2**, 214–221.
430. Boley, A., Müller, W.R. and Haider, G. (2000) Biodegradable polymers as solid substrate and biofilm carrier for denitrification in recirculated aquaculture systems. *Acquacultural Engineering*, **22**, 75–86.
431. Horiba, Y., Khan, S.T. and Hiraishi, A. (2005) Characterization of the microbial community and culturable denitrifying bacteria in a solid-phase denitrification process using poly(-caprolactone) as the carbon and energy source. *Microbes and Environments*, **20**, 25–33.
432. Yang, Q., Shang, H.T., Wang, X.L. *et al.* (2006) Anaerobic degradation of tetrachloroethylene using different co-substrates as electron donors. *Biomedical and Environmental Sciences*, **19**, 73–76.
433. Atsuta, Y., Shinbo, T., Ishihara, K. *et al.* (2009) Effectiveness of recycling system of poly(lactic acid) with hydrothermal reaction technology by life cycle assessment (in Japanese). *Environmental Science*, **22**, 11–18.
434. Landis, A.E. (2010) Cradle to gate environmental foot print and life cycle assessment of poly(lactic acid), in *Poly(Lactic Acid): Synthesis, Structures, Properties, Processing, and Applications* (Wiley Series on Polymer Engineering and Technology) (eds R. Auras, L.-T. Lim, S.E.M. Selke and H. Tsuji), John Wiley & Sons, Inc., NJ, pp. 431–441.

9

Other Polyesters from Biomass Derived Monomers

Daan S. van Es, Frits van der Klis, Rutger J. I. Knoop, Karin Molenveld, Lolke Sijtsma, and Jacco van Haveren

Wageningen University and Research Centre – Food and Biobased Research, the Netherlands

9.1 Introduction

In the transition from a fossil-based to a bio-based economy the introduction of bio-based chemicals can be achieved via two distinctly different approaches. The first approach is based on the conversion of bio-mass into existing (petro)chemicals; that is, the 'drop-in' approach. The main benefit of this approach is that it can make optimal use of the existing knowledge base and infrastructure. For example, the development of bio-based terephthalic acid, will allow for the production of bio-based PET (with reduced material carbon footprint), which can be processed under the same conditions as conventional PET. A drawback of this approach is that extensive, energy consuming defunctionalization of the biomass is required in order to obtain 'drop-in' chemicals. The second approach is based on the development of bio-based chemicals with a unique structure and functionality. In this approach the biomass is selectively defunctionalized to increase stability and reduce the number of functional groups in order to enhance selectivity, while retaining (part of) the unique structural characteristics of the biomass feedstock. This approach follows the thermodynamically most favourable path from feedstock to building block. However, one of the other benefits of this approach (new monomers and polymers with new properties) can also be considered a drawback because, both for the monomers as well as for the polymers, new synthesis and processing technologies will have to be developed. In their now famous 2004 study for the US Department of Energy (DoE) called 'Top Value Added Chemicals from Biomass', Werpy *et al.* listed the top 12 bio-based chemicals that can be considered as the main examples for the second approach [1].

Bio-Based Plastics: Materials and Applications, First Edition. Edited by Stephan Kabasci.

In their follow-up study from 2010, Bozell and Petersen revised this list based on the level of technological development and commercialization [2]. Nevertheless, in both studies only three types of bio-based difunctional building blocks were included: that is, succinic acid, furan-2,5-dicarboxylic acid (2,5-FDA or FDCA) and 1,4 : 3,6-dianhydrosorbitol (isosorbide).

In this chapter we will discuss four types of biomass derived monomers and their application in polyesters; three building blocks with a more or less unique structure from the 2004 DoE report (i.e. succinic acid, FDCA, and isosorbide), and one true 'drop-in' replacement (i.e. bio-terephthalic acid). For each type of building block the chemical structure, raw material source, and synthesis will be described. The synthesis and properties of the polyesters will then be discussed and, depending on the type of material, current and future commercial applications will be described.

9.2 Isohexide Polyesters

9.2.1 Introduction

Isohexides, also known as 1,4 : 3,6-dianhydrohexitols, are a group of chiral, rigid secondary diols based on C6-sugar alcohols. Depending on the C6 sugar source, three different isomers can be obtained by successive reduction of the aldose to the alditol, followed by acid catalysed cyclodehydration to the isohexide. The three major isomers, that is, isosorbide (1,4 : 3,6-dianhydrosorbitol), isoidide (1,4 : 3,6-dianhydroiditol) and isomannide (1,4 : 3,6-dianhydromannitol), are bicyclic structures composed of two cis-fused tetrahydrofuran rings with a 120° angle between the rings, which differ only by the orientation of the hydroxyl groups at positions 2 and 5 [3]. Whereas two isomers are symmetrical with both hydroxyl groups in the endo (isomannide) or exo (isoidide) position, the third isomer (isosorbide) has a 2-exo, 5-endo configuration (Figure 9.1).

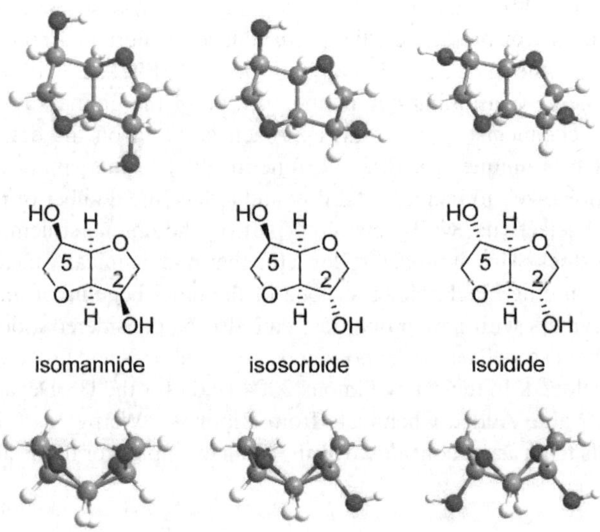

Figure 9.1 Chemical structure and spatial orientation of the three major isohexide isomers; top-view (top) and side-view (bottom) are based on DFT calculations (B3LYP-TZV).

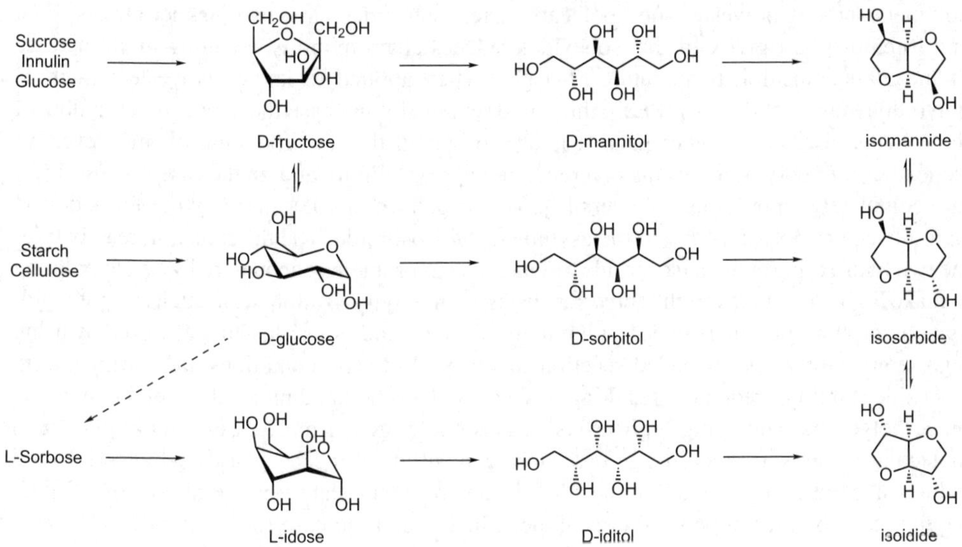

Figure 9.2 *Schematic overview of synthetic routes to isohexides.*

The position of the hydroxyl groups has significant influence on the reactivity, depending on the type of chemistry involved. Whereas the exo-hydroxyl group is less sterically hindered, the endo-hydroxyl group is more nucleophilic due to intramolecular hydrogen bonding. This difference in reactivity is an intrinsic feature in isosorbide chemistry.

Of the three isomers mentioned, only isosorbide is produced on industrial scale, which is mainly due to the limited accessibility of the precursors of the other two isomers. Isosorbide is produced from glucose in two steps in high yields, which makes the synthetic process efficient, while the feedstock glucose is abundantly available (Figure 9.2). Whereas currently starch is the major source of glucose, developments in lignocellulosic bio-refineries will ultimately result in the production of glucose from nonfood sources, thereby preventing competition with the food chain. Isomannide is prepared from mannitol, which is made by hydrogenation of fructose in 50% yield. Fructose is less commonly available than glucose, so the price and availability of fructose is significantly higher. Sources of fructose are, for example, sucrose or inulin, while also enzymatic isomerization of glucose to fructose is performed on industrial scale. In contrast to sorbitol, the cyclodehydration of mannitol to isomannide is also less efficient [3]. The third isomer, isoidide, is even less accessible since the parent hexose, that is, L-idose, is rarely found in nature and hence can only be prepared chemically. A recent patent application describing the synthesis of iditol in five steps starting from glucose, underlines the difficulty of obtaining this isomer [4]. On the other hand, according to the same authors, the cyclodehydration of iditol to isoidide is highly efficient. An alternative approach to obtaining isohexide isomers is the epimerization as described by Wright and Brandner, which involves catalytic inter-conversion of all three isomers in hydrogen atmosphere [5, 6].

Apart from various other applications such as pharmaceuticals, plasticizers and solvents, isohexides are also interesting building blocks for polymers. Due to their chirality and intrinsic rigidity isohexides have been explored for applications such as for example, liquid crystalline polymers (LCPs), powder coating resins, and engineering plastics. When built into step growth polymers such as polyesters and polycarbonates isohexides have two distinct effects. First and foremost the rigidity of the isohexides reduces chain mobility, resulting in an increase of the glass transition temperature (T_g) [7]. When applied to known polyesters, such as poly(ethylene terephthalate) (PET), this can dramatically increase the application window of these materials. On the other hand it is also observed that incorporation of high levels of isohexides into polyesters results in a reduction of crystallinity and, in the case of isosorbide, into completely amorphous polymers [8]. This is believed to be caused by the formation of stereo-irregular polymers due to the asymmetry of isosorbide. The difference in reactivity of the two hydroxyl groups in isosorbide often results in problems encountered when attempting to make high molecular weight polyesters using melt polymerization. Molecular weight build-up of amorphous polymers under melt polymerization conditions is usually achieved by using high temperatures and extended reaction times. Such drastic conditions are detrimental to isohexide stability, frequently resulting in severe colouration and molecular weight decrease. Since polyesters containing high levels of isosorbide are, in most cases, amorphous, high molecular weights are required in order to attain sufficient mechanical properties. Standard industrial practice to increase molecular weight by solid state post condensation (SSPC) requires the polymer to be semi-crystalline, which is not the case for isosorbide polyesters. Hence application of isosorbide in high molecular weight engineering plastics has not been successful yet. Of the other two isomers isomannide has been shown to be rather ineffective for obtaining polyesters with industrially relevant properties. In contrast, isoidide, due to its symmetry, gives stereo-regular polymers, resulting in semi-crystalline materials, albeit at low degrees of crystallinity [8]. Also the higher reactivity of the two exo-hydroxyl groups generally results in significantly reduced colour formation. Unfortunately, due to its limited availability data on isoidide polyesters are scarce compared to the commercially available isosorbide.

Despite the broad industrial and academic interest only a limited number of review papers have been published on isohexides. The older review by Stoss *et al.* [3] deals with isohexides in a broad context, while others focus more on their application in polymers [9]. Especially a recent review by Fenouillot *et al.* gives a convenient overview of the work published in the last two decades [8]. In this chapter we aim to introduce the reader to the most significant scientific as well as industrial developments in the area of isohexide polyesters.

9.2.2 Semi-Aromatic Homo-Polyesters

The first papers on isohexide based semi-aromatic homo-polyesters were published by Thiem *et al.* in 1984 [10, 11]. They prepared polyesters from all three isohexides by melt condensation with terephthaloyl chloride at elevated temperatures (Figure 9.3). Colourless, brittle oligomers were obtained with M_n 3000–8000. These materials exhibited very high T_gs – that is,

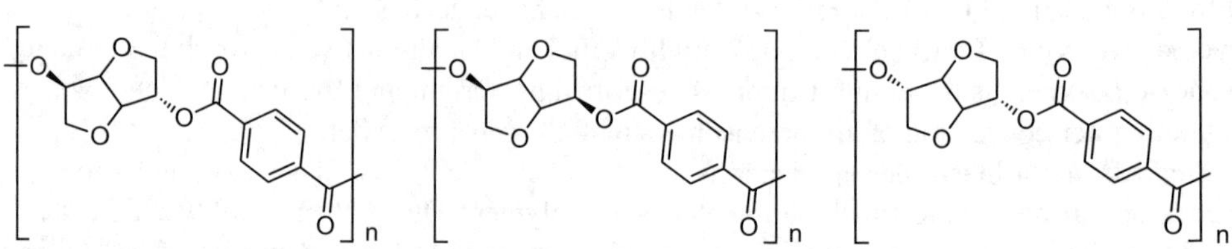

Figure 9.3 Poly(isohexide terephthalate)s (PIT) from isosorbide, isomannide and isoidide respectively.

153 °C for isosorbide and 155 °C for isoidide. The synthesis of isomannide-based polyesters was unsuccessful.

Later, Braun *et al.* reported the synthesis of a broad range of polyesters from isosorbide/isomannide and acid dichlorides *via* melt condensation [12]. The acid chlorides used include linear C4–C16 alkanoyl, ortho/iso/terephthaloyl, and 1,8- and 2,6-naphthalenoyl. Relatively high M_w aliphatic polyesters were obtained, with M_w up to 60 000 (GPC) for poly(isosorbide sebacate). Contrary to the aliphatic polyesters, the semi aromatic polyesters are highly viscous at high temperatures, hampering polycondensation. The T_gs recorded for polyesters obtained by melt polymerization of isosorbide or isomannide with terephthaloyl chloride were 147 and 156 °C, respectively.

Storbeck and Ballauff prepared polyesters from isohexides and terephthaloyl dichloride by solution polymerization in pyridine, giving colourless, fibrous materials [13]. Isomannide and isoidide yielded semi-crystalline materials; however, crystallinity could not be recovered after annealing since their glass transition temperature and melting temperature are too close to each other for them to be crystallized from the melt. Thermogravimetric analysis (TGA) of poly(isosorbide terephthalate) (PIT) showed thermal stability up to 360 °C under a nitrogen atmosphere.

Kricheldorf reported PIT synthesis by transesterification with dimethyl terephthalate, giving a polymer with a T_g of 197.5 °C [14] (Table 9.1).

Despite the broad range of molecular weights obtained with the different polymerization methods, and hence the spread in observed T_g values, it is clear that poly(isosorbide terephthalate), or PIT, is a high T_g amorphous material.

The absence of crystallinity in isosorbide polyesters can be attributed to the lack of symmetry in this building block. Storbeck showed that polyesters based on the symmetrical isoidide are semi-crystalline [13]. Under normal polymerization conditions the presence of the two different OH groups in isosorbide gives rise to the formation of stereo-irregular polyesters (Figure 9.4).

Thiem was the first to show that stereo regular poly(isosorbide terephthalate) can be prepared via a stepwise synthesis using modified isosorbide monomers [10]. This resulted in a material with a higher molecular weight (M_n 8000) and higher T_g (174 °C) compared to the irregular analogue (T_g 155 °C). The material was described as partially crystalline, however without reference to any actual measurements.

Table 9.1 *Properties of poly(isohexide terephthalate)s.*

Diol	M_n (g/mol)	T_g (°C)	T_m (°C)	Polymerization method	Reference
isosorbide	3000	155	–	Melt	[10,11]
	n.d.	147	–	Melt	[12]
	n.d.	197	–	Solution	[13]
	25 600	205[a]	–	Solution	[13]
	n.d.	197.5	–	Transesterification	[14]
isomannide	n.d.	n.d.	–	Melt	[10,11]
	n.d.	156	–	Melt	[12]
	n.d.	n.d.	–	Melt	[13]
isoidide	3800	153	–	Melt	[10,11]
	14 500	209[b]	261	Solution	[13]

Notes: n.d.: not determined. [a]Amorphous according to WAXS analysis. [b]Semi-crystalline according to WAXS analysis.

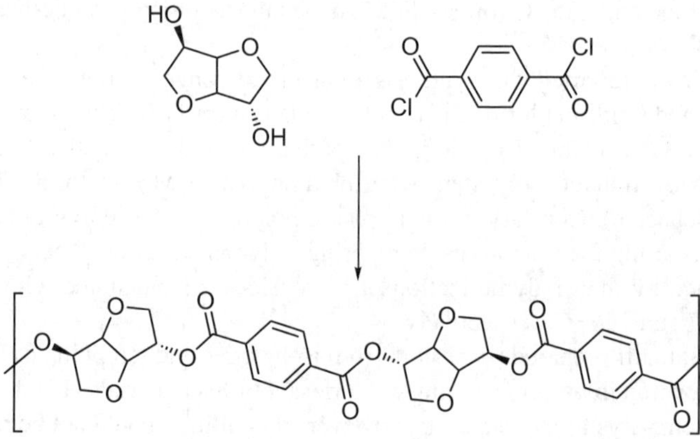

Figure 9.4 *Random /stereo-irregular poly(isosorbide terephthalate) (PIT).*

More recently Feng *et al.* reported on the same materials via comparable routes [15]. Both possible AB-type isosorbide analogues were prepared (one with a free *exo*-OH and the other with a free *endo*-OH). Differences in reactivity were observed, although both starting materials finally yielded the same stereo-regular polyester. The monomer with the free *exo*-OH yielded a polymer with a T_g of 145 °C, while the monomer with the free *endo*-OH gave a polymer with multiple T_gs, indicating a mixture of compounds. Also, the degree of polymerization was lower. However, after annealing both materials the final products showed a T_g of 150 °C and were partially crystalline. Scrambling of the stereo-regularity (during the melt) was not observed by NMR.

Another class of interesting isohexide polyesters contains furan-2,5-dicarboxylic acid (2,5-FDA or FDCA, Figure 9.5) as the aromatic diacid component. Since FDCA can be obtained from carbohydrates via various routes, this opens up possibilities for fully bio-based semiaromatic polyesters.

Storbeck and Ballauff prepared polyesters from isohexides and furan-2,5-dicarboxylic acid dichloride via solution polymerization [16]. The colourless fibrous materials had relatively high molecular weights: M_n values (membrane osmometry) for isosorbide/isomannide/isoidide were 25 000/20 400/21 500 respectively (see Table 9.2). These polyesters also had very high T_gs of 194/191/196 °C. According to WAXS analysis all materials exhibited very low degrees of crystallinity. TGA showed that these materials were stable up to temperatures in excess of 300 °C. More recently, Gomes *et.al.* prepared the same isosorbide and isoidide furanoate polyesters via similar methods, albeit with lower molecular weights and T_gs (Table 9.2) [17].

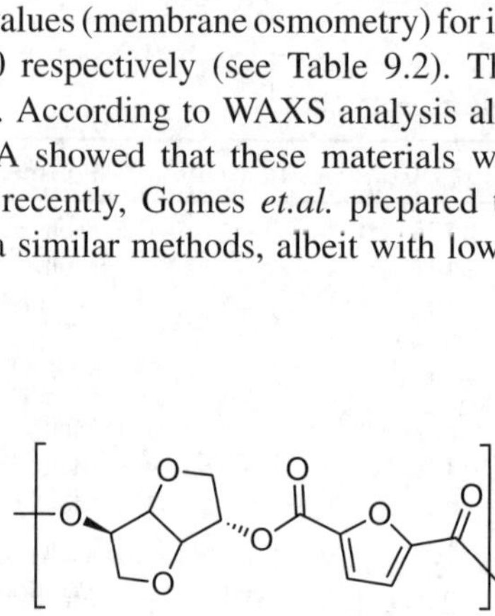

Figure 9.5 *Poly(isosorbide furan-2,5-dicarboxylate).*

Table 9.2 properties of poly(isohexide furanoates).

Diol	M_n	T_g (°C)	Reference
isosorbide	25 000	194	[16]
	13 750	180	[17]
isoidide	21 500	196	[16]
	5670	140	[17]
isomannide	20 400	191	[16]

9.2.3 Semi-Aromatic Co-Polyesters

Whereas the semi-aromatic isohexide homo-polyesters are very interesting from a scientific point of view and hold a promise for high performance applications, their extremely high glass transition temperatures, relatively low molecular weights and elaborate methods of preparation thus far have precluded any commercial applications. The use of isohexides in semiaromatic co-polyesters on the other hand has been extensively explored, mainly by industrial research groups, leading to a wealth of patent literature.

The first patents in which isosorbide based polyesters are described are from the 1960s. Most of these were aimed at fibre applications and involved co-polyesters with terephthalic acid and other monomers [18–21].

Surprisingly, it took nearly 40 years until, in 1999, DuPont started filing an extensive suit of patents dealing with the preparation and the application of PEIT, poly(ethylene-co-isosorbide terephthalate) (Figure 9.6).

Incorporation of isosorbide into poly(ethylene terephthalate) (PET) increases the T_g, allowing for various high-temperature applications, like hot-fill containers [22] and optical discs [23]. Furthermore, a range of other applications, such as films [24], sheets [25] and fibres [26], was described. Various other patent applications were filed on the preparation of isosorbide containing polyesters [27, 28]. Later the general methodology was expanded to other polyesters such as poly(1,3-propylene terephthalate) or PPT [29]. The formation of colour and of diethylene glycol (DEG) during prolonged post-condensation was also addressed [30]. Both issues are related to the relatively low reactivity of the secondary hydroxyl groups in isosorbide, giving rise to byproduct formations under melt polymerization conditions. A strategy to overcome these reactivity issues was also described [31]. In this approach isosorbide containing oligo/polyesters were prepared with a high (>40%) isosorbide content, which were subsequently reacted in the melt with a PET homo-polymer. The result is a higher isosorbide incorporation combined with high M_w. For example PIT mixed with PET yields PEIT with 26.3 mol% isosorbide content (30% theoretical) and $T_g = 139\,°C$. This material approaches the range of polycarbonate (T_g 148–150 °C). Since the extensive work by DuPont and Roquette on the development of PEIT many other patent applications on PEIT and related polyesters

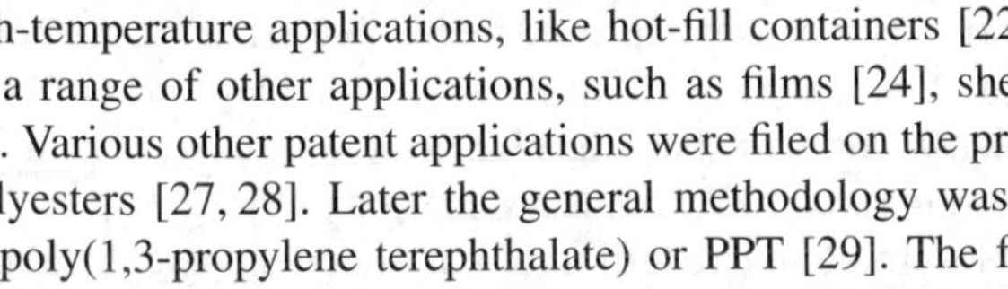

Figure 9.6 Ideal structure of PEIT.

Table 9.3 Influence of isosorbide content on PEIT properties [32]. (Data from ref. [32]. Copyright © John Wiley & Sons, Inc., 1996.)

Isosorbide (%)	T_g (°C)	T_m (°C)
0 (PET)	85	261
10.3	90	231
20.0	103	221
34.3	115	–
50.6	140	–
81.1	178	–
100 (PIT)	197	–

have been published. Although for the PEIT-type co-polyesters many patents mention or claim isohexides in general, most examples are only given for isosorbide. Data on the actual incorporation of isomannide or isoidide in PET are extremely limited, even in the scientific literature.

In the late 1990s, Storbeck and Ballauf reported the synthesis of PEIT by solution polymerization (toluene/pyridine) with terephthaloyl dichloride [32]. They showed that the T_g increased linearly with isosorbide content (Table 9.3), while T_m decreased with increasing isosorbide content with concomitant decrease in crystallinity. In general PEIT is an amorphous material when isosorbide incorporation levels exceed 20% relative to the total amount of diol. The materials obtained by Storbeck showed no signs of degradation after 15 min heating at 280 °C. M_n values were not reported but the authors mention that the degree of polymerization is in the order of 50, which is sufficient to ensure minimization of the influence of end groups, in particular when discussing T_g.

Recently Bersot *et al.* described an improved melt polymerization method for PEIT by using antimony catalysts in combination with other metals (Al, Mg or Li) [33]. Although higher polymerization efficiency and less (but still high) colouration were achieved, the maximum incorporation of isosorbide was 20% with M_n values up to 9600.

Apart from PET, isosorbide incorporation in poly(butylene terephthalate) (PBT) was also investigated. Kricheldorf *et al.* observed that in the case of melt polymerization of dimethyl terephthalate, isosorbide and 1,4-BDO insufficiently high molecular weights were obtained [14]. By changing to terephthaloyl chloride in solution, increased molecular weights were obtained, but the actual degree of incorporation of isosorbide into the polyesters was lower than was expected based on the feed composition. As in the case of PEIT the incorporation of isosorbide in PBT resulted in an increase in T_g (from 55 °C at 6% isosorbide to 92 °C at 42%), and concomitant decrease of T_m, resulting in completely amorphous material at an isosorbide content of 42% (Table 9.4). In an alternative approach Sablong *et al.* used solid-state post-condensation (SSPC) to incorporate isosorbide in PBT [34]. The block-like structures that were obtained allow for crystallization to take place, enabling SSPC. While co-polyesters with M_n values of approximately 11 000 were obtained from melt polymerization, SSPC yielded materials with $M_n > 90\,000$.

9.2.4 Aliphatic Polyesters

Aliphatic isohexide polyesters have been studied with various types of applications in mind. Noordover *et al.* described co-polyesters of succinic acid with combinations of isosorbide/isoidide and various other diols for powder coating application [35]. Depending on

Table 9.4 Influence of isosorbide content on PBIT properties.

Percentage isosorbide in feed	Percentage isosorbide incorporated	T_g (°C)	T_m (°C)	Ref
0 (PBT)	0	–	226	[14]
0 (PBT)	0	43	211–220	[34]
10	6	55	208	[14]
14	11	59	207	[34]
20	18	58	191	[14]
20	17	69	199	[34]
28	25	78	194	[34]
30	24	84	162	[14]
40	30	78	155	[14]
50	42	92	–	[14]
100 (PIT)	100	197	–	[14]
100 (PIT)	100	186	–	[34]

the isohexide content, and the type of second diol, materials with T_g's varying between −18 and 56 °C were obtained: in this case again T_g's increased with higher isohexide content. Higher M_n values (approx. 6000 versus 3000) were obtained for the isoidide based polyesters compared to the isosorbide based polyesters.

Homo-polyesters of isohexides and linear aliphatic diacids have been reported by various groups (see Table 9.5). Braun *et al.* reported the synthesis of polyesters from isosorbide or isomannide with a variety of linear aliphatic acid dichlorides (C4–C16) *via* melt condensation, giving relatively high molecular weight polymers [12]. Comparable polyesters from all three isohexides were prepared by Okada [36–38], Noordover *et al.* also described the synthesis

Table 9.5 Properties of the homo-polyesters derived from isohexides and various aliphatic diacids.

Diol	Diacid	M_n	T_g (°C)	T_m (°C)	Reference
isosorbide	Succinic (C4)	–	77	–	[40]
	Succinic (C4)	7500	65	–	[12]
	Succinic (C4)	2400	57	–	[35, 39]
	Succinic (C4)	7400	36	–	[36]
	Adipic (C6)	13 000	25	–	[12]
	Adipic (C6)	26 000	21	–	[37]
	Sebacic (C10)	23 000	0	–	[12]
	Sebacic (C10)	20 000	−10	61	[37]
	cis-CHDA	11 000	146	–	[40]
isomannide	Succinic (C4)	10 000	75	175	[36]
	Succinic (C4)	6000	60	–	[12]
	Adipic (C6)	8800	30	–	[12]
	Adipic (C6)	16 000	28	–	[36]
	Sebacic (C10)	13 700	0	–	[12]
	Sebacic (C10)	18 000	−8	–	[38]
	cis-CHDA	–	133	–	[40]
isoidide	Succinic (C4)	4 200	73	171	[35, 39]
	Adipic (C6)	34 000	45	164	[38]
	Sebacic (C10)	28 000	0	134	[38]
	cis-CHDA	–	115	–	[40]

and properties of isohexide succinates [35, 39]. They showed that while poly(isosorbide succinate) is amorphous, the isoidide based polyester is semi-crystalline. In a series of papers Okada *et al.* reported on the synthesis and biodegradability of various aliphatic isohexide polyesters [36–38]. The work of Kricheldorf *et al.* shows that higher T_gs can be obtained when a more rigid aliphatic diacid like cis-cyclohexane dicarboxylic acid (CHDA) is used [40]. The homo-polyester of isosorbide and cis-CHDA was prepared by three different methods, but only polycondensation of isosorbide and CHDA dichloride yielded satisfactory molecular weights ($M_n = 11{,}000$). MALDI-TOF-MS analysis revealed a high content of cyclics. The homo-polyesters of CHDA and all three isohexides displayed high T_gs (146 °C for isosorbide, 133 °C for isomannide and 115 °C for isoidide).

The case of poly(isosorbide succinate) (PIS) shows that it is difficult to compare results from different groups. Differences in methods of preparation, measured molecular weight distribution, and methods to obtain molecular weight distributions lead to a broad range of T_g values, apparently not related to the value of M_n.

9.2.5 Modified Isohexides

The relatively low reactivity of the secondary hydroxyl groups in isosorbide has posed serious challenges to obtaining low colour, high molecular weight polyesters, especially in the case of semi-aromatic (co-)polyesters. A way to solve these issues is to increase the reactivity of the bicyclic secondary diol. By reacting isosorbide with ethylene oxide, the diethylene glycol ether (i.e. bis(2-hydroxyethyl)isosorbide) can be obtained as described by DuPont [41]. This primary diol was significantly more reactive, which allowed for quantitative incorporation into co-polyesters. Unfortunately, the high flexibility of the ethylene glycol ether units resulted in an actual decrease of the T_g comparable to the incorporation of diethylene glycol in PET.

More recently Wu *et al.* reported an alternative approach based on a 1-carbon extension of the isohexide skeleton, yielding a new family of isohexide based building blocks (Figure 9.7) [42]. By directly attaching the primary hydroxyl bearing groups to the bicyclic ring structure the authors expect that rigidity is more likely to be maintained. Detailed investigations into the effects on polymer properties of these building blocks have shown that whereas 1-carbon extension results in some decrease in T_g, reactivity increases dramatically. Furthermore, due to symmetry and higher mobility, incorporation of 1-carbon extended isohexides results in semi-crystalline polymers, which allows for mild solid state post-condensation (SSPC) to further increase molecular weight [7, 43, 44].

Overall we can conclude that isohexides, and in particular isosorbide, are unique renewable rigid diols, which are capable of dramatically altering polymer properties when incorporated into polyesters. Whereas the beneficial effects of the intrinsic rigidity are obvious, and of high industrial importance, various other issues, like reduced reactivity, colour formation and reduced crystallinity have so far hampered large-scale industrial implementation. The wealth of both academic and industrial knowledge obtained with this type of molecule suggests that it should be possible to find solutions to use these unique and interesting building blocks effectively in polyesters. This can be achieved by developing new polymerization methods specifically suited to meet the requirements of bio-based building blocks such as isosorbide, or by using more reactive isohexide species such as isoidide or the 1-carbon extended derivatives. In order for the latter two to become successful more economical, industrially viable routes to these monomers will have to be developed, so as to unlock their full potential.

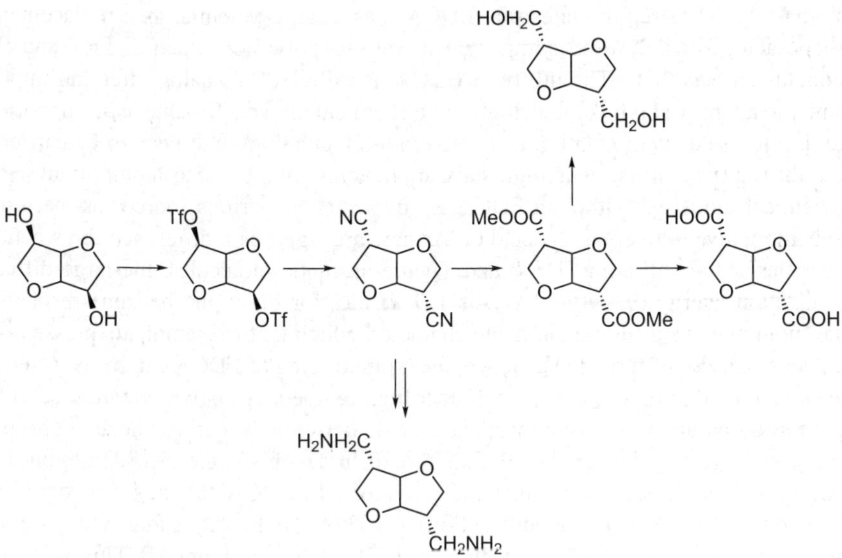

Figure 9.7 *Novel family of 1-carbon extended isohexide building blocks.*

9.3 Furan-Based Polyesters

9.3.1 Introduction

Furan derivatives have been investigated as potential renewable building blocks for polymers for over a century. Initially, thermosetting resins based on furfural or furfuryl alcohol found use in a range of applications [45]. Although not yet commercially available on a large scale, furan-2,5-dicarboxylic acid (2,5-FDA or FDCA, Figure 9.8) is probably one of the most interesting furan-based monomers (based on the properties of the monomer as well as on the characteristics of FDCA based polymers) [17, 46–50]. In the 2004 DOE study mentioned above, Werpy *et al.* rated FDCA as second in their top 12 value-added chemicals

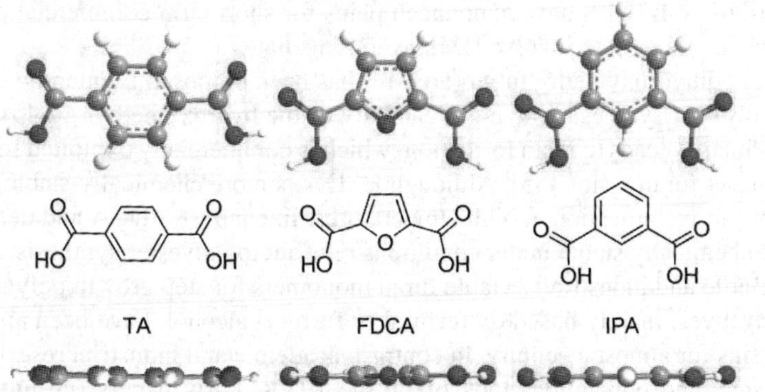

Figure 9.8 *Structural comparison between terephthalic acid (TA), furan-2,5-dicarboxylic acid (FDCA), and isophthalic acid (IPA); top-view (top) and side-view (bottom) are based on DFT calculations (B3LYP-TZV).*

from biomass [1]. The report states that FDCA ‘has a large potential as a replacement for terephthalic acid, a widely used component in various polyesters, such as PET and PBT’. Their conclusion was that: ‘The utility of FDCA as a PET/PBT analog offers an important opportunity to address a high volume, high value chemical market. To achieve this opportunity, R&D to develop selective oxidation and dehydration technology will need to be carried out. However, the return on investment might have applicability of interest to an important segment of the chemical industry.’ Although FDCA is advocated by various sources as the ultimate bio-based alternative to terephthalic acid (TA), there are significant differences between these two molecules. Although both FDCA and TA are aromatic molecules, the large difference in delocalization energy (92 kJ/mol versus 151 kJ/mol for furan and benzene respectively) gives the furan nucleus a significant diene character, which is, for example, expressed by the Diels–Alder reactivity of furan [51]. Hence, the aromatic ring of FDCA can be expected to be more reactive than that of TA, especially under high temperature polymerization conditions. Another relevant parameter for polymers is the angle between the carboxylic acid groups that determines the degree of linearity of the monomer. In TA this angle is 180°, giving highly linear, rigid, and often easily crystallizable polymers. In FDCA this angle is significantly smaller ranging from 129.4° (X-ray diffraction) to 133.6° (DFT) [42], and hence closer to the value of 120° found for isophthalic acid (IPA), as is illustrated in Figure 9.8. This difference in angles will have profound effects on for example, crystallization kinetics, crystal structure, and degree of crystallinity [46, 52]. Furthermore, the difference in dipole moment will influence polymer properties [53].

Furan-based monomers for polyesters can be obtained from two major classes of bio-based feedstocks; that is, pentosan based and hexose based [54]. Pentosans are obtained from hemicellulose containing agricultural waste streams such as corn cobs, oat hulls, or wood chips, giving furfural (furan-2-carbaldehyde) as primary intermediate, which, by means of carbonylation or hydroxymethylation, can be transformed into the platform intermediate 5-hydroxymethylfurfural (HMF) (Figure 9.9). Hexoses, mainly D-glucose and D-fructose, are readily available by hydrolysis of polysaccharides such as starch, cellulose or inulin respectively and can be directly transformed into HMF by cyclodehydration (Figure 9.9). From HMF, by either oxidation or reduction of the functional groups, three types of monomers relevant for polyester synthesis can be obtained: a diol (2,5-dihydroxymethylfuran or DHMF), a hydroxyacid (5-hydroxymethylfuroic acid or HMFA), and a diacid (FDCA). In spite of its versatility as a platform molecule, HMF is currently not produced on an industrial scale, although various companies have announced plans for short term commercial production of FDCA, which are thought to involve HMF as intermediate.

Whereas 2,5-dihydroxymethylfuran (DHMF) has been proposed frequently as a diol component for polyester synthesis, the high reactivity of the hydroxymethyl groups under acidic conditions primarily leads to resin formation, which is commercially exploited for the production of thermoset furan resins [55]. Although HMFA is more chemically stable than DHMF, it is also prone to resinification at high temperatures. In contrast, FDCA and derivatives have proven to be chemically stable under conditions relevant to polyester synthesis, making them the most versatile and industrially viable furan monomers for step growth polymers.

Furan derivatives, mainly based on furfural or furfuryl alcohol, have been applied in thermosetting resins for almost a century. In contrast, academic and industrial research into furan based polyesters and polyamides took off in the 1950s. Early papers (up until the 1970s) are usually rather qualitative, generally just describing the polymer synthesis, combined with rudimentary property analysis. Since the 1980s, more quantitative scientific papers have been

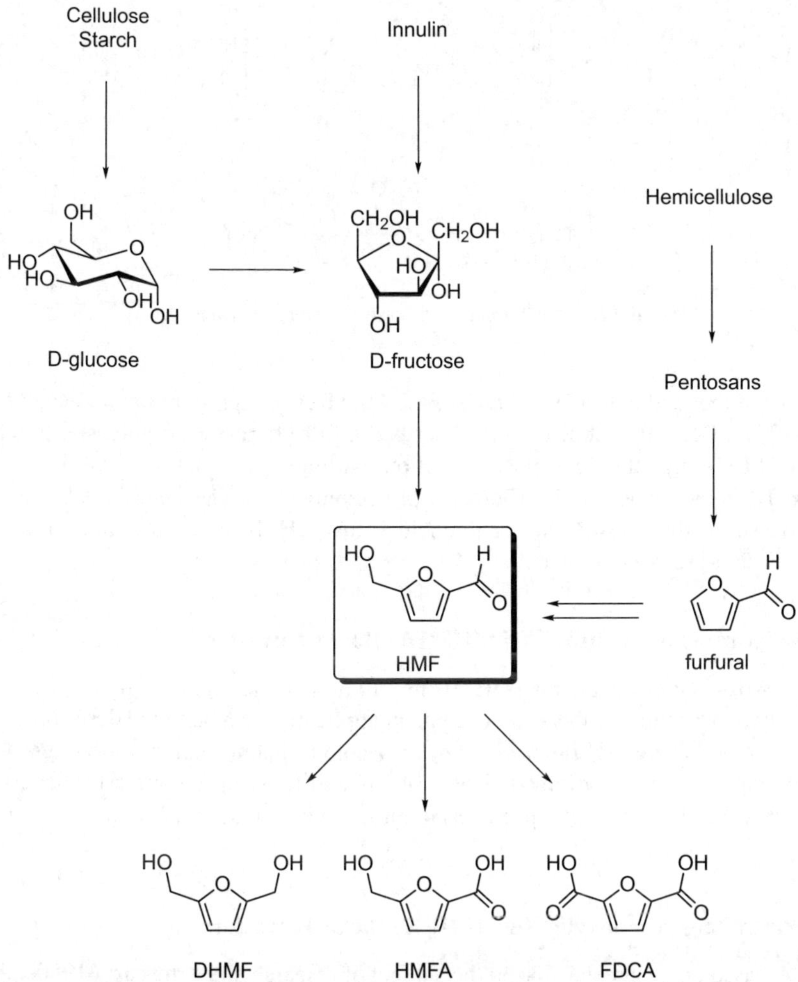

Figure 9.9 *Routes from polysaccharides to furan building blocks via HMF.*

published, including extensive property analyses with a focus on industrial applicability. More recently, in the wake of the substantial developments in bio-fuels and bio-refineries a strong growth in both scientific papers as well patent applications can be seen, with a focus on the synthesis and polymer applications of FDCA.

In this chapter attention is focused mainly on 2,5-FDCA. However, also DHMF and HMFA based polyesters will be discussed briefly in order to gain a better overview of the full potential of furan based polyesters.

9.3.2 2,5-Dihydroxymethylfuran (DHMF)-Based Polyesters

The first reported attempts to prepare DHMF based polyesters, in combination with FDCA (Figure 9.10), are from the mid-1970s, however with inconclusive results [56]. Later, Kelly and

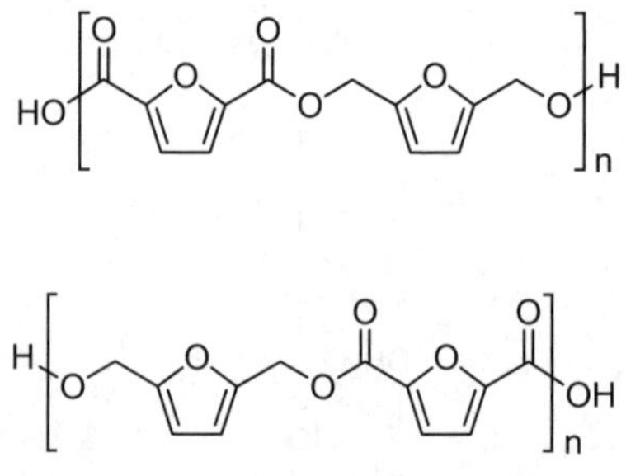

Figure 9.10 *Ideal structure of polyester from DHMF and FDCA.*

Moore reacted the acid chloride of FDCA with DHMF, yielding only low molecular weight polyesters [57]. Recently, Gomes *et al.* described a fully furan based polyester *via* the acid chloride of FDCA by interfacial polymerization, with an M_n of 3800 $g.mol^{-1}$, and a T_g of 83 °C (no T_m is mentioned) [17]. The onset of decomposition was determined to be around 205 °C. Based on these results we can conclude that DHMF is too unstable to be used as monomer in step-growth polymers.

9.3.3 5-Hydroxymethylfuroic Acid (HMFA) Based Polyesters

Although HMFA is more stable than DHMF, its still limited thermal stability is reflected in the fact that only two scientific papers and one patent application can be found describing HMFA-based polyesters. Whereas Hirai *et al.* were only able to obtain macrocyclic oligo-esters by solution polymerization in pyridine [58, 59], Moore and Kelly used trans-esterification of the methyl ester of HMFA, also giving materials in low yield and with low molecular weights [60].

9.3.4 Furan-2,5-Dicarboxylic Acid (FDCA) Based Polyesters

The FDCA based polyesters have been the subject of research since the early 1950s. A patent to the Celanese Corporation in 1951 is one of the first publications on poly(ethylene furan-2,5-dicarboxylate) (PEF, Figure 9.11, n = 2) [61]. This was obtained by using either the free diacid or the dimethylester, having a melting point of 205–210 °C. Later that decade, Hachihama *et al.* reported the preparation of polyesters based on FDCA dimethylester and aliphatic diols ranging from ethylene glycol to 1,6-hexanediol, for which only melting points were recorded [62]. A Czech patent from 1959 only gives a basic description of the preparation of PEF [63], while another Czech patent from 1960 describes co-polyesters from dimethylterephthalate, FDCA, and ethylene glycol (PEFT), having a melting point of 200 °C and a molecular weight of 20 000 [64]. Somewhat later Heertjes and Kok also reported the synthesis of PEF as well

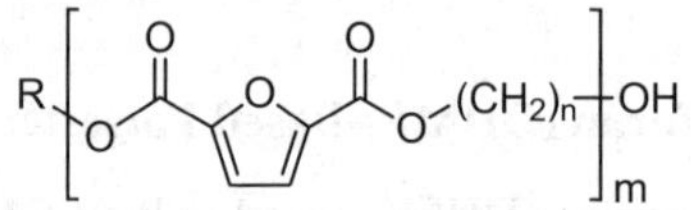

Figure 9.11 *Structure of polyesters of FDCA with linear α-ω-diols; R = H or Me, n = 2, 3, 4, 6, 8, 12, 18.*

as a polyester based on FDCA and bisphenol A [65]. Again only very rudimentary properties were given. The paper by Moore *et al.*, describing polyesters prepared from FDCA dimethyl ester and a series of different diols including DHMF, ethylene glycol and 1,6-hexanediol, fits the same trend, focusing on polymer synthesis rather than properties [57].

Storbeck and Ballauff were the first to report on polyesters based on FDCA and dianhydrohexitols [16]. Here, the acid chloride of FDCA was reacted with all three dianhydrohexitol isomers, in 1,1,2,2-tetrachloroethane in the presence of pyridine, giving white, fibrous polyesters. This work has been discussed in detail in the preceding section on isohexide polyesters (section 9.2.2). Later, Gandini and co-workers performed some kinetic studies on the transesterification of the FDCA dimethyl ester [66–68]. However, little attention was paid to polymer properties. More recently a series of papers on FDCA polyesters has been published by various groups, indicating the renewed interest in this subject.

Grosshardt *et al.* prepared a series of furan based polyesters, by reacting FDCA dimethyl ester with 1,3-propanediol, 1,6-hexanediol, 1,12-dodecanediol and 1,18-octadecanediol, using calcium acetate and antimony(III) oxide as catalysts [47]. Molecular weights and thermal properties were determined for all polyesters and are summarized in Table 9.6.

Gandini and co-workers followed a different approach by first separately synthesizing the (ethylene glycol)-diester of FDCA, which was subsequently polymerized to make PEF [46]. This semi-crystalline material had a T_g of 75–80 °C and a T_m at 210–215 °C. TGA analysis showed that the material was stable up to 300 °C. Since molecular weight analysis was performed by hydroxyl end-group functionalization with pentafluorobenzoyl chloride, the reported degree of polymerization of 250–300 was probably overestimated. Later, the same group reported some preliminary results on FDCA polyesters, including diols such as ethylene glycol, 1,3-propanediol, isosorbide and bis(1,4-hydroxymethyl)benzene [48]. In a follow-up paper the authors systematically compared different polymerization procedures for FDCA and

Table 9.6 *Combined data of various diols in FDCA based polyesters.*

Sample	M_n (g.mol^{-1})	M_w (g.mol^{-1})	PDI	T_g (°C)	T_m (°C)	E-modulus (MPa)	σ_{max} (MPa)	ε_{break} (%)
PEF (C2)	22.400[a]	44.500[a]	1.99	80	215			
	45.500[b]	25.200[c]		90	210	2.070	66.7	4.2
	105.300[c]							
PPF (C3)	21.600[a]	27.600[a]	1.28	50	174			
	49.000[b]	8.9800[c]	1.49	58	177	1.550	68.2	46
	6.0200[c]	30.600[d]	2.2	53				
	13.900[d]							
PBF (C4)	1.7800[c]	4.200[c]	2.38	31	172	1.110	19.8	2.8
PHF (C6)	3.2100[c]	6.6700[c]	2.08	28	148	493	35.5	210
	13.400[d]	22.300[d]	1.67		147			
POF (C8)	2.0700[c]	4.7500[c]	2.29	22	149	340	20.3	15
PET				76[e]	250[e]			
PBT				49[f]	227[f]			

Notes: [a]Measured by SEC, 70/20/10 (v/v/v) mixture of dichloromethane/chloroform/1,1,1,3,3,3-hexafluoro-2-propanol (HFIP) 70/20/10 was used as the eluent, compared to polystyrene [17]. [b]Measured by elemental analysis after hydroxyl end-group functionalization with pentafluorobenzoyl chloride [17]. [c]Measured by SEC, HFIP as eluent, compared to polymethylmethacrylate [49]. [d]Measured by SEC, HFIP as eluent, compared to polymethylmethacrylate [47]. [e]Ref. [70]. [f]Refs. [14, 34, 70].

the previous series of diols, now also including isoidide, DHMF and hydroquinone [17]. An overview of their results is included in Table 9.6.

Ma *et al.* investigated the copolymerization reactivity of ethylene glycol and 1,4-butanediol with FDCA catalysed by $Ti(OC_4H_9)_4$ [50]. Besides this reactivity study, they also performed a thermal analysis of the different co-polyesters. All polyesters were obtained as semi-crystalline materials with T_g's varying from 83 °C to 36 °C for PEF and PBF respectively. The melting temperatures were determined as 201 °C and 169 °C for PEF and PBF respectively. The reactivity of both diols towards FDCA was also evaluated and a large difference in reactivity was observed. Assuming first order kinetics, rate constants of 3.1×10^{-3} min^{-1} and 4.7×10^{-2} min^{-1} were determined. Molecular weights for the obtained polyesters were determined by Ubbelohde capillary viscosity measurements using phenol/ 1,1,2,2-tetrachloroethane (3/2 (m/m)) as solvent at four different concentrations. M_vs of 26 000 and 24 000 $g.mol^{-1}$ were obtained. However, for the calculation the Mark Houwink parameters for PET were used because these values are not yet known for 2,5-FDCA based polymers. Unfortunately, no further mechanical analysis of the various block co-polyesters was performed.

Jiang *et al.* reported the direct esterification of FDCA with a range of diols from ethylene glycol up to 1,8-octanediol, using tetrabutyl titanate as catalyst [49]. All polymers were obtained as brown to yellow products, depending on the maximum polymerization temperature. The authors described several structural and mechanical analyses used in order to evaluate the full potential of furan based polyesters. It appeared that all obtained polyesters were semi crystalline polymers, however, the sharp peaks of PEF, PHF (n = 6) and POF (n = 8) indicate a high macromolecular order. A relation between specific viscosity (η_{sp}) of the polyesters of 0.5 g dL^{-1} in 1,1,2,2-tetrachloroethane/phenol (1 : 1, w/w) and molecular weights was estimated. These molecular weights were determined by SEC in HFIP using PMMA standards (Table 9.6). Furthermore, densities and contact angles were determined. The tensile moduli of all polyesters were determined as well as the mechanical properties as function of temperature (DMTA). The data are included in Table 9.6. DMTA shows a clear loss of mechanical properties above the T_g for all polyesters indicating that these samples were amorphous.

Gruter *et al.* from the Avantium company evaluated different catalysts for the polycondensation of FDCA or its dimethyl ester with ethyleneglycol and 1,4-butanediol. They concluded that starting from the dimethyl ester of FDCA higher reaction rates can be achieved, and hence shorter reaction times and concomitant reduction in colour [69]. In addition, a patent by the same authors describes the synthesis of high molecular weight PEF by solid state polymerization [70]. Also recently Dam *et al.* reported some application properties of PEF, like gas barrier properties and mechanical properties after recycling [71]. Gas permeation measurements showed that the barrier properties of PEF are superior to those of PET: 2 times better for H_2O, >2 times better for CO_2, and >6 times better for O_2. According to the authors, this may allow for new packaging applications that are currently unattainable for PET, such as for beer, wine, vegetable oils, and so on.

9.3.5 Future Outlook

The ongoing developments with regard to biorefineries and the bio-based economy have sparked renewed interest in the known bio-based building block FDCA. Although academic research so far has shown that FDCA-based polyesters have interesting properties, and are in many ways comparable to terephthalic acid based polyesters, various challenges remain to be overcome in order to come to full-scale industrial production. First of all, large-scale cost-effective production of resin grade FDCA needs to become a reality. According to Dam *et al.* 'Due to its superior performance, compared to pTA, the potential market size of FDCA can exceed 50 million tons per year. The main price drivers for bio-based FDCA are the feedstock-price and economy of scale. However, at >300 000 t/a scale we believe that the price of FDCA will be <EUR 1000 per ton, and therefore competitive with pTA produced at the same scale' [71]. Secondly, despite the growing knowledge base on FDCA polyester (i.e. PEF) properties, there is still a considerable need for more insight into properties and behaviour, necessary to identify potential fields of application, such as packaging, textile or industrial fibres, coatings, and so on. Furthermore, industrial processing of FDCA polyesters by, for example, extrusion, injection moulding, fibre spinning and so on will need to be developed in order to come to viable production processes for polyester products. Last but not least, it should be clear that the transition of a petrochemical based material to a bio-based material must not only be economically feasible but also ecologically sound. In this respect, a recent paper by Eerhart *et al.* on the energy and greenhouse gas (GHG) balance of the production of poly(ethylene furanoate) (PEF) shows, in principle, that PEF can be produced sustainably [72]. The authors concluded that the production of PEF can reduce the nonrenewable energy use (NREU) by approx. 40% to 50%, while the GHG emissions can be reduced by approx. 45%–55% compared to PET. However, as the authors mention, the assessment is based on modelled process data, so the technology must first be proven in demonstration plants, followed by an update of the environmental assessment, if necessary.

Hence it can be concluded that, provided that most of the issues mentioned above are addressed, FDCA based polyesters have a bright future.

9.4 Poly(Butylene Succinate) (PBS) and Its Copolymers

The interest in poly(butylene succinate) (PBS) and its copolymers finds its origin in the 1980s, mainly based on its biodegradability. However, PBS also combines excellent processability with good mechanical properties. Poly(butylene succinate) is synthesized from succinic acid and 1,4-butanediol (BDO). Via copolymerization with other dicarboxylic acids or diols, the properties of PBS can be varied over a wide range. Compared to biopolymers such as poly(lactic acid) (PLA) and polyhydroxyalkanoates (PHAs) the properties of PBS are most similar to commodity plastics like polyethylene (PE) and polypropylene (PP). Currently, there is renewed interest in PBS, this time with a focus on developing bio-based PBS. Apart from environmental issues, these developments are also cost driven as it is expected that the C4 building blocks could become more cost effective when produced from biological origin.

9.4.1 Succinic Acid

At present succinic acid is a specialty chemical with an annual production volume of about 30 000 tons worldwide. Fossil-based succinic acid is most commonly prepared via hydrogenation of maleic anhydride (by oxidation of *n*-butane or benzene) [73]. In the field of bio-based chemicals and building blocks succinic acid is considered to be one of the most important platform chemicals [1, 74, 75], and as a result of the introduction of biosuccinic acid the production volume is expected to double or triple within years. Several fermentation processes have been described to produce bio-based succinic acid. Common feedstocks for these processes include glucose, starch and xylose [76]. The commercial potential for bio-succinic acid is illustrated by the numerous initiatives by companies that are working towards, or already are, at pilot scale production. Most prominent are the activities of Bioamber, a joint venture of Diversified Natural Products Inc. (DNP, United States) and Agroindustrie Recherches et Dévelopements (ARD, France). The Bioamber pilot plant started producing bio-succinic acid based on wheat-derived glucose, using an *Escherichia coli* on a 2000 metric ton/year scale in January 2010.

A partnership between DSM (the Netherlands) and Roquette Freres (France) is producing bio-succinic acid on a 300 ton/year scale, starting from starch and using an innovative enzyme-based fermentation technology. Scale-up towards 10 000 ton/year is expected for 2012 and this facility in Cassano Spinola (Italy) will then be Europe's largest bio-based succinic acid facility. BASF (Germany) and Purac (the Netherlands) announced the construction of a 25 000 ton/year facility in the proximity of Barcelona (Spain). This plant should be operational in 2013 according to a press release from August 2011. Specific features of the BASF/Purac process are the use of the bacterial strain *Basfi succiniproducens,* which was developed by BASF, in a highly efficient fermentation process that uses glucose or glycerine as renewable substrates.

Myriant (United States) received a grant from the US Department of Energy for their activities on the production of bio-succinic acid. These activities include building of a 20 000 litre reactor in Louisiana, modification of a leased facility of 15 000 ton/year and the construction of a greenfield plant also with a capacity of 15 000 ton/year. The Myriant process uses *E. coli* and unrefined sugar as feedstock.

9.4.2 1,4-Butanediol (BDO)

The annual production of BDO is about 1 million metric tons worldwide. BDO is used as a solvent, as a building block for polymers (e.g. PBT), but most importantly for the production of tetrahydrofuran (THF). Several industrial processes based on fossil feedstocks are used to produce BDO, for instance:

- The Reppe process, in which acetylene reacts with two equivalents of formaldehyde to form 1,4-butynediol, which is subsequently converted to BDO via hydrogenation. The Reppe process is the most widely used process for example by BASF.
- LyondelBasell, the second largest producer of BDO, uses its own process that consists of three major steps; isomerization of propylene oxide to allyl alcohol followed by hydroformylation to 4-hydroxybutyraldehyde and subsequent hydrogenation to BDO.
- Mitsubishi produces BDO and/or THF via consecutive acetoxylation, hydrogenation and hydrolysis of 1,3-butadiene.
- A fourth route is the Davy Process, which consists of the oxidation of benzene or butane to maleic anhydride followed by hydrolysis to maleic acid and hydrogenation to BDO.

The main routes for the production of bio-based 1,4-butanediol are:

- Catalytic reduction of bio-based succinic acid to BDO.
- Direct fermentation of sugars to BDO using *E. coli.*

This last process was developed and patented by Genomatica (USA) [77–79]. Genomatica has genetically engineered *E. coli* via a genome-scale metabolic model. This *E. coli* strain can produce BDO from glucose, xylose, sucrose and biomass-derived mixed sugar streams [80]. Genomatica has produced BDO at a demonstration scale since June 2011 together with Tate & Lyle (United States). Scale up is foreseen with Tate & Lyle and with other partners.

Genomatica and Tate & Lyle are working on a plant with a capacity of 45 000 ton/year. Together with Novamont (Italy) a facility of approximately 18 000 ton/year is constructed in Adria (Italy). Production is expected to start in 2013. Moreover, together with Mitsubishi bio-BDO production on a commercial scale is planned in Asia.

9.4.3 Poly(Butylene Succinate) (PBS)

Poly(butylene succinate) belongs to the poly(alkylene dicarboxylate) family that can be obtained by polycondensation of α,ω-diols such as ethylene glycol and 1,4-butanediol, with aliphatic dicarboxylic acids, such as succinic and adipic acid. PBS is commonly prepared via esterification of succinic acid and BDO or transesterification of dimethyl succinate and BDO to oligomers followed by a subsequent polycondensation reaction, removing excess BDO. Catalysts include $SnCl_2$ [81], p-toluenesulfonic acid [82], tetrabutyltitanate [83] and lanthanide triflates [84]. To produce PBS with sufficiently high molar mass often chain-extenders are used. Examples are the use of diisocyanates [83, 85–87], bisoxazoline [88] and biscaprolactamates [89].

Poly(butylene succinate) is a white, highly crystalline thermoplastic polymer [83] (Table 9.7). The equilibrium melting point of single crystals of PBS is 132 °C [90, 91] and the T_g is −38 °C. Poly(butylene succinate) exhibits multiple melting behaviour, and using differential scanning calorimetry (DSC) melting points are usually found around 100–120 °C. The heat distortion temperature (HDT) of PBS is 70–90 °C. The mechanical properties of PBS are in between those of polyethylene and polypropylene and the material exhibits good toughness provided the molar mass is sufficiently high. Elongation at break is about 350% and the tensile strength is 30–35 MPa. Depending on the degree of crystallinity the E-modulus is around 700 MPa.

9.4.4 PBS Copolymers

The properties of PBS can be varied over a wide range via copolymerization with other dicarboxylic acids or diols. Amongst the co-monomers studied are adipic acid [83, 93], terephtalic acid [94, 95], ethylene glycol [96–98], 1,3-propanediol [99–101] and lactic acid [100, 102]. Of these co-monomers, the introduction of adipic acid for the production of PBSA has received most attention. Elongation at break and impact strength can be significantly

Table 9.7 *Properties of PBS as compared to other plastics [92]. (Data from ref. [92]. Copyright © John Wiley & Sons, Inc., 2008.)*

	PLA (Ingeo)[a]	PBS (GSPla)[b]	PBSA (GSPla)	PP	LDPE
T_g (°C)	55–60	−38	−45	−20	−120
T_m (°C)	150–170	115	90	170	110
HDT (°C)[c]	55	70–90	60	90	50
E-modulus (MPa)	3500	550–700	350	1100–1600	150–250
Tensile strength (MPa)	70	40	40	30–70	20
Elongation at break (%)	6	350	800	150–700	100–1000
Notched impact (kJ/m^2)	2	5–10	35	3–15	>40
Density (kg/m^3)	1240	1260	1240	900	920

Notes: [a]Ingeo Datasheet (Natureworks). [b]GSPla Datasheet (Mitsubishi Chemical company). [c]The Heat Distortion Temperature (HDT) is determined by the test procedure according to ASTM D648.

improved with the introduction of adipate co-monomer [83]. However, melting points and degree of crystallinity decrease. Tserki *et al.* report a melting point of 70 °C and a degree of crystallinity of 27% of a PBSA copolymer containing 40% adipate units and 60% succinate units [83]. Contrary to other observations, Xu and Guo report a higher degree of crystallinity for PBSA with 5–15% mol% adipate building blocks and state this could be due to co-crystallization of butylene adipate units within the crystal lattice of PBS [92].

9.4.5 PBS Biodegradability

The interest in PBS originates from the biodegradability of this aliphatic polyester [85, 103–105]. The biodegradability of PBS and copolymers depends on several factors including the chemical structure, degree of crystallinity and biodegradation environment [81]. The biodegradation rate of PBSA is much higher than the degradation rate of PBS since PBSA has a lower degree of crystallinity. Nevertheless, commercial PBS grades are also certified to be compostable according to EN13432.

9.4.6 PBS Processability

Poly(butylene succinate) has excellent processability, superior to biopolymers like PLA and polyglycolic acid (PGA) [104]. Like all polyesters, during processing at elevated temperatures PBS is sensitive to hydrolysis and needs to be dried prior to use. Maximum processing temperatures are around 200–230 °C. To improve melt strength long-chain branches can be introduced using peroxides or small amounts of trifunctional monomers such as trimethylol propane (TMP) [106, 107]. Improving the melt strength of PBS is necessary for applications like film blowing, stretched-blown bottles and extrusion foams. To allow fast processing, for example in injection moulding, nucleating agents like talcum and α-cyclodextrin can be used [108].

9.4.7 PBS Blends

Poly(butylene succinate) is often used in blends with other biodegradable polymers like starch [109], PLA [110–114] and PHAs [115,116]. In many cases PBS is added to other biopolymers to improve properties like heat stability (heat distortion temperature) and impact resistance, and to improve processing behaviour. Although PBS and PLA are immiscible, compatibility is sufficient to allow preparation of blends with good mechanical properties [92, 105]. Peroxides can be used to improve the compatibility of the blend leading to improved impact strength [117].

9.4.8 PBS Markets and Applications

The first commercial PBS polymers were produced by Showa High Polymer (Japan) under the trade name Bionolle ™ [118] (Table 9.8). Mitsubishi Chemical (Japan) introduced PBS into the market in 2003 under the name GSPla™. In 2006 Hexing Chemical Anhui (China) started producing PBS and in 2007 Xinfu Pharmaceutical (China) built a PBS-production line.

Poly(butylene succinate) is currently used in fast-food packages, bottles, disposable plastic bags, mulch films and so on [85, 104, 105]. In these applications PBS is used as a homo

Table 9.8 Commercial activities in the field of PBS(A) and succinic acid production.

Company	Production Location	Product	Capacity 2009 kt/year	Capacity planned kt/year
BASF and Purac	Germany/Spain	Succinic acid	R&D	25
Bioamber and ARD	France	Succinic acid	2	65
Changsha May Shine	China	PBS	6	
DSM and Roquette	France/Italy	Succinic acid	0.3	10
DuPont de Nemours	United States	1,4-BDO	R&D	50
Hexing Chemical	China	Succinic acid, PBS	10	
Ire Chemical	Korea	PBS	3.5	
Kingfa	China	PBSA	1	30
Mitsubishi Chemical	Japan/Thailand	PBS, PBSA	3	20
Myriant	United States	Succinic acid	Pilot scale	50
Showa High Polymers	Japan	PBS, PBSA	3	
Sinoven Biopolymers Inc.	China/United States	mPBS (PBS nanocomposites)	Market development	
Xinfu	China	PBS	3	

polymer or in blends with other biopolyesters. Sinoven has announced the production of modified PBS (PBS nanocomposites) with a HDT above 100 °C.

Poly(butylene succinate) is also used as biomedical material; for example, a promising substance for bone and cartilage repair. Its processability is better than that of PGA or PLA. It has better mechanical properties than PE or PP. Its insufficient biocompatibility could be enhanced by plasma treatment [119].

The current market for PBS is small (10 000–15 000 ton/year) at a price of 3–4 €/kg. However, the market and production volume are expected to increase rapidly with the development of bio-based PBS. At the same time, with the development of bio-based succinic acid, the price of PBS is expected to drop.

9.4.9 Future Outlook

In the near future, PBS will become a (partly) bio-based polymer with an improved carbon footprint. Compared to other biopolyesters, PBS combines excellent processability with a high HDT and good impact properties, making it a very versatile material. It is expected that the introduction of bio-based building blocks will also lead to a price decrease and this will help to further expand the market for PBS.

9.5 Bio-Based Terephthalates

9.5.1 Introduction

In addition to bio-based polyesters such as poly(lactic acid) (PLA), polyhydroxyalkanoates (PHAs), and poly(ethylene furanoate) (PEF), all based upon biomass-derived building blocks that have a structure different from today's commercial petrochemical-based polyesters, bio-based polyesters can be developed having an identical structure to well-known petrochemical based polyesters. A very important class of such 'drop-in' type bio-based polyesters are represented by polyesters based upon either isophthalic acid or terephthalic acid, such as PET,

PPT and PBT. For this particular class of polyesters there is little need to develop alternative polymer chemistry or to optimize polymer properties and processing parameters but focus needs to be on developing efficient synthetic pathways to the constituting monomers – that is, diols like ethylene glycol (EG), 1,3-propanediol (1,3-PDO) and 1,4-butanediol (1,4-DBO) and aromatic diacids like isophthalic acid and terephthalic acid. Moreover, significant attention has to be paid to develop methods for obtaining polymer grade – that is, high-purity monomers. As PET is the most dominant commercially produced polyester in terms of volume, this chapter focuses on the possibilities for obtaining bio-based PET.

The huge interest in the development of renewable packaging materials using either PEF or bio-based PET is being reflected by announcements of companies like Coca Cola, Pepsi Cola and Danone that they are working towards 100% bio-based PET beverage bottles. These developments are being pursued by means of partnerships with technology providers, such as Virent, Gevo and Avantium (www.icis.com, accessed 22 June 2013).

9.5.2 Bio-Based Diols: Ethylene Glycol, 1,3-Propanediol, 1,4-Butanediol

The route to bio-based ethylene glycol (EG) starting from ethanol is well known and relies on dehydrating bio-ethanol to ethylene, followed by oxidation to ethylene oxide and subsequent rehydration to EG. This method was practised until the 1960s when it was taken over by petrochemical routes. Today's production of bioethylene by companies like Braskem is reviving the possibility of obtaining bio-based glycol. Indian glycols presently converts first-generation ethanol into EG [120]. The likely future establishment of second generation bioethanol production from lignocellulosic biorefineries will make this route more sustainable. Nonetheless, it will not be an ideal route in the longer term, as first all oxygen functionality is removed from the biomass and then reintroduced. A more direct route to ethylene glycol will be preferred. Such an alternative approach to bio-based glycol is based on hydrogenolysis of either glycerol or sorbitol, leading to mixtures of ethylene glycol and propylene glycol. A patent application by Werpy *et al.*, describes the conversion of biomass, including glycerol, into propylene glycol (PG) [121,122] (Figure 9.12). Using a Ni-Re bi-metallic catalyst, at 60% conversion of glycerol, a selectivity of 78–88% towards the production of propylene glycol was obtained, together with the production of lactate (8–10%) and ethylene glycol (12%). ADM (Archer Daniels Midland) recently announced the start of the pilot plant production of bio-based glycols using a similar process (www.adm.com, accessed 22 June 2013).

Increasing selectivity towards EG production could render a more direct process towards bio-based EG. A direct route to produce EG from cellulose has, for example, been described by Ji *et al.*, who reported the direct conversion of cellulose into EG with high yields (up to 60%) using nickel-promoted tungsten carbide catalysts [123].

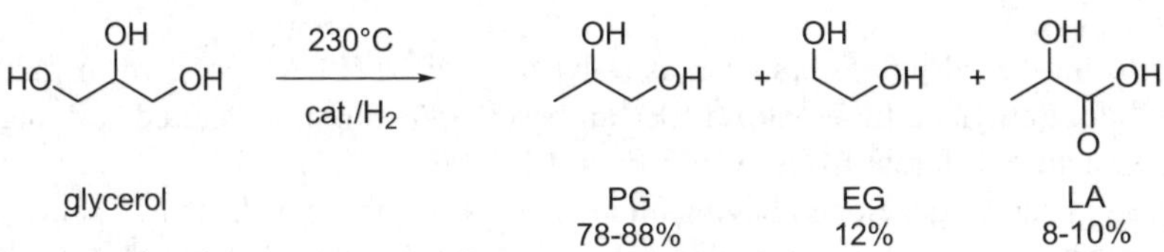

Figure 9.12 Conversion of glycerol to EG and PG according to Werpy et al. [122]. (Reproduced from ref. [122].)

The production of 1,3-PDO from biomass has been commercialized by DuPont, which has established biotechnological routes for the production of 1,3-PDO from sugars or glycerol [124]. 1,3-PDO is currently used for the production of the PPT fibre commercialized under the brand name Sorona and being used in applications like clothing, residential carpets, and automotive fabrics. Polyethers based upon 1,3-PDO are being used in the production of polyols for polyurethanes [125].

Routes towards based 1,4-BDO are currently under development. The production of succinic acid from biomass is currently being commercialized by companies like for example, BioAmber and the DSM-Roquette joint venture Reverdia [126]. Succinic acid is a potential feedstock for the production of 1,4-BDO via reduction of the diacid to the diol [127]. The US based company Genomatica claims it has been successful in developing technology for the direct biotechnological production of 1,4-BDO (see www.genomatica.com, accessed 22 June 2013; see also section 9.4.2).

9.5.3 Bio-Based Xylenes, Isophthalic and Terephthalic Acid

The dominant current commercial process for the production of isophthalic acid (IA) and terephthalic acid (TA) is the catalytic oxidation of *m*-xylene and *p*-xylene (PX). Xylenes are oxidized to IA or TA, respectively, in the so called Amoco process at 110–205 °C and 15–30 bar and in the presence of 95% acetic acid [128]. Furthermore, Co and Mn need to be added as catalysts and NH_4Br and tetrabromoethane as co-catalysts. It is important to realize that some routes towards bio-based IA and TA may involve the production of the xylenes as an intermediate, whereas others might not. Furthermore, production routes may rely on technologies generating a slate of products comprising mostly fuel components next to chemicals like PX, whereas other technologies rely on the dedicated production of PX.

9.5.3.1 Production of Mixtures of Bio-Aromatics including P-Xylene (PX) from Biomass

Methods for the conversion of biomass into a slate of product, also comprising chemicals like p-xylene, are those promoted by US-based companies Anellotech and Virent. In Anellotech's 'Biomass to Aromatics' process, biomass such as corn, sugar beet and so on is dried and milled, and then injected into a fluidized bed in the presence of a cheap zeolite, based on ZSM-5. The process takes place at high temperatures (600 °C) at a short period of time. Coking initially was considered as a drawback in this process. Annellotech claims to be able to produce 190 L product out of 1 ton biomass (www.anellotech.com, accessed 22 June 2013). So far the process has been performed on a small scale only and scale-up needs to be realized.

Recently, Vispute *et al.* showed that pyrolysis oils can be converted into a mixture of industrial commodity chemical feedstocks (benzene, toluene, xylene, ethylbenzene, ethylene, propylene, butylene) [129]. These authors used an integrated catalytic approach that involves hydroprocessing of the bio-oils over supported metal catalysts, followed by conversion over zeolite catalysts. The hydroprocessing increases the intrinsic hydrogen content of the pyrolysis oil (and reduces the oxygen content), producing polyols and alcohols. The zeolite catalyst then converts these hydrogenated products into light olefins and aromatic hydrocarbons in a yield of about 60%. The reaction can be steered by changing the temperature, amount of added hydrogen and zeolite properties. In this process, drawbacks associated with the prior bio-oil hydrogenation processes were overcome by operating at moderate temperatures (≤250 °C) at which no catalyst coking or reactor plugging was observed.

The Virent company has developed a comparable process in which plant derived sugars using a 'drop-in' process are being converted into a mixture of components, including PX (BioFormPX™) (www.virent.com, accessed 22 June 2013). The continuous process uses a heterogeneous catalysts to remove oxygen from the biomass feedstock, involving a series of reactions at 175–300 °C and 10–90 bar of hydrogen. The process consists of: (i) hydrogen production, (ii) dehydrogenation of sugars and alcohols/hydrogenation of carbonyl groups, (iii) deoxygenation reactions, (iv) hydrogenolysis and (v) condensation and cyclization. Use is being made of zeolite and aluminosilicate based catalysts, including adapted ZSM5 catalysts. As the process requires significant amounts of hydrogen, part of the biomass is being converted into hydrogen, using an aqueous phase reforming (APR) process [130, 131]. Mixtures of products including, amongst others paraffins, aromatics and olefins are being formed. This product slate can be used in traditional crude oil refining processes and subsequently be used for the production of PX. The company has announced to start production capacity at a scale of 37 000 L/year.

9.5.3.2 Dedicated Approaches Towards P-Xylene (PX) and Terephthalic Acid (TA)

9.5.3.2.1 TA from Terpenes One of the possible approaches described in literature to arrive at bio-based TA is using terpenes as feedstock. Use of terpenes such as limonene and α- and γ-terpinenes offers potential for the production of aromatic compounds such as, for example, substituted phenols or TA. The aromatization of terpenes such as limonene using zeolites has been described in the literature [132] (Figure 9.13). Yields of *p*-cymene described so far were moderate and next to α- and γ-terpinenes, substantial amounts of unidentified products were formed. A patent application to Sabic describes the synthesis of TA from *p*-cymene using nitric acid as a stoichiometric oxidant [133]. Following this approach, Colonna *et al.* recently described the synthesis of fully bio-based terephthalate polyesters, including radiocarbon evidence [134]. The preparation of *p*-cymene from other terpenes is also reported in the literature [135]. In particular, pinene – that is, the major component of turpentine oil and available in larger amounts with respect to limonene – can be isomerized to limonene or directly converted in *p*-cymene [136]. Terpenes can be obtained from biomass resources such as orange peels and soft woods. Current production volumes of terpenes are in the range of hundreds of thousands of tonnes instead of the million tonnes needed to substitute a significant amount of aromatics production. Also prices of limonene and related terpenes are still relatively high compared to aromatics. Therefore, from an economical point of view,

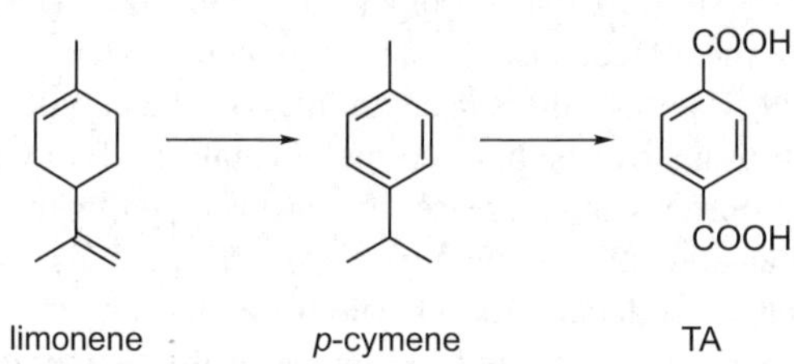

Figure 9.13 *Potential production of TA from limonene.*

production volumes of terpenes should be significantly increased and terpenes should become available at lower prices to be an attractive feedstock for bulk aromatics production.

9.5.3.2.2 Production of PX via a Combined Biotechnological Chemo-Catalytic Approach An alternative approach to bio-based TA is being advocated by the US-based company GEVO (www.gevo.com, accessed 22 June 2013). In this particular approach a route to bio-based TA is envisaged by a combination of biotechnological and chemo-catalytic steps. Using genetically modified organisms, sugars are being fermented into isobutanol by a recombinant yeast in which the pathway for conversion of pyruvate to isobutanol is added. In order to enhance the isobutanol production, enzyme activities in competing pathways, with glycerol-3-phospate dehydrogenase (GPD) and pyruvate decarboxylase (PDC) as the most important enzymes, were reduced [137]. Isobutanol can be dehydrated to isobutylene, which subsequently, using essentially known chemistry, can be dimerized and cyclized into a mixture of xylenes from which PX and *m*-xylene can be separated and used for the production of IA and TA (Figure 9.14). Gevo described a PX yield of 19% on a weight by weight basis. The main other products formed were isobutylene (26%) and isooctene (19%) [138]. The yield of PX may be enhanced by reutilization of the intermediate [138]. This particular approach can rely on abundantly available sugars but needs two molecules of sugar in order to produce one molecule of IA and TA. The economic viability of the process is, at present, difficult to assess as a result of the lack of public data on fermentation parameters like space-time yields of isobutanol.

The French company Global Bioenergies, is pursuing another approach by aiming at the direct biotechnological production of isobutylene, rather than isobutanol. Their technology relies on using *Clostridia* bacteria in which the aceton pathway is optimized; aceton is intracellular converted into hydroxyisovalerate, which is subsequently decarboxylated into isobutylene [139–141]. The merits of this process will be that, as isobutylene is a gaseous product, the downstream processing costs will be reduced compared to producing isobutanol.

9.5.3.2.3 Production of TA via Cycloaddition Reactions A number of other approaches towards bio-based TA have in common that a cycloaddition reaction of two components is being used as one of the critical steps in deriving bio-based aromatics (Figure 9.15).

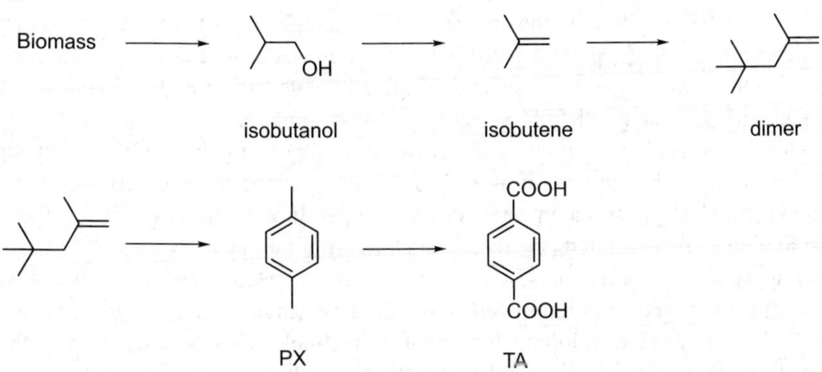

Figure 9.14 *The GEVO process.*

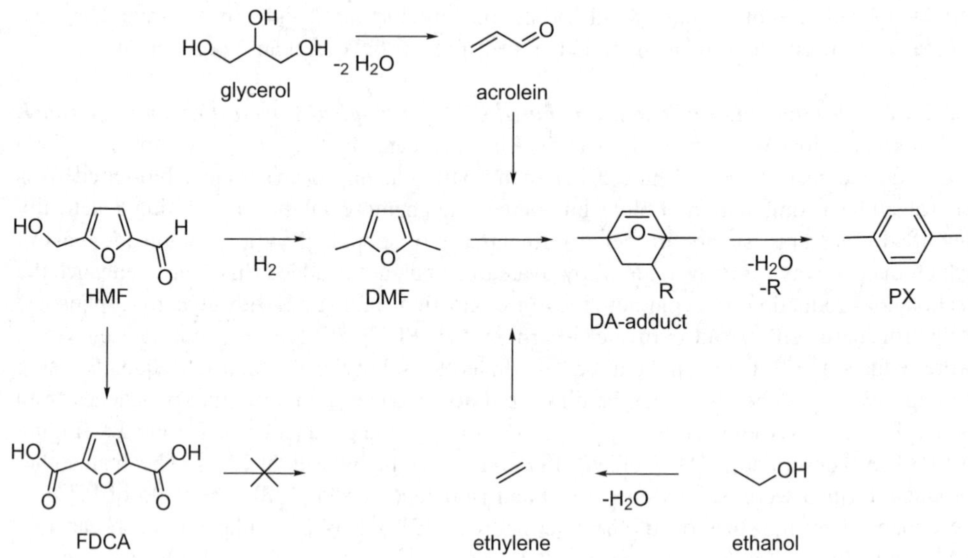

Figure 9.15 Cycloaddition routes to TA.

One of the first reports to mention a catalytic tandem Diels-Alder aromatization reaction to synthesize aromatic bulk chemicals from furans (i.e. FDCA) is a review paper by Okkerse and van Bekkum [142]. Unfortunately, no experimental details are given. A patent application to BP mentions a yield of less than 1% of TA following this approach [143]. This is to no surprise as in normal electron demand Diels–Alder reactions frontier molecular orbital theory requires the diene to have electron releasing substituents and the dienophile to have electron withdrawing substituent in order for the reaction to proceed. Since FDCA has electron withdrawing substituents, only inverse electron demand Diels–Alder reactions are possible, for which ethylene is unsuitable. Instead of using FDCA as furan-based diene, 2,5-dimethylfuran (DMF), obtained by hydrogenation of HMF, has recently been shown to react with ethylene to PX in a one-pot consecutive cycloaddition-aromatization sequence [144]. The UOP patent application claims a 30% yield per pass, at 200–330 °C and 30–45 bar of ethylene pressure. The efficacy of the Diels–Alder reaction is reflected by the fact that this method gives PX in good yields (up to 92%), although the process requires a high pressure of ethylene, as well as a high reaction temperature. A paper by Williams *et al.*, further describes this reaction, showing that using various zeolite type catalysts and temperatures up to 573 K and 75 bar of ethylene pressure PX can be formed with 75% selectivity [145].

In another approach, Shiramizu and Toste describe a method for converting 2,5-dimethylfuran into TA by cycloaddition with acrolein followed by aromatization and decarbonylation [146]. Using Lewis acid catalysts and temperatures as low as −55 °C, the authors were able to obtain (non-isolated) yields of the Diels–Alder adducts in up to 75%.

Another approach to produce bio-TA via a cycloaddition reaction as one of the key steps, is promoted by the Draths corporation (recently acquired by Amyris (www.amyris.com, accessed 22 June 2013)). They are developing fermentation technology to produce renewable-based monomers from muconic acid. Amongst others, the company's product portfolio includes bio-based TA (Figure 9.16).

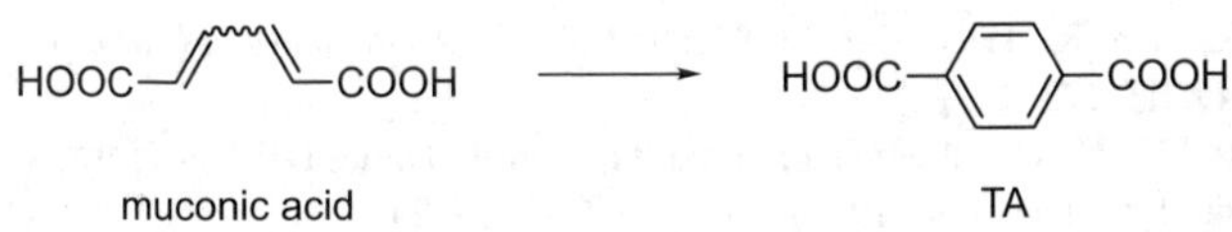

Figure 9.16 TA from muconic acid.

In this process, sugars, obtained from biomass, are fermented at low pH into cis-muconic acid. The process of microbial muconic adic formation was already described by Frost and co-workers, who developed *E. coli* WN1/pWN2.248 that synthesized 36.8 g/L of *cis,cis*-muconic acid in 22% (mol/mol) yield from glucose after 48 h of culturing under fed-batch fermentation conditions [147]. This strain did not possess the aroE encoded shikamate dehydrogenase preventing the cells to convert 3-dehydroshikimic acid into shikimic acid which is available for production of cis,cis-muconic acid. Optimization of microbial *cis,cis*-muconic acid synthesis required expression of three enzymes not typically found in *E. coli*. A recent patent application by Bui *et al.* describes a productivity of 59 g/L cis muconic acid from 248 g/L glucose by a modified *E. coli*. in a 20 L fermenter in 88 h.

In another patent application Frost *et al.* claim the synthesis of dimethyl cyclohex-2-ene-1,4-dicarboxylate starting from *trans,trans*-dimethyl muconate, using high pressures ethylene at 150–170 °C. According to the patent application, dimethyl cyclohex-2-ene-1,4-dicarboxylate can be successfully converted into TA by reacting it over Pt/C in the presence of oxygen and acetic acid [148].

In conclusion, with regard to bio-based TA or IA based polyesters, it can be stated that routes to most of the important diols are already established and even commercialized, whereas various routes towards bio-based TA or IA are currently under development. The significant progress that has recently been made in this area is a strong indication that bio-based TA can become viable in the near future.

9.6 Conclusions

In this chapter we have given an overview of four types of biomass-derived monomers and their applications in polyesters. Despite the fact that most of these monomers have been known for over a century, the development of efficient industrial routes to the monomers, as well as the development of polymer applications (isosorbide, FDCA, and succinate) is being intensively pursued by both industry and academia. All of the materials discussed in this chapter have unique and commercially relevant properties, so we can expect significant scientific breakthroughs and commercialization in the years to come.

References

1. Werpy, T. and Petersen, G. (2004) Top Value Added Chemicals from Biomass: Volume I – Results of Screening for Potential Candidates from Sugars and Synthesis Gas. DOE/GO-102004-1992, http://www1.eere.energy.gov/bioenergy/pdfs/35523.pdf (accessed 7 July 2013).
2. Bozell, J.J. and Petersen, G.R. (2010) Technology development for the production of biobased products from biorefinery carbohydrates-the US Department of Energy's 'Top 10' revisited. *Green Chemistry*, **12** (4), 539–554.

3. Stoss, P. and Hemmer, R. (1991) 1,4 : 3,6-Dianhydrohexitols. *Advances in Carbohydrate Chemistry and Biochemistry*, **49**, 93–173.
4. Fuertes, P. (2006) Process for the preparation of high-purity L-iditol, EP1647540.
5. Wright, L.W. and Brandner, J.D. (1962) Isoiodide, US3023223.
6. Wright, L.W. and Brandner, J.D. (1964) Catalytic isomerization of polyhydric alcohols. II. The isomerization of isosorbide to isomannide and isoidide. *Journal of Organic Chemistry*, **29** (10), 2979–2982.
7. Wu, J. (2012) Carbohydrate-Based Building Blocks and Step-growth Polymers; Synthesis, Characterization and Structure-Property Relations. PhD thesis, Eindhoven Technical University.
8. Fenouillot, F., Rousseau, A., Colomines, G. *et al.* (2010) Polymers from renewable 1,4 : 3,6-dianhydrohexitols (isosorbide, isomannide and isoidide): a review. *Progress in Polymer Science*, **35** (5), 578–622.
9. Kricheldorf, H.R. (1997) 'Sugar diols' as building blocks of polycondensates. *Journal of Macromolecular Science, Reviews in Macromolecular Chemistry and Physics*, **C37** (4), 599–631.
10. Thiem, J. and Lueders, H. (1984) Synthesis and directed polycondensation of starch-derived anhydroalditol building blocks. *Starch/Staerke*, **36** (5), 170–176.
11. Thiem, J. and Lueders, H. (1984) Synthesis of polyterephthalates derived from dianhydrohexitols. *Polymer Bulletin (Berlin)*, **11** (4), 365–369.
12. Braun, D. and Bergmann, M. (1992) Polymers from 1,4 : 3,6-dianhydrosorbitol. *Journal Fur Praktische Chemie-chemiker-zeitung*, **334** (4), 298–310.
13. Storbeck, R., Rehahn, M. and Ballauff, M. (1993) Synthesis and properties of high-molecular-weight polyesters based on 1,4 : 3,6-dianhydrohexitols and terephthalic acid. *Makromolecular Chemistry*, **194** (1), 53–64.
14. Kricheldorf, H.R., Behnken, G. and Sell, M. (2007) Influence of isosorbide on glass-transition temperature and crystallinity of poly(butylene terephthalate). *Journal of Macromolecular Science, Part A. Pure and Applied Chemistry*, **44** (7), 679–684.
15. Feng, X., East, A.J., Hammond, W.B. *et al.* (2011) Overview of advances in sugar-based polymers. *Polymers for Advanced Technologies*, **22** (1), 139–150.
16. Storbeck, R. and Ballauff, M. (1993) Synthesis and properties of polyesters based on 2,5-furandicarboxylic acid and 1,4 : 3,6-dianhydrohexitols. *Polymer*, **34** (23), 5003–5006.
17. Gomes, M., Gandini, A., Silvestre, A.J.D. and Reis, B. (2011) Synthesis and characterization of poly(2,5-furan dicarboxylate)s based on a variety of diols. *Journal of Polymer Science Part A: Polymer Chemistry*, **49** (17), 3759–3768.
18. Jenkins, V.F., Mott, A. and Wicker, R.J. (1963) Unsaturated polyesters and coatings therefrom, GB927786 (1963).
19. Tate, C.W. and Scruggs, J.G. (1962) Polyolefin blends for textiles, BE618363.
20. Le Maistre, J.W. and Ford, E.C. (1965) Fibrous products bound with polyester resins, GB1012563.
21. Collins, J.R. (1967) Polyesters, GB1079686.
22. Charbonneau, L.F. and Johnson, R.E. (1999) Polyester containers and method for making them, WO9954533.
23. Khanarian, G., Charbonneau, L.F. and Witteler, H.B. (1999) Optical articles comprising isosorbide polyesters and method for making same, WO9954378.
24. Khanarian, G., Charbonneau, L.F., Kitchens, C. and Shen, S.S. (1999) Polyester films, and their manufacture, US5958581.
25. Khanarian, G., Charbonneau, L.F., Witteler, H.B. *et al.* (1999) Sheets formed from polyesters containing isosorbide, WO9954129.
26. Charbonneau, L.F., Khanarian, G., Johnson, R.E. *et al.* (2000) Isosorbide-containing polyester fiber and methods for making same, US6063495.
27. Charbonneau, L.F., Johnson, R.E., Witteler, H.B. and Khanarian, G. (1999) Polyesters including isosorbide as a comonomer and methods for their production, US5959066.
28. Khanarian, G., Charbonneau, L.F., Witteler, H.B. and Johnson, R.E. (1999) Isosorbide containing polyesters and methods for making same, WO9954119.

29. Adelman, D.J., Greene, R.N. and Putzig, D.E. (2003) Melt polymerization process for the preparation of poly(1,3-propylene-co-isosorbide) terephthalate, US20030232960.
30. Charbonneau, L.F. (2006) Processes for making low-color poly(ethylene-co-isosorbide) terephthalate polymers in the presence of antioxidants, WO2006032022.
31. Adelman, D.J., Charbonneau, L.F. and Ung, S. (2003) Process for making poly(ethylene-co-isosorbide) terephthalate polymer, US6656577.
32. Storbeck, R. and Ballauff, M. (1996) Synthesis and thermal analysis of copolyesters deriving from 1,4 : 3,6-dianhydrosorbitol, ethylene glycol, and terephthalic acid. *Journal of Applied Polymer Science*, **59** (7), 1199–1202.
33. Bersot, J.C., Jacquel, N., Saint-Loup, R. *et al.* (2011) Efficiency increase of poly(ethylene terephthalate-co-isosorbide terephthalate) synthesis using bimetallic catalytic systems. *Macromolecular Chemistry and Physics*, **212** (19), 2114–2120.
34. Sablong, R., Duchateau, R., Koning, C.E.G. *et al.* (2008) Incorporation of isosorbide into poly(butylene terephthalate) via solid-state polymerization. *Biomacromolecules*, **9** (11), 3090–3097.
35. Noordover, B.A.J., Sablong, R.J., Duchateau, R. *et al.* (2008) Process for the production of a dianhydro-hexitol based polyester, WO2008031592.
36. Okada, M., Okada, Y. and Aoi, K. (1995) Synthesis and degradabilities of polyesters from 1,4 : 3,6-dianhydrohexitols and aliphatic dicarboxylic acids. *Journal of Polymer Science Part A: Polymer Chemistry*, **33** (16), 2813–2820.
37. Okada, M., Tsunoda, K., Tachikawa, K. and Aoi, K. (2000) Biodegradable polymers based on renewable resources. IV. Enzymatic degradation of polyesters composed of 1,4 : 3,6-dianhydro-D-glucitol and aliphatic dicarboxylic acid moieties. *Journal of Applied Polymer Science*, **77** (2), 338–346.
38. Okada, M., Okada, Y., Tao, A. and Aoi, K. (1996) Biodegradable polymers based on renewable resources: polyesters composed of 1,4 : 3,6-dianhydrohexitol and aliphatic dicarboxylic acid units. *Journal of Applied Polymer Science*, **62** (13), 2257–2265.
39. Noordover, B.A.J., van Staalduinen, V.G., Duchateau, R. *et al.* (2006) Co- and terpolyesters based on isosorbide and succinic acid for coating applications: synthesis and characterization. *Biomacromolecules*, **7** (12), 3406–3416.
40. Garaleh, M., Yashiro, T., Kricheldorf, H.R. *et al.* (2010) (Co-)Polyesters derived from isosorbide and 1,4-cyclohexane dicarboxylic acid and succinic acid. *Macromolecular Chemistry and Physics*, **211** (11), 1206–1214.
41. Hayes, R. and Brandenburg, C. (2003) Bis(2-hydroxyethyl isosorbide), preparation, polymers, and use, US6608167.
42. Wu, J., Eduard, P., Thiyagarajan, S. *et al.* (2011) Isohexide derivatives from renewable resources as chiral building blocks. *ChemSusChem*, **4** (5), 599–603.
43. Wu, J., Eduard, P., Jasinska-Walc, L. *et al.* (2012) Fully isohexide-based polyesters: synthesis, characterization, and structure–properties relations. *Macromolecules*, **46** (2), 384–394.
44. Wu, J., Eduard, P., Thiyagarajan, S. *et al.* (2012) Semi-crystalline polyesters based on a novel renewable building block. *Macromolecules*, **45** (12), 5069–5080.
45. Gandini, A. (1977) The behaviour of furan derivatives in polymerization reactions. *Advances in Polymer Science*, **25**, 47–92.
46. Gandini, A., Silvestre, A.J.D., Pascoal Neto, C. *et al.* (2009) The furan counterpart of poly(ethylene terephthalate): an alternative material based on renewable resources. *Journal of Polymer Science Part A: Polymer Chemistry*, **47** (1), 295–298.
47. Grosshardt, O., Fehrenbacher, U., Kowollik, K. *et al.* (2009) Synthesis and characterization of polyesters and polyamides from 2,5-furandicarboxylic acid. *Chemie Ingenieur Technik*, **81** (11), 1829–1835.
48. Gandini, A., Coelho, D., Gomes, M. *et al.* (2009) Materials from renewable resources based on furan monomers and furan chemistry: work in progress. *Journal of Materials Chemistry*, **19** (45), 8656–8664.
49. Jiang, M., Liu, Q., Zhang, Q. *et al.* (2012) A series of furan-aromatic polyesters synthesized via direct esterification method based on renewable resources. *Journal of Polymer Science Part A: Polymer Chemistry*, **50** (5), 1026–1036.

50. Ma, J., Pang, Y., Wang, M. *et al.* (2012) The copolymerization reactivity of diols with 2,5-furandicarboxylic acid for furan-based copolyester materials. *Journal of Materials Chemistry*, **22** (8), 3457–3461.
51. Katritzky, A.R., Jug, K. and Oniciu, D.C. (2001) Quantitative measures of aromaticity for mono-, bi-, and tricyclic penta- and hexaatomic heteroaromatic ring systems and their interrelationships. *Chemical Reviews*, **101** (5), 1421–1450.
52. Zamorsky, Z. and Justo, G. (1960) Crystallization kinetics of the polyethylene derivative of dehydromucic acid. *Chemicky Prumysl*, **10**, 492–495.
53. Moore, J.A. and Bunting, W.W. (1985) Polyesters and polyamides containing isomeric furan dicarboxylic acids. *Polymer Scienceand Technology, Advanced Polymer Synthesis*, **31**, 51–91.
54. Lichtenthaler, F.W. (2002) Unsaturated O- and N-heterocycles from carbohydrate feedstocks. *Accounts of Chemical Research*, **35**, 728–737.
55. Gandini, A. (1977) The behavior of furan derivatives in polymerization reactions. *Advances in Polymer Science*, **25**, 47.
56. Ivanov, A., Primelles Alberteris, E. and Aluija Pereda, H. (1975) Synthesis and physical-chemical properties of poly(2,5-furandimethylene 2,5-furandicarboxylate). *Sobre los Derivados de la Cana de Azucar*, **9** (2), 48–64.
57. Moore, J.A. and Kelly, J.E. (1978) Polyesters derived from furan and tetrahydrofuran nuclei. *Macromolecules*, **11** (3), 568–573.
58. Hirai, H. (1984) Oligomers from hydroxymethylfurancarboxylic acid. *Journal of Macromolecular Science – Chemistry*, **A21** (8–9), 1165–1179.
59. Hirai, H., Naito, K., Hamasaki, T. *et al.* (1984) Syntheses of macrocyclic oligoesters from 5-hydroxymethyl-2-furancarboxylic acid. *Makromolecular Chemistry*, **185** (11), 2347–2359.
60. Moore, J.A. and Kelly, J.E. (1984) Polyhydroxymethylfuroate [poly{2,5-furandiylcarbonyl-oxymethylene}]. *Journal of Polymer Science – Polymer Chemistry Edition*, **22** (3), 863–864.
61. Drewit, J.G.N. and Lincoln, J. (1951) Polyesters from heterocyclic components, US2551731.
62. Hachihama, Y., Shono, T. and Hyono, K. (1958) Syntheses of polyesters containing a furan ring. *Technology Reports of the Osaka University*, **8**, 475–480.
63. Lukes, R. and Janda, M. (1959) Ethylene glycol polyester, CS87340.
64. Zamorsky, Z. (1960) Mixed polyesters, CS96326.
65. Heertjes, P.M. and Kok, G.J. (1974) Polycondensation products of 2,5-furandicarboxylic acid. *Delft Progress Reports, Series A,* **1** (2), 59–63.
66. Khrouf, A., Abid, M., Boufi, S. *et al.* (1998) Polyesters bearing furan moieties. Part 2. A detailed investigation of the polytransesterification of difuranic diesters with different diols. *Macromolecular Chemistry and Physics*, **199** (12), 2755–2765.
67. Khrouf, A., Boufi, S., El Gharbi, R. *et al.* (1996) Polyesters bearing furan moieties. Part 1. Polytransesterification involving difuranic diesters and aliphatic diols. *Polymer Bulletin (Berlin)*, **37** (5), 589–596.
68. Khrouf, A., Boufi, S. El Gharbi, R. and Gandini, A. (1999) Polyesters bearing furan moieties. Part 3. A kinetic study of the transesterification of 2-furoates as a model reaction for the corresponding polycondensations. *Polymer International*, **48** (8), 649–659.
69. Gruter, G.-J.M., Sipos, L. and Dam, M.A. (2012) Accelerating research into bio-based FDCA-polyesters by using small scale parallel film reactors. *Combinatorial Chemistry and High Throughput Screening*, **15** (2), 180–188.
70. Sipos, L. (2010) Preparation of polyesters containing 2,5-furandicarboxylate moieties within the polymer backbone, WO2010077133.
71. Dam, M.A., Gruter, G.-J.M., Sipos, L. *et al.* (2011) 2,5-Furandicarboxylic acid; a versatile building block for a very interesting class of polyesters. EUROTEC® 2011, Barcelona, Society of Plastics Engineers.
72. Eerhart, A.J.J.E., Faaij, A.P.C. and Patel, M.K. (2012) Replacing fossil based PET with biobased PEF; process analysis, energy and GHG balance. *Energy and Environmental Science*, **5** (4), 6407–6422.
73. Kanakam, R., Pathy, M.S.V. and Udupa, H.V.K. (1967) Electroreduction of maleic and fumaric acids at a rotating cathode. *Electrochimica Acta*, **12** (3), 329–332.

74. Bechthold, I., Bretz, K., Kabasci, S. *et al.* (2008) Succinic acid: A new platform chemical for biobased polymers from renewable resources. *Chemical Engineering and Technology*, **31** (5), 647–654.
75. McKinlay, J.B., Vieille, C. and Zeikus, J.G. (2007) Prospects for a bio-based succinate industry. *Applied Microbiology and Biotechnology*, **76** (4), 727–740.
76. Song, H. and Lee, S.Y. (2006) Production of succinic acid by bacterial fermentation. *Enzyme and Microbial Technology*, **39** (3), 352–361.
77. Burk, M.J., Van Dien, S.J., Burgard, A. and Niu, W. (2008) Compositions and methods for the biosynthesis 1,4-butanediol and its precursors, WO08115840.
78. Burk, M.J., Burgard, A.P., Osterhout, R.E. and Sun, J. (2010) Microorganisms for the production of 1,4-butanediol, WO10030711.
79. Van Dien, S.J., Burgard, A.P., Haselbeck, R. *et al.* (2010) Microorganisms for the production of 1,4-butanediol and related methods, WO10141920.
80. Yim, H., Haselbeck, R., Niu, W. *et al.* (2011) Metabolic engineering of *Escherichia coli* for direct production of 1,4-butanediol. *Nature Chemical Biology*, **7** (7), 445–452.
81. Zhu, C., Zhang, Z., Liu, Q. *et al.* (2003) Synthesis and biodegradation of aliphatic polyesters from dicarboxylic acids and diols. *Journal of Applied Polymer Science*, **90** (4), 982–990.
82. Song, D.K. and Sung, Y.K. (1995) Synthesis and characterization of biodegradable poly(1,4-butanediol succinate). *Journal of Applied Polymer Science*, **56** (11), 1381–1395.
83. Tserki, V., Matzinos, P., Pavlidou, E. *et al.* (2006) Biodegradable aliphatic polyesters. Part I. Properties and biodegradation of poly(butylene succinate-co-butylene adipate). *Polymer Degradation and Stability*, **91** (2), 367–376.
84. Takasu, A., Oishi, Y., Iio, Y. *et al.* (2003) Synthesis of aliphatic polyesters by direct polyesterification of dicarboxylic acids with diols under mild conditions catalyzed by reusable rare-earth triflate. *Macromolecules*, **36** (6), 1772–1774.
85. Fujimaki, T. (1998) Processability and properties of aliphatic polyesters, 'BIONOLLE', synthesized by polycondensation reaction. *Polymer Degradation and Stability*, **59** (1–3), 209–214.
86. Han, Y.K., Kim, S.R. and Kim, J. (2002) Preparation and characterization of high molecular weight poly(butylene succinate). *Macromolecular Research*, **10** (2), 108–114.
87. Tserki, V., Matzinos, P., Pavlidou, E. and Panayiotou, C. (2006) Biodegradable aliphatic polyesters. Part II. Synthesis and characterization of chain extended poly(butylene succinate-co-butylene adipate). *Polymer Degradation and Stability*, **91** (2), 377–384.
88. Huang, C.Q., Luo, S.Y., Xu, S.Y. *et al.* (2010) Catalyzed chain extension of poly(butylene adipate) and poly(butylene succinate) with 2,20-(1,4-phenylene)-bis(2-oxazoline). *Journal of Applied Polymer Science*, **115** (3), 1555–1565.
89. Zhao, J.B., Li, K.Y. and Yang, W.T. (2007) Chain extension of polybutylene adipate and polybutylene succinate with adipoyl- and terephthaloyl-biscaprolactamate. *Journal of Applied Polymer Science*, **106** (1), 590–598.
90. Wang, X., Zhou, J. and Li, L. (2007) Multiple melting behavior of poly(butylene succinate). *European Polymer Journal*, **43** (8), 3163–3170.
91. Miyata, T. and Masuko, T. (1998) Crystallization behaviour of poly(tetramethylene succinate). *Polymer*, **39** (6–7), 1399–1404.
92. Xu, J. and Guo, B.H. (2010) Poly(butylene succinate) and its copolymers: Research, development and industrialization. *Biotechnology Journal*, **5** (11), 1149–1163.
93. Nikolic, M.S. and Djonlagic, J. (2001) Synthesis and characterization of biodegradable poly(butylene succinate-co-butylene adipate)s. *Polymer Degradation and Stability*, **74** (2), 263–270.
94. Luo, S., Li, F., Yu, J. and Cao, A. (2010) Synthesis of poly(butylene succinate-co-butylene terephthalate) (PBST) copolyesters with high molecular weights via direct esterification and polycondensation. *Journal of Applied Polymer Science*, **115** (4), 2203–2211.
95. Nagata, M., Goto, H., Sakai, W. and Tsutsumi, N. (2000) Synthesis and enzymatic degradation of poly(tetramethylene succinate) copolymers with terephthalic acid. *Polymer*, **41** (11), 4373–4376.

96. Cao, A., Okamura, T., Nakayama, K. *et al.* (2002) Studies on syntheses and physical properties of biodegradable aliphatic poly(butylene succinate-co-ethylene succinate)s and poly(butylene succinate-co-diethylene glycol succinate)s. *Polymer Degradation and Stability*, **78** (1), 107–117.
97. Deng, L.M., Wang, Y.Z., Yang, K.K. *et al.* (2004) A new biodegradable copolyester poly(butylene succinate-co-ethylene succinate-co-ethylene terephthalate). *Acta Materialia*, **52** (20), 5871–5878.
98. Mochizuki, M., Mukai, K., Yamada, K. *et al.* (1997) Structural effects upon enzymatic hydrolysis of poly(butylene succinate-co-ethylene succinate)s. *Macromolecules*, **30** (24), 7403–7407.
99. Papageorgiou, G.Z. and Bikiaris, D.N. (2007) Synthesis, cocrystallization, and enzymatic degradation of novel poly(butylene-co-propylene succinate) copolymers. *Biomacromolecules*, **8** (8), 2437–2449.
100. Xu, Y., Xu, J., Guo, B. and Xie, X. (2007) Crystallization kinetics and morphology of biodegradable poly(butylene succinate-co-propylene succinate)s. *Journal of Polymer Science Part B: Polymer Physics*, **45** (4), 420–428.
101. Xu, Y., Xu, J., Liu, D. *et al.* (2008) Synthesis and characterization of biodegradable poly(butylene succinate-co-propylene succinate)s. *Journal of Applied Polymer Science*, **109** (3), 1881–1889.
102. Shibata, M., Inoue, Y. and Miyoshi, M. (2006) Mechanical properties, morphology, and crystallization behavior of blends of poly(l-lactide) with poly(butylene succinate-co-l-lactate) and poly(butylene succinate). *Polymer*, **47**(10), 3557–3564.
103. Vroman, I. and Tighzert, L. (2009) Biodegradable polymers. *Materials*, **2** (2), 307–344.
104. Ishioka, R., Kitakuni, E. and Ichikawa, Y. (2002) Aliphatic polyesters: 'Bionolle', in *Biopolymers* (eds Y. Doi and A. Steinbüchel), Wiley-VCH, Weinheim, Germany, pp. 275–297.
105. Xu, J. (2010) Microbial succinic acid, its polymer poly(butylene succinate), and applications, in *Plastics from Bacteria: Natural Functions and Applications, Microbiology Monographs* (ed G.Q. Chen), Springer-Verlag, Heidelberg-Berlin, Germany, pp. 347–388.
106. Kim, E.K., Bae, J.S., Im, S.S. *et al.* (2001) Preparation and properties of branched polybutylenesuccinate. *Journal of Applied Polymer Science*, **80** (9), 1388–1394.
107. Yoshikawa, K., Ofuji, N., Imaizumi, M. *et al.* (1996) Molecular weight distribution and branched structure of biodegradable aliphatic polyesters determined by s.e.c.-MALLS. *Polymer*, **37** (7), 1281–1284.
108. Dong, T., He, Y., Zhu, B. *et al.* (2005) Nucleation mechanism of α-cyclodextrin-enhanced crystallization of some semicrystalline aliphatic polymers. *Macromolecules*, **38** (18), 7736–7744.
109. Lai, S.M., Huang, C.K. and Shen, H.F. (2005) Preparation and properties of biodegradable poly(butylene succinate)/starch blends. *Journal of Applied Polymer Science*, **97** (1), 257–264.
110. Chuai, C.Z., Zhao, N., Li, S. and Sun, B.X. (2011) Study on PLA/PBS blends. *Advanced Materials Research Guilin*, **197–198**, 1149–1152.
111. Harada, M., Ohya, T., Iida, K. *et al.* (2007) Increased impact strength of biodegradable poly(lactic acid)/poly(butylene succinate) blend composites by using isocyanate as a reactive processing agent. *Journal of Applied Polymer Science*, **106** (3), 1813–1820.
112. Park, J.W. and Im, S.S. (2002) Phase behavior and morphology in blends of poly(L-lactic acid) and poly(butylene succinate). *Journal of Applied Polymer Science*, **86** (3), 647–655.
113. Yokohara, T., Okamoto, K. and Yamaguchi, M. (2010) Effect of the shape of dispersed particles on the thermal and mechanical properties of biomass polymer blends composed of poly(L-lactide) and poly(butylene succinate). *Journal of Applied Polymer Science*, **117** (4), 2226–2232.
114. Yokohara, T. and Yamaguchi, M. (2008) Structure and properties for biomass-based polyester blends of PLA and PBS. *European Polymer Journal*, **44** (3), 677–685.
115. Qiu, Z., Ikehara, T. and Nishi, T. (2003) Miscibility and crystallization behaviour of biodegradable blends of two aliphatic polyesters. Poly(3-hydroxybutyrate-co-hydroxyvalerate) and poly(butylene succinate) blends. *Polymer*, **44** (24), 7519–7527.
116. Qiu, Z., Ikehara, T. and Nishi, T. (2003) Poly(hydroxybutyrate)/poly(butylene succinate) blends: Miscibility and nonisothermal crystallization. *Polymer*, **44** (8), 2503–2508.
117. Wang, R., Wang, S., Zhang, Y. et al. (2009) Toughening modification of PLLA/PBS blends via in situ compatibilization. *Polymer Engineering and Science*, **49** (1), 26–33.

118. Takashi, F. (1998) Processability and properties of aliphatic polyesters, 'BIONOLLE', synthesized by polycondensation reaction. *Polymer Degradation and Stability*, **59** (1–3), 209–214.
119. Wang, H., Ji, J., Zhang, W. *et al.* (2009) Biocompatibility and bioactivity of plasma-treated biodegradable poly(butylene succinate). *Acta Biomaterialia*, **5** (1), 279–287.
120. de Jong, E., Highson, A., Walsh, P. and Wellish, M. (2009) Biobased Chemicals; value added products from biomass. IEA Bioenergy, Task 42 report.
121. Haveren, J., Scott, E.L. and Sanders, J. (2008) Bulk chemicals from biomass. *Biofuels, Bioproducts and Biorefining*, **2** (1), 41–57.
122. Werpy, T.A., Frye, J.G.J., Zacher, A.H. and Miller, D.J. (2003) Hydrogenolysis of 6-Carbon Sugars and Other Organic Compounds, WO2003035582.
123. Ji, N., Zhang, T., Zheng, M. *et al.* (2008) Direct catalytic conversion of cellulose into ethylene glycol using nickel-promoted tungsten carbide catalysts. *Angewandte Chemie*, **120** (44), 8638–8641.
124. Burch, R.R., Laffend, L.A., Nagarajan, V. *et al.* (2007) Bioconversion of a fermentable carbon source to 1,3-propanediol by a single microorganism, US 7169588.
125. Sunkara, H.B. and Manzer, L.E. (2001) Production of Polytrimethylene Ether Glycol and Copolymers Thereof, WO0144348.
126. Cukalovic, A. and Stevens, C.V. (2008) Feasibility of production methods for succinic acid derivatives: a marriage of renewable resources and chemical technology. *Biofuels, Bioproducts and Biorefining*, **2** (6), 505–529.
127. Prichard, W.W. (1979) Conversion of furan to 1,4-butanediol and tetrahydrofuran, US4146741.
128. Wang, Q., Cheng, Y., Wang, L. and Li, X. (2007) Semicontinuous studies on the reaction mechanism and kinetics for the liquid-phase oxidation of p-xylene to terephthalic acid. *Industrial and Engineering Chemistry Research*, **46** (26), 8980–8992.
129. Vispute, T.P., Zhang, H., Sanna, A. *et al.* (2010) Renewable chemical commodity feedstocks from integrated catalytic processing of pyrolysis oils. *Science*, **330** (6008), 1222–1227.
130. Davda, R.R., Shabaker, J.W., Huber, G.W. *et al.* (2005) A review of catalytic issues and process conditions for renewable hydrogen and alkanes by aqueous-phase reforming of oxygenated hydrocarbons over supported metal catalysts. *Applied Catalysis B*, **56** (1–2), 171–186.
131. Huber, G.W. and Dumesic, J.A. (2006) An overview of aqueous-phase catalytic processes for production of hydrogen and alkanes in a biorefinery. *Catalysis Today*, **111** (1–2), 119–132.
132. Lopes, C., Lourenco, J., Pereira, C. and Marcelo-Curto, M.J. (2002) Aromatization of limonene with zeolite Y, in *Natural Products in the New Millennium: Prospects and Industrial Application* (eds A.P. Rauter, F.B. Palma, J. Justino *et al.*), Kluwer Academic publishers, Dordrecht, The Netherlands, pp. 429–436.
133. Berti, C., Binassi, E., Colonna, M., *et al.* (2010) Biobased Terephthalate Polyesters, WO2010078328.
134. Colonna, M., Berti, C., Fiorini, M. *et al.* (2011) Synthesis and radiocarbon evidence of terephthalate polyesters completely prepared from renewable resources. *Green Chemistry*, **13** (9), 2543–2548.
135. Gandini, A. (2011) The irruption of polymers from renewable resources on the scene of macromolecular science and technology. *Green Chemistry*, **13** (5), 1061–1083.
136. Corma, A., Iborra, S. and Velty, A. (2007) Chemical routes for the transformation of biomass into chemicals. *Chemical Reviews*, **107** (6), 2411–2502.
137. Feldman, R.M., Gubawardena, U., Urano, J. *et al.* (2011) Yeast organism producing isobutanol at a high yield, US2011183392.
138. Peters, M.W., Taylor, J.D., Jenni, M. *et al.* (2010) Integrated process to selectively convert renewable isobutanol to p-xylene, WO2011044243.
139. Marliere, P. (2009) Production of Alkenes by Enzymatic Decarboxylation of 3-Hydroxyalkanoic acids, EP2304040.
140. Marliere, P. (2010) Method for the Enzymatic Production of 3-Hydroxy-s-Methylbutyric Acid from Acetone and Acetyl-CoA., WO2011032934.
141. Marliere, P. Delcourt, M., Anissimova, M. and Tallon, R. (2011) Production of Alkenes by Combined Enzymatic Conversion of 3-HydroxyAlkanoic Acids, WO2012052427.

142. Okkerse, C. and van Bekkum, H. (1999) From fossil to green. *Green Chemistry*, **1** (2), 107–114.
143. Gong, W.H. (2007) Terephthalic acid composition and process for the production thereof, US7385081.
144. Brandvold, T.A. (2010) Carbohydrate Route to para-Xylene and Terephthalic Acid, WO2010151346.
145. Williams, C.L., Chang, C.-C., Do, P. *et al.* (2012) Cycloaddition of biomass-derived furans for catalytic production of renewable p-Xylene. *ACS Catalysis*, 935–939.
146. Shiramizu, M. and Toste, F.D. (2011) On the diels–alder approach to solely biomass-derived polyethylene terephthalate (PET): Conversion of 2,5-dimethylfuran and acrolein into p-Xylene. *Chemistry: A European Journal*, **17** (44), 12452–12457.
147. Niu, W., Draths, K.M. and Frost, J.W. (2002) Benzene-free synthesis of adipic acid. *Biotechnology Progress*, **18** (2), 201–211.
148. Frost, J.W., Miermont, A., Schweitzer, D. and Bui, V. (2010) Novel terephthalic and trimellitic based acids and carboxylate derivatives thereof, WO2010148081.

10

Polyamides from Biomass Derived Monomers

Benjamin Brehmer
Evonik Industries AG, Marl, Germany

10.1 Introduction

10.1.1 What are Polyamides?

Polyamides are probably better known under their colloquial name nylon, originating from DuPont's widely successful introduction of PA66 for women's pantyhose (stockings) in the 1940s. While the name 'nylon' was originally limited to the registered trademark of PA66, it has since come to be synonymous with the nomenclature of all polyamides. Even though the words 'nylon' and 'polyamide' have since become interchangeable, in this chapter the official chemical terminology of polyamides (PA) will be exclusively used.

Polyamide resins are linear condensation polymers with a high degree of crystallinity with repeating amide (–CO–NH–) links in their molecular chain. The number of carbon atoms between the amide bonds in the main backbone designates the number after the prefix 'PA'. There are, however, two distinct types of polyamides with the following chemical formulas [1]:

$$\text{Monadic (AB):}\quad \mathrm{H-[NH-C_nH_{2n-z}R_z-CO]_p-OH}$$

$$\text{Dyadic (AABB):}\quad \mathrm{H-\left[NH-C_nH_{2n-z}}R'_z\mathrm{-NHCO-C_nH_{2n-z}}R''_z\mathrm{-CO\right]_p-OH}$$

Monadic (or AB) polyamides are based upon a single repeating lactam with, on the one end an amine reactive group, and on the other end a carboxylic acid group. The lactam thus reacts with itself and the number describes the lactam involved. Dyadic (or AABB) polyamides are based on the reaction of a diamine and a diacid. The diamine and the diacid react alternatively with each other in a salt solution. In regards to the nomenclature, the first number describes

Bio-Based Plastics: Materials and Applications, First Edition. Edited by Stephan Kabasci.

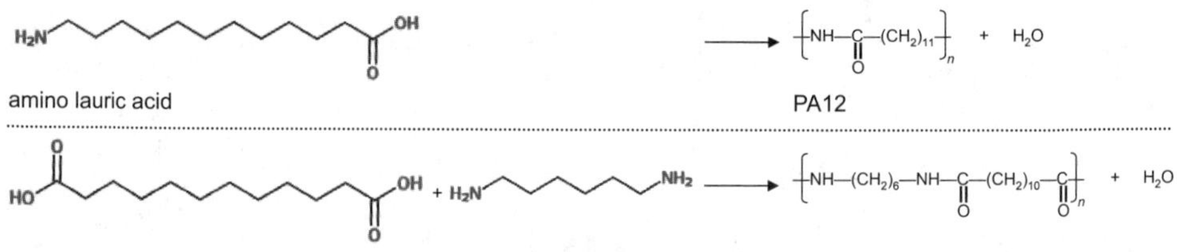

Figure 10.1 Visual depiction of monadic (AB) and dyadic (AABB) polyamides.

the diamine while the second number describes the diacid. Depending on the familiarity of the AABB-polyamide involved, the numbers can be written with or without a comma (e.g. PA612 or PA6,12).

Visually, the classification of polyamides is easier to comprehend (see Figure 10.1). Monadic (AB) is a single monomer system reacting upon itself, while dyadic (AABB) is a dual component system.

Although, the lactams are limited to certain geometrically stable ring structures, in theory nearly any combination of an amide and carboxylic acid monomer could create a polyamide. In practice, however, only several major petro-based and bio-based types have been established.

10.1.2 What is the Polymer Pyramid?

The world of polymers has spawned such success and diversification that grouping and classification is beneficial. A tool frequently used is the polymer pyramid. Listed below (Figure 10.2) is a typical polymer pyramid. The base of the pyramid is labelled as commodity polymers, seeing that they are produced in ample quantities (indicated by the width of the

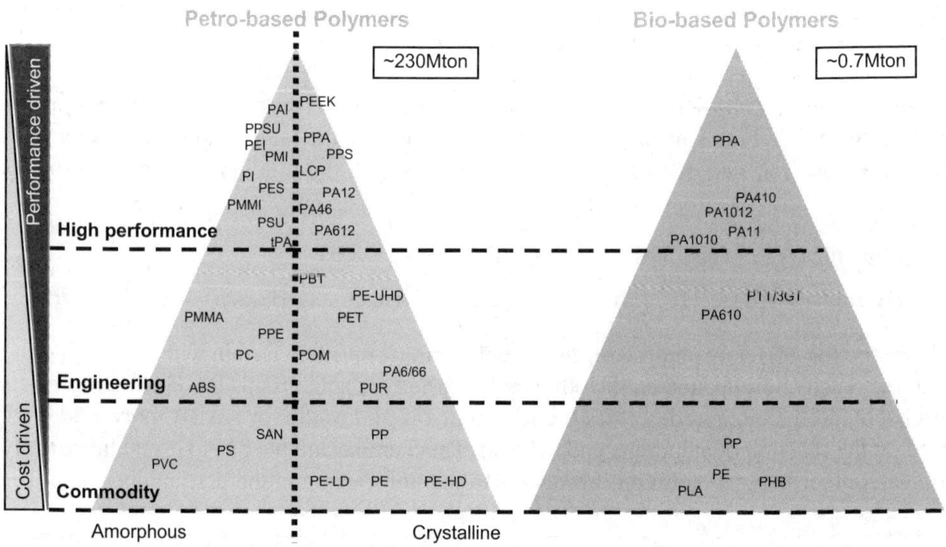

Figure 10.2 Cost vs. performance pyramid of petro-based and bio-based polymers.

pyramid) and their use is more cost driven than performance driven. Those polymers that are firstly, less frequent and secondly, more performance driven are located at the peak of the pyramid. Polyamides are generally classified as either engineering or high performance, although their exact location is rather subjective as each major petrochemical company will construct their own pyramid. The implications of performance are discussed in the next section. What is important here is to draw parallels between the traditional petro-based polyamides and the novel bio-based polyamides. Unlike most available biopolymers, the polyamides from biomass derived monomers are located in the upper half of the pyramid.

The classification of all polymers and their consolidated production capacities is presented at http://www.plasticseurope.org/documents/document/20130627092230-plasticseurope,_statistical_monitoring,_pemrg,_2013-06a_-_summay.pdf and http://en.european-bioplastics.org/market/ (accessed 25 July 2013). Focused solely on their global production capacities, polyamides are rightfully positioned in the upper portion of the pyramid. Total demand of polyamides in 2006 was 2.29×10^6 tons, which was forecasted to reach 2.79×10^6 tons by 2011 (less than 1% of the total polymer market) as more recent data is not yet available [2]. This relates to an annual growth rate of about 4%. Of this modest quantity, the 'engineering' types (PA6 and PA66) combined represent the vast majority of this global polyamide market. Between 1996 and 2006, only 5% to a maximum of 15% of the polyamides were not PA6 or PA66. This is of course even less for the bio-based polyamides; as recently as 2010, they achieved a total global capacity of 40 000 tons, of which more than half was allocated to the production of PA11. In this regard, polyamides, especially bio-based polyamides, are niche products selected mainly for their performance as opposed to their cost position.

10.1.3 Where do Polyamides from Biomass Derived Monomers Fit?

The monomers for the bio-based polyamides are derived partially or fully from renewable feedstocks. Hence, their carbon backbone is sourced from a biomass resource. It can be argued that a strong reduction in greenhouse gas emissions or the inclusion of such bio-sourcing is an additional performance factor. This would facilitate the fusion of the two pyramids (as displayed in Figure 10.2) into one. The bio-based polymers would receive an 'eco-bonus' and be displayed at a higher level than their petro-based counterparts or alternatives. On a purely technical level, this is of course not correct. However, from a marketing perspective, this may very well be a key performance factor that justifies higher pricing. Therefore, the central question remains how to evaluate the polyamides produced from biomass derived monomers subjectively.

10.2 Technical Performance of Polyamides

10.2.1 How to Differentiate Performance

As classified above, these high-performance polymers are characterized by their exceptional hardness, good impact strength and excellent abrasion resistance. Their fibres are also generally associated with silky quality due to their natural elasticity. This combination of properties differentiates the polyamides as a separate category against most other polymer materials. Such categorization is routed in the characteristic amide functional links, which facilitate an internal hydrogen bond between the different polymer chains. Especially in regards with the thermal and mechanical properties, the shorter polyamides, with their high proportion of amide links, outperform their long aliphatic relatives. On the other hand, the amide links also

Table 10.1 *Selection of properties for typical base petro-based and bio-based polyamides.*

Polymer	Bio-sourcing	LCA*	Temperatures (°C)			Tensile (MPa)		Water adsorption
	Percentage of C-atom	$kgCO_{2eq}/kg$	T_g	T_s	T_m	Strength	Modulus	Uptake (%)
PA6	0	9.1	47	179	218	80	3000	10.5
PA66	0	7.9	50	230	258	72	2500	8.2
PA610	63	4.6	48	196	206	61	2100	2.9
PA1010	100	4.0	37	171	191	54	1800	1.8
PA1012	45	5.2	49	151	181	40	1400	1.6
PA10T	50	6.9	125	280	285	73	2700	3.0
PA11	100	4.2	42	154	183	34	1100	1.9
PA12	0	6.9	138	151	176	45	1400	1.5

Notes: All values are averages. MATWEB Material Property Data (www.matweb.com, accessed 22 June 2013) and Evonik internal technical product datasheets. *LCA: lifecycle assessment, based on PE International database (http://www.gabi-software.com/databases/, accessed 22 June 2013), internally certified studies and GaBi-based calculations and industrial datasheets [2, 3, 4]. T_g: Glass transition temperature, as measured by ISO11357.; T_s: Vicat softening temperature as measured by ISO306; T_m: melting-point temperature as measured by ISO11357; tensile strength and modulus as measured by ISO527 1/2; water adsorption at saturation based on full water contact, internal procedure.

lead to a higher degree of water uptake and chemical attack meaning that the longer aliphatic polyamides are inheritably more chemical and hydrolysis resistant while displaying a lower water uptake. In many demanding applications, these properties are a more decisive criterion than mechanical strength and thermal stability alone. Table 10.1 lists an overview of some of the key properties of typical petro-based and bio-based polyamides.

The level of bio-sourcing of the carbon backbone is listed in the first column and as the nomenclature suggests, the bio-based polyamides have a mid to long chain length. This enables these bio-based polyamides (PA610, PA1010, PA1012, and PA11) with straight chains to outperform the standard short chain polyamides (PA6, PA66) in terms of water uptake and to a degree, chemical resistance – yet, due to the sheer complexity and quantity of chemical solutions and testing options, the term 'chemical resistance' must be generalized and therefore is not listed. As bio-based polyamides are also shorter than their long-chain relatives (PA12) they outperform them in terms of mechanical strength and heat resistance. In essence, the bio-based polyamides lie between the short-chain and long-chain polyamides and depending on the desired application can be either superior or inferior to their petro-based relatives.

In the polyamide business, base polymers (as the properties above adhere to) are rarely applied. In addition to this – as standard practice – the base resins are usually altered by controlling the resulting molecular weight, producing copolymers or simply by blending. The values listed in Table 10.1 are thus averages, implying that such alterations can create large deviations between the different base polymer types. To make the matter of comparisons even more precarious, polyamides destined for high demanding niche markets are further tailored through compounding to achieve the desired end-use properties. A wide range of compounding options are available, such as adding thermal stabilizers, anitoxidants, impact modifiers, plasticizers, light stabilizers, flame retardants and lubricating additives. Therefore, direct comparison of the different polyamides based on their technical performance is not as straightforward as Table 10.1 might suggest.

The categories of bio-sourcing and lifecycle assessment will be separately discussed in the following sections. Price performance ratios or even indications cannot be considered as prices are not listed and specially tailored compounds are frequent in the high-performance segment of polymers.

10.2.2 Overview of Current Applications

While a layman or laywoman will likely associate polyamides (here nylons) with panties, other apparel items and/or fishing lines, they are in fact employed in a diverse range of applications. Furthermore, the public perception is now far from the truth as the extrusion of filaments and fibres for nontechnical applications such as fishing nets and lines is currently only a minor niche field. To better categorize and describe the application fields in which polyamides are applied, segregation based on downstream processing technologies is beneficial to provide a good overview.

In general, polyamides can be broken down into two large processing segments; injection moulding and extrusion processing [5]. Other smaller processing fields do exist but are not significant compared to these broad processing segments. As a niche processing technology, powders (especially powder coating) stand out and are applied in noticeable quantities but this will not be mentioned here in further detail. Listed below is an overview of this process determined segregation with their application fields and several product examples as well as their corresponding production ratios.

As briefly mentioned in section 10.2.1, polyamides are selected for their performance and thus specific properties with only a minor price pressure. In recent years, the automotive market has evolved to become the single largest application field for moulded polyamides. They are selected for their excellent resistance to chemicals (such as gasoline and other petroleum products) and the environment (such as water and light) plus their relatively high temperature stability. Much of the growth mentioned in section 10.1.2 is derived directly from this market to the point that the total polyamide consumption is stipulated and almost dependent on the success of new car and truck production.

Injection moulding (76% of total)

- Automotive and truck parts: tanks, reservoirs, covers, tubing, piping.
- Industrial machinery parts: parts for small engines; gears, caps, housings.
- Consumer articles: valves for aerosol cans, wheels, toys, shoes, ski boots.
- Electrical and electronic parts: connector clips, fasteners, guiders, castings.
- Appliance parts: dishwashers, refrigerators, freezers, irons, blenders.
- Hardware and furniture parts: fasteners and rollers.

Extrusion (23% of total)

- Film and extrusion coating: lamination, foils, barrier layers, sealants
- Wire and cable: protective layers, insulation
- Tubing and pipe: pressurized pneumatic systems, brake cables, gas
- Nontextile monofilament: paintbrushes, toothbrushes
- Stock shapes: precision parts, bottles.

Other (1% of total)

It is conceivable that the bio-based polyamides might also find many suitable applications in the broad area of automotive and truck parts, but due to their current price situation and low capacities it is highly unlikely that this will be the dominant segment. As seen in Table 10.1, the bio-based polyamides have a different set of properties compared to the more price sensitive short-chain polyamides. To gain market entrance, it would be wise to focus on application

Table 10.2 Overview of major[a] producers of bio-polyamides.

Company	Trade Name Basis	PA410	PA610	PA1010	PA1012	PA10T	PA11
Arkema	Rilsan						•
BASF	Ultramid		•				
Dongchen	n/a		•				
DSM	EcoPaxx	•					
EMS	Grilamid		•	•		•	
Evonik	Vestamid Terra		•	•	•	•	
Solvay	Technyl		•				
Suzhou Hipro	Hiprolon		•	•	•		
Wuxi Xingda	n/a		•	•	•		

Notes: [a]Notable capacity: i.e. larger than 1 kton/a. Solvay includes previous activities of Rhodia; Wuxi includes previous activities of DuPont; Arkema recently acquired the activities of Hipro.

areas that demand their specific set of properties. The emerging trend of LOHAS (lifestyle of health and sustainability) enables the bio-sourcing and ecological aspects to become actual performance factors for some application segments. Generally, the closer products are to the end customer (actively practising LOHAS), the greater the willingness to pay an eco or bio premium as a key performance indicator. This will likely shift the focus of bio-polyamides to consumer articles and other fields currently held by other petro-based polymers. Well differentiated petrochemical/polymer companies offering a multitude of polyamides including the novel bio-based variants (such as Evonik, see Table 10.2) will benefit particularly strongly by providing customers with the specific properties required in the various application areas. The global trend of LOHAS and product personalization will further support this reasoning, opening new application fields for bio-based polyamides.

As a final note, it is also conceivable that several petro-based and bio-based polyamides will be copolymerized or blended to tailor to meet the needs of certain application areas. In essence, this would create an ideal combination of properties to facilitate an ideal price-to-performance ratio. It should then be mentioned that this will further expand the product portfolio and make listed comparisons, such as in Table 10.1, nearly impossible to construct. The future of bio-polyamides may be in consolidation and a clear-cut focus on certain defined applications areas demanding their specific properties. The actual development remains to be seen.

10.2.3 Typical Association of Biopolymers

Biopolymers can mean different things to different people. In many regions, there seems to be even a bit of confusion in the industry and amongst key stakeholders. Much like how nylon is associated with pantyhose, biopolymers are innately associated with inferior performance and biodegradability. The common attribute of this analogy is that both associations can be traced back to the first major success stories. Yet, unlike with the case of polyamides, to date the biodegradable and low performance biopolymers still dominate the global production. In Figure 10.3, the distribution of the current 700 000 tons global biopolymer production capacity is shown [6].

Even this listing style assists in rousing confusion towards the classification of biopolymers. Biodegradable polymers, even those 100% based on the traditional petrochemical production route are including in this listing. In some cases, polymers with a petrochemical feedstock, but a biotechnological intermediary step (such as fermentation or enzymatic catalysis) are

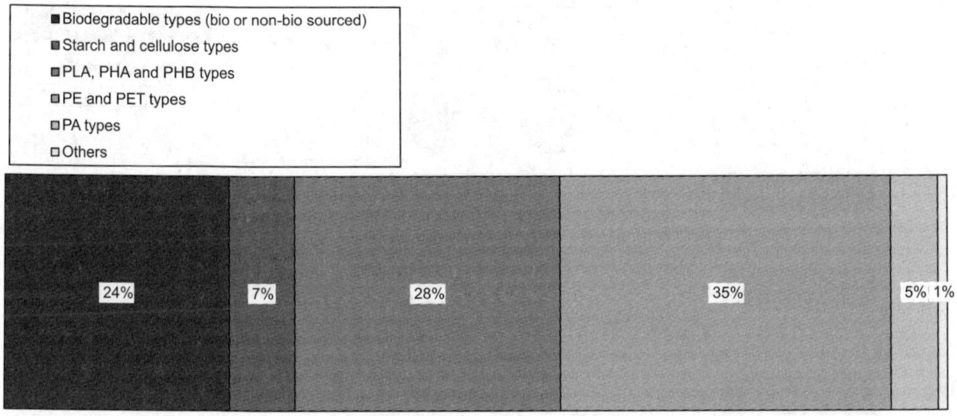

Figure 10.3 *Global biopolymer production capacity distribution.*

included in the semantic 'biopoylmers'. The correct definition of biopolymers would be those polymers synthesized from monomers that are fully or partially based on (or sourced from) biomass feedstocks. Terms such as renewably sourced, vegetable based or bio-based are all synonymous with the prefix 'bio' as in biopolymer. Sourcing only refers to the carbon backbone originating from a biomass (renewable plant) source.

Glancing at the list, the second association of inferior performance becomes apparent. The two single largest biopolymers on the market are PE/PET types and PLA/PHA types; PE and PET are pure commodity products and even their durable petrochemical equivalents are not necessarily associated with any particularly high performance criteria. They are primarily price driven. Starch and cellulose, as well as PLA/PHA, are a new breed of polymers, which mostly have even worse properties than existing olefins.

'Bio' does not necessarily have to denote low quality or inferior let alone biodegradable. As mentioned in section 10.2.1, polyamides are true high performance polymers, yet seeing that only 5% of the current biopolymer market is served by PA-types it may take a while to modify the public perception and association. In this respect, the title of this chapter has been well chosen. Furthermore, several of the main bio-based polyamide producers have opted to use the trade names originally deemed for their main petro-based polyamides to uphold the notion of high performance, for example Arkema with Rilsan or Evonik with Vestamid (see Table 10.2).

10.3 Chemical Synthesis

10.3.1 Castor Bean to Intermediates

Castor bean is a rarity amongst the oilseed crops. Firstly, the bean has an abnormally high oil content of 40–60%. Secondly, the chemical composition of the castor oil is highly defined and concentrated on a single fatty acid. Between 85 and 90% of the castor oil is the C18 fatty acid ricinoleic acid found in the form of a triglyceride ester, hence the binomial nomenclature (Latin name), *Rizinus cumminus*.

Although the cultivation practices and supply chain will be described in the following section, the cultivation methods and locations apparently have little influence on the chemical

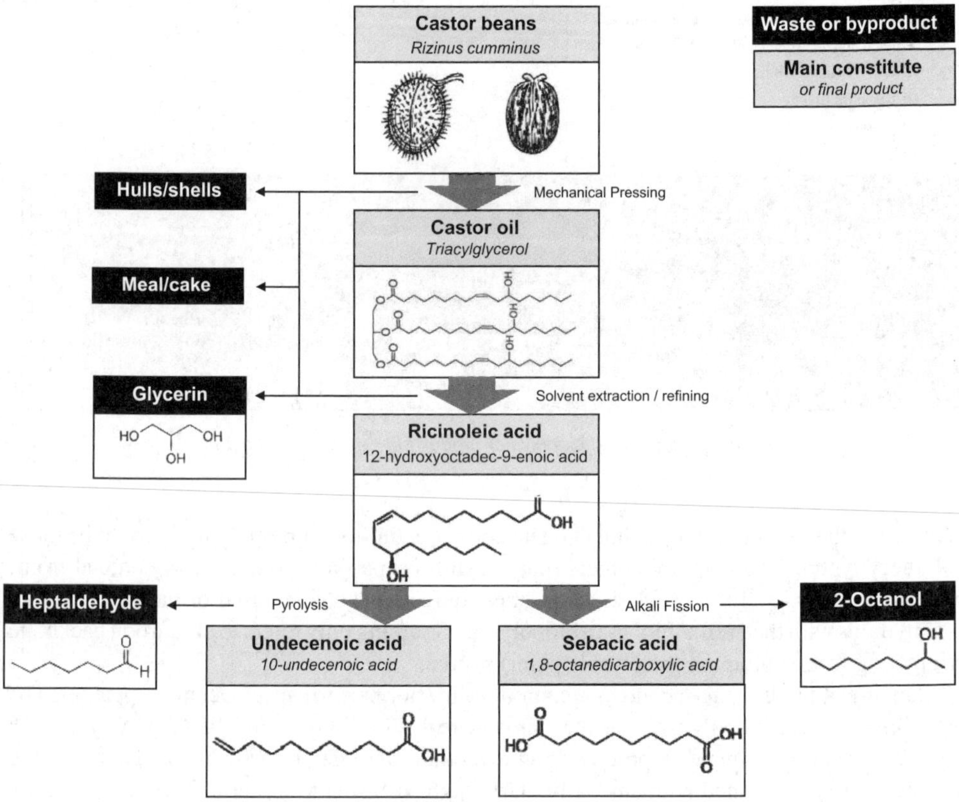

Figure 10.4 Production pathway for monomers based derived from castor bean.

composition of the castor bean. This robust trait leads to a reproducible quality that enables bean mixing and the option of broad sourcing of the material. The initial oil pressing and refinery steps can thus operate with a wide array of sourced castor beans.

The main production pathways for the monomers used in polyamides are illustrated in Figure 10.4. Ricinoloeic acid ($C_{18}H_{34}O_3$) is the essential building block and starting point for the two distinct production routes; undecenoic acid and sebacic acid. The mechanical pressing and solvent extraction methods employed to release and separate the fatty acids (oils) are standard techniques. Nonetheless, there is a wide degree of technologies employed and some modern castor specific techniques have also been explored and incorporated [7, 8].

As mentioned, castor oil is a triacylglycerol composed of various fatty acids and glycerol with an extraordinary high concentration of the C18 fatty acid ricinoleic acid. The oil pressing and refinery steps are employed to extract the C18 fatty acid and purify it for further use as a building block. As seen in Figure 10.4, several side products and residues are produced. Firstly, the beans are physically or mechanically separated from the outer shell or husks; they are typically burnt for their calorific value. Secondly, after the mechanical pressing and subsequent solvent extraction a residue, called the castor cake, remains. Untreated, the castor cake contains toxic compounds such as ricin and ricinine and allergens. A detoxifying step is therefore required to facilitate its use as a fodder, feed or as an agricultural fertilizer. There are numerous physical and chemical methods available to convert the cake to a meal [7]. Thirdly, to obtain single fatty acid esters, the triglycerides (oils) are subjected to either a transesterification step, common in biodiesel processing, or a saponification step unique to castor-oil processing. Both steps involve reacting the triglycerides with an agent to break the ester bond. As a side product glycerine is produced. In regards to the transesterification, an alcohol (usually methanol) is used, which results in the intermediate molecule, methyl ricinoleate. In regards to the saponification step, an alkali (usually sodium hydroxide) is used, which results in pure sodium ricinoleate. Not displayed is the waste-water stream resulting from the neutralization of the alkaline solution related to this process. Some processes are being studied using physical, as opposed to chemical, treatment to reduce the waste water stream. Lastly, and for both routes, the residual fatty acids and other impurities are left untreated in the meal or in the waste water streams.

10.3.2 Undecenoic Acid Route

The monomer used for the AB polyamide PA11 is 11-aminoundecanoic acid. There is a multitude of process steps involved to arrive at this monomer, which have been termed here collectively as the undecenoic acid route.

As displayed in Figure 10.4, the ricinoleic acid (here the transesterification product, methyl ricinoleate) is subjected to a pyrolysis step. More specifically, the cracking pyrolysis takes place at 500–600 °C in presence of water vapour but in the absence of air to release undecenoic acid (methyl undecenoate) and heptaldehyde. The side product heptaldehyde is a source of several seven carbon-containing co-products such as heptanoic acid or heptanol. The main intermediate, methyl undecenoate, is further treated [9]. Firstly, it is hydrolysed to yield undecenoic acid. Then, it is reacted with hydrogen bromide in a nonpolar solvent to enable the reverse addition reaction, which forms bromoundecanoic acid. Finally, upon ammonia treatment, the crystalline solid 11-aminoundecanoic acid is formed and separated.

The pyrolysis step is not particularly efficient at yielding methyl undecylenate [7]: straightforward techniques lead to low yields of 17–35% whereas newer techniques can reach values between 45–50%. By including these process values and starting chemical composition, the overall yield to undecenoic acid is rather low: 15–20% of the original castor bean mass.

The formation of PA11 from 11-aminoundecanoic acid is conducted using the standard water solution polycondensation technique with subsequent reactant water extraction. The reactant water typically contains trace monomers and other impurities and is therefore treated.

10.3.3 Sebacic Acid Route

The diacid (BB) monomer used for the AABB polyamide PA610 and PA1010 is sebacic acid (1,8-octanedicarboxylic acid). While the process to arrive at this monomer is straightforward, it too will receive a collective term, the sebacic acid route.

As displayed in Figure 10.4, the ricinoleic acid (here the saponification product, pure ricinoleic acid) is subjected to a further alkaline treatment. At elevated temperatures (180–270 °C) the ricinoleic acid is subjected to a strong soda (with NaOH or KOH) treatment to cleave 2-octanol and sebacic acid. The 2-octanol side product could be a source for various eight carbon products, but is currently marketed in low-end applications areas such as plasticizers, if not directly combusted.

The alkali steps are not particularly efficient at yielding sebacic acid [7]: straightforward techniques lead to low yields of 23–42% whereby newer techniques can reach values up to 72–76%. By including the standard process values and starting chemical composition, the overall yield to sebacic acid is rather low: 10–15% (30–45% under new technologies) of the original castor-bean mass. This value is however very subjective to the technology involved as the several small and large sebacic acid producers employ strikingly different degree of technological advancements.

The diamine (AA) monomer used for the PA610 AABB polyamide is hexamethylene diamine (HMDA), which is produced petrochemically from butadiene.

AABB polymerization has an extra element of complexity compared to AB polymerization. Slight deviations in the proportion between the diacid and diamine can lead to premature chain termination causing low molecular weights and an uneven distribution. In order to achieve the desired molecular weight (chain length) and equal distribution a so-called 'nylon salt' solution is initially prepared. This crystalline solid structure is formed at room temperature using an exact stoichiometric ratio of 1 to 1. Afterwards, at elevated temperatures (250–300 °C) the salt solution reacts to form polyamide. Chain length termination is controlled by bringing the solution out of equilibrium via either diacid or diamine dosing. Thus, the formation of PA610 from sebacic acid and hexamethylene diamine is conducted using the nylon salt polycondensation technique with subsequent reactant water extraction. The reactant water typically contains trace monomers and other impurities and is therefore treated.

10.3.4 Decamethylene Diamine Route

The diamine (AA) monomer used for the AABB polyamide PA1010, PA1012, and PA10T is decamethylene diamine (DMDA or 1,10-diaminodecane). As DMDA is derive from sebacic acid it belongs to the abovementioned sebacic acid route.

There are several processing steps involved to form DMDA from sebacic acid [10]. Firstly, sebacic acid is exposed and reacts with a nitrogen source such as gaseous ammonia to form diammonium sebacate. Upon continuous dehydration at elevated temperatures (200–220 °C) the dinitrile octamethylene dicyanide ($C_{10}H_{16}N_2$) is formed which, after purification, is hydrogenated (80–100 °C, 2.5 MPa) in the presence of KOH and a nickel catalyst. Pure DMDA is subsequently obtained through vacuum distillation.

The formation of PA1010 from sebacic acid and decamethylene diamine, like other AABB polyamides, is conducted using the nylon salt polycondensation technique. PA1012 is formed by the petro-based dodecanoic diacid (DDDA) and DMDA. PA10T is formed by DMDA and terephthalic acid, a diacid with a benzene ring as a backbone.

10.4 Monomer Feedstock Supply Chain

10.4.1 Description of Supply Chain

While the castor bean plant is native to tropical Asia and Africa and grows in the wild in many regions, it has found widespread cultivation success in only a few isolated regions (http://www.hort.purdue.edu/newcrop/default.html, accessed 22 June 2013). These regions correspond well with the ideal agricultural conditions to obtain maximum oilseed yields. Throughout the growth season, the castor oil responds well to temperatures between 20 and 26 °C with a low humidity and grows best in loamy soils with medium texture. Nonetheless, castor beans are renowned as a low maintenance crop with the ability to be cultivated especially on marginal lands and can tolerate various weather conditions. The vast majority of castor bean cultivation is nevertheless still confined to the Indian state of Gujarat (northwest India). This area is ideal to achieve high castor oilseed yields as can be seen by comparing the official harvested yield figures [11]. The following lists the most recent (2006–2010) four-year average of the main castor bean producing regions:

1. India: 12.6 ton/ha yield and 71.6% total harvest capacity;
2. China: 8.6 ton/ha yield and 13.2% total harvest capacity;
3. Brazil: 6.4 ton/ha yield and 7.2% total harvest capacity.

There are also several smaller players/regions (the residual 8.1%) who are increasing their production and might play a role in the near future. At the moment, the market is clearly dominated by India with yields nearly twice as high as in Brazil.

Castor bean is indeed cultivated at commercial operations in India, but cannot yet be considered as a plantation crop. Most cultivation plots are operated by small family farms and in many cases as a secondary (crop) source of income. Its low maintenance nature and Gujarat's ideal growing conditions facilitate this behaviour. In this respect, weather patterns and more particularly the onset of the monsoon season, are critical for predicting the season's harvest. While it cannot be classified as such, to a certain extent castor oil is already being treated as a commodity crop and is influenced by speculation [12]. The impact of pricing will be covered in the next section.

In 2010, the worldwide castor oilseed production was over 1×10^6 tons while the castor oil production amounted to around 600 000 tons [6, 12]. Most of the oil is destined for motor lubrication and for use in the cosmetics area. After distribution to the other handful of application fields, only a mere 150 000 tons are allocated for the production of sebacic acid. This quantity leads to approx. 70 000 tons of sebacic acid, of which only roughly 20 000 tons are currently dedicated for polyamides. Even with the additional circa 50 000 tons of castor oil needed for PA11, polyamides only represent about 10% of the crop harvest.

The castor bean is stable and stores well, which gives the option of transporting the oilseed for further processing. Nonetheless, most castor-oil production is decentralized and takes place closer to the cultivation areas. On the other hand, sebacic acid and undecenoic acid production is generally centralized by a few players operating in the vicinity of their distribution markets. The polymerization facilities, if not linked directly to the acid processing plants, are also centralized at key petrochemical hubs.

10.4.2 Pricing Situation

Listed in Figure 10.5 is the monthly price history of a selection of raw materials (in US$) [12, 13]. In bold and black, the highly volatile yet clearly increasing price of crude oil can be seen. Crude oil is the key raw material source for the petrochemical industry. Nearly superimposed is the dotted grey line representing sebacic acid. Sebacic acid on the Chinese spot market seems to more or less adhere to the world crude oil price. This is logical as its derivatives (such as plasticizers, stabilizers and paints) and its already mentioned use as a raw material for bio-polyamides compete directly with petro-based alternatives. Sugar, another raw material used for bio-based materials and fuels (such as bio-ethanol and bio-polyethylene) as well as food, has also risen during the period 2004–2012, but more stable and with a lesser volatility compared to crude oil or sebacic acid. This is influenced by a lot of factors, of which

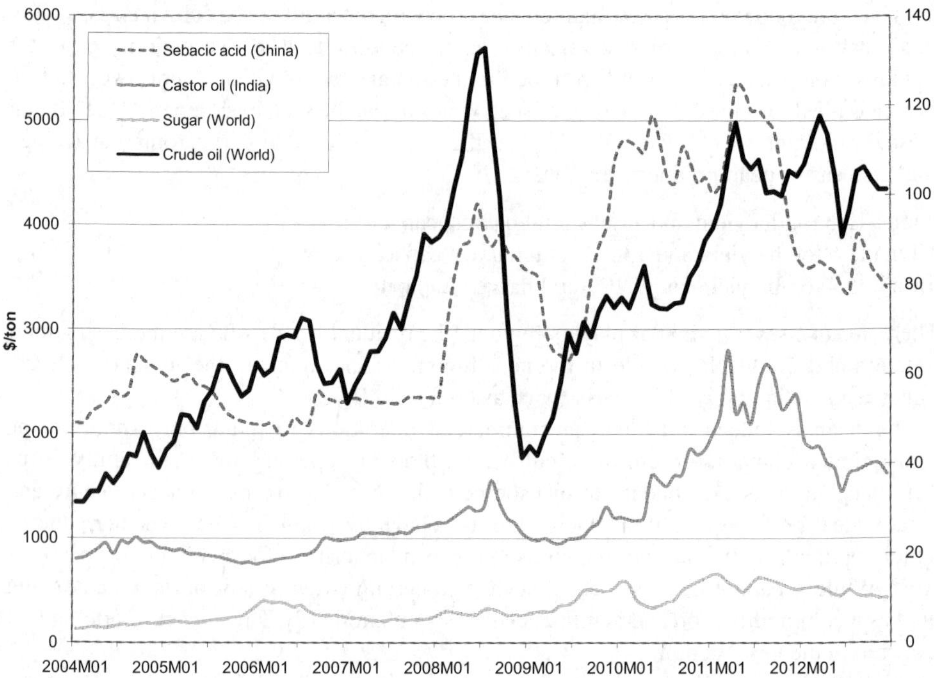

Figure 10.5 *Selected raw material price history.*

many are debatable. Yet, as opposed to speculating or hypothesizing on the reasons, the key issue shown in Figure 10.5 is the price change of castor oil between 2009 and 2012 in relation to the other raw materials. Compared to sugar or crude oil, the 2011 peak of castor oil is substantially higher than the previous peak of 2008 (or for sugar 2010). Naturally, due to the direct use of castor oil, sebacic acid has experienced that same peak trend, whereby it is still more closely related to crude oil price developments. To understand any product sourced from castor oil, it is important to understand such price trends and the recent discrepancies with the global trend.

The key difference between sugar and castor oil, for example, is marked in the level of professionalism in the cultivation practices and the scale of cultivation. Sugar is a true world commodity and is cultivated in many regions using state-of-the-art technology. The relationship between supply and demand (and thus price) cannot be severely influenced by a bad crop harvest of an isolated area. As mentioned above, castor oil on the other hand is cultivated on a relatively limited scale and as a low maintenance (hence nonplantation) crop does not incorporate all the amenities of modern agriculture practice. In the case of castor oil, the relationship between supply and demand (and thus price) can be greatly influenced by a bad crop harvest in an isolated area. This is the reason why monsoon predictions for the Gujarat region have become an integral part of market data pertaining to castor oil. Market speculation or a bad harvest or a slight yearly difference between supply and demand cannot be easily compensated. While polyamides only represent 10% of the market share, their market presence and perceived success, and public interest may have driven both speculation and, at least temporarily, the supply and demand balance to higher prices. With the announcement and inauguration of several new Chinese polyamide producers (see next section), the unusually high price increase between 2009 and 2011 can be partially explained by these phenomena.

As was expected, the higher prices of 2010 and 2011 led many farmers (in all suitable regions) to plant more castor beans, which increased the supply in the following years. Prices in 2012 dropped heavily due to an exceptional year and the cultivation efforts of the previous years. Especially in 2012, the prices were not in direct correlation with the crude oil prices. What remains to be seen is if that supply and the rising demand will balance the equation or if speculation and inefficient trading will dictate spot prices.

Another factor stipulating the recent (speculative) price development of castor compared to other oil crops (such as soy bean, rape seed or palm oil – prices not shown) is castor oil's consistently high concentration of C18 fatty acid. This makes replacement with other oil crops difficult for existing industries based on this chemistry as there are no realistic crop alternatives.

Presently, all polyamides produced from biomass derived monomers are based on those monomers derived from castor oil. To a large extent, the price of bio-polyamides will thus be directly influenced by the castor oil price. Such price formulas in addition to the moderate economy-of-scale options and logistics will likely keep bio-polyamides in the upper price range.

More information on pricing of biomass and raw materials for bio-based polymers can be found in Chapter 13 of this book.

10.5 Producers

Table 10.2 lists the currently available bio-polyamide types and their notable producers and trade names. No producer offers all variants, whereby Evonik has the largest selection. Arkema is the sole worldwide producer of PA11 and of the undecenoic acid process. DSM is the only company pursuing PA410. Aside from DSM and Arkema, all other producers active in bio-polyamides offer PA610. PA610 is generally the market entry point for the Chinese producers due to their cost positioning. Several smaller Chinese producers not mentioned in Table 10.2 produce PA610 for their own semi-finished product consumption/production.

This table does not give any indication of grades or compounding options. As mentioned in section 10.2.1, rarely is a polymer, let alone a polyamide marketed pure (as a base polymer). Especially in high performance applications, tailoring, modifications and certain quality features are needed and sometimes essential. To ascertain the possibility of employing a bio-polyamide for demanding applications, joint customer accompanied research and development is advised.

10.6 Sustainability Aspects

10.6.1 Biosourcing

Biosourcing is a relatively new term conjured to describe materials based partially or complete on renewable biomass. Biomass is biological material from living or recently living organisms; crops, trees and castor beans are prime examples. As the title of this article implies and as the last few sections explained in detail, bio-polyamides are biosourced to various degrees (see Table 10.1).

All organic and biologically based materials, fresh and fossilized, share one fundamental chemical feature, which is that the carbon atom (C) forms the backbone. The field of organic chemistry is based for the large part on the three elements carbon, hydrogen (H) and oxygen (O). In fact, this is why life forms on earth are considered carbon-based. Bio-sourcing is an attempt to differentiate between chemicals or products based on fossil fuels (petro-based) against those based on renewable biomass (bio-based). It is a relatively crude yet universally understood concept to quantify, in simple terms, the level of renewability or how 'green' a chemical or product is. This classification applies to all organic chemicals and, more specifically, all polymers. Inorganic chemicals such as chlorine or silica cannot be quantified using bio-sourcing: quantification is mass-based.

An obvious limitation to bio-sourcing is that of functional groups. Many organic chemicals and polymers contain other elements aside from C, H, and O, albeit typically in smaller quantities. This is irrelevant for inorganic materials such as chlorine or sulfur as bio-sourcing does not apply; however, for nitrogen functionalities it should apply. Adding nitrogen functionality to a chemical or monomer (as described in section 10.3.4) involves chemically reacting it with a nitrogen-based molecule, such as ammonia or dinitrile. These nitrogen-based molecules are directly sourced from fossil fuels, most notably natural gas. Polyamides and bio-polyamides by definition contain nitrogen functionality in the form of their amine bonds. Bio-sourcing is thus limited to quantifying the carbon source. For this reason, an additional lifecycle assessment (LCA) is advised to quantify the actual level of sustainability for all environmental criterions.

10.6.2 Lifecycle Assessments

While it is implied that a bio-sourced product (or polymer) does contain carbon from a biomass origin in its backbone, it does not necessarily imply an environmentally sound product. Since the Brundtland Commission of 1987, 'environmentally sound' has been encompassed by the broader term 'sustainable'. There are many aspects of sustainability, some can be quantified whereas others are still more subjective.

Amongst the currently subjective aspects, social issues and responsibilities are frequently being associated with the broad term 'sustainable': issues such as work conditions, exposure to toxic chemicals, destruction of rainforests, or the more passionate issue with regards to the bio-based economy; food versus fuel. The 'food versus fuel' debate arose when large quantities of the food chain (such as wheat, sugar and corn) were allocated or planned to be allocated for the production of biofuels or bioenergy. The validity of this subjective issue is superfluous for bio-polyamides such as castor bean (as briefly mentioned in section 10.4.1). Castor bean is sometimes referred to as a wonder tree because the oilseed contains toxic substances that restricts its direct use in the food chain, it is grown mainly in arid to marginal lands, it requires little or no fertilizer or other industrial chemicals and it has been cropped in the existing cultivation regions for centuries [10]. It therefore does not compete with food or cause rainforest destruction.

The currently quantifiable aspects of sustainability are analysed using an LCA according to ISO14040:2006. Figure 10.6 illustrates the general LCA structure with the various impact categories and types (scope). While exposure to toxic chemicals cannot be quantified, an LCA allows, for example, the release of toxic substances to be quantified as a separate impact category. There are some aspects that must be clarified as they are directly relevant to biopolymers and thus bio-polyamides.

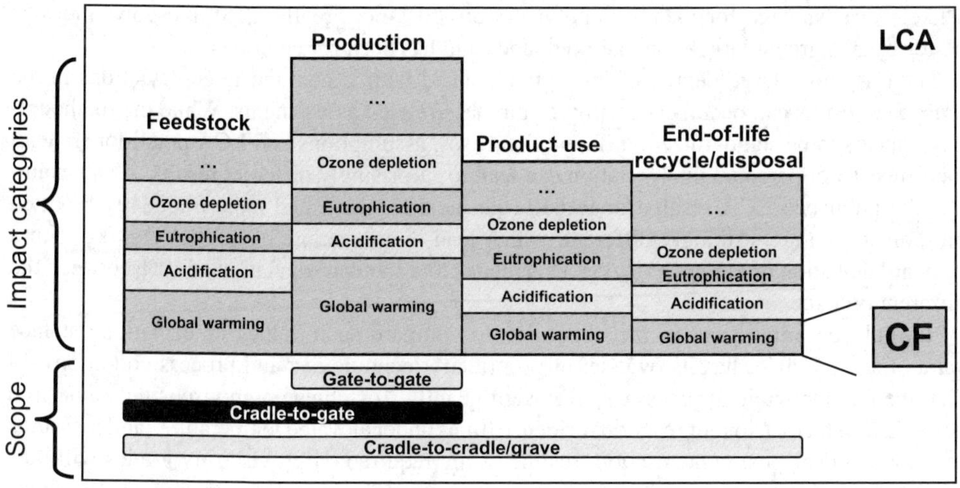

LCA: Life cycle assessment
CF: carbon footprint, greenhouse gas emissions CO_{2eq} over 100years

Figure 10.6 *Structure of lifecycle assessments.*

Firstly, scope. Listed in Figure 10.6 are three scope variants: gate-to-gate, cradle-to-gate and cradle-to-grave. The cradle-to-gate scope is preferred for polymers with the gate set at the delivery of pellets. This scope entails all process requirements and environmental impacts involved to acquire feedstock and produce the polymer. As mentioned previously, polymers, especially high performance bio-polyamides, are not marketed pure (base) but are tailored (compounded) to meet specific requirements of high-demand application areas. This makes product use and end-of-life disposal exceedingly customer specific and too complex and broad for the polymer producer to assess.

Secondly, impact categories. Figure 10.6 lists just a few select examples: global warming, acidification, eutrophication and ozone depletion. A full LCA facilitates dozens of separate impact categories. The global warming potential or otherwise frequently called 'CF: carbon footprint' is preferred not only by polymer producers, but by their customers. Often when an LCA has been conducted, it is the carbon footprint (CO_{2eq}) that is requested by customers and end customers. All other impact categories are simultaneously calculated and can be documented; it would appear, however, that CO_{2eq} has become the sole universal and quantifiable impact category to represent sustainability.

Conducting LCAs is a recent development and they are evolving over time, with data accuracy and calculation methods constantly improving, while adapting new scientific standpoints. Land-use change has quickly emerged out of the subjective and into the quantifiable realm of sustainability. Clearing a rainforest for palm-oil plantations or for cattle pastures are typical examples of land-use change. The CF impact category has been unofficially agreed upon to express global warming potential over the course of 100 years. Attempts are now being made to quantify land-use change by calculating the biogenic carbon lost through the conversion of fertile lands or forests to biomass cultivation plots. In newer LCA databanks and calculations land-use change is included in the CF impact category for biomass crops. This will present a likely disadvantage for many bio-polymers but does not greatly affect bio-polyamides as castor bean is grown largely on marginal lands and has been for centuries.

There is still a large degree of uncertainty linked with LCAs and large deviations in the presented results can occur, especially for biomass-related assessments. While the methodology appears to be standardized, different databases, assumptions and LCA practitioners with their own set of personal interpretation can lead to surprisingly different results. As an example, the published LCA results for PA610 (the most common and most frequently reported bio-polyamide) are strikingly different, which lead to an average of 3.3 $kgCO_{2eq}/kg$ with a standard deviation of 1.3 $kgCO_{2eq}/kg$. This relates to a standard deviation of 40% between the different producers.

It would be unfair, and in fact incorrect, to compare such values or deviations at face value. Each producer has its own set of proprietary technologies and process conditions. In fact, the polymerization process can vary significantly. To achieve high viscosities (a desired performance factor), longer reaction residence times under elevated temperatures and pressures as well as longer post-condensation treatments are required. High viscosity grades will thus result in higher CO_{2eq} values. The energy source involved in the polymerization will also have an impact; nuclear energy has a lower CF than coal. Despite these valid issues, the indicated differences with these PA610 results are mainly attributed to the currently inaccurate nature of an LCA. Values are rarely accompanied with a sensitively analysis.

To continue with the PA610 example, in Figure 10.7 the cumulative energy demand (CED) and greenhouse gas emission cycle are visualized [14]. Almost all bio-based or bio-sourced products have a lower CF due to the locked atmospheric (biogenic) carbon in the product. In some cases it can even be negative. As previously described, the components of PA610 are sebacic acid ($C_{10}H_{18}O_4$) and HMDA ($C_6H_{16}N_2$). As the sebacic acid is bio-sourced, its carbon content is of biomass origin. These interlocked carbon atoms in the resulting bio-polyamide

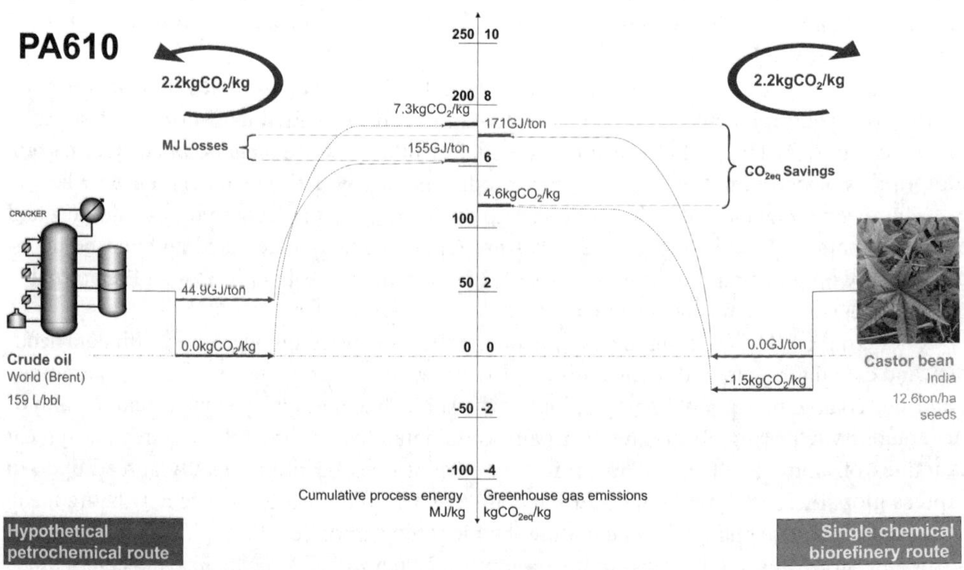

Figure 10.7 *Pictorial representation of PA610 petrochemical and biorefinery route.*

would result in approximately 1.5 kg of CO_{2eq} release upon combustion or degradation at the end-of-life. A total of 2.2 kg of CO_{2eq} would be released for PA610. On the right-hand of the graph, the −1.5 kgCO_2/kg represent this biogenic carbon. The 4.6 kgCO_2/kg expresses the total CF involved to procure the raw material and process the polymer minus the biogenic carbon of the sebacic acid. Conversely, carbon attributed to crude oil or naphtha is not included. On the left-hand side of the graph, the 0.0 kgCO_2/kg represents the allocation of the interlocked carbon originating from crude oil. The 7.3 kgCO_2/kg expresses the total CF should PA610 be produced exclusively from petrochemicals. Polymers are robust materials but typically reach their end-of-life before 100 years. The 2.2 kg of CO_{2eq} will inevitably be released within the 100 year timeframe.

Comparing the CO_{2eq} results at the 'intermediary' factory gate would lead to a savings of 37% for the bio-polyamide. However, when including the interlocked carbon backbone release, the savings are reduced to 28%. This interpretation is not the case with the CED calculations, for the energy content locked in the crude oil is included. Solar energy used for biomass growth is of course free and not a primary energy source. Therefore, the associated CED of the petrochemical raw material is 44.9 MJ/ton and 0.0 GJ/ton for the biorefinery route. The energy contained in PA610 and released upon combustion is 36.9 MJ/ton. Seeing that this 'end-of-life' value has already been accounted for on the petrochemical route, and to compare by equal measures, this calorific value should be deducted from the 'intermediary' factory gate for the biorefinery route. In this case, the displayed energy loss of 16 MJ/ton becomes a saving of 21 MJ/ton. To be consistent, either the released biogenic carbon should be included or the calorific energy should be deducted from the bio-polymer in question.

This issue, visually depicted in Figure 10.7, is common for all LCAs involving the comparison between petro-based and bio-based products. It has been elaborately described to stress the inconsistency and to an extent where inaccuracies can arise in performing LCAs. The entire methodology is a work in progress and is evolving to address such issues. Lifecycle assessments are nonetheless vital to measure sustainability. Currently, however, the results must be taken with a grain of salt. Furthermore, as will be described, bio-based chemistry is relatively new and many technological advancements are foreseeable in the near future. The sustainability of bio-polyamides and their CF (LCA) values will inevitably improve.

10.6.3 Labelling and Certification

The topics of biosourcing and lifecycle assessment, mentioned above, are perfect candidates for labelling and certification for bio-polyamides. The bio-based content of products such as polymers can be tested and verified using radiocarbon dating. Carbon originating from fossil fuel sources compared to biomass source will display a different set of C^{14} isotopes. The DIN ISO 10694:1996 uses such testing methods and is authorized to issue biosourcing labels. Other government bodies and organizations (such as the USDA) are also pursuing biosourcing labelling. As more end customers request 'green' products it is becoming increasingly important to guarantee authenticity and use the labelling as a marketing tool to promote biomass derived products. Efforts are being made to certify LCAs as specialized institutions offer service to verify ones methodology and set of assumptions. It is a first step to address the above mentioned issues of personal interpretation. The validity of such a certified LCA would nonetheless be limited as improvements or drastic operational changes will require amendments and constant monitoring. In this respect, the LCA certification provides a reliable guide to sustainability whereas the biosourcing labelling provides a quality guarantee.

10.7 Improvement and Outlook

The collective value and supply chain of bio-based products is relatively new. While some bio-polyamides have been on the market longer than others (read: PA11 vs. PA10T), they are still novel polymers with plenty of room for improvement.

On the farm level, as mentioned in section 10.3.1, new processing technologies have already been envisioned and tested, which when employed could increase yield, lower waste water and generally improve the sustainability factors. Yet, due to the market structure and geographical location of the castor oil to intermediates, it is unlikely that any groundbreaking improvements will occur in the near future. Gradual improvements and more consolidated cultivation is, however, more foreseeable. In any case, improvements and investments will be driven by castor oil's primary buyers, who are not bio-polyamide producers.

On the petrochemical site level, there are several direct improvement options that the bio-polyamide producers can pursue. Although, confined to the AABB types, direct melt polymerization is one such option. Process control is more difficult, meaning that polymer knowhow and technological finesse are prerequisites. The system would translate into lower energy consumption, reduced waste water treatment, shortened drying times and higher throughputs. According to internal LCA calculations it would be possible to reduce CF and CED by another 43 and 12% respectively.

Confined to PA1012, the diacid (BB) DDDA can be bio-sourced. Facilities do exist that can process palm kernel oil into a lauric acid derivative and consequently into DDDA. At the moment, the sourcing of such a raw material is not competitive or reliable. The financial incentive of the eco-bonus performance factor might steer this development.

Thus far this chapter has focused upon bio-polyamides existing on the market. There are, in fact, several research projects, either bent on developing a bio-based monomer alternative for a traditional polyamide or on investigating completely new polyamides. Research work has begun to create a new 4-diamine (AA) production route for PA46 or possibility for PA410 using novel biotechnological techniques [15]. Biomass (like sugar) is fed to cyanophycin bacteria with the resulting product enzymatically cleaved to aspartic acid and arginine. At the current stage of development, the researchers were successful at stabilizing and immobilizing the *Bacillus subtilis arginase* enzyme needed to create ornithine from arginine. The 4-diamine, specifically 1,4-diaminobutane, is produced from ornithine. In this case, the nitrogen functionality originally produced in the cyanophycin bacteria is upheld, which could yield energetic and economic benefits [9]. Other research work has begun exploring the options of succinic acid as a completely new 4-diacid (BB) type or as another source for the 4-diamine (AA) [16]. Succinic acid is used for a variety of industrial applications and has recently aroused further industrial interest with the announcement of capacity expansions (see http://www.dsm.com/nl_NL/html/dcn/03_05_2011_news.htm, accessed 25 July 2013). Succinic acid is produced via fermentation of sugar. As a diacid, multistep chemical treatment with a nitrogen source can convert it to the diamine much like sebacic acid to DMDA. The novel PA44 can be synthesized from succinic acid and butylmethylene diamine (BMDA). It is a very high temperature polyamide (M_p circa 300 °C). Due to the high concentration of amide links, it is expected that PA44 will possess an unacceptably high water uptake for most applications.

Many of the wonders and promises of biotechnology remain to be seen if they are transferrable and applicable to bio-polyamides. Those bio-polyamides on the market are currently all sourced from castor oil. The notion of discovering and cultivating other suitable biomass feedstock material or using biotechnology to enable simple sugars to become feedstock is promising. Future bio-polyamides will surely incorporate technological or biomass-based improvements.

As a final word, bio-polyamides are a clear example showing that bio-based and high performance can correlate and, considering that the original definition of sustainable is 'something that lasts', bio-polyamides are truly sustainable.

References

1. Keim, W. (2006) *Polyamide in Kunststoffe: Snythese, Herstellungsverfahren, Apparaturen*, Wiley-VCH, Weinheim, Chapters 4 and 8.
2. Arkema (2009) *RILSAN® PA 11: Created from a Renewable Source* (product data sheet), Arkema, Puteeaux.
3. Häger, H. (2010) *VESTAMID Terra, Lifecycle Assessment of Biobased Polyamides* (Evonik product data sheet), Evonik Degussa GmbH, Marl.
4. DSM (2011) *EcoPaXX: The Green Performer* (product data sheet), DSM Engineering Plastic, Geleen.
5. Ormonde, E.V. and Mori, H. (2007) *Nylon Resins—CEH Marketing Research Report*, SRI Consulting, Menlo Park, CA.
6. Internal market analysis of bio-PA market based on sebacic acid trends.
7. Mutlu, H. and Meier, M.A.R. (2010) Castor oil as a renewable resource for the chemical industry. *European Journal of Lipid Science and Technology*, **112**, 10–30.
8. Ogunniyi, D.S. (2006) Castor oil: a vital industrial raw material. *Bioresource Technology*, **97**, 1086–1091.
9. Scott, E., Peter, F. and Sanders, J. (2007) Biomass in the manufacture of industrial products—the use of proteins and amino acids. *Applied Microbiology and Biotechnology*, **75**(4), 751–762.
10. Wang, M.S. and Huang, J.C. (1994) Nylon 1010 properties and applications. *Journal of Polymer Engineering*, **13**(2), 155–174.
11. FAOSTAT (2006–2009) *FAO Data Based on Imputation Methodology*, FAO, Rome.
12. Fuerst Day Lawson Market Intelligence (2011) *Market Report—Castor Oil and Industrial Chemicals*, FDL, London.
13. World Bank Commodity Price Data (Pink Sheet) http://econ.worldbank.org/WBSITE/EXTERNAL/EXTDEC/EXTDECPROSPECTS/0,,contentMDK:21731023~menuPK:538203~pagePK:64165401~piPK:64165026~theSitePK:476883~isCURL:Y,00.html (accessed 25 July 2013).
14. Brehmer, B., Boom, R.M. *et al.* (2009) Maximum fossil fuel feedstock replacement potential of petrochemicals via biorefineries. *Chemical Engineering Research and Design*, **87**(9), 1103–1119.
15. Könst, P.M., Franssen, C.R. *et al.* (2011) Stabilization and immobilization of Trypanosoma brucei ornithine decarboxylase for the biobased production of 1,4-diaminobutane. *Green Chemistry*, **13**, 1167–1174.
16. Kabasci, S. and Bretz, I. (2012) Succinic acid: synthesis of biobased polymers from renewable resources, in *Renewable Polymers* (ed. V. Mittal), Scrivener Publishing LLC, ISBN: 978-0-470-93877-S.

11

Polyolefin-Based Plastics from Biomass-Derived Monomers

R.J. Koopmans
Dow Europe GmbH, Switzerland

11.1 Introduction

'Although theoretically possible, the utilization of biofuels as a primary feedstock for production of commodity chemicals will most likely be constrained by a shortage of cropland, limited capital, and the availability of lower cost alternatives. Absent unforeseen technological innovations or significant government mandates, this situation is unlikely to change on a wholesale basis in the coming decades' [1]. This conclusion, published in a paper by Banholzer *et al.* in 2008, based on environmental, economic, and technical facts and arguments, captures the challenge of obtaining biomass derived versus fossil fuel derived monomers for the production of polyolefin-based plastics. It is however a singular petrochemical industry perspective. This industry has boomed since the 1940s, developing extremely efficient and highly integrated processes for the mass production of chemicals and materials, including plastics, essential to the functioning of society today. Alternative approaches beyond using fossil fuel have just begun to make inroads into the established paradigm. Biotechnology and the concept of biorefinery making use of renewable feedstock are receiving industrial attention. However, significant scientific and technological challenges remain to be solved [2, 3].

Polyethylene and polypropylene, as generic classes of plastics, are by far the most dominant. Together they account for approximately 70% of the world plastics production of about 230×10^6 tons in 2010 (www.plasticeurope.org, accessed 22 June 2013). Their production volumes reflect their versatility as materials in multiple applications.

While society's demand for these plastics continues to rise at a long-term rate of about 4% (www.plasticeurope.org, accessed 22 June 2013), awareness for alternatives to fossil resources derived plastics is growing. In addition, more emphasis is placed on sustainability of plastics

Bio Based Plastics: Materials and Applications, First Edition. Edited by Stephan Kabasci.

from an end-use perspective. The current economic paradigm, however, is not inductive to bio(mass)-based alternatives [1] and current engineering technology does not enable immediate replacement of existing plastics to deal with the growing concern [4] about specific uses and end-of-life issues of their various applications [5].

A holistic approach and a new paradigm are required to consolidate the multiple drivers that define the need for more sustainable products. For example, the cradle-to-cradle concept of McDonough and Braungart [5] proposes a more nature-inspired product design principle versus the more opportunistic commerce driven design of products and manufacturing systems having grown out of the Industrial Revolution. However, 'bio' – typically the plant-alternative to fossil feedstock technology – needs to be examined for its true environmental footprint [6]. System approaches such as the BASF AgBalance^{TM} programme (http://www.agro.basf.com, accessed 22 June 2013) or the University of Stuttgart GaBi^{TM} software (http://www.lbp-gabi.de/5-0-GaBi-Software.html and http://www.gabi-software.com, both accessed 22 June 2013) consider all aspects of crop cultivation in relation to economic, environmental and societal impact enabling more sustainable decision making. Furthermore, the bio-alternatives cannot be seen as independent from the required or desired final product performance. For example, making ethylene (ethene) from sugarcane or corn via fermentation and dehydration of resulting ethanol does not lead to a different kind of polyethylene from that produced from fossil feedstock [1]. Making lactic acid from corn/maize to produce poly(lactic acid) does not provide a plastic having the same performance characteristics as polyethylene nor can it be used in the same or similar applications at identical cost. Moreover, a plastic that can be composted does not require a plant-based feedstock [7]. Growing plants requires energy as well and additional effects have to be considered, for example: the effective harvesting of the desired chemical, the disposal or use of the remaining biomass – possibly as 'bio-fuel' -, and the management of the water issues of lignocellulosic waste [8]. Similarly, a biotechnological approach, using micro-organisms to produce chemicals, has energy issues impacting the economic, environmental and social value in terms of sustainability and relevance.

In this context, looking for and finding viable options to manufacture polyolefin-based plastics from bio(mass)-derived monomers, needs to be questioned in terms of its necessity given the current petrochemical technology and infrastructure. The functionality of the produced products and the relative limited fossil fuel used for their production should be balanced with efforts to find avenues for alternative materials and applications stepping beyond the current ones. Next-generation materials based on building blocks that nature uses may sidestep the issue of trying to decouple the fossil fuel for energy from fossil fuel for chemicals and materials just as plastics created a novel functionality enabling applications beyond metals.

11.2 Polyolefin-Based Plastics

'Plastics' is a catch-all term for materials composed mainly of polymers and other molecules that enhance the use and functionality towards specific applications. Polymers are large (macro) molecules obtained by connecting the same or similar small molecules (monomers). Polymeric materials consist of a mixture of macromolecules being chemically the same that may have different molar masses within a specific molar mass distribution range.

Polyolefin plastics are essentially polymers that are composed out of hydrocarbon molecules. These are typically obtained from fossil fuel that is, petroleum. After a relatively simple heat treatment in a refining (distillation) process, specific distillate fractions, being hydrocarbon

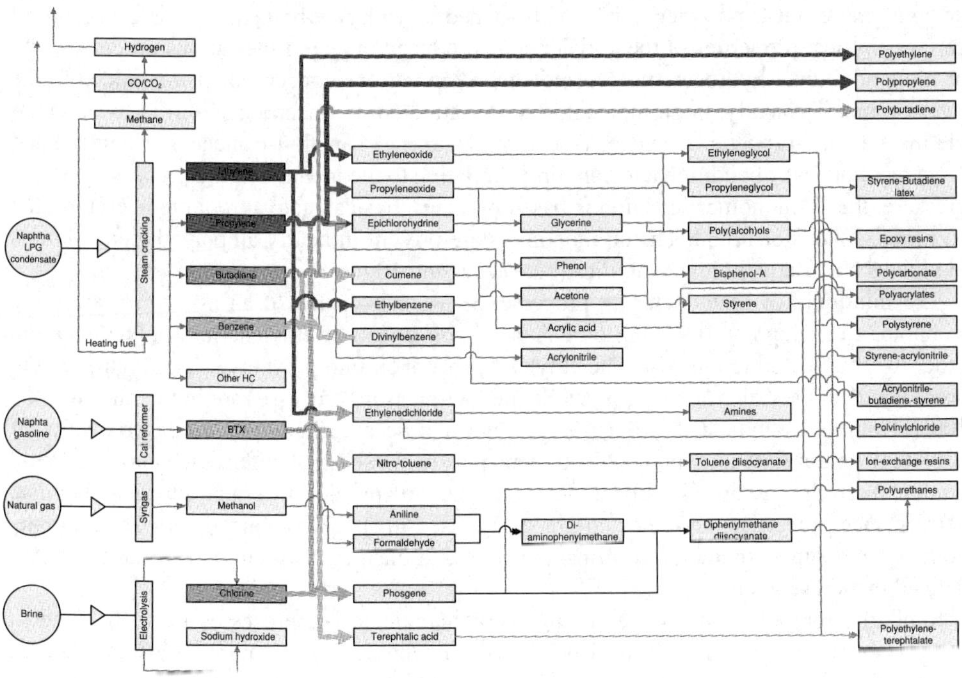

Figure 11.1 *A schematic of the production chain from fossil feedstock to final polymers.*

mixtures referred to as naphtha, ethane or liquefied petroleum gas, are 'cracked'. This 'cracking' process exposes the hydrocarbon mixtures for a very short time to high heat in a furnace, breaking down the molecules into very pure streams of chemicals such as ethylene and propylene (1-propene). These unsaturated hydrocarbon molecules – that is, monomers – can be easily reacted to form polyethylene and polypropylene polymers. Also produced are butadienes, a mixture of aromatic compounds (benzene, toluene, xylene), and hydrogen, each functioning as feedstock for a complex chain of so-called 'intermediates' and additional polymers [8]. The entire production chain from feedstock to final products indicates a strongly integrated number of processes generating multiple intermediates, which are optimized for maximum efficiency and minimal energy consumption at the lowest possible cost and are being continuously engineered for optimal performance (Figure 11.1).

In the case of ethylene, several large-scale manufacturing processes exist to polymerize the monomers [9]. A general distinction can be made between polyethylene manufacturing processes in terms of their operational pressure, being either high (2000–3000 bar) or low (1–100 bar). A further subdivision is made for high-pressure processes as being autoclaves or tubular processes. For the low-pressure processes a subdivision is made into gas-phase, solution or slurry technology. For each subdivision there exist multiple manufacturing processes that have been designed and optimized for specific purposes and proprietary reasons. A key understanding is that, for the multitude of processes, each will generate a different class of polymeric materials in terms of the 'mixture of macromolecules' (molar mass distribution) and the detailed chemical composition of the macromolecules. High-pressure processes generate macromolecules having a very branched chemical structure, whereas low-pressure processes generate more linear polymers, either unbranched or with few branches. In order to control certain physical properties of the polyethylene produced in one of the various processes, one or more additional hydrocarbons (co-monomers) in various concentrations are added to the reaction mix. Typically, these molecules are unsaturated hydrocarbons referred to as alpha-olefins that is, 1-propene, 1-butene, 1-pentene, 1-hexene, 4-methyl-pentene-1, 1-heptene and 1-octene. For ease of addition and handling in relation to the manufacturing process operation pressure, the co-monomer selection is based on their physical state at room temperature that is, being gaseous or liquid. The co-monomers are present in nearly all polyethylene plastics in various but relatively low quantities (between 0 and 10 mol%).

The multitude of manufacturing processes is a consequence of an evolving processing technology development focusing on enhanced conversion of ethylene into polyethylene in order to reduce production cost. The ethylene conversion into polyethylene of high-pressure processes is in the order of 15–30%, while the low-pressure processes are in the range of 80–100%. However, as indicated earlier, the multitude of process technologies for the production of polyethylene translate into many polyethylene plastic classes and, within each class, different grades. Accordingly, not all polyethylene plastics will provide the same physico-chemical performance properties, and selected grades are used in specific applications – for example, bottles, bags, pipes, membranes, films and fibres, to ensure maximum performance for the defined market segment.

Similarly, for polypropylene, a multitude of manufacturing processes exist. In contrast to polyethylene, however, industrial polymerization processes are more decisively defined by the catalyst system [10]. Polypropylene is, chemically speaking, a stereo-specific polymer. Depending on the spatial positioning of the methyl group of the subsequent propylene molecules in the macromolecule, three distinct polypropylenes are produced – isotatic, syndiotactic and atactic. The latter is typically an amorphous polymer, generated as a side product of the isotactic polypropylene production. Syndiotactic polypropylene production requires very specific catalyst technology and, like isotactic polypropylene, it is a so-called semi-crystalline polymer [11]. Semi-crystalline refers to a partial organization of the macromolecules leading to a diffraction pattern in an X-ray analysis experiment, very similar to inorganic crystals. The tacticity critically defines the ability of the macromolecules to organize themselves (crystallize) in mesoscopic structures. The type and degree of crystallinity determines the final material properties. Similarly polyethylene has a semi-crystalline nature. The degree of semi-crystallinity is measured in terms of the density of the grades, indicating that low density polyethylene has much less macromolecular organization then high density polyethylene, making the latter a harder and stiffer material. In contrast, for polypropylene, levels and type of tacticity are used to differentiate grades and not density. The isotactic arrangement is the most preferred in terms of performance versatility and ability to manufacture on a large scale. Catalyst and process conditions are selected to ensure a specific tacticity dominates. Just as with ethylene, polymerization of propylene can be done in a liquid, gaseous or dissolved state. The associated manufacturing plants are referred to as bulk, gas-phase and slurry processes, all operated at relatively low-pressures [10]. Irrespective of the process, the main co-monomer used is ethylene to either produce a so-called random- or block-copolymer for enhancing the performance properties of the homo (isotactic) polypropylene.

The polyolefin manufacturing technologies have taken advantage of 'engineering of scale' indicating polyolefin plastics production in units capable of producing several hundred thousand tons on a yearly basis, reaching global levels of about $80 \cdot 10^6$ tons each – polyethylene and polypropylene – in 2010 (www.plasticeurope.org., accessed 22 June 2013). All processes require very high investment cost (several hundred millions of euro per production site), making the plastics-producing industry one with a high risk and a high financial entry barrier. The industry also requires a significant feedstock position for operation in terms of energy and basic chemicals. Only proximity to feedstock, refineries and crackers, combined with large-scale polymerization production facilities, will lead to a cost advantage, avoiding expensive transportation – pipelines, super-tankers. Part of the success of 'synthetic' polyolefin plastics has been precisely their very low production cost and their ability to replace, as well as most often significantly improve on the performance of existing, natural materials such as glass, ceramic, wood, metal, and natural fibres (e.g. wool, cotton). Thus, a significant petrochemical industry has been created where engineers and scientist have worked on, and still work on, optimizing the effectiveness and efficiency of processes for reducing energy use. However, as long as the fossil fuel for energy cost is coupled to the fossil fuel for chemicals, the existing value chain needs reconsidering. Some challenges to sustain the present petrochemical industry are: that feedstock is typically not found where its products are being consumed while there exists a huge petrochemical infrastructure close to the consumer; that economics drive the choice of transportation of either feedstock to production facilities close to consumers or products transportation to consumers from production facilities close to the feedstock; that very few alternative technologies exist to produce chemicals and materials that can compete efficiently with the existing petrochemical ones. Above all, the turning point for a paradigm shift towards a bio-based, renewable and sustainable business model remains somewhat elusive in its practical outlook from a petrochemical industry perspective. Indeed, the need for energy for heating and transportation is the major driver for the fossil-fuel industry. The petrochemical industry is engaged with a relatively minor fraction of about 4–8% product wise but majorly coupled to the energy pricing. A decoupling may provide a more sustainable petro-chemical industry future. One approach is the bio-refinery concept that focuses mostly on local and specific biomass feedstock with relatively low production volumes. However, this relinquishes the advantage of large-scale production units affecting the economics unfavourably. Furthermore, significant technology improvement in terms of separation and purification are required. Alternative chemicals and materials from biomass may be more opportune to consider providing a different vantage point in terms of what can be done beyond what existing petrochemicals and materials bring, just as stones were replaced by metals and metals by plastics because of the application versatility and the opportunities beyond what stone or metal can provide.

11.3 Biomass

Biomass availability is estimated to be 180×10^9 ton per year [3], having a significantly more complicated chemical makeup than fossil feedstock. Biomass refers to a wide range of materials that are mostly carbohydrates (~75%), followed by lignin (~20%) and smaller amounts of proteins, fatty acids, alkaloids, terpenoids, and nucleic acids comprising together about 5% [3]. A chemically obvious route to hydrocarbons would be to dehydrate the carbohydrates. However, this is a thermodynamically unfavourable route [1]. The materials comprising biomass can be plant or animal-based, by-products from other processes or waste products [2, 3]. The premise of biomass lies with its renewability in the sense that it can be 're-grown or re-made' in a relatively short period of time. The key insight is the time window within which sufficient energy and associated chemicals can be generated to sustain the demand of society, as it is known today.

Table 11.1 Typical times for biomass renewal indicating a variety of time scales [14].

Biomass base	Recycle / renewable time
Algae	1 month
Agricultural Crops	3 months to 1 year
Grasses	1 year
Shrubs	1–5 years
Trees	5–80 + years
Oil, gas and coal	200×10^6 years

The sun is the ultimate driver for biomass generation and will be around for a few billion years [12]. The present practice of harvesting fossil fuel is banking on the stored reserves of biomass of eons ago. The challenge at hand is not just generating feedstock in a form that enables the sustenance of products available today but more of finding potentially new ones with improved attributes and final performance. For 'old' monomers or plastics such as ethylene and propylene, respectively polyethylene and polypropylene, very few options are available and many have been researched. Accordingly, biomass as a renewable feedstock for the production of the same as fossil fuel chemicals and materials entails several challenges [13]:

- Biomass harvesting is typically far more labour intensive then pumping fossil raw materials, which – despite the incurred societal advantage in the biomass producing regions – means an economic drawback for the bio-based product.
- The biological way of producing chemicals (plant growing) is relatively slow (Table 11.1) [14] versus fossil raw materials 'harvesting' via pumping a well, the distillation and cracking process.
- The yield of the desired chemicals per unit land area is typically low.
- The chemicals desired are biomass byproducts, complicating the separation and purification and associated economics.
- The biomass composition tends to be varied and inconsistent.
- Biomass production on a large scale can lead to competition with 'food-crop' land.
- Biomass production on a large scale is climate-zone specific and may only be regionally practised at optimal levels with a potential need for adjustable genetically modified plants.
- Biomass handling typically requires new infrastructural investment on energy, water and waste handling.
- Biomass harvesting and processing technologies of scale have generally not (yet) lived up to their promise in terms of efficiency, overall economics, and final performance of the plastics.

11.4 Chemicals from Biomass

Chemicals from biomass is a practice that is far from new as it was more the norm than the exception before the twentieth century although the quest for hydrocarbons is a more recent consequence of the fossil feedstock debate. The premise for producing hydrocarbons has been documented abundantly; the practical implementation has been lagging however [2, 3,

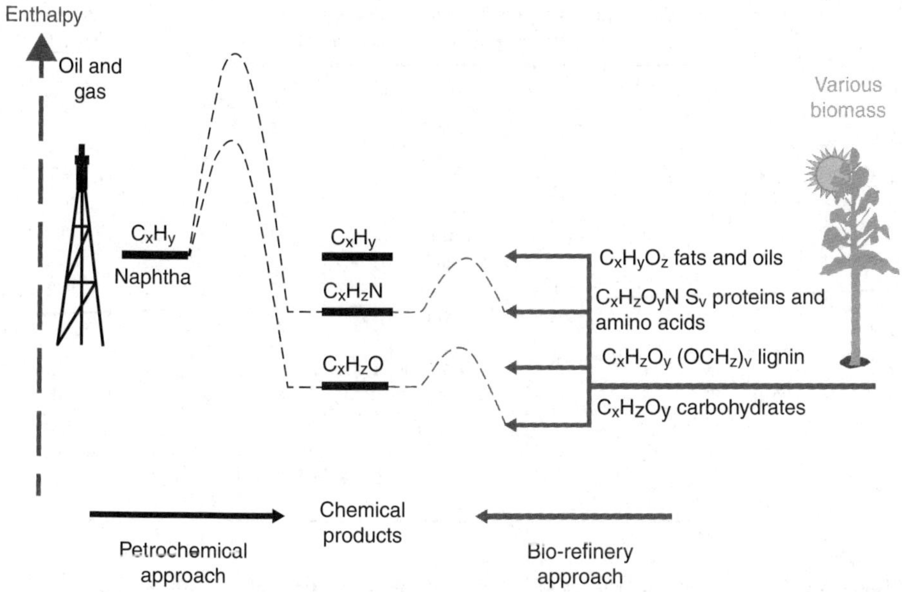

Figure 11.2 *Sketch of thermodynamic considerations for the production of desired functional chemicals from biomass as compared to petrochemical route. (Reproduced with permission from [16]. Copyright © 2007, Springer.)*

14–18]. The scientific and technical challenge can best be understood when considering the enthalpy of combustion or the calorific value of the feedstock. The petrochemical industry uses high-enthalpy, low oxidation-state starting materials to produce products of lower enthalpy and higher oxidation state through the use of additional process energy [1, 16]. In contrast, biomass is composed of many components that have an even lower enthalpy value and higher oxidation state, indicating that it costs significant net energy to synthesize desired monomers such as those obtained by cracking distillates from fossil fuels and, in particular, polyolefins [1]. Figure 11.2 illustrates that unless 'free or advantaged' energy is available – most probably via solar or wind technology – to be incorporated into the overall energy balance equation, the use of biomass is thermodynamically unfavourable for producing basic chemicals such as ethylene and propylene manufactured today.

However, some advantages for biomass are evident when chemical structures with higher functionality that are hard to synthesize from fossil based chemicals can be harvested and used directly. Furthermore, when the required 'process' energy is low for producing the desired chemical or material, significant benefits can be anticipated. Some natural products exemplified in amino acids, peptides and proteins can generate base chemicals and, by taking advantage of self-assembly processes, build functional materials without substantial 'process' energy [16]. Ultimately, innovative catalyst technology may be needed to tip the balance in favour of biomass as feedstock for a much broader bio-based chemical production [1] although the use of enzymes may offer significant overall advantages eventually [17].

Another challenge, beyond the thermodynamic one, is that the adoption of chemicals from biomass, whether produced intentionally or as by-products, is strongly connected to the

Table 11.2 *A selection of chemicals derived from biomass and categorized according carbon number.*

Carbon number	Bio-based chemical
C1	Methanol
C2	Ethanol
C3	Glycerol
C4	Succinic acid
	Fumaric acid
	Malic acid
	Furan
C5	Xylitol
	Furfural
	Itaconic acid
	Levulinic acid
C6	Sorbitol
	Citric acid
	Furan-2,5-dicarboxylic acid
C6 + x	Fatty acids

eventual successful (technically efficient and economically affordable) production of biofuels. This raises the issue of the amount of available land and types of waste or crops to be used for the production of biofuels versus food, chemicals and materials. As indicated before, fossil raw materials are mostly used for transportation fuels and heating/cooling and less than 10% is used for the production of chemicals and materials [4, 18, 19] (see also http://www.bp.com/en/global/corporate/about-bp/statistical-review-of-world-energy-2013/energy-outlook-2030-.html, accessed 8 July 2013).

Some of the more important relevant chemicals already available from biomass for producing plastics are summarized in Table 11.2. One approach for classifying biomass derived chemicals is according to their carbon number as some can be produced via multiple routes [2, 3, 8]. Starting with C1, methanol can be manufactured via syngas derived from biomass. The next is C2, being primarily ethanol, which can be obtained in a similar way from syngas as methanol and by fermentation processes from carbohydrates. Both C1 and C2 chemicals are sources for ethylene using proven large-scale technology. The most relevant bio-derived C3 chemical is glycerin. It could be a source for propylene but that is technically far more challenging than it is to produce ethylene from methanol or ethanol. The synthesis would involve first producing 1,2 propanediol from glycerol, a further dehydration to 2-propanol and eventually propylene [20,21]. Higher carbon numbers are relating to bio derived chemicals more relevant to non-polyolefin plastics.

11.5 Chemicals from Biotechnology

Industrial biotechnology aims to provide solutions to chemicals and materials production by using micro-organisms [2,3, 17]. Fermentation processes are operational and can produce both methanol and ethanol. A key challenge for using microorganisms to produce hydrocarbons is the incompatibility with the aqueous environment in that fermentation typically takes place. To date, even genetically modified microorganisms are no match for the synthetic, petrochemical approach. Here again, a focus on higher value than bulk chemicals is most probably a more sustainable model.

11.6 Plastics from Biomass

A plastic from biomass is again nothing new. In the nineteenth century, rubber, basically polyisoprene, and cellulose were directly harvested from a plant source and saw their first industrial uses. In fact, any kind of biomass is a 'plastic' in the sense that it is a mixture of macromolecules consisting of proteins, polysaccharides, polynucleotides and some smaller molecules such as lipids, water and inorganic elements. The biomass challenge, however, is the definition of viable unit operations to obtain clean chemical streams that enable a direct harvesting of the desired plastic or of the chemical that leads to it. This view leans towards a multitude of options for natural 'plastics' development of which many have not been explored, at least from a scientific perspective, let alone a technology scale up. The natural polymers, grown in nature, such as proteins, polysaccharides and polynucleotides as well as lipids offer an unimagined ability to form structure, smart functionality, and energy and information storage. This goes far beyond the more pragmatic outlook of biodegradable polymers, which have been pursued in the last few decades. Their perceived advantage is compostability and the possibility to either be obtained directly from plants or microorganisms or synthetically made either from fossil- or bio-based chemicals. Some combinations of fossil- and bio-based are also being researched [22]. Alternatively, the same chemicals can produce noncompostable bio-based plastics. Various global capacity/volume projections for those plastics anticipate 1.5 to 2.5×10^6 tons in the next decade [23], less than 1% of the current synthetic fossil fuel based plastics capacity. Some of these plastics will be discussed more extensively in this book.

However, the 'Zeitgeist' seems to be that although the rise of the large-scale production synthetic plastics will slow down [24], they are not about to disappear as yet or in the foreseeable future. Near term, the five major plastics polyethylene, polypropylene, poly(vinyl chloride), poly(ethylene terephthalate), polystyrene and styrene based plastics that make up the bulk of all synthetic plastics will be difficult to replace in view of their versatility, performance and low cost. Ultimately there may be no need until better materials replace their functionality or the application becomes redundant. For example, poly(vinyl chloride) is no longer used for music long-player disks, and floppy disks and videocassettes from polystyrene do not exist any more because novel technologies made the applications redundant. This also suggests that the search for novel polymers based on physico-chemical insights of structure forming properties of natural polymers can lead to a completely novel materials paradigm, bypassing the 'fossil fuel for energy' dilemma.

11.7 Polyolefin Plastics from Biomass and Petrochemical Technology

In view of the significant investment cost for building an integrated chemical complex, the prevailing petrochemical industry business logic – at least as a short (5 year) to medium (10 year) term strategy – is to move facilities closer to the fossil feedstock. Another view is exiting the business or exploring technology that provides the same or similar feedstock from different renewable resources and takes advantage of existing infrastructure that is, green chemistry.

Just as the fossil fuel and petrochemicals industries are closely connected, the production of the same but bio-based chemicals will be closely related to the manufacturing of biofuels. The focus of commercialization by using biomass as a feedstock is mainly on: ethanol, biodiesel, butanol, hydrogen, Fischer–Tropsch fuels, methanol, methane, and MTBE/ETBE (methyl- or ethyl-tert-butylether). These products can be used either as biofuels or chemicals for further converting. In the context of this chapter, only routes for the ethylene and propylene

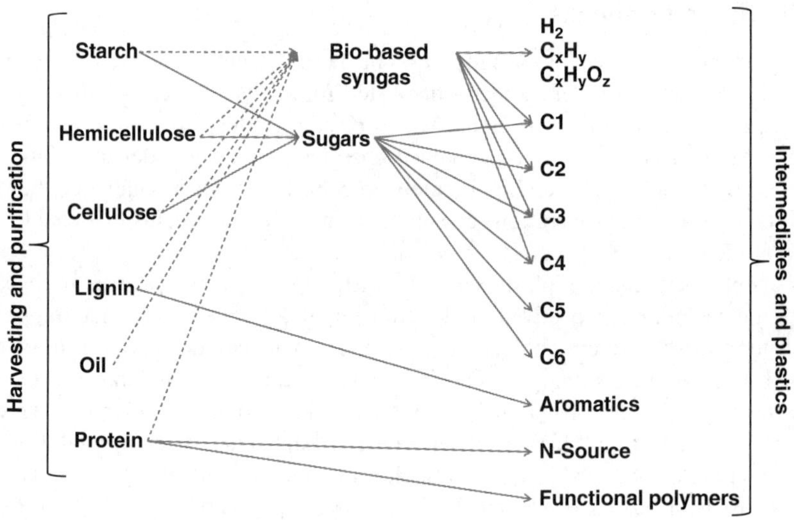

Figure 11.3 *Bio-based polyolefin plastics can potentially be obtained from several purified natural products.*

monomer production are discussed, as there exist multiple technologies to convert biomass into hydrocarbon feedstock. Moreover, there are huge amounts and multiple potential sources of biomass available [25]. Considering the carbon numbers as identifier building blocks to generate ethylene and propylene the following approaches are explored. Figure 11.3 gives a schematic representation of biomass sources and their potential for obtaining chemicals to produce intermediates and plastics.

11.7.1 One-Carbon Building Blocks

Since the 1920s the Fisher–Tropsch process (see http://www.fischer-tropsch.org, accessed 22 June 2013) [26] is the one most commercially exploited as a route to hydrocarbons. Through gasification of biomass (or other carbon sources such as coal or natural gas) a synthesis gas (syngas) composed mainly of carbon monoxide, carbon dioxide and hydrogen can be produced.

$$\text{biomass} + (O_2,\ H_2O) \rightarrow \text{syngas}(CO,\ CO_2,\ H_2) + \text{other chemicals}$$

$$\text{isolated syngas: nCO} + (2n+1)H_2 \rightarrow C_nH_{(2n+2)} + nH_2O$$

The ultimate production of hydrocarbons thus allows existing petrochemical processes to produce the required intermediates, including ethylene. Alternatively, partial pyrolysis of biomass allows for the production of organic molecules such as methanol. This thermal process for making wood alcohol has long been replaced by more efficient natural gas (methane) reforming processes [27]. The chemical approach of producing methanol has significant industrial potential, as addressed in the book by George Olah *et al.*, propagating a 'methanol economy' [28]. This inspired research for alternative routes to methanol such as methane oxidation with homogeneous catalysts in sulfuric acid media [29], ethane bromination followed by hydrolysis of the obtained bromomethane [30, 31], direct oxidation of methane with oxygen [32, 33] and microbial or photochemical conversion of methane [34].

Irrespective of the technology, methanol can be converted catalytically into ethylene and propylene. These methanol to olefin (MTO) technologies use highly selective catalysts to yield very efficient (>95%) hydrocarbon-producing processes [35–37].

11.7.2 Two-Carbon Building Blocks

Ethylene is a nontoxic, natural product that affects the growth, development, ripening and ageing of all plants [38]. It is produced in small quantities by fruits and vegetables. The quantities, however, are certainly not in the order of magnitude that justifies any industrial process development or polyethylene production based on naturally produced ethylene. Beside the MTO route, ethylene can also be obtained from ethanol. The dehydration of ethanol is, in fact, a reversible reaction of the synthetic direct hydration of ethylene [39–42].

Besides bio-based ethylene additional bio-based alpha-olefin co-monomers will be required in order to tailor the various polyethylene grades to a required performance. It will require bio-based ethylene to be di-, tri- or tetramerized to form 1-butene, 1-hexene or 1-octene [43–47].

11.7.3 Three-Carbon Building Blocks

Propylene is, next to ethylene, the most important basic chemical to produce not only polypropylene but also other intermediates for example propylene oxide and acrylonitrile. Just like ethylene, propylene can be produced via a hydrocarbon feedstock produced from a biomass [35–37]. Bio-glycerol produced as a byproduct of biodiesel can be dehydrogenated to produce propylene [48]. Bio-based ethylene can be dimerized to produce n-butene, which can then react with remaining ethylene via metathesis to produce propylene [49]. The use of fermentation products of biomass such as 1-butanol [50] enables the formation of n-butene, followed by a subsequent methathesis [49]. Alternatively, hydrothermal carboxylate reforming of fermentation products such as butyric acid or 3-hydroxybutyrate is also proposed as a viable option to propylene [51].

Other direct fermentation products such as 1-propanol and 2-propanol [52] can, in principle, be used for the production of propylene by dehydration but little research has been carried out [16, 18].

Another route is the use of vegetable oils or fatty acids that can be fed into an enhanced fluid catalytic cracker (FCC) unit to produce propylene (see www.chemsystems.com, accessed 22 June 2013).

11.8 Polyolefin Plastics from Biomass and Biotechnology

Bio-based ethanol is obtained from fermenting carbohydrate biomass such as starches (grains, corn, potatoes, root crops), sugars (molasses, cane juice, sugar beets, fruits) and cellulose (wood, waste from paper and pulp operations, waste paper). The viability of the bio-based ethanol route to ethylene is defined by the issue of scale and overall process economics. In regions such as Brazil, with a significant sugarcane industry and an ethanol fuel economy, the ethylene and ultimate polyethylene production becomes economically attractive. In combination with co-energy regeneration using the lignocellulose (bagasse) waste, the overall process and greenhouse gas emissions can be called sustainable [1]. The technical challenge for the fermentation process lies with the compositional complexity and variability of the biomass feedstock, requiring pre-treatments before enabling the micro-organisms to convert carbohydrates to ethanol. The cane sugar conversion to ethanol is a relatively simple process, however corn or wheat starch need first an enzymatic hydrolysis (amylases) to produce glucose prior to the fermentation process [53]. The release of fermentable carbohydrates from plant cellulose and hemicelluloses comprising respectively 23–53% and 20–35% of plant material requires even more extensive pretreatments affecting the ethanol production cost [53].

The fermentation – that is, biotechnological – approach to feedstock chemicals is divided into four classes depending on the metabolic pathways followed [54]. Fermentative short-chain alcohols are obtained using genetically engineered yeasts and bacteria that produce mainly ethanol and 1-butanol [55]. Non fermentative short-chain alcohols are synthesized using enzymes for decarboxylating 2-keto acid intermediates from an amino acid biosynthesis followed by reducing the resulting aldehyde to an alcohol [54]. Isoprenoid-derived hydrocarbons, such as isoprene and isoprenoid alcohols and olefins, are converted from biosynthetic phosphates and pyrophosphates intermediates using specific enzymes [54]. Longer olefins can be obtained from the enzymatic decarbonylation of unsaturated fatty acids [54]. Overall, the yield of biosynthetic processes remains low, indicating that the value of this approach lies with the synthesis of chemicals beyond the focus of the current-day petrochemical industry.

Considered as the third generation biomass – indicating the direct photosynthetic production of feedstock molecules versus fermentation and chemical technologies – algae hold the promise of being a high-quality biofuel production option [56, 57]. Using sunlight, the organisms turn carbon dioxide and water into lipids that can form the source of biodiesel, ethanol and other chemicals such as ethylene.

Propylene is also naturally produced and research has been carried out for a direct fermentation method producing propane and propylene from a glucose medium. Multiple micro-organism strains were tested but the efficiency remained extremely low [58].

11.9 Bio-Polyethylene and Bio-Polypropylene

Considering the state of the art in technology combined with economic attractive nonfossil feedstock, the production of, in the first instance, bio-polyethylene is feasible. The National Fuel Alcohol Program established in Brazil in the 1970s and aimed at managing the sugar industry overcapacity as well as ensuring energy feedstock independence, has led to an industry that has maximized the use of sugarcane. The productivity of sugarcane in Brazil is approximately 82.5 tons per hectare. Brazil has 340×10^6 hectares of arable land, 65% occupied for cattle and only 1% of the land is devoted to ethanol production from sugarcane (for the sugarcane industry see www.unica.com.br, accessed 22 June 2013 – in Portuguese). The majority of sugarcane (~70%) is produced in the Sao Paulo province. In September 2010, the Brazilian petrochemical company Braskem started the production of 200 000 tons of ethylene from ethanol to make 200 000 tons of polyethylene. This takes approximately 65 000 hectares of sugar cane to provide around 460×10^6 litres of ethanol as an intermediate. The Braskem 'green PE' portfolio consists mainly of high density polyethylene with >96% renewable content aimed at packaging applications (www.braskem.com.br, accessed 22 June 2013). The validation of the renewable content is possible via carbon-14 isotope determination based on ASTM D6866 [59]. The environmental footprint for 'green', bio-based polyethylene is considered favourable, for 1 ton polyethylene production will have the effect of capturing 2.5 tons of carbon dioxide. Moreover the overall energy balance is calculated to be neutral

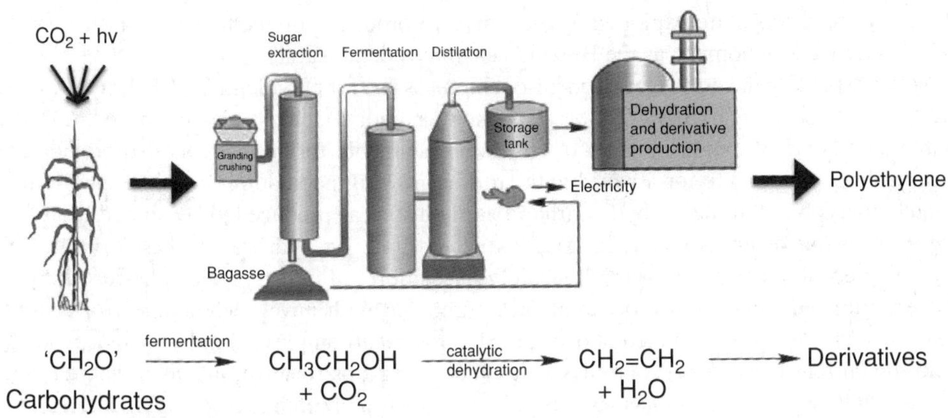

Figure 11.4 *Schematic flow diagram of the production of bio-polyethylene from sugarcane glucose via the fermentation into ethanol and subsequent dehydration into ethylene.*

when considering appropriate water management and energy recovery via the residual biomass (mainly bagasse).

In addition, Braskem formed a partnership with Novozymes, the global leader in the production of industrial enzymes, to expand research for the development of propylene made from sugarcane for the production of 30 000 tons of 'green polypropylene' by 2013.

In Brazil, for similar reasons, the Dow Chemical Company and Mitsui & Co Ltd are engaging in a 50/50 joint venture to cultivate sugarcane for the production of ethanol and subsequently ethylene. Initially a 350 000 tons ethanol production unit is planned, eventually leading to polyethylene production by 2015. This will take about 120 000 hectares of sugar cane aimed at providing a cheaper and less economically volatile ethanol feedstock (see http://www.biofuelsdigest.com/bdigest/2012/03/22/dow-mitsui-to-finalize-brazilian-bio-based-polyethylene-jv/, accessed 8 July 2013).

The technology used for the production of ethylene is schematically represented in Figure 11.4. The final conversion from ethylene into polyethylene will make use of different processes being for Braskem a slurry and gas phase process, while the Dow/Mitsui joint venture may aim for a solution process. Furthermore, ethylene can also be used for other basic chemical derivatives production besides polyethylene. In 2013 the 2007 started project, however, is put on hold in view of changed Ethanol economics (see http://www.worldindustrialreporter.com/dow-mitsui-postpone-1-5b-sugarcane-to-plastic-brazil-plant/, accessed 22 August 2013).

11.10 Perspective and Outlook

The quest for polyolefin-based plastics from biomass-derived monomers reduces essentially to developing nonfossil fuel-based options for producing ethylene and propylene, including some higher alpha olefins. A multitude of technologies are available and industrially scalable for using biomass to generate olefins and the subsequent polymers and plastics. A large-scale use of biomass, however, is hampered by the need for scaling and the socio-economic impact in relation to the petrochemical alternative. The direct use of micro-organisms in a fermentation process for either a direct synthesis of olefins (let alone polyolefin plastics) is in its infancy and will require significant technology developments. Replacing polyolefin plastics with other types of chemistries is extremely difficult and may not be essential. Regional alternative feedstocks are exploited as incentives to produce polyolefin-based plastics from biomass-derived monomers, as the Brazilian examples demonstrate.

Notwithstanding the odds against an all-out biomass use for polyolefin-based plastics, eventually, the petrochemical, fossil fuel-based economy paradigm may need to be shifted. A more viable alternative is sought in order to ensure a sustainable future that can also provide an acceptable quality of life for a world with an ever growing population. It can be questioned if the biomass-based routes to hydrocarbon-based plastics as produced today should be abandoned in favour of novel materials. The fossil feedstock for hydrocarbon-based plastics is becoming economically unfavourable, mainly in relation to fossil-fuel dependence. Alternative energies reducing the use of fossil fuel, combined with novel technologies for energy efficiency, can allow a continued and sustainable hydrocarbon plastics industry independent from the current fossil fuel economics. Ultimately the paths forward, just as in nature, will be the intelligent use of biomass-based polymers such as proteins and polysaccharides as material building molecules providing advanced solutions beyond the current hydrocarbon plastics uses.

References

1. Banholzer, W.F., Watson, K.J. and Jones, M.E. (2008) How might biofuels impact the chemical industry? *Chemical Engineering Progress Magazine*, **104**(3), 7–14.
2. Kamm, B., Gruber, P. and Kamm, M. (eds) (2006) *Biorefineries – Industrial Processes and Products*, vol. **1**, Wiley-VCH Verlag GmbH, Weinheim.
3. Kamm, B., Gruber, P. and Kamm, M. (eds) (2006) *Biorefineries – Industrial Processes and Products*, vol. **2**, Wiley-VCH Verlag GmbH, Weinheim.
4. Freinkel, S. (2011) *Plastics. A Toxic Love Story*, Houghton Mifflin Harcourt Publishing Company, New York.
5. McDonough, W. and Braungart, M. (2002) *Cradle to Cradle*, North Point Press, New York.
6. Searchinger, T., Heimlich, R., Houghton, R. *et al.* (2008) Use of U.S. croplands for biofuels increases greenhouse gasses through emissions from land use change. *Science*, **29**, 1238–1240.
7. Rudnik, E. (2008) *Compostable Polymer Materials*, Elsevier, Amsterdam.
8. US Department of Energy (2004) *Top Value Added Chemicals From Biomass Volume 1 - Results of Screening for Potential Candidates From Sugars and Synthesis Gas*, http://www.osti.gov/bridge/product.biblio.jsp?osti_id=15008859 (accessed 8 July 2013).
9. Vasile, C. and Seymour, R. (eds) (1993) *Handbook of Polyolefins*, Marcel Dekker, New York.
10. Karger-Kocsis, J. (ed.) (1999) *Polypropylene. An A-Z Reference*, Kluwer Academic Publisher, Dordrecht.
11. Moore, E. (ed.) (1996) *Polpropylene Handbook*, Hanser Publishers, Munich.
12. Schröder, K.-P. and Smith, R. (2008) Distant future of the Sun and Earth revisited. *Monthly Notices of the Royal Astronomical Society*, **386**(1), 155.
13. Davenport, R. (2007) Chemicals from Biomass, SRI, Menlo Park, CA.
14. Liu, S., Amidon, T., Francis, R. *et al.* (2006) From forest biomass to chemicals and energy: biorefinery initiative in New York. *Industrial Biotechnology*, **2**(2), 113–120.
15. Sanders, J., Scott, E. and Weusthuis, R. (2007) Bio-refinery as the bio-inspired process to bulk chemicals. *Macromolecular Bioscience*, **7**(2), 105–117.
16. Scott, E., Peter, F. and Sanders, J. (2007) Biomass in the manufacture of industrial products—the use of proteins and amino acids. *Applied Microbiology and Biotechnology*, **75**(4), 751–762.
17. Praeve, P., Faust, U., Sittig, W. and Sukatch, D. (eds) (1994) *Handbuch der Biotechnologie*, 4th edn, Oldenburg Verlag, Vienna.
18. Ren, T., Daniels, B., Patel, M. and Blok, K. (2009) Petrochemicals from oil, natural gas, coal and biomass: production costs in 2030–2050. *Resources, Conservation and Recycling*, **53**, 653–663.

19. Marshall, A. and Alaimo, P. (2010) Useful products from complex starting materials: common chemicals from biomass feedstocks. *Chemistry – A European Journal*, **16**, 4970–4980.
20. Vanhaveren, J., Scot, E. and Sanders, J. (2008) Bulk chemicals from biomass. *Biofuels, Bioproducts and Biorefining*, **2**, 41–57.
21. Sardessi, A. and Lee, S. (2005) Alternative source of propylene. *Energy Sources*, **27**, 489–500.
22. Koopmans, R. (ed.) (2009) *Engineering Aspects of Self-Organising Materials, Advances in Chemical Engineering*, vol. **35**, Elsevier, Amsterdam.
23. Shen, L., Haufe, J. and Patel, M. (2009) *Product Overview and Market Projection of Emerging Bio-Based Plastics,* downloadable from http://www.epnoe.eu/research/Life-Cycle-Analysis (accessed 8 July 2013).
24. Lemstra, P. (2008) Introduction – Synthetic versus natural polymers. European polymer Federation workshop on Bioplastics: Crossing the border between synthetic and natural polymers – May 30–31, Paris.
25. Kim, S. (2004) Global potential bioethanol production from wasted crops and crop residues. *Biomass and Bioenergy*, April, 361–375.
26. Kreutz, T., Larson, E., Liu, G. and Williams, R. (2008) Fischer-Tropsch Fuels from Coal and Biomass Princeton Environmental Institute. Princeton University 25th Annual International Pittsburgh Coal Conference, 29 September–2 October.
27. Keil, F. (1999) Methanol-to-hydrocarbons: process technology. *Microporous and Mesoporous Materials*, **29**, 49–66.
28. Olah, G., Goeppert, A. and Surya Prakash, G. (2006) *Beyond Oil and Gas: The Methanol Economy*, Wiley-VCH, New York.
29. Periana, R., Bhalla, G., Tenn, W. *et al.* (2004) Perspectives on some challenges and approaches for developing the next generation of selective, low temperature, oxidation catalysts for alkane hydroxylation based on CH activation reaction. *Journal of Molecular Catalysis A: Chemical*, **220**(1), 7–25.
30. Xu, H., Wang, K., Li, W. and Zhou, X. (2005) Dimethyl ether synthesis from methane by non syngas process. *Catalysis Letters*, **100**(1–2), 53–57.
31. Zhou, X., Lorkovic, I. and Sherman, J. (2002) Integrated process for synthesizing alcohols, ethers, and olefins from alkanes, US Patent 6,486,368.
32. Hickman, D. and Schmidt, L. (1993) Production of syngas by direct catalytic oxidation of methane. *Science*, **259**(5093), 343–346.
33. Foster, N.R. (1985) Direct catalytic oxidation of methane to methanol: a review. *Applied Catalysis*, **19**(1) 1–11.
34. Sipma, J., Henstra, A., Parshina, S. *et al.* (2006) Microbial CO conversions with applications in synthesis gas purification and bio-desulfurization. *Critical Reviews in Biotechnology*, **26**(1), 411–65.
35. Wunder, F. and Leopold, E. (1980) A selective pathway from methanol to ethylene and propene. *Angewandte Chemie International Edition*, **19**(2), 126–127.
36. Chen, J., Bozzano, A., Glover, B. *et al.* (2005) Recent advancements in ethylene and propylene production using the UOP/Hydro MTO process. *Catalysis Today*, **106**(1–4), 103–107.
37. Sardesai, A. and Lee, S. (2005) Alternative source of propylene. *Energy Sources*, **27**(6), 489–500.
38. Arteca, R. (1996) *Plant Growth Substances: Principles and Applications*, Chapman & Hall, New York.
39. Britton, R. Direct hydration of ethylene to ethanol, US 3686334, 22 August 1972.
40. Zsao, U. and Zasloff, H. Production of ethylene from ethanol, US 4134926, 16 January 1979.
41. Jacobs, J., Jacobs, P. and Uyttterhoeven, J. Process for containing ethylene from ethanol US 4670620, 2 June 1987.
42. Megumu, I., Kazuhisa, M., Isao, T. and Inoue, K-I. (2011) Production of olefins from ethanol by Fe and/or P-modified H-ZSM-5 zeolite catalysts. *Journal of Chemical Technology and Biotechnology*, **86**(1), 95–104.
43. Kuhlmann, S., Paetz, C., Haegeli, C. *et al.* (2009) Chromium catalyzed tetramerization of ethylene in a continuous tube reactor – proof of concept and kinetic aspects. *Journal of Catalysis*, **262**, 83–91.
44. McGuiness, D., Rucklidge, A., Tooze, R. and Slawin, A. (2007) Cocatalyst influence in selective oligomerization: effect on activity, catalyst stability, and 1-hexene/1-octene selectivity in the ethylene trimerization and tetramerization reaction. *Organometallics*, **26**, 2561–2569.

45. McGuiness, D., Overett, M., Tooze, R. *et al.* (2007) Ethylene tri- and tetramerization with borate cocatalysts: effects on activity, selectivity, and catalyst degradation pathways. *Organometallics*, **26**, 1108–1111.
46. Al-Sa'doun, A. (1993) Dimerization of ethylene to butene-1 catalyzed by Ti(OR′)4—AIR3. *Applied Catalysis A: General*, **105**, 1–40.
47. Sohn, J., Park, W. and Shin, D. (2002) Characterization and catalytic activity for ethylene dimerization of nickel sulfate supported on zirconia. *Applied Catalysis A: General*, **230**, 11–18.
48. Yu, C., Ge, Q., Xu, H. and Li, W. (2006) Propane dehydrogenation to propylene over Pt-based catalysts. *Catalysis Letters*, **112**(3–4), 197–201.
49. Chiusoli, G. and Maitlis, P. (eds) (2006) *Metal-catalysis in Industrial Organic Processes*, Royal Society of Chemistry, Cambridge, United Kingdom, pp. 201–217.
50. Jones, D. and Woods, D. (1986) Acetone-butanol fermentation revisited. *Microbiological Reviews*, **50**(4), 484–524.
51. Fischer, C., Peterson, A. and Tester, J. (2011) Production of C3 hydrocarbons from biomass via hydrothermal carboxylate reforming. *Industrial and Engineering Chemistry Research*, **50**, 4420–4424.
52. Hanai, T., Atsumi, S. and Liao, J. (2007) Engineered synthetic pathway for isopropanol production in *Escherichia Coli*. *Applied and Environmental Microbiology*, **73**(24), 7814–7818.
53. Mielenz, J. (2001) Ethanol production from biomass: technology and commercialization status. *Current Opinion in Microbiology*, **4**, 324–329.
54. Rude, M. and Schirmer, A. (2009) New microbial fuels: a biotech perspective. *Current Opinion in Microbiology*, **12**, 274–281.
55. Lin, Y. and Tanaka, S. (2005) Ethanol fermentation from biomass resources: current state and prospects. *Applied Microbiology and Biotechnology*, **69**(6), 627–642.
56. Lynch, J. (1972) Identification of substrates and isolation of microorganisms responsible for ethylene production. *Nature*, **240**, 45–46.
57. Greenwell, H., Laurens, L., Shields, R. *et al.* (2010) Placing microalgae on the biofuels priority list: a review of the technological challenges. *Journal of the Royal Society Interface*, **7**(May), 703–726.
58. Fukuda, H., Kawaoke, Y., Fuji, T. and Ogawa, T. (1987) Production of a gaseous saturated hydrocarbon mixture by *Rhizopus japonicus* under aerobic conditions. *Agricultural and Biological Chemistry*, **51**, 1529–1534.
59. American Standard Test Method (ASTM) D6866 – 11: Standard Test Methods for Determining the Biobased Content of Solid, Liquid, and Gaseous Samples Using Radiocarbon Analysis, http://www.astm.org/DATABASE.CART/HISTORICAL/D6866-11.htm (accessed 22 August 2013).

12

Future Trends for Recombinant Protein-Based Polymers: The Case Study of Development and Application of Silk-Elastin-Like Polymers

Margarida Casal[1], António M. Cunha[2], and Raul Machado[1]

[1] *CBMA (Centre of Molecular and Environmental Biology), Department of Biology, University of Minho, Portugal*

[2] *IPC (Institute of Polymers and Composites), Department of Polymer Engineering, University of Minho, Portugal*

12.1 Introduction

Materials science and engineering emerged and developed throughout the twentieth century with a primary focus on metals, ceramics and petroleum-based polymers. However, since the 1990s they also expanded to study biological materials, due to the significant progress made in this field. In fact, the recognition of the mechanisms linking nano- and micro-scale structures with macromolecular assembly and organization enabled molecular biologists to understand Nature's refined ways of creating high performance structural materials.

Structural proteins are abundant biomacromolecules and the natural building blocks for all the living organisms, representing an utmost case of function specialization and high-performance materials found in Nature. Throughout evolution, Nature has designed specialized proteins, by the association of amino acid sequences, creating refined assemblies with impressive properties, including outstanding mechanical behaviour [1]. These specialized polymeric

Bio Based Plastics: Materials and Applications, First Edition. Edited by Stephan Kabasci.

Table 12.1 *Amino acids repeating units of some structural proteins found in Nature. (Adapted from Tatham and Shewry, 2000 [6] and Sanford and Kumar, 2005 [7].)*

Protein family	Monomeric unit sequence
Elastin	GVGVP, VPGG, APGVGV
Silk fibroin	GAGAGS
Byssus	GPGGG
Flagelliform silk	GPGGX
Dragline silk	GPGQQ, GPGGY, GGYGPGS
Collagen	GAPGAPGSQGAPGLQ, GAPGTPGPQGLPGSP
Keratin	AKLKLAEAKLELA
Titin (I-band)	PPAKVPEVPEPKKPVPEEKVPVPVPKKPEA
Abductin	GGFGGMGGGX

Note: [a]Sequences are given in single-letter amino acid code, where X is any amino acid.

biomolecules evolved by the association of repetitive units of amino acid sequences to carry out a wide variety of functions (Table 12.1).

A key feature of biological systems is the ability to precisely control the structure of proteins at the molecular level. Conservative blocks of amino acid sequences are arranged in a way that creates flexible, rigid or tough domains, which are responsible for the remarkable mechanical properties found in proteins like elastin, resilin, abductin, mussel byssus or silk (Figure 12.1). The repetitive amino-acid sequences propagate through the protein, encoding a basic structural feature that is responsible for the properties of the polymeric material.

As natural polymers provide challenging examples for materials scientists, although the respective structure-properties relationships are not fully understood, there is already an important demand for translating biopolymers' conceptual architectures into bio-inspired materials. The interdisciplinary synergy between materials science engineers and biologists implies a convergence of knowledge to allow *de novo* creation of materials, with refined sequences and tailored for specific applications. Each protein has a proper amino acid sequence that specifies its unique three-dimensional conformation. The use of molecular genetics techniques provides the necessary tools to fill the gap between the concept and the synthesis of the material. By exploiting the 20 naturally occurring amino acids, researchers are able to select the required amino acid building blocks to meet the properties desired in the synthetic structural protein and program cell synthesis for its production. In the development of such functional materials, either for understanding the organizational concept or for its biotechnological application, there is the need to precisely control the macromolecular architecture. Recombinant protein-based polymers (rPBPs) are an example of such advanced materials derived from any macromolecule, designed and produced by the specific association of amino acids, and synthesized by the use of recombinant DNA technologies.

This overview will mainly focus on those polymers found in natural structural proteins, particularly the silk-elastin-like polymers (SELPs) of recombinant DNA technology origin.

12.2 Production of Recombinant Protein-Based Polymers (rPBPs)

According to the *Glossary of Basic Terms in Polymer Science* [5], a *polymer molecule* or *macromolecule* is defined as a molecule of high relative molecular mass, the structure of which essentially comprises the multiple repetition of units derived, actually or conceptually,

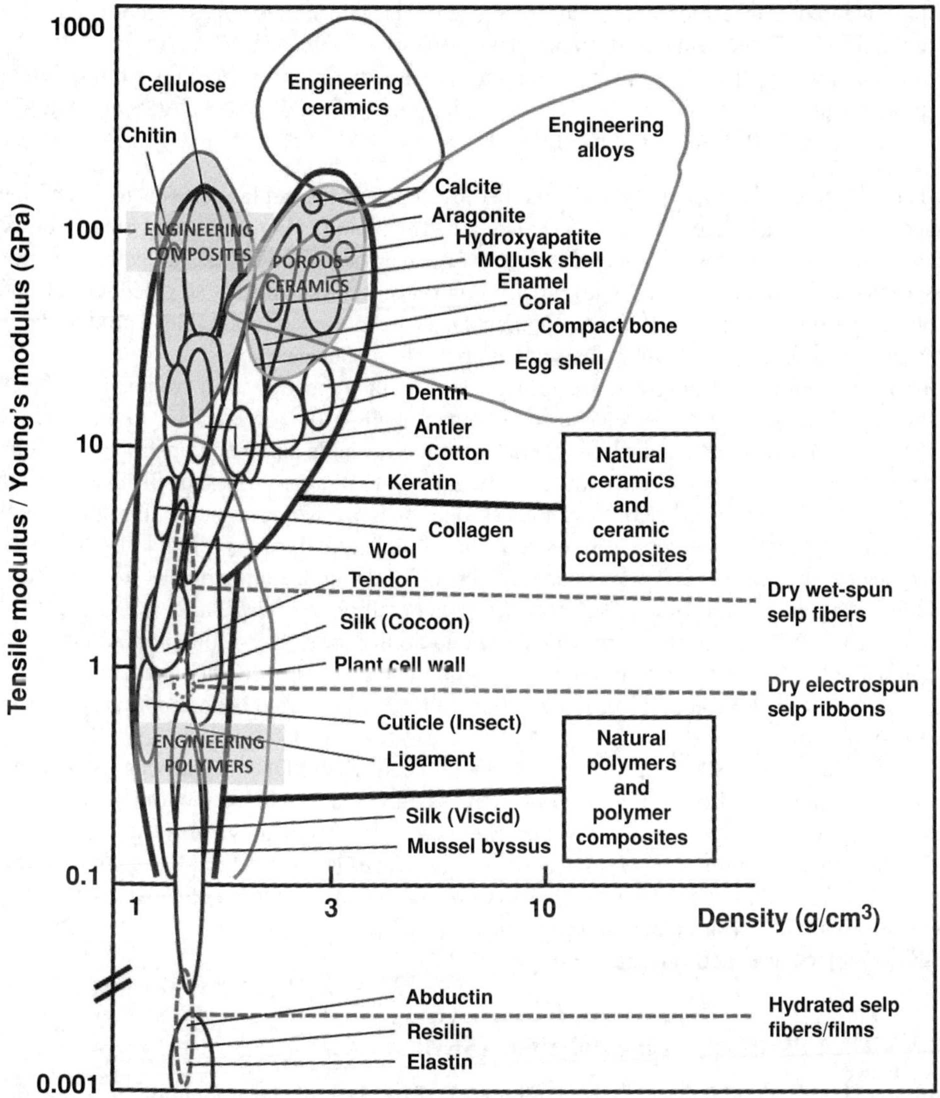

Figure 12.1 Tensile modulus as a function of density for several biological materials overlaid in comparison with synthetic materials. (Adapted with permission from [73], Copyright © Elsevier, 2008. Also special credit to Ashby (1989) [74] and Wegst and Ashby (2004) [75] with permission from Elsevier.)

from molecules of low relative molecular mass. According to this definition, and integrating the concept of recombinant protein-based polymers (rPBPs) in polymer science context, they can be defined as high molecular weight molecules comprising multiple *tandem* repetitions of a unique amino acid, or blocks of amino acids as units. Recombinant protein-based polymers can comprise unlimited primary structural arrangements, such as those that are made by the:

- polymerization of a sole amino acid, in particular, the monomeric molecule, for example, poly-L-lysine, poly-L-proline;

- polymerization of a monomeric unit of the same protein family type, as those sequence units listed in Table 12.1, for example, poly-GVGVP or poly-APGVGV; or
- the copolymerization of a block comprising at least two monomeric units from different protein family types (Table 12.1), for example, poly-[(GVGVP)*n*(GAGAGS)*m*], where *n* and *m* represent the number of repetitions.

The rPBPs are synthetic polymers obtained by standard molecular genetics tools and the associated fermentative techniques for production. The biological production of protein-based polymers presents several advantages over conventional chemical synthesis allowing complex macromolecules to be obtained with precise control on their sequence, stereochemistry and molecular weight. Moreover, with recombinant DNA technology it is possible to produce large amounts of recombinant polymers by scaling up the fermentation process.

In the class of rPBPs, polymeric materials based on the repeating units from silk, collagen and elastin are among the most characterized materials [8–10]. However, due to their unique properties in terms of mechanical, biodegradable, thermostable and biocompatible behaviour, the interest has been mostly concentrated in two major rPBPs families: the silk-like [11] and the elastin-like polymers [12,13]. Nevertheless, one can find other examples with compositions based on conservative amino acid sequence units of mussel adhesive protein [14–16], wheat glutenin [17], resilin [18–23] or even copolymers comprising repeating blocks from different proteins [24]. Regarding biological production, generally, *Escherichia coli* is the preferred micro-organism [25], due to its easy handling and ability to be grown on an industrial scale at relatively low cost; but other biological systems such as yeasts [26–29], filamentous fungi [30], plants [31,32], insects [8] or mammalian cells [33] have also been reported for rPBPs expression.

Several strategies have been proposed to assemble synthetic DNA sequences [34–38] but generally it follows a 'Lego' approach in which, a synthetic DNA block monomer, the 'Lego' basic DNA unit, is first produced by either chemical synthesis or polymerase chain reaction, and then subjected to multimerization of the 'Lego' pieces (Figure 12.2). Although an initial investment in molecular biology is needed, once the genetically modified organism is obtained, the fast production of an added value product will rapidly compensate the costs associated with the molecular genetics steps.

12.3 The Silk-Elastin-Like Polymers (SELPs)

With the aid of recombinant DNA technology for rPBP design and production, it is possible to genetically assemble conservative blocks of amino acid sequences from different structural proteins, creating an assortment of materials with improved mechanical properties (Figure 12.1). Silk-elastin-like polymers (SELPs) are recombinant copolymers with composition based on silk-like blocks combined with elastin-like blocks. By combining the elasticity and high resilience of an ideal elastomer like elastin [40], with the mechanical and tensile strength of silk fibroin [41], it is possible to create copolymers that in theory will exhibit the properties of both proteins. The rationale is based on the periodic inclusion of elastomeric sequences that reduces the overall crystallinity of the system by disrupting the silk-like blocks and, as a consequence, increases the flexibility, as well as aqueous solubility.

The silk-like unit characterized by the sequence GAGAGS (G – glycine, A – L-alanine and S – L-serine) is prevalent in silk fibroin from the silkworm *Bombyx mori*; it assembles into packed antiparallel β-sheet structures stabilized by hydrogen bonding [42]. The elastin-like

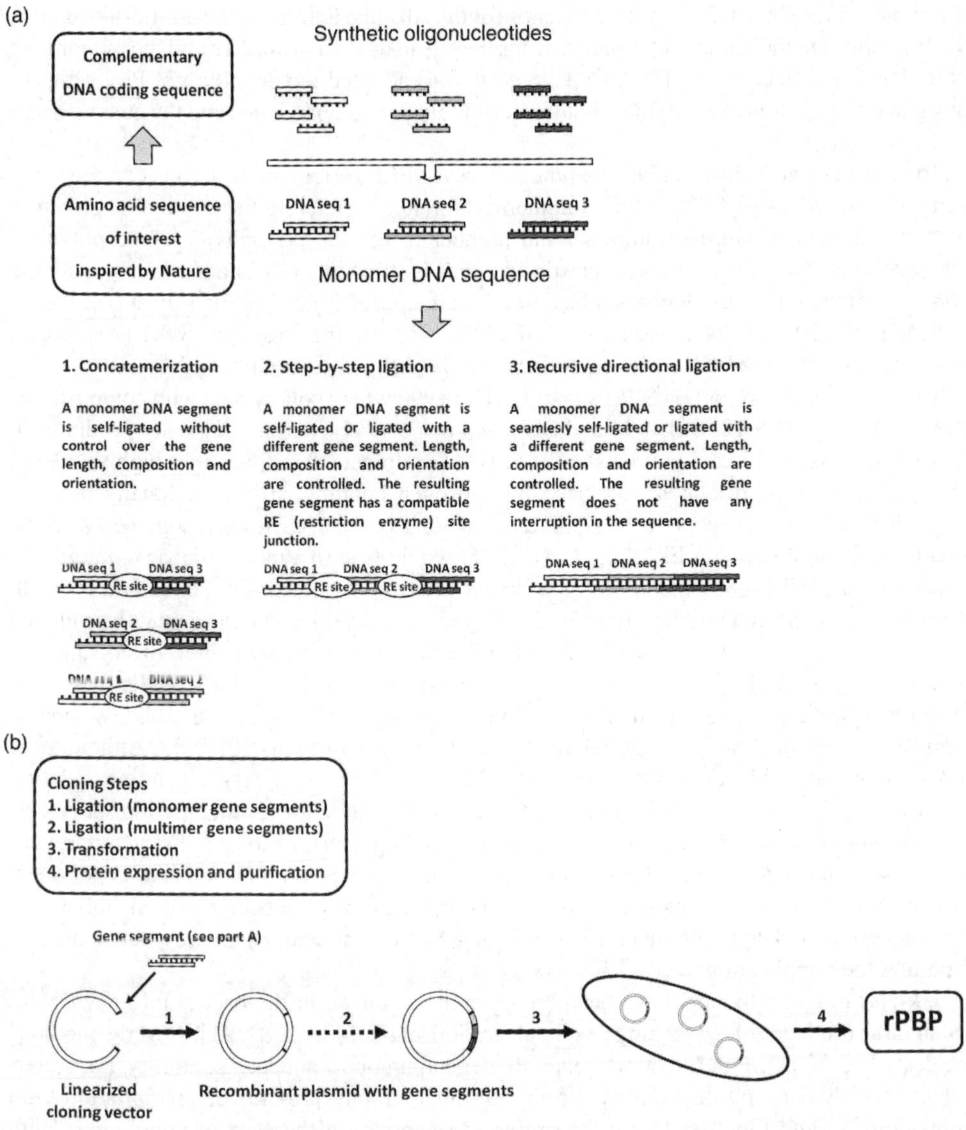

Figure 12.2 *Recombinant DNA approaches for gene multimerization (a) and cloning steps involved in rPBPs expression (b). RE represents the restriction enzyme site for compatible sticky-ends. (Adapted with permission from [76]. Copyright © 2011, American Chemical Society.)*

unit, on the other hand, with the main sequence VPGXG (V – L-valine, P – L-proline, X is the guest residue and represents any amino acid except for proline) is a flexible component that has been reported as displaying either the tendency to form flexible β-turns [12] or an amorphous structure with no defined conformation [43]. Although there is no consensus about the secondary structure of the elastin-like pentamer, it is generally accepted that it adopts a highly flexible conformation [12, 43]. The formulation of these copolymers is represented by the nomenclature [(*S*)*x*(*E*)*y*]*n*, where *S* stands for the silk-like unit, *E* for the elastin-like unit; *x*, *y* and *n* represent the number of repetitions for the silk-like, the elastin-like and the monomeric unit blocks, respectively. In 1998, Capello *et al.* [44] adopted the definition of ProLastins to designate silk-elastin-like polymers but this terminology was abandoned and the generic name SELP was adopted.

To allow a clear distinction between the different SELP formulations, in this overview, the generic name SELP is followed by a numeric reference *xy* relating the copolymer structure with the number of silk-like units (*x*) and number of elastin-like units (*y*). In copolymers where the guest residue of a single elastin-like unit is substituted by another amino acid, the corresponding amino acid letter is added to the numeric reference. In this way, a SELP with structure (S)4(E)8 will be designated as SELP-48 whereas the structure (S)4(E)4(EK)(E)3, where EK stands for VPGKG (K – L-lysine), will be designated as SELP-47K.

The first scientific report on SELPs goes back to 1990 where copolymers with compositions (S9E4)14 and (S8E8)12, SELP-94 and SELP-88 respectively were biologically produced in *E. coli* [45]. As mentioned previously, the GAGAGS domain assembles into highly ordered antiparallel β-sheet structures, stabilized by hydrogen bonding, so it is rationally possible to predict that a higher content of silklike units (total of 126 repetitions) together with the smaller interrupting elastin-like units in SELP-94 would lead to a more compact and ordered structure than SELP-88, which comprises larger interrupting elastin-like units and less silk content (total of 96 repetitions). In [45], the degree of crystallinity in unoriented lyophilized powders was analysed by X-ray diffraction and effectively, a higher degree of crystallinity was obtained for SELP-94 when compared with SELP-88 [45]. Similarly, FTIR dichroism of polymer films cast from formic acid solutions onto silver chloride glass slides, showed a higher content of β-sheet structures for SELP-94 in opposition to SELP-88. Although the inclusion of elastin-like units provided water solubility, after several days in solution, gelation occurred for both copolymers, initially for SELP-94 and after a longer time period for SELP-88. This work represented a first insight into understanding how the molecular structure relates to the macromolecular material properties. Since then, silk-like and elastin-like copolymers have been extensively characterized and several combinations have been reported. Table 12.2 summarizes an update of all the currently existing SELP formulations and respective articles reporting their applications.

Silk-elastin-like polymers have been biologically produced in the bacterium *E. coli* with volumetric productivities reaching levels up to 500 mg/L [48–50, 81, 91]. Several methods are described for SELP purification, generally depending on its aqueous solubility. For water-soluble copolymers, purification by affinity chromatography [48–50] or precipitation with ammonium sulfate [36, 81, 91] are the preferred methods. In the case of copolymers with larger silk-like units content, or less interrupting elastin-like units, purification is achieved by dissolving the insoluble pellet with lithium bromide 9 M, followed by dialysis [82]. Generally, affinity chromatography is the chosen method for purification but although it allows very pure polymer fractions to be obtained, it is a very laborious and expensive method. Consequently, it is not surprising that most of the applications with these copolymers have been directed for biomedical applications, especially when considering the high versatility of SELPs together with the high impact of health sciences. As base materials for biomedical purposes, their use includes a wide variety of applications, namely matrix-mediated delivery systems for gene therapy [49, 51–57, 73, 74], drug delivery systems [44, 58], (bio)functional materials [36], stimuli (pH, temperature or ionic strength) sensitive materials [46, 48] or as scaffolds for the encapsulation and chondrogenesis of human mesenchymal stem cells [59].

Table 12.2 *Update of all the SELP formulations found in literature and their developments.*

Code	Schematic formulation	Molecular Weight (kDa)	References
SELP-18	[(S) (E)8]11	47.6	[46]
SELP-28	[(S)2 (E)8]18	80.5	[47]
SELP-48	[(S)4 (E)8]13	70.0	[44]
SELP-68	[(S)6 (E)8]12	80.3	[47]
SELP-88	[(S)8 (E)8]12	84.3	[45]
SELP-812	[(S)8 (E)12]8	79.6	[47]
SELP-816	[(S)8 (E)16]7	84.6	[44]
SELP-94	[(S)9 (E)4]14	89.0	[45]
SELP-17E-11mer	[(S) (E)4 (EE) (E)3]11	47.6	[46]
SELP-17E-16mer	[(S) (E)4 (EE) (E)3]16	66.1	
SELP-37E	[(S)2 (E)4 (EE) (E)3 (S1)]16	66.1	[48]
SELP-27K	[(S)2 (E)4 (EY) (E)3]18	80.7	[44]
SELP-37K	[(S)3 (E)4 (EK) (E)3]12	60.1	[44]
SELP-47K	[(S)4 (E)4 (EK) (E)3]13	69.8	[49–70]
SELP-415K-6mer	[(S)4 (E)4 (EK) (E)11]6	55.1	[49, 50, 54, 56, 57, 70, 71]
SELP-415K-8mer	[(S)4 (E)4 (EK) (E)11]8	71.5	
SELP 415K-10mer	[(S)4 (E)4 (EK) (E)11]10	87.9	
SELP-67K	[(S6)(E4)(EK)(E3)]13	70.0	[72]
SELP-815K-3mer	[(S8) (E4) (EK) (E11)]3	35.6	[50, 57, 70, 73–77]
SELP-815K-4mer	[(S8) (E4) (EK) (E11)]4	45.5	
SELP-815K-5mer	[(S8) (E4) (EK) (E11)]5	55.5	
SELP-815K-6mer	[(S8) (E4) (EK) (E11)]6	65.4	
SELP-815K-MMPRS-6mer	SELP-815K fused with a matrix-metallo-proteinase sensitive peptide sequence	≈ 69	[78]
SELP-17Y	[(S) (E)4 (EY) (E)3]14	55.7	[79]
SELP-27Y	[(S)2 (E)4 (EY) (E)3]12	53.0	[79]
SELP-47Y	[(S)4 (E)4 (EY) (E)3]9	47.8	[79]
SELP-27K	[(S)2 (E)4 (EK) (E)3]12	≈ 53	[80]
MBI- SELP-47K	SELP-47K fused with cecropin-bee melittin	87.6	[36]
CAM-SELP-47K	SELP-47K fused with cecropin A melittin	80.6	
CBP-SELP-47K	SELP-47K fused with CBP	78.8	
P4-SELP-47K	SELP-47K fused with P4	76.7	
SELP-59-A	[(S5) (EA)9]9	≈55	[81,91]
SELP-520-A	[(S5) (EA)20]5	≈55	
SELP-1020-A	[(S10) (EA)20]4	≈55	
SELP-109-A	[(S10) (EA)9]7, where EA represents VPAVG	≈55	

12.3.1 SELPs for Biomedical Applications: Hydrogels for Localized Delivery

The most critical feature when designing SELP copolymers is the length of both the silk-like and the elastin-like blocks, which will determine the macromolecular properties of the material. Depending on such structural parameters (mainly the length/number of silk blocks), with the adequate composition (structure and concentration), SELPs are able to undergo an irreversible sol-gel transition at physiological temperatures by means of an exothermic event [44]. The formation of SELP hydrogels occurs through crystallization of the silk-like blocks via hydrogen-bonding and can be controlled by manipulating the copolymer architecture. In the same conditions, SELPs with larger silk-like blocks display higher gelation rates. For instance, SELP-27K, SELP-47K and SELP-816 are copolymers with different silk-like block lengths;

SELP-27K and SELP-47K share elastin-like blocks with the same length but a different silk to elastin block ratio while SELP-47K and SELP-816 exhibit the same silk to elastin ratio (1:2). When analysed by solution viscosimetry at 37 °C, copolymer solutions of 20% (w/w) of SELP-816 and SELP-47K gelled at rates reflecting the increase in their silk-like block length. For the same viscosity value of 3 Pa s, while SELP-816 reached this value in about 50 minutes, SELP-47K took almost twice the time (105 minutes) [44]. In the case of SELP-27K, no increase in viscosity was observed during the 125 minutes time period of the experiment, suggesting that the length of the silk-like blocks plays a critical role in the sol-gel transition event [44]. The elastin-like block is a factor that cannot be underestimated and indeed it influences the sol-gel transition. If large enough, together with small silk-like blocks (low silk to elastin ratio), the size of the flexible elastin-like block can hamper the formation of interchain contacts by increasing the distance between the silk-like units. In a study comparing the gelation of SELP-47K and SELP-415K-8mer, both with the same silk-like block length, but with different sizes of elastin-like units, SELP-47K formed firm gels with a concentration as low as 6 wt%, after 4 hours of cure time at 37 °C, while SELP-415K-8mer formed robust gels only at a concentration of 12 wt% [49]. The lower silk-to-elastin ratio of SELP-415K (1:4) presents a lower number of hydrogen-bonding sites and therefore it requires more polymer chains (higher critical gelation concentration) to self-assemble.

The gelation process is not exclusively determined by molecular factors related to copolymer architecture. External factors like concentration, temperature and additives also affect the sol-gel transition. Increasing the concentration or the temperature implies faster gelation rates, while the addition of chaotropic agents, such as urea, practically eliminates the crystallization event associated with the gelation process due to hydrogen bonding disruption [44]. On the other hand, the addition of a nucleating agent, like precrystallized SELP to the copolymer solution, accelerates the process of crystallization and the subsequent gelation [44].

Typically, SELP materials are extremely biocompatible [44, 60] and biodegradable. The resorption rate depends on the number of silk-like blocks, because an increasing number of silk blocks decreases the rate of resorption [44, 57]. When injected in guinea pigs, a SELP-47K solution quickly formed a stable gel without material migration into adjacent tissues with no signs of immunological reaction or inflammation [44]. These aspects reflect the very high potential of these copolymers, especially for biomedical applications, where the high level of customization represents a major advance in the research of tailored biomaterials. In the particular aspect of localized release, it is quite remarkable that gelation of SELPs occurs in mild conditions with physiologically biocompatible solvents and without the need for chemical crosslinking agents. This means that compounds can be stably incorporated in the polymer solution without chemical changes or loss of activity [44, 52, 53]. In fact, human mesenchymal stem cells were incorporated into SELP solutions that, after gelation, generated a bioactive matrix with living cells [59]. This suggests that SELPs can be used for the integration of organelles or intact tissues, creating bioactive scaffolds. As drug-delivery devices, SELPs exhibit a release profile matching a Fickian type of diffusion [44, 61] with a release rate dependent on formulation, concentration, cure time and molecular weight of the loaded compound, varying between a couple of days or several weeks for low molecular weight (<10 kDa) and high molecular weight (>40 kDa) compounds, respectively [44].

As mentioned previously, the architectural design strongly influences the gelation process due to molecular network rearrangement. Increasing the concentration or cure time leads to increased polymer-polymer interactions resulting in a higher degree of physical contacts. Increasing the cure time allows the silk units to align and configure the three-dimensional network to form more and / or stronger physical contacts [50]. This produces a dense network with reduced mean pore size and, as a consequence, leads to slower compound release. In fact, the cumulative percentage release of plasmid DNA from SELP-47K hydrogels has been shown to be inversely proportional to the concentration or cure time, for the reasons described above [51].

In conclusion, SELPs with larger silk-like blocks or with higher silk-like block content gel faster and this is accelerated by increasing the temperature or the concentration or by addition of nucleating agents. In gels with a higher level of chain entanglement, the pores are closely tightened, decreasing the release rate of the loaded compounds.

12.3.2 Mechanical Properties of SELP Hydrogels

The formation of hydrogen bonds between the silk-like blocks is the major driving force behind gelation, resulting in interlocking points for physical chain entanglement. The density of the chain network is a function of polymer concentration and structure as well as of the cure conditions that overall significantly influence the mechanical properties of hydrogels. As viscosity is directly proportional to the density of chain entanglement, the rheological properties can be manipulated easily by controlling both molecular structure and concentration.

Higher concentrations imply a great number of molecules able to establish physical contacts, decreasing the distance between hydrogen-bonding molecules. In fact, transmission electron microscopy images have shown a decrease in network spacing with increasing SELP-47K concentration [54] (Figure 12.3).

The concentration effect was demonstrated in the viscoelastic behaviour of SELP-47K hydrogels, cured for 4 hours at 37 °C. The storage modulus determined by dynamic mechanical analysis (DMA) was 75.4 kPa for SELP-47K at 4 wt%, whereas a greater value of 1600 kPa was obtained for 12 wt%. Macroscopically, the 4 wt% hydrogels showed to be translucent, soft and easily deformable, while the 12 wt% hydrogels were opaque and firm [54].

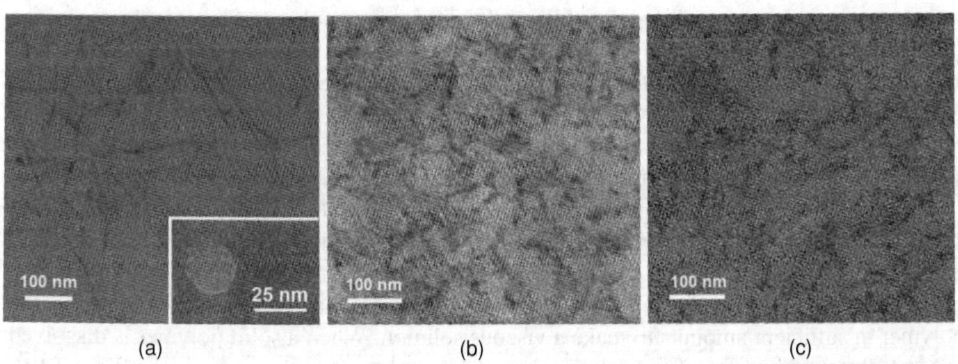

Figure 12.3 Transmission electron micrograph of 4 wt% SELP-47K, (inset) icosahedral Ad-CMV-Luc virion (a), 12 wt% SELP-47K (b), and 12 wt% SELP-415K (c). The Ad-CMV-Luc virion was included in image A for size comparison purposes. Figure A shows a relatively open structure with well-defined fibrils spaced about 93 nm apart. In Figure B, the micrograph shows a highly packed structure with small spacing between well-defined fibrils. Fibrils in image C are poorly defined, thicker and less regular. (Adapted with permission from [47]. Copyright © 2011, American Chemical Society.)

The molecular architecture also influences the mechanical behaviour of SELPs, as the block size is a parameter that affects the critical gelation concentration as well as the time for gelation [49, 54]. For a cure time of 4 hours at 37 °C, physically robust gels were obtained with SELP-47K at concentrations as low as 6 wt%, while for SELP-415K-8mer (structure in Table 12.2), only at a concentration of 12 wt% was it possible to form a defined three-dimensional structure. The mechanical properties of these copolymers was shown to be clearly related to the structural architecture, because both display similar molecular weight, the same silk-like block length, but the elastin-like block size doubles in SELP-415K-8mer. SELP-47K 12 wt% hydrogels failed with a strain to failure between 1.7 and 1.8% while SELP-415K-8mer remained intact up to a strain to failure of 2% [50]. With a strain to failure of 1.5%, the storage modulus was 5.09 MPa and 0.52 MPa for SELP-47K and SELP-415-8mer, respectively [50]. Interestingly, the storage modulus of 12 wt % SELP-415K-8mer hydrogels was similar to that of 4 wt% SELP-47K of about 70 kPa [54]. The complex shear modulus (G^*) for solutions of SELP-47K and SELP-415K-8mer has revealed a dramatic change at the gelation point reaching 608 and 93.9 kPa, with maximum values at the end of the experiment of 1.33 and 0.337 MPa, respectively [49]. The longer elastin-like block of SELP-415K-8mer reduces the molecular network packing, increasing the elasticity and reducing the stiffness of the hydrogel.

In summary, increasing the silk-like block length increases the number of contact points for potential physical interchain junctions. By varying the length of the elastin-like block, the distance between silk-like blocks for contacts will also vary. The presence of longer elastin blocks determines the spacing of the junction areas in the hydrogel network while the silk blocks provide mechanical stiffness and structural rigidity [50]. Consequently, with larger interrupting elastin-like blocks, both the critical gelation concentration and the time for gelation increase [49].

12.3.3 Spun Fibres

Electrospinning is a process of fibre formation that relies on electrostatic forces to control its production (Figure 12.4).

Due to its ease of implementation, electrospinning has received a lot of attention as a technique to produce nanofibres [83]. When the diameter of polymer fibre materials shrinks from the microscale to the submicro or nanoscale, several new characteristics appear, such as enhanced surface area-to-volume ratio and a superior mechanical performance [84]. Therefore, biopolymer nanofibrous mats show great potential to be used as particle filters, nanocomposite reinforcing fibres, protective clothing and in biomedical applications like wound dressings, sutures, tissue engineering scaffolds, implantable devices and drug delivery [83–85].

The morphology of the spun structures is influenced by the properties of the polymer solution, namely: viscosity, surface tension and density. As the stability of the jet depends of the viscous and viscoelastic properties of the polymer solution, the choice of solvent is a critical factor. Therefore, the solvent should be chosen according to its ability to dissolve the polymer in sufficient amounts to make a viscous solution. When a solid polymer is dissolved in a solvent, the viscosity of the solution is proportional to the polymer concentration. Thus, higher polymer concentration produces larger fibre diameter, whereas a diluted solution leads to breakage of the polymer fibre into droplets due to the effects of surface tension [84].

Water [62], or water with addition of a polyethylene oxide-sodium dodecyl sulfate (PEO-SDS) complex [72], and formic acid [60, 63, 72] have been used to dissolve SELP copolymers for fibre spinning. Water acts as a plasticizer in spun-fibres, inhibiting the β-sheet conversion

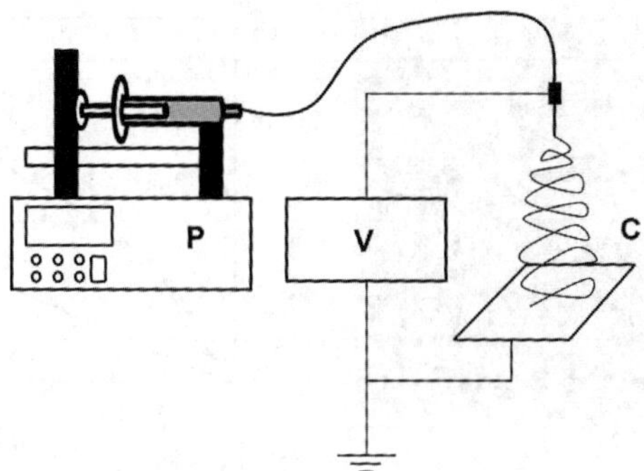

Figure 12.4 Schematic representation of electrospinning setup. A syringe is assembled into a metering pump with a digital controller (P) and connected to a needle by a silicon capillary. The electrode of a power voltage supply (V) is placed in contact with the needle. The other electrode is in contact with the collector (C) and grounded. The polymer solution is forced through the syringe and as a result of the surface tension there is the formation of pendant droplets. As the fluid charges into a high potential, the repulsion between charges at the free surface deforms the droplet into a conical shape, the Taylor cone. At a critical charge density, the repulsive electrostatic forces overcome the surface tension, the cone gets unstable and a jet fluid forms, moving towards the collector, which has the opposite electrical charge. During the time the jet takes to reach the collector, the solvent evaporates and an interconnected web of polymer fibres is deposited.

[64], and thus as-spun SELP fibres are highly water soluble. However, water insolubility can be rendered by inducing the crystallization of the silk-like blocks through the formation of β-sheets. Usually, this structure stabilization is achieved via chemical vapour treatment with methanol [60, 62] or glutaraldehyde (GTA) [60], by using a methanol coagulant bath [63] or autoclaving as a chemical-free process [64]. While methanol induces crystallization in a noncovalent way, interfibre crosslinking can be obtained by using GTA that reacts with the lysine (K) residues present in the SELP-47K.

Nanoribbons with average diameters of 238 nm, 472 nm and 535 nm were obtained by electrospinning aqueous solutions of SELP-47K at concentrations of 12, 15 and 18% (w/w), respectively [62], thus indicating that an increase in solution concentration leads to broader fibre-size distributions. Although SELP-47K aqueous solutions can irreversibly form a gel leading to nonelectrospinnable solutions, the authors claim that no gelation was observed during the time of the experiment (3 hours) at room temperature. As for mechanical properties, dry electrospun ribbons with average sample thickness of 20 μm showed an average ultimate tensile strength of 30.8 MPa, a strain to failure of 7.9% and a tensile modulus of 0.88 GPa [62]. However, these values can be improved by further inducing β-sheet crystallization in the fibre meshes.

As discussed in the previous section, the length of the silk-like block exerts a great influence in the sol-gel transition, where copolymers with larger silk-like blocks generally display higher gelation rates. This sol-gel transition can be delayed by decreasing the solution concentration. However, the low concentration might not be sufficient to confer the adequate solution viscosity for the electrospinning process. To overcome this problem Nagarajan *et al.* [72] employed a polymer-micelle complex of PEO-SDS as an aid for SELP-67K electrospinning in aqueous solution at a concentration of 6.7 wt%, producing nanofibres of 200–300 nm diameter.

As lyophilized SELPs exhibit limited water solubility, formic acid has been used as a solvent [60, 72]. Fibres of 200–300 nm diameter have been produced by electrospinning a SELP-67K

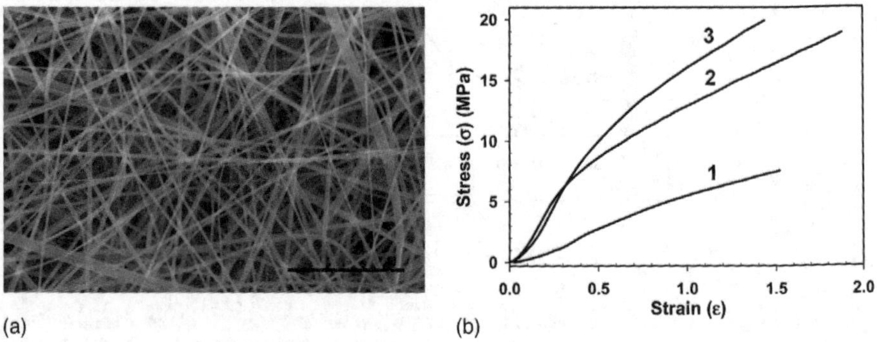

Figure 12.5 Scanning electron micrograph of as-spun SELP-47K fibres (a) and representative tensile stress-strain curves (b) of MeOH- (1), GTA- (2), and MeOH-GTA-treated (3) SELP-47K nanofibrous scaffold. Scale bar: 5 μm. (Adapted with permission from [64], Copyright © 2010, American Chemical Society and [56], Copyright © 2011 American Institute of Physics.)

solution at a concentration of 15 wt% in formic acid, whereas a decrease in the concentration to 8 wt% produced defective fibres due to the low solution viscosity [72]. The electrospinning of a SELP-47K formic acid solution, at a concentration of 15 wt%, produced a nonuniform fibre population ranging from 50 to 200 nm (Figure 12.5a) [60]. Chemical vapour treatment with methanol, GTA, or a combination of both affected the mechanical properties of the hydrated meshes (Figure 12.5b).

The methanol-treated sample revealed an ultimate tensile strength of 7.2 ± 2.3 MPa, a strain to failure of 130 ± 30 % and a tensile modulus of 3.7 ± 1.2 MPa; whereas the GTA-treated sample showed values of 14.1 ± 3.8 MPa, 130 ± 40 % and 18.2 ± 3.3 MPa for ultimate tensile strength, strain to failure and tensile modulus, respectively [60]. The samples treated with both chemicals showed mechanical properties similar to the GTA-treated sample. Preconditioning the chemical treated electrospun scaffolds with six cycles of stretching with a strain of about 33 %, allowed an increase in resilience to 80–87%, in contrast to the 46–60% before preconditioning [60]. Water insolubility of SELP-47K formic acid solution electrospun fibres was promoted upon autoclaving [64]; the mechanical analysis evidenced a tensile modulus of 1.02 ± 0.28 MPa, an ultimate tensile strength of 0.34 ± 0.04 MPa and a strain to failure of 29 ± 3%. After six cycles of mechanical preconditioning, the autoclaved nanofibres displayed an increase of resilience to 90.7 ± 3% [64].

Another process for obtaining SELP fibres is based on the wet spinning technique. In this process, the spinnerets are submerged in a coagulant bath and, as the filaments emerge, they precipitate from solution and solidify. Fibres that are metres in length and have diameters ranging from less than 10 μm to over 60 μm, were obtained by wet-spinning a SELP-47K solution at a concentration of 25% (w/v) in formic acid into a coagulant bath of methanol [63]. Mechanical analysis of dry fibres revealed a strong dependence on fibre diameter, with fibres with smaller diameter showing higher tensile modulus and thus higher tensile strength. For these dry fibres, the strain to failure was less than 2%, with the ultimate tensile strength varying between 20 and 80 MPa and a tensile modulus in the range of 1–5 GPa. Upon hydration in PBS (phosphate buffered saline), the strain to failure increased several hundred-fold although accompanied with a decrease in both ultimate tensile strength and tensile modulus to values of about 1 and 0.2 MPa, respectively.

Table 12.3 *Mechanical properties of SELP-47K fibres, produced by wet-spinning or electrospinning, and stabilized by methanol (MeOH), glutaraldehyde (GTA), MeOH-GTA or autoclaving.*

Fibre state	Technique	Solvent	Stabilization Treatment	Strain to failure (%)	UTS (MPa)	Tensile modulus (MPa)
Dry	Electrospin	Water	n.a.	7.9	30.8	880
Hydrated	Electrospin	Formic acid	MeOH vapour	130 ± 30	7.2 ± 2.3	3.7 ± 1.2
Hydrated	Electrospin	Formic acid	GTA vapour	130 ± 40	14.1 ± 3.8	18.2 ± 3.3
Hydrated	Electrospin	Formic acid	MeOH-GTA vapour	100 ± 30	15.8 ± 5.3	13.8 ± 3.7
Dry	Wet-spin	Formic acid	MeOH bath	<2	20–80	1000–5000
Hydrated	Wet-spin	Formic acid	MeOH bath followed by GTA vapour	323 ± 201	8.1 ± 6.06	3.64 ± 1.6
Hydrated	Electrospin	Formic acid	Autoclaving	29 ± 3	0.34 ± 0.04	1.02 ± 0.28

Note: UTS refers to ultimate tensile strength. n.a. (not applied). For references consult the text.

The resilience of preconditioned SELP-47K fibres, either autoclaved or chemical cross-linked, approached that of native elastin, which is 90% [60]. It is worth noticing that mechanical properties of SELP-47K fibres change dramatically between the dry state and the full hydrated state (Table 12.3). This is likely due to the plasticizing effect of water, which gives the silk-like blocks conformational flexibility by decreasing their crystallinity.

As a remarkable example to demonstrate how hydration greatly affects the mechanical properties of protein-based proteins, it is worth mentioning the case of the spider major ampullate silk, which in the dry state shows a tensile modulus of 7–10 GPa, but upon hydration it decreases by a hundredfold [86].

12.3.4 Solvent Cast Films

Although there are several SELP formulations described in literature (Table 12.2), the most described and widely used is SELP-47K. Besides the ability to produce hydrogels, self-assembled structures and electrospun fibres, it is also possible to fabricate thin films by solvent evaporation at room temperature [65, 66]. The major difference between hydrogels and thin films resides in the fabrication process and the resulting microstructure. Thin films of SELP-47K were made by dissolving the lyophilized powder in water and letting the solvent evaporate from the casting moulds [65–67]. Film thickness is controlled by the solution concentration and the volume used to cast.

Recombinant protein-based polymers are rarely reported as being optically transparent, however, SELP-47K thin films are optically transparent to visible light with notable transmittance in the wavelength range of 350–800 nm [65, 66]. Increasing the thickness (from 16 to 92 μm) had a slight effect in the transmittance with a reduction of about 3% [65]. This optical transparency is maintained even after methanol-induced crystallization, with only a slight reduction in the transmittance from 95 to 92% at 800 nm.

The cast SELP-47K films are not mechanically stable after hydration [65, 67] and it is therefore necessary to stabilize its structure. Methanol induces the crystallization of the silk-like blocks, leading to aqueous insolubility, and improves the mechanical properties by the conversion of less-ordered conformations into more ordered anti-parallel beta-sheet structures [87, 88]. Under uniaxial tensile stress-strain, fully hydrated treated films exhibited an ultimate tensile strength of 2.5 ± 0.4 MPa, a strain to failure of 190 ± 60% and a tensile modulus of 1.57 ± 0.3 MPa [65].

The mechanical properties of hydrated SELP-47K films resemble those of native elastin from bovine ligament, which displays a tensile modulus of 1.1 MPa, ultimate tensile strength of 2 MPa and deformability of 150% [89]. Preconditioning the methanol-treated SELP-47K films to 10 repetitive cyclic strains with deformations of 45% showed an increase in the resiliency from 66 ± 4% to 86 ± 4% [67], possibly due to the stabilization of the microstructures of the elastin-like blocks and crystallized silk-like blocks. When fully hydrated, the methanol-treated films absorb approximately 100% of its weight in water, meaning that hydrated SELP-47K films possess 50% (w/w) of water content [65]. Remarkably, after full hydration, the methanol-treated films still maintained their dry-state dimensions with a slight variation of less than 5% in length and width [65]. With particular regard to the application of SELP-47K for contact lenses, the surface of methanol-treated SELP-47K films showed a mean surface roughness of 1.421 ± 0.271 nm and a contact angle of 67.8 ± 2.2°, which is similar to those of contact lens materials such as Asmoficon (71.2 ± 1.5°), Balafison (71.5 ± 1.1°) and Enfilcon (68.3 ± 1.5°) [65]. Considering the refractive index (RI), a key property for a material to be considered as suitable for contact lenses application, SELP-47K films displayed similar RI values (1.5300 ± 0.002 and 1.5345 ± 0.0039 for non- and methanol-treated, respectively) to those of commonly used materials such as PMMA (1.490 ± 0.001), Etafilcon A (1.412 ± 0.0017) and Lotrafilcon B (1.424 ± 0.0003) [65].

As in the case of hydrogels, SELP-47K films have been demonstrated to be extremely biocompatible while allowing cell proliferation [65]. This biocompatible feature, together with mechanical properties, allows SELPs to be explored for several applications, namely contact lenses, synthetic corneas, intraocular lenses, ophthalmic drug delivery matrices, and so on. Although methanol is routinely used to render aqueous insolubility as well as to improve the mechanical properties, autoclaving has been shown to stabilize electrospun nanofibres and therefore can be an alternative nonchemical method for the stabilization of the film structure.

12.4 Final Considerations

Until now, the research done with rPBPs has been almost exclusively directed towards biomedical applications. However, the exploitation of their remarkable properties in general polymer and materials science provides opportunities as it opens an excellent window for the creation and refinement of both new and currently existing materials. The use of rPBPs, especially in the case of SELPs, provides novel answers to the limitations of current polymers, constituting a new generation of materials with the ability to be tailored towards specific purposes. For the SELPs described in this chapter, the range of tensile moduli is depicted by dashed lines in Figure 12.1. Accordingly, and in terms of mechanical properties, these recombinant copolymers display a tensile modulus similar to their natural counterparts in both dry and hydrated states. As in the natural systems, hydration plays a crucial role in mechanical behaviour as water acts as a 'softening' agent.

By exploiting the use of recombinant DNA technology it is possible to redesign the molecular structure of rPBPs in order to combine bioactive domains and in this way achieve the creation of highly customable functionalized materials. Materials with antimicrobial properties can be obtained by fusion with antimicrobial peptides for biodefensive purposes [36]. By combining the rPBPs with UV protective peptides it is possible to create a photo-protective material for the prevention of photoaging in human skin [36]. The incorporation of cellulose binding peptides allows the development of smart textile materials [36]. These three cases demonstrate the high versatility of recombinant DNA technology in the development of tailored-made performance materials and, ultimately, the possibility of creating a highly advanced material exhibiting the required properties. In fact, a new set of novel copolymers based on plastic elastin blocks with sequence VPAVG has been recently described [81].

In natural structural proteins, the repetitive blocks of amino acid sequences are not arranged in a precise regular frequency. Amino acid substitutions occur throughout the protein. This is possibly due to a natural process of minimizing the repetitiveness of the genetic material template encoding such proteins. However, by means of recombinant DNA technology and biotechnology tools it is possible to design a multimerized DNA template and produce rPBPs with the predefined amino acid sequence. The produced rPBPs will therefore consist of a 'perfect' *tandem* repeat of the amino acid units. Although we are able to produce such 'perfect' protein-based macromolecules, often the rPBPs exhibit weaker mechanical properties when compared to their natural counterparts. In order to create materials with similar properties found in natural polymers, further developments are still needed to clearly understand the complexity underlying the material processing mechanisms that occur in the biological systems. According to Bensaude–Vincent [90] this implies the emergence of a new paradigm where the current processing techniques are inadequate for such materials: 'Nature works with minimal energy, at low temperature, with raw materials; it is too versatile, too impure to meet the requirements and standards of industrial production.'

The recent advances in nano(bio)technology emerged from the increase of knowledge in structural and molecular biology combined with the use of recombinant DNA technology, facilitating the advent of a new class of synthetic biomacromolecules, the rPBPs. Through evolution, Nature created and refined proteins for a broad range of functions, working as structural components, as molecular motors or as signalling molecules; amino acid sequences fold and combine into precise 3D-patterns to create materials, able to perform work when submitted to specific stimuli (e.g. silks, bioadhesives, collagen and elastin). The biosynthesis of rPBPs with an absolute control of their composition, sequence and length, opens the opportunity of creating novel biological compositions engineered towards a desired application.

References

1. Meyers, M.A., Chen, P.-Y., Lin, A.Y.-M. and Seki, Y. (2008) Biological materials: structure and mechanical properties. *Progress in Materials Science*, **53**, 1–206.
2. Chen, P.Y., Lin, A.Y.M., Lin, Y.S. *et al.* (2008) Structure and mechanical properties of selected biological materials. *Journal of the Mechanical Behavior of Biomedical Materials*, **1**, 208–226.
3. Ashby, M.F. (1989) On the Engineering Properties of Materials. *Acta Metallurgica*, **37**, 1273–1293.
4. Wegst, U.G.K. and Ashby, M.F. (2004) The mechanical efficiency of natural materials. *Philosophical Magazine*, **84**, 2167–2181.
5. Jenkins, A.D., Kratochvil, P., Stepto, R.F.T. and Suter, U.W. (1996) Glossary of basic terms in polymer science. *Pure and Applied Chemistry*, **68**, 2287–2311.
6. Tatham, A.S. and Shewry, P.R. (2000) Elastomeric proteins: biological roles, structures and mechanisms. *Trends in Biochemical Sciences*, **25**, 567–571.
7. Sanford, K. and Kumar, M. (2005) New proteins in a materials world. *Current Opinion in Biotechnology*, **16**, 416–421.
8. Vendrely, C. and Scheibel, T. (2007) Biotechnological production of spider-silk proteins enables new applications. *Macromolecular Bioscience*, **7**, 401–409.

9. Chow, D., Nunalee, M.L., Lim, D.W. *et al.* (2008) Peptide-based biopolymers in biomedicine and biotechnology. *Materials Science and Engineering R-Reports*, **62**, 125–155.
10. Gomes, S., Leonor, I.B., Mano, J.F. *et al.* (2012) Natural and genetically engineered proteins for tissue engineering. *Progress in Polymer Science*, **37**, 1–17.
11. Shao, Z. and Vollrath, F. (2002) Surprising strength of silkworm silk. *Nature*, **418**, 741.
12. Urry, D.W. (2006) *What Sustains Life? Consilient Mechanisms for Protein-based Machines and Materials*, Springer, New York, N.Y.
13. Rodriguez-Cabello, J.C., Prieto, S., Reguera, J. *et al.* (2007) Biofunctional design of elastin-like polymers for advanced applications in nanobiotechnology. *Journal of Biomaterials Science-Polymer Edition*, **18**, 269–286.
14. Hwang, D.S., Yoo, H.J., Jun, J.H. *et al.* (2004) Expression of functional recombinant mussel adhesive protein Mgfp-5 in *Escherichia coli*. *Applied and Environmental Microbiology*, **70**, 3352–3359.
15. Hwang, D.S., Gim, Y. and Cha, H.J. (2005) Expression of functional recombinant mussel adhesive protein type 3A in *Escherichia coli*. *Biotechnology Progress*, **21**, 965–970.
16. Kim, D., Hwang, D.S., Kang, D.G. *et al.* (2008) Enhancement of mussel adhesive protein production in *Escherichia coli* by co-expression of bacterial hemoglobin. *Biotechnology Progress*, **24**, 663–666.
17. Feeney, K.A., Wellner, N., Gilbert, S.M. *et al.* (2003) Molecular structures and interactions of repetitive peptides based on wheat glutenin subunits depend on chain length. *Biopolymers*, **72**, 123–131.
18. Elvin, C.M., Carr, A.G., Huson, M.G. *et al.* (2005) Synthesis and properties of crosslinked recombinant pro-resilin. *Nature*, **437**, 999–1002.
19. Kim, M., Elvin, C., Brownlee, A. and Lyons, R. (2007) High yield expression of recombinant pro-resilin: lactose-induced fermentation in *E. coli* and facile purification. *Protein Expression and Purification*, **52**, 230–236.
20. Lyons, R.E., Lesieur, E., Kim, M. *et al.* (2007) Design and facile production of recombinant resilin-like polypeptides: gene construction and a rapid protein purification method. *Protein Engineering Design and Selection*, **20**, 25–32.
21. Nairn, K.M., Lyons, R.E., Mulder, R.J. *et al.* (2008) A synthetic resilin is largely unstructured. *Biophysical Journal*, **95**, 3358–3365.
22. Lyons, R.E., Nairn, K.M., Huson, M.G. *et al.* (2009) Comparisons of recombinant resilin-like proteins: repetitive domains are sufficient to confer resilin-like properties. *Biomacromolecules*, **10**, 3009–3014.
23. Li, L., Teller, S., Clifton, R.J. *et al.* (2011) Tunable mechanical stability and deformation response of a resilin-based elastomer. *Biomacromolecules*, **12**, 2302–2310.
24. Bracalello, A., Santopietro, V., Vassalli, M. *et al.* (2011) Design and production of a chimeric resilin-, elastin-, and collagen-like engineered polypeptide. *Biomacromolecules*, **12**, 2957–2965.
25. Wang, J.J., Sivashanmugam, A., Murray, V. *et al.* (2009) Practical protocols for production of very high yields of recombinant proteins using *Escherichia coli*. *Protein Science*, **18**, 936–948.
26. Fahnestock, S.R. and Bedzyk, L.A. (1997) Production of synthetic spider dragline silk protein in *Pichia pastoris*. *Applied Microbiology and Biotechnology*, **47**, 33–39.
27. Fahnestock, S.R., Yao, Z. and Bedzyk, L.A. (2000) Microbial production of spider silk proteins. *Journal of Biotechnology*, **74**, 105–119.
28. Sallach, R.E., Conticello, V.P. and Chaikof, E.L. (2009) Expression of a recombinant elastin-like protein in *Pichia pastoris*. *Biotechnology Progress*, **25**, 1810–1818.
29. Schipperus, R., Teeuwen, R.L., Werten, M.W. *et al.* (2009) Secreted production of an elastin-like polypeptide by *Pichia pastoris*. *Applied Microbiology and Biotechnology*, **85**, 293–301.
30. Herzog, R.W., Singh, N.K., Urry, D.W. and Daniell, H. (1997) Expression of a synthetic protein-based polymer (elastomer) gene in *Aspergillus nidulans*. *Applied Microbiology and Biotechnology*, **47**, 368–372.
31. Scheller, J., Henggeler, D., Viviani, A. and Conrad, U. (2004) Purification of spider silk-elastin from transgenic plants and application for human chondrocyte proliferation. *Transgenic Research*, **13**, 51–57.
32. Xu, J., Dolan, M.C., Medrano, G. *et al.* (2012) Green factory: plants as bioproduction platforms for recombinant proteins. *Biotechnology Advances*, **30**, 1171–1184.

33. Lazaris, A., Arcidiacono, S., Huang, Y. *et al.* (2002) Spider silk fibers spun from soluble recombinant silk produced in mammalian cells. *Science*, **295**, 472–476.
34. McMillan, R.A., Lee, T.A.T. and Conticello, V.P. (1999) Rapid assembly of synthetic genes encoding protein polymers. *Macromolecules*, **32**, 3643–3648.
35. Meyer, D.E. and Chilkoti, A. (2002) Genetically encoded synthesis of protein-based polymers with precisely specified molecular weight and sequence by recursive directional ligation: examples from the elastin-like polypeptide system. *Biomacromolecules*, **3**, 357–367.
36. Kumar, M., Sanford, K.J., Cuevas, W.A. *et al.* (2006) Designer protein-based performance materials. *Biomacromolecules*, **7**, 2543–2551.
37. Mi, L.X. (2006) Molecular cloning of protein-based polymers. *Biomacromolecules*, **7**, 2099–2107.
38. McDaniel, J.R., MacKay, J.A., Quiroz, F.G. and Chilkoti, A. (2010) Recursive directional ligation by plasmid reconstruction allows rapid and seamless cloning of oligomeric genes. *Biomacromolecules*, **11**, 944–952.
39. Rabotyagova, O.S., Cebe, P. and Kaplan, D.L. (2011) Protein-based block copolymers. *Biomacromolecules*, **12**, 269–289.
40. Urry, D.W., Hugel, T., Seitz, M. *et al.* (2002) Elastin: a representative ideal protein elastomer. *Philosophical Transactions of the Royal Society B: Biological Sciences*, **357**, 169–184.
41. Altman, G.H., Diaz, F., Jakuba, C. *et al.* (2003) Silk-based biomaterials. *Biomaterials*, **24**, 401–416.
42. Fossey, S.A., Némethy, G., Gibson, K.D. and Scheraga, H.A. (1991) Conformational energy studies of β-sheets of model silk fibroin peptides. I. Sheets of poly(Ala-Gly) chains. *Biopolymers*, **31**, 1529–1541.
43. Rodríguez-Cabello, J.C., Alonso, M., Díez, M.I. *et al.* (1999) Structural investigation of the poly(pentapeptide) of elastin, poly(GVGVP), in the solid state. *Macromolecular Chemistry and Physics*, **200**, 1831–1838.
44. Cappello, J., Crissman, J.W., Crissman, M. *et al.* (1998) In-situ self-assembling protein polymer gel systems for administration, delivery, and release of drugs. *Journal of Controlled Release*, **53**, 105–117.
45. Cappello, J., Crissman, J., Dorman, M. *et al.* (1990) Genetic engineering of structural protein polymers. *Biotechnology Progress*, **6**, 198–202.
46. Nagarsekar, A., Crissman, J., Crissman, M. *et al.* (2003) Genetic engineering of stimuli-sensitive silkelastin-like protein block copolymers. *Biomacromolecules*, **4**, 602–607.
47. Megeed, Z., Cappello, J. and Ghandehari, H. (2002) Genetically engineered silk-elastin-like protein polymers for controlled drug delivery. *Advanced Drug Delivery Reviews*, **54**, 1075–1091.
48. Nagarsekar, A., Crissman, J., Crissman, M. *et al.* (2002) Genetic synthesis and characterization of pH- and temperature-sensitive silk-elastin-like protein block copolymers. *Journal of Biomedical Materials Research*, **62**, 195–203.
49. Haider, M., Leung, V., Ferrari, F. *et al.* (2005) Molecular engineering of silk-elastin-like polymers for matrix-mediated gene delivery: Biosynthesis and characterization. *Molecular Pharmaceutics*, **2**, 139–150.
50. Dandu, R., Von Cresce, A., Briber, R. *et al.* (2009) Silk-elastin-like protein polymer hydrogels: Influence of monomer sequence on physicochemical properties. *Polymer*, **50**, 366–374.
51. Megeed, Z., Cappello, J. and Ghandehari, H. (2002) Controlled release of plasmid DNA from a genetically engineered silk-elastin-like hydrogel. *Pharmaceutical Research*, **19**, 954–959.
52. Megeed, Z., Haider, M., Li, D.Q. *et al.* (2004) In vitro and in vivo evaluation of recombinant silk-elastin-like hydrogels for cancer gene therapy. *Journal of Controlled Release*, **94**, 433–445.
53. Hatefi, A., Cappello, J. and Ghandehari, H. (2007) Adenoviral gene delivery to solid tumors by recombinant silk-elastin-like protein polymers. *Pharmaceutical Research*, **24**, 773–779.
54. Cresce, A.V., Dandu, R., Burger, A. *et al.* (2008) Characterization and real-time imaging of gene expression of adenovirus embedded silk-elastin-like protein polymer hydrogels. *Molecular Pharmaceutics*, **5**, 891–897.
55. Greish, K., Araki, K., Li, D. *et al.* (2009) Silk-elastin-like protein polymer hydrogels for localized adenoviral gene therapy of head and neck tumors. *Biomacromolecules*, **10**, 2183–2188.

56. Hwang, D., Moolchandani, V., Dandu, R. *et al.* (2009) Influence of polymer structure and biodegradation on DNA release from silk-elastin-like protein polymer hydrogels. *International Journal of Pharmaceutics*, **368**, 215–219.
57. Gustafson, J., Greish, K., Frandsen, J. *et al.* (2009) Silk-elastin-like recombinant polymers for gene therapy of head and neck cancer: from molecular definition to controlled gene expression. *Journal of Controlled Release*, **140**, 256–261.
58. Dinerman, A.A., Cappello, J., Ghandehari, H. and Hoag, S.W. (2002) Swelling behavior of a genetically engineered silk-elastin-like protein polymer hydrogel. *Biomaterials*, **23**, 4203–4210.
59. Haider, M., Cappello, J., Ghandehari, H. and Leong, K.W. (2008) In vitro chondrogenesis of mesenchymal stem cells in recombinant silk-elastin-like hydrogels. *Pharmaceutical Research*, **25**, 692–699.
60. Qiu, W., Huang, Y., Teng, W. *et al.* (2010) Complete recombinant silk-elastin-like protein-based tissue scaffold. *Biomacromolecules*, **11**, 3219–3227.
61. Dinerman, A.A., Cappello, J., Ghandehari, H. and Hoag, S.W. (2002) Solute diffusion in genetically engineered silk–elastin-like protein polymer hydrogels. *Journal of Controlled Release*, **82**, 277–287.
62. Ner, Y., Stuart, J.A., Whited, G. and Sotzing, G.A. (2009) Electrospinning nanoribbons of a bioengineered silk-elastin-like protein (SELP) from water. *Polymer*, **50**, 5828–5836.
63. Qiu, W.G., Teng, W.B., Cappello, J. and Wu, X. (2009) Wet-spinning of recombinant silk-elastin-like protein polymer fibers with high tensile strength and high deformability. *Biomacromolecules*, **10**, 602–608.
64. Qiu, W., Cappello, J. and Wu, X. (2011) Autoclaving as a chemical-free process to stabilize recombinant silk-elastin-like protein polymer nanofibers. *Applied Physics Letters*, **98**, doi: 10.1063/1.3604786
65. Teng, W., Huang, Y., Cappello, J. and Wu, X. (2011) Optically transparent recombinant silk-elastin-like protein polymer films. *Journal of Physical Chemistry B*, **115**, 1608–1615.
66. Teng, W., Cappello, J. and Wu, X. (2011) Physical crosslinking modulates sustained drug release from recombinant silk-elastin-like protein polymer for ophthalmic applications. *Journal of Controlled Release*, **156**, 186–194.
67. Teng, W., Cappello, J. and Wu, X. (2009) Recombinant silk-elastin-like protein polymer displays elasticity comparable to elastin. *Biomacromolecules*, **10**, 3028–3036.
68. Megeed, Z., Cappello, J. and Ghandehari, H. (2004) Thermal analysis of water in silk-elastin-like hydrogels by differential scanning calorimetry. *Biomacromolecules*, **5**, 793–797.
69. Dinerman, A.A., Cappello, J., El-Sayed, M. *et al.* (2010) Influence of solute charge and hydrophobicity on partitioning and diffusion in a genetically engineered silk-elastin-like protein polymer hydrogel. *Macromolecular Bioscience*, **10**, 1235–1247.
70. Anumolu, R., Gustafson, J.A., Magda, J.J. *et al.* (2011) Fabrication of highly uniform nanoparticles from recombinant silk-elastin-like protein polymers for therapeutic agent delivery. *ACS Nano*, **5**, 5374–5382.
71. Hwang, W., Kim, B.H., Dandu, R. *et al.* (2009) Surface induced nanofiber growth by self-assembly of a silk-elastin-like protein polymer. *Langmuir*, **25**, 12682–12686.
72. Nagarajan, R., Drew, C. and Mello, C.M. (2007) Polymer−micelle complex as an aid to electrospinning nanofibers from aqueous solutions. *Journal of Physical Chemistry C*, **111**, 16105–16108.
73. Greish, K., Frandsen, J., Scharff, S. *et al.* (2010) Silk-elastin-like protein polymers improve the efficacy of adenovirus thymidine kinase enzyme prodrug therapy of head and neck tumors. *Journal of Gene Medicine*, **12**, 572–579.
74. Gustafson, J.A., Price, R.A., Greish, K. *et al.* (2010) Silk-elastin-like hydrogel improves the safety of adenovirus-mediated gene-directed enzyme-prodrug therapy. *Molecular Pharmacology*, **7**, 1050–1056.
75. Chang, J., Peng, X.-F., Hijji, K. *et al.* (2011) Nanomechanical stimulus accelerates and directs the self-assembly of silk-elastin-like nanofibers. *Journal of the American Chemical Society*, **133**, 1745–1747.
76. Price, R., Gustafson, J., Greish, K. *et al.* (2012) Comparison of silk-elastin-like protein polymer hydrogel and poloxamer in matrix-mediated gene delivery. *International Journal of Pharmaceutics*, **427**, 97–104.
77. Johnson, S., Ko, Y.K., Varongchayakul, N. *et al.* (2012) Directed patterning of the self-assembled silk-elastin-like nanofibers using a nanomechanical stimulus. *Chemical Communications*, **48**, 10654–10656.

78. Gustafson, J.A., Price, R.A., Frandsen, J. *et al.* (2013) Synthesis and characterization of a matrix-metalloproteinase responsive silk-elastin-like protein polymer. *Biomacromolecules*. doi: 10.1021/bm3013692
79. Xia, X., Xu, Q., Hu, X. *et al.* (2011) Tunable self-assembly of genetically engineered silk-elastin-like protein polymers. *Biomacromolecules*, **12**, 3844–3850.
80. Sun, Z., Qin, G., Xia, X. *et al.* (2013) Photoresponsive retinal-modified silk-elastin copolymer. *Journal of the American Chemical Society*. doi: 10.1021/ja312647n
81. Machado, R., Azevedo-Silva, J., Correia, C. *et al.* (2013) High level expression and facile purification of recombinant silk-elastin-like polymers in auto induction shake flask cultures. *AMB Express*. doi: 10.1186/2191-0855-3-11
82. Rogers, G.E. (2000) Genetic engineering for novel fibres. *Journal of the Textile Institute*, **91**, 24–31.
83. Schiffman, J.D. and Schauer, C.L. (2008) A review: electrospinning of biopolymer nanofibers and their applications. *Polymer Reviews*, **48**, 317–352.
84. Huang, Z.M., Zhang, Y.Z., Kotaki, M. and Ramakrishna, S. (2003) A review on polymer nanofibers by electrospinning and their applications in nanocomposites. *Composites Science and Technology*, **63**, 2223–2253.
85. Sill, T.J. and von Recum, H.A. (2008) Electrospinning: applications in drug delivery and tissue engineering. *Biomaterials*, **29**, 1989–2006.
86. Shao, Z. and Vollrath, F. (1999) The effect of solvents on the contraction and mechanical properties of spider silk. *Polymer*, **40**, 1799–1806.
87. Wilson, D., Valluzzi, R. and Kaplan, D. (2000) Conformational transitions in model silk peptides. *Biophysical Journal*, **78**, 2690–2701.
88. Hu, X., Kaplan, D. and Cebe, P. (2006) Determining beta-sheet crystallinity in fibrous proteins by thermal analysis and infrared spectroscopy. *Macromolecules*, **39**, 6161–6170.
89. Gosline, J., Lillie, M., Carrington, E. *et al.* (2002) Elastic proteins: biological roles and mechanical properties. *Philosophical Transactions of the Royal Society of London Series B-Biological Sciences*, **357**, 121–132.
90. Bensaude-Vincent, B. (2007) Reconfiguring nature through syntheses: From plastics to Biomimetics, in *The Natural and the Artificial. An Evolving Polarity* (eds B. Bensaude-Vincent and W.R. Newman), MIT Press, Cambridge, MA.
91. Collins, T., Azevedo-Silva, J., da Costa, A., Branca, F., Machado, R. and Casal, M. (2013) Batch production of a silk-elastin-like protein in *E. coli* BL21(DE3): key parameters for optimisation. *Microbial Cell Factories*, **12**, 21.

13

Renewable Raw Materials and Feedstock for Bioplastics

Achim Raschka, Michael Carus, and Stephan Piotrowski
nova-Institut GmbH, Germany

13.1 Introduction

To produce bio-based plastics there is a broad spectrum of feedstock options. Today, bio-based plastics are mainly based on sugar, starch, vegetable oil and natural rubber, the so-called first-generation feedstock.

Because of potential competition with food and animal feed, politicians and scientists have introduced, since the early 2000s, the idea of using lignocellulose feedstock by transforming it into fermentable sugar (whether this will have less impact on food security will be discussed below). Lignocellulose means wood, short rotation coppice, such as poplar, willow or miscanthus, or lignocellulose containing agricultural byproducts. Another option is to use byproducts from agriculture and forestry or different biomass waste streams which contain sugar and starch (Figure 13.1).

13.2 First- and Second-Generation Crops: Advantages and Disadvantages

Table 13.1 indicates the advantages and disadvantages for different feedstock options for bio-based plastics. It shows that there is no easy answer to the question of whether second-generation feedstock really is a better solution for the production of bio-based products. The answer depends on the criteria one is looking for.

Industrial biotechnology as a whole, as well as the production of bioplastics, has given rise to an ongoing public, political and industrial debate on the competition between food, feed and industrial markets for agricultural raw materials, with wide-reaching implications. This has created a lot of confusion and insecurity within the bioplastics industry. For example, the

Bio Based Plastics: Materials and Applications, First Edition. Edited by Stephan Kabasci.

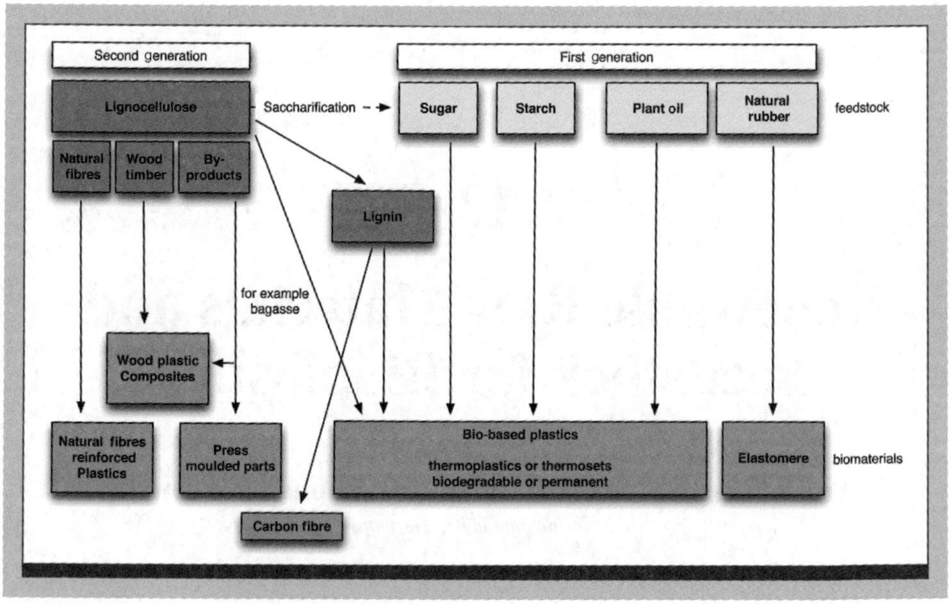

Figure 13.1 *Feedstock for bio-based plastics and composites. (© nova-Institut GmbH, 2011.)*

German automotive industry decided not to use bioplastics based on potential foodstuffs like sugar, starch or edible oil. The bioplastics industry should be well prepared for this discussion on feedstock.

Forced by public discussion during the food crisis in 2008, politics and industry gave a very simple answer to the potential food-versus-industry conflict: industry should only use nonfood crops as feedstock [1].

But the question of food versus nonfood crops for industry is itself oversimplified and misleading. The real questions and conflicts are different [1]:

- Question 1. Are there – worldwide, in the EU, in a single state or in a region under consideration – free agricultural areas left, which are not necessary for food and animal feed, domestic use and export? If yes (and in many regions the answer is yes), continue.
- Question 2 (the real question). How can we use these free areas for industry with the highest resource efficiency and the highest climate protection?

In many cases food crops will best fulfil these criteria – just because they have been especially cultivated to produce maximum yields over many, many years and all the logistical requirements have been established.

Food crops for industry can also serve as an emergency reserve for food and feed supply – second-generation lignocellulose cannot. This is exactly what was happening in 2011 in Brazil. The Brazilian government had reduced the bioethanol fuel quota to save sugar for the demanding food and feed market. That's why 'no food crops for industry' can lead to a misallocation of agriculture resources. We need a comprehensive concept for feedstock for food, feed, industrial material use and bioenergy.

Table 13.1 A comparison of first and second generation crops [1]. (Reproduced with permission from European Bioplastics, Germany.)

Criteria	First generation (sugar, starch, oil, natural rubber)	Second generation (lignocellulose – wood and short rotation coppice)	Byproducts from agriculture and forestry
Yield per hectare in terms of fermentable sugar equivalents	Broad range but more or less on the same level		If the byproducts have no markets yet, this means an extra yield
Greenhouse gas emissions for biomass derived bio-based plastics (cradle to factory gate)	Broad range, but more or less on the same level		Very low, because of the methodology of LCA
Technical maturity	Very high	Still a lot to do	Depending on the content of the by-product
Economically competitive	Not yet (except for specific properties)	Not yet	Not yet, but close
Secure supply with raw materials at a reasonable price	Competing for food, animal feed and bioenergy	Competition with bioenergy and traditional industrial material use	Still a huge potential for inefficiently used byproducts, or even those not used at all
Direct competition with food and feed	Yes	No	No
Indirect competition to food and feed	Yes – on land use		No
Emergency reserve for food and feed	Yes	No	No

13.3 The Amount of Land Needed to Grow Feedstock for Bio-Based Plastics

There are 3300×10^6 ha of rainfed arable land available on this planet, which are used for cropland (circa 1500×10^6 ha), residential areas, road and rail (100×10^6 ha), protected areas (330×10^6 ha) and potential forest land (800×10^6 ha) – so there are still 570×10^6 ha left (Figures 13.2 and 13.3). Those areas are in Russia, Kazakhstan, Africa and South America – often far away from any agricultural infrastructure. Until 2020, huge areas will be activated for cropland, but still 360 million ha are expected to remain 'free' for agricultural uses. To activate this potential, huge investments and reforms in rural areas will be necessary. Even in the European Union, circa 8×10^6 hectares are free and could be used for bioenergy or biomaterials. Most of this land is located in the member states in Eastern Europe.

Today, most of the arable land is used for animal feed (69%), followed by food (17%), material use (7%, including bio-based plastics) and finally bioenergy (3.5%).

Of the worldwide total of roughly 13×10^9 tonnes of biomass harvested in the form of crop and forest biomass, energy and materials as renewable resources account for only about 3% each and wood altogether for 21%. Animal feed (55%) and food (15%) production account for the vast majority (Figure 13.4). This classification is made according to the primary production objective and the figures therefore include byproducts that can also be put to other uses.

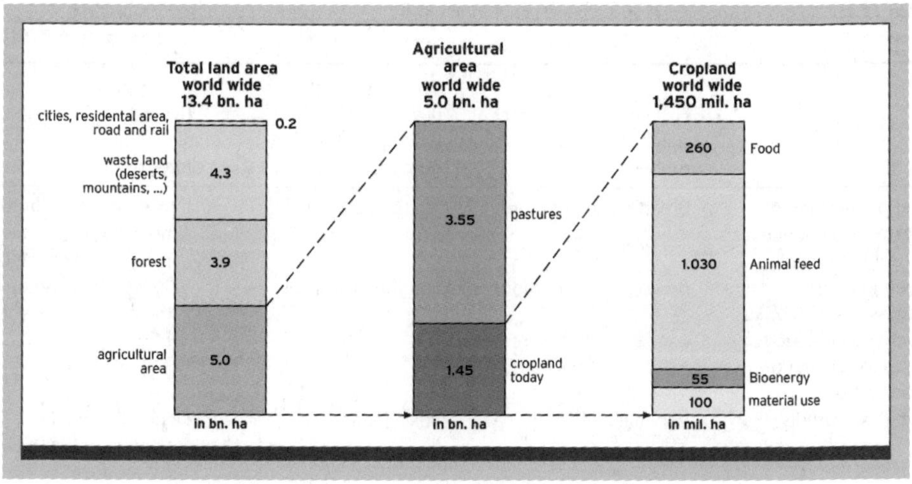

Figure 13.2 *Global land use for food production and renewable resources 2008. (© nova-Institut GmbH, 2012.)*

If one relates the use only to agricultural biomass, then the division becomes even clearer. While both the energy and materials sectors use of renewable crops from arable land represent about 4% of, their overall production, animal feed accounts for 74% – or almost three-quarters – of the total volume and 18% of harvested agricultural biomass is used directly for food (Figure 13.5). The feed fraction is predominantly required for meat and milk production; grass from pastureland for grazing or feed production is not included in these numbers.

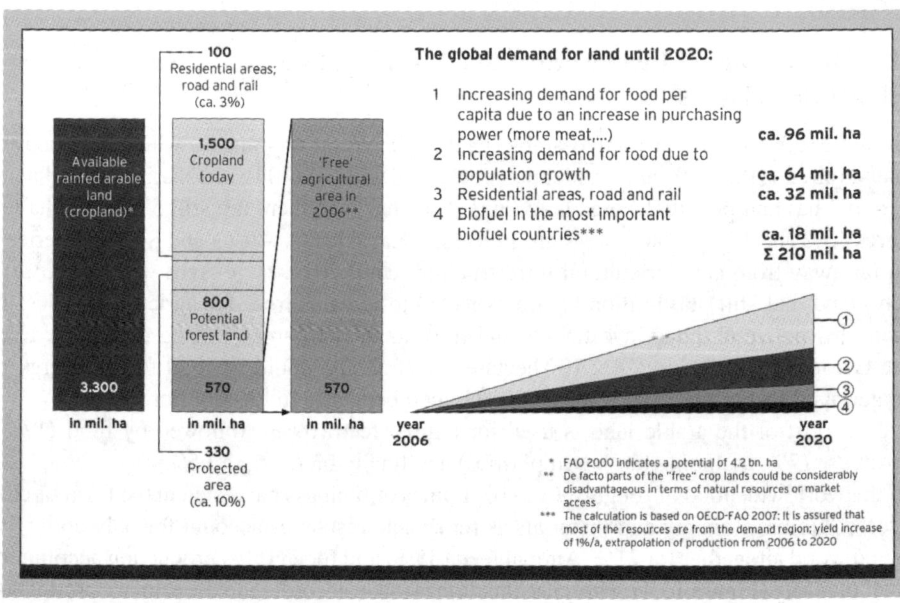

Figure 13.3 *'Free' agricultural area in 2006 and the global demand for area in 2020. (© nova-Institut GmbH, 2012.)*

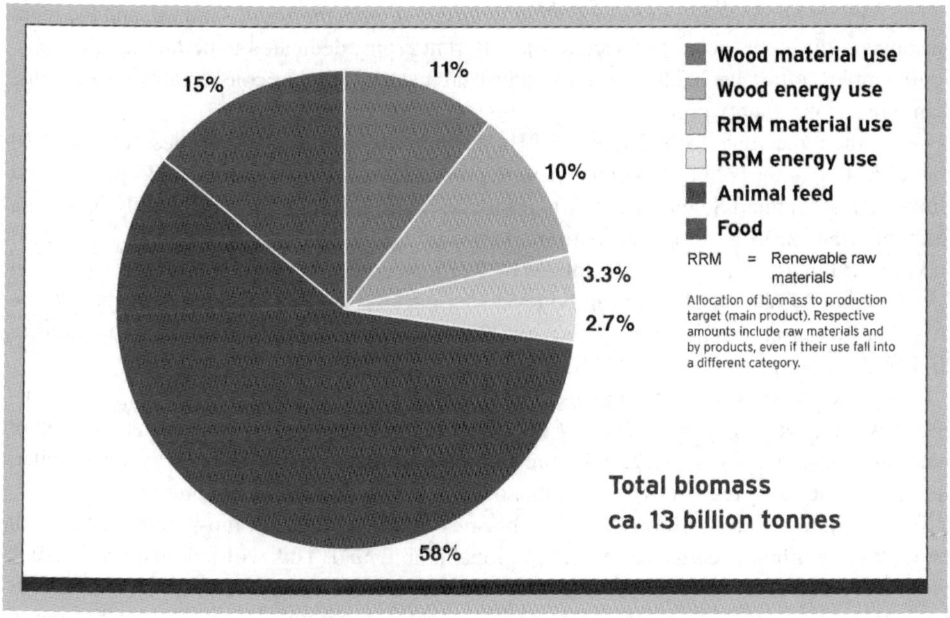

Figure 13.4 *Use of harvested forestal and agricultural biomass worldwide 2008. (© nova-Institut GmbH, 2011.)*

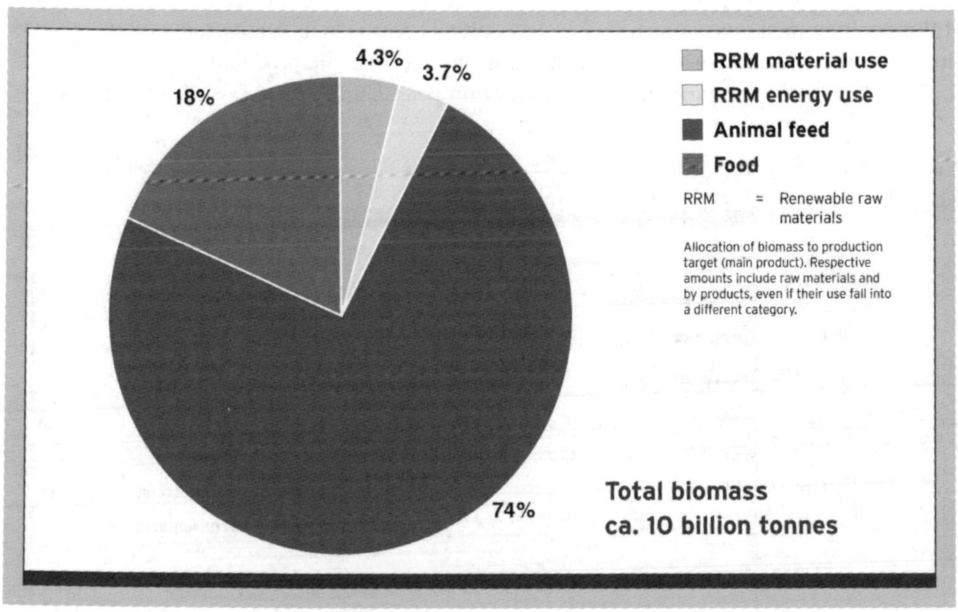

Figure 13.5 *Use of harvested agricultural biomass worldwide 2008. (© nova-Institut GmbH, 2011.)*

The data on biomass in tonnes look slightly different from the arable land use and the main reasons are different yields per hectare for different crops, dedicated to certain applications. For example, cotton, the leading crop for industrial material use, has considerably lower yields than most of the energy crops.

Using the recent data from Endres (FH Hannover) and European Bioplastics, today (2010) about 724 000 tonnes of bio-based plastics are produced worldwide (European Bioplastics and University of Applied Sciences and Arts Hanover: Market development of biopolymers, data from http://en.european-bioplastics.org/market/market-development/, accessed 22 June 2013) and this will increase to 1.71×10^6 tonnes by 2015. According to a rough average estimation 2.5 tonnes of bio-based plastics can be produced per hectare and per year. This means that crops for bio-based plastics were grown on 290 000 hectares (0.02% of global crop land) in 2010 and will be grown on 684 000 hectares (0.05%) in 2015 [1].

Compared to other raw materials used around the world, renewable resources and in particular wood have a very large share of the whole and are comparable in mass terms to other materials such as steel and concrete (Figure 13.6). If one takes material and energy use together, the use of renewable resources exceeds that of all other raw materials in volume terms.

To substitute all 250 million tonnes of plastics in the world with bio-based plastics will demand 100 million hectares or 7% of the global arable land. This will only happen if crude oil is really scarce and very expensive. Then solar and wind energy will also be taking over the energy sector, including bioenergy, so these arable areas will be set free for bio-based chemicals and plastics.

13.4 Productivity and Availability of Arable Land

Still more important than activating new agricultural areas is to increase productivity on areas already in use. Modern agricultural processing can increase the productivity up to ten times compared to traditional farming. Even in the European Union, there is still much scope for

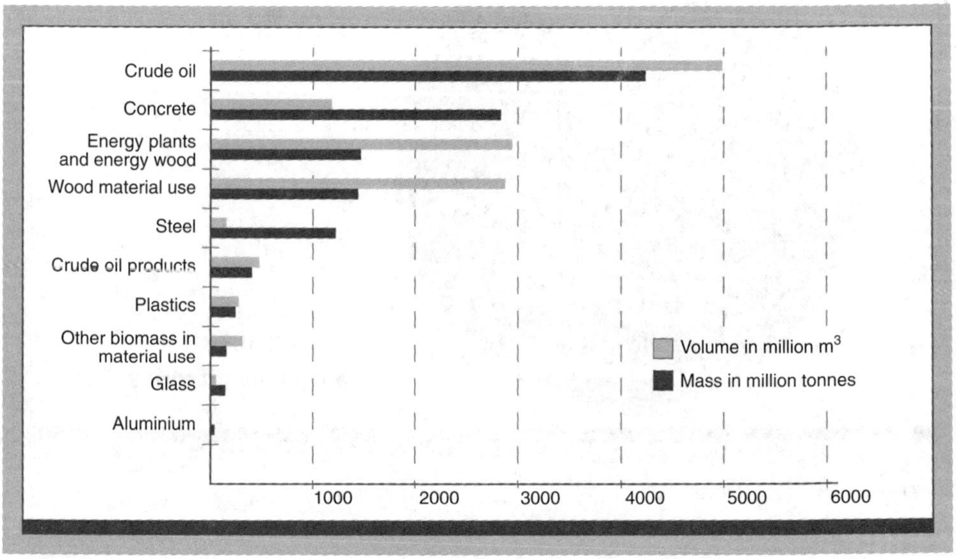

Figure 13.6 Relation of selected raw materials in material and energy use worldwide 2008. (© nova-Institut GmbH, 2011.)

productivity increases. In Romania, for example, yields for most crops are less than 50% of the corresponding average yields in the former EU-15, despite good quality soils.

According to a recent FAO work using longer term population and income projections, global food production needs to increase more than 40% by 2030 and 70% by 2050 compared to average 2005–2007 levels [2]. As the OECD and FAO state: 'Finally, over the longer term, agricultural supply is facing increased uncertainties and limitations to the amount of new land that can be taken into cultivation. Public and private investments in innovation and increasing agricultural productivity, particularly in developing countries, would greatly improve supply prospects by helping to broaden the production base and lessen the chance of recurring commodity price spikes.' [3]

In July 2009, the world leaders pledged to commit $\$20 \cdot 10^9$ over three years for a 'food security initiative' to develop sustainable agriculture in poor countries. Addressing the G8, FAO Director-General Jacques Diouf said 'I am convinced that you will "walk the talk" not only for natural ethical considerations but also for sound economic reasons and, last but not least, to ensure peace and security in the world.' [4] This commitment will trigger further investments in agriculture and will ease the supply situation.

Increased food production can either come from productivity growth or area expansion. On the one hand, there is theoretically room for expansion of agricultural areas. According to the OECD/FAO Agricultural Outlook, based on work of Fischer *et al.* 2002 [5], some 1.6×10^9 ha could be added to the current 1.4×10^9 ha of cropland. Over half of the additionally available land is found in Africa and in Latin America. These regions account for most of the available land that has the highest suitability class for rain-fed crop production. But historical expansion of arable land has been slow, and bringing more marginal land into production can involve considerable investment and lower average yields, while possibly incurring social and environmental costs [3].

Also historically, expansion of arable land has since the 1960s mainly taken place in Africa and Asia while it has continually declined in Europe (Figure 13.7).

However, the following two figures show that the total global agricultural area has in reality stagnated in the 1990s and has even decreased since 2001 (Figure 13.8). Furthermore,

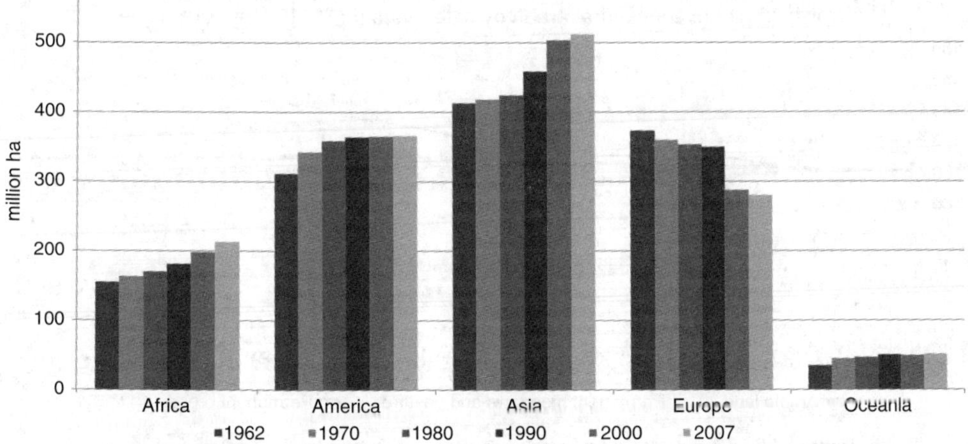

Figure 13.7 *Arable land by region. (© nova-Institut GmbH, 2012.)*

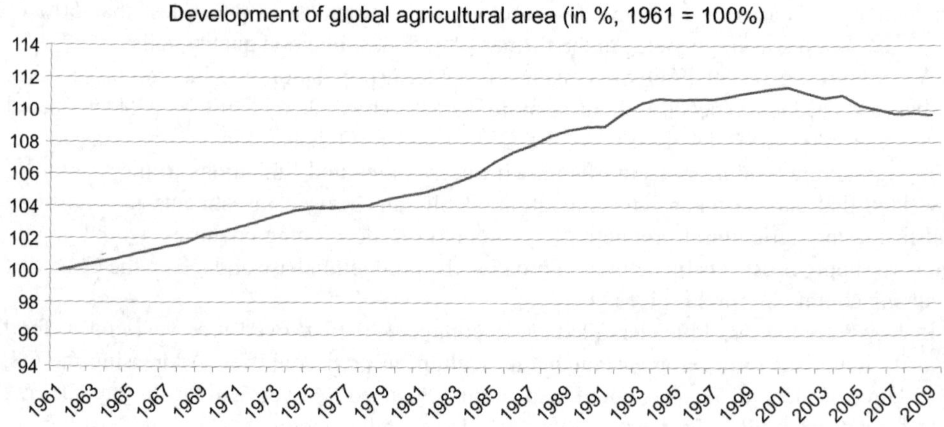

Figure 13.8 *Global agricultural area. (© nova-Institut GmbH, 2012.)*

the increase in agricultural area has been due mainly to a tremendous increase of land for permanent crops (including oil palm, cocoa, coffee, natural rubber etc.) and not arable land [6] (Figure 13.9).

Impact on food security is one of the core social factors to be considered in the development of the use of renewable resources for biofuels and material use like the use in biorefineries. The growing market (today mainly biofuel, in future biorefineries as well) represents a new source of demand for agricultural commodities, which must be supplied from available land resources without negative impacts on the food and feed sector [6].

13.5 Research on Feedstock Optimization

Feedstock was and is constantly adapted to the needs of the modern (agricultural) industry. One focus on national and international research concerning the availability of feedstock for

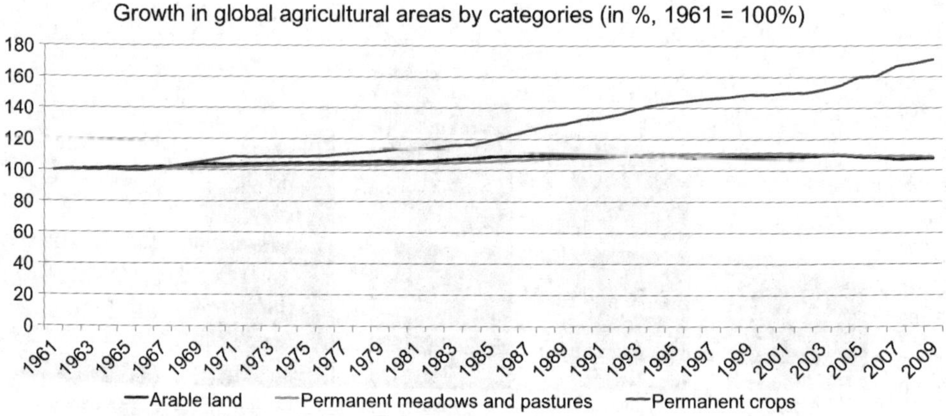

Figure 13.9 *Global agricultural area by categories. (© nova-Institut GmbH, 2012.)*

a bio-based economy is the optimization of the yields and contents of crops for different uses. There are different ways to achieve these aims. This includes the technological optimization of agricultural processes as well as the direct optimization of crops (and wood for short rotation plantations) via conventional breeding, refinements of conventional breeding or biotechnological methods. Modern breeding methods offer a wide repertoire of different approaches. In addition to traditional selection procedures and genetic engineering, (green) biotechnology for feedstock optimization is increasingly receiving attention, although experts, politicians and the general public regard its benefits as controversial.

Increasing demand among chemical companies for renewable resources for a bio-based economy requires adjustments in plant breeding and production. The focus is on the increase in the production of the desired plant ingredients or the total biomass production by direct biomass increase and development of resistance breeding to increase crop yields. Conventional breeding of all crops is used and has been used for hundreds of years to optimize crop yields for the food industry and to produce plants with high contents of, for example, starch or sugar.

Due to the sensitive implications of expansion of agricultural areas (considerable investment and lower average yields, while possibly incurring social and environmental costs), the alternative to increasing global food production is to further increase land productivity. From the 1960s on, global cereal yields have continually increased.

The crucial question is whether the trend of increasing global yields will continue at the same pace. The tremendous potential for increasing yields in developing countries is hindered by lack of technology and infrastructure on the one hand and unfavourable agricultural policies on the other. Both areas are not easy to change in the short term.

While yields for major crops still grow, the growth rate has markedly slowed down for many crops. Bringezu (2009) showed that the global crop yields for all main crops grew slower than in the past and that there was very little optimization through conventional breeding [7].

Figure 13.10 shows that the 5-year moving average of wheat-yield growth slowed down from about 4% in the beginning of the 1960s to less than 1%. The same trend is visible in Asia and Europe. Only in the case of Africa and the Americas is yield growth about constant, but with large variations over the years.

13.6 Advanced Breeding Technologies and Green Biotechnology

A second focus is the optimization of the plant ingredients – for example, through a change of the starch molecule for technical uses of the potato (Amflora potato from BASF and TILLING potato, both with starch consisting of only amylopectin) or the change in the composition of fatty acids in rapeseed, sunflower or crambe. The German Plant Breeders Association (BDP) declares as the main breeding objectives for example, 'increasing the income and improving the quality of the harvested crop, an improved resistance to harmful organisms, reducing the costs of pesticides, improving tolerance to abiotic stress factors in order to reduce agronomic efforts and securing yields under changing environmental conditions, the optimization of the content of plant ingredients and an improved intake of nutrients for a reduced use of fertilizers' (Bundesverband Deutscher Pflanzenzüchter e.V., 'Züchtungsziele', at http://www.bdp-online.de/de/Pflanzenzuechtung/Zuechtung_und_Forschung/Zuechtungsziele/, accessed 22 June 2013).

Conventional breeding is dedicated to new technological developments in addition to the development of new breeding objectives. Due to this, the concept of the use in cascades and product chains was established and optimized and breeding aims are defined accordingly.

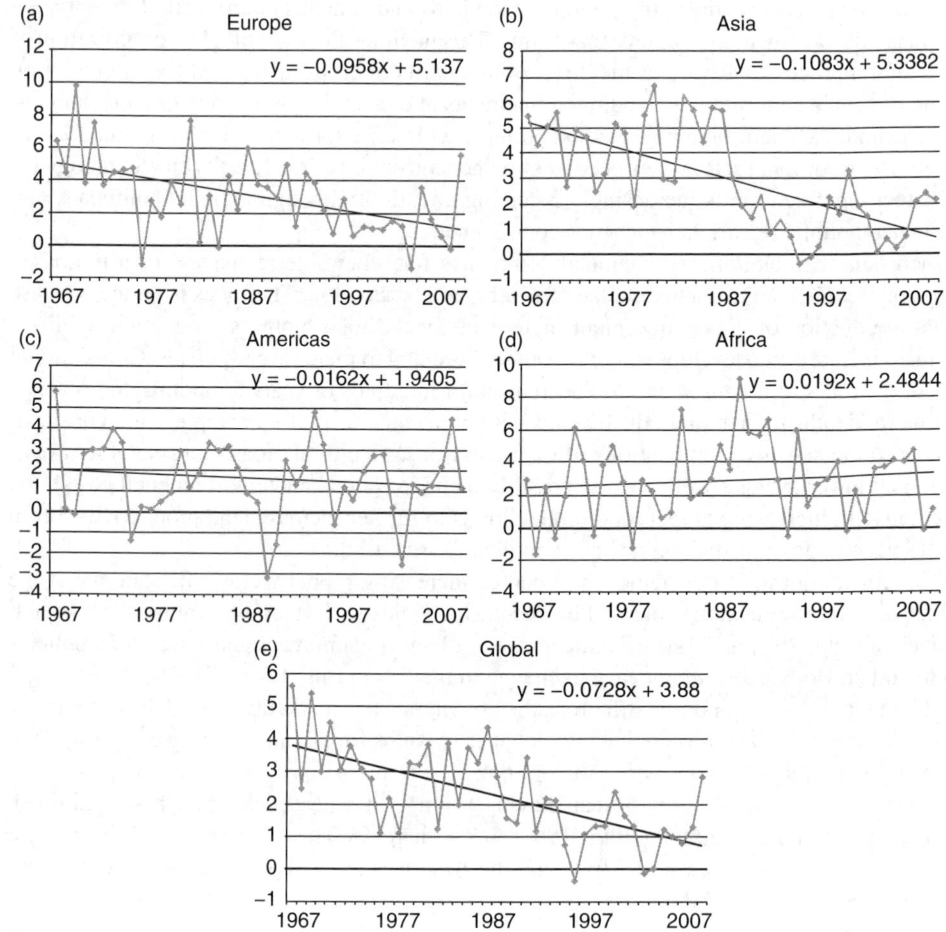

Figure 13.10 *Wheat yield growth by world regions (5-year moving averages in percentages). (© nova-Institut GmbH, 2012.)*

Examples for this may be the breeding for sugar beets with high sugar content or the improvement of the harvest index for rapeseed. The output of these efforts should also be used for energy and material purposes. Another level of development is the smart breeding technology (smart meaning 'Selection with Markers and Advanced Reproductive Technologies'). In this process, the gene responsible for a specific property is identified using molecular methods in the first step (DNA sequencing and next-generation sequencing, PCR, molecular markers). In the breeding process this is used to analyse the progeny of a crossbreed very early in regard to the presence of the feature and to select it. Time savings of up to 50% can be the result of this process. Examples of smart breeding can be found, for instance, for rice [8].

Cisgenetics is a similar approach, where genes of a species are identified, isolated and then reintroduced into the same species to optimize it. With this technology, breeding objectives can also be achieved much faster, so these techniques are discussed as tools of a second green revolution [9]. Another advancement is the TILLING technology (Targeting Induced

Local Lesions In Genomes), where previously very laboratory-intensive mutation breeding has been expanded and improved in efficiency [10], or EcoTILLING, where natural alleles are identified and used in breeding [11]. The newest developments also should lead to accurate and efficient breeding. One of those is homologous recombination [12], where only genes of certain features are targeted to avoid undirected and therefore unwanted genetic changes.

Some of the developments in breeding can be understood as examples of green biotechnology. Other developments can be mentioned, for example, embryo-rescue, anther cultures and *in vitro* culture/micropropagation to increase clones. Another example green biotechnology is genetic engineering, which in some countries, like China, Brazil and India, is regarded as a promising and important technology and is widely used [9].

13.7 Some Facts about Food Prices and Recent Food Price Increases

Compared to other raw materials the price increase for agriculture raw materials was moderate since 2008. In inflation-adjusted terms, price levels in 2008 were even much lower than in the 1970s. This is reported, for example, in the OECD/FAO Agriculture Outlook:

> The commodity price spikes witnessed in the last couple of years, and particularly most recently, are exceptional when viewed from the perspective of the last decade or so but not so much so when seen in a longer historical context ... the recent price spike is neither the only, nor even the most important, one to occur in the last 30-plus years. In inflation adjusted terms, today's prices fall well short of peaks achieved in the early 1970s, and neither current maize nor wheat prices are averaging much above levels achieved as recently as the mid-1990s [3].

Until now, biofuels have had only small effects on world food prices but, although it is smaller than the increase in food and feed, biofuel demand is the largest source of new demand in recent decades and a strong factor underpinning the upward shift in agriculture commodity prices. The medium-term impact of biofuels on crop markets, at least until 2017, however, should not be overestimated, having had an influence on cereal and oilseed prices of 3% to, at most, 10% [13].

Recently, the impact of bioplastics on land use has been approximately 250 times lower than the impact of biofuels; consequently the price influence can be estimated to be lower than 0.1%. Therefore, the impact of bioplastics on the world food market is negligible. Producing biofuels or bioplastics also means, in most cases, producing high-value protein-rich byproducts that can be used as feed.

The main driver for the price increase of agricultural products is the fast growing demand for meat and milk products (Figure 13.11). According to a special *Biofuels Digest* report, 'Fat vs Fuel', 70% of US corn and soy production is devoted to feed, not food or fuel: feed for animals to provide meat, dairy and other livestock byproducts. According to the FAO and the USDA, US meat consumption has increased to 62 kg per person per year since the 1950s, with a resulting increase in grain usage of 170 kg per person (the grain fed to pigs, cattle and poultry). Cheese consumption has increased faster than milk's decline, and Americans consume an extra 81 kg of milk, which uses up another 29 kg of grain [14].

High prices for agricultural raw materials are good for some and bad for others. Unpredictable movements in food prices can still provide problems in the future. With high prices, the consequences in terms of hunger or malnutrition, especially in poor urban areas, will surface. But with low prices, consequences for poor farmers will be disastrous. Until recently, hundreds of millions of farmers could not lift themselves out of poverty because of low food

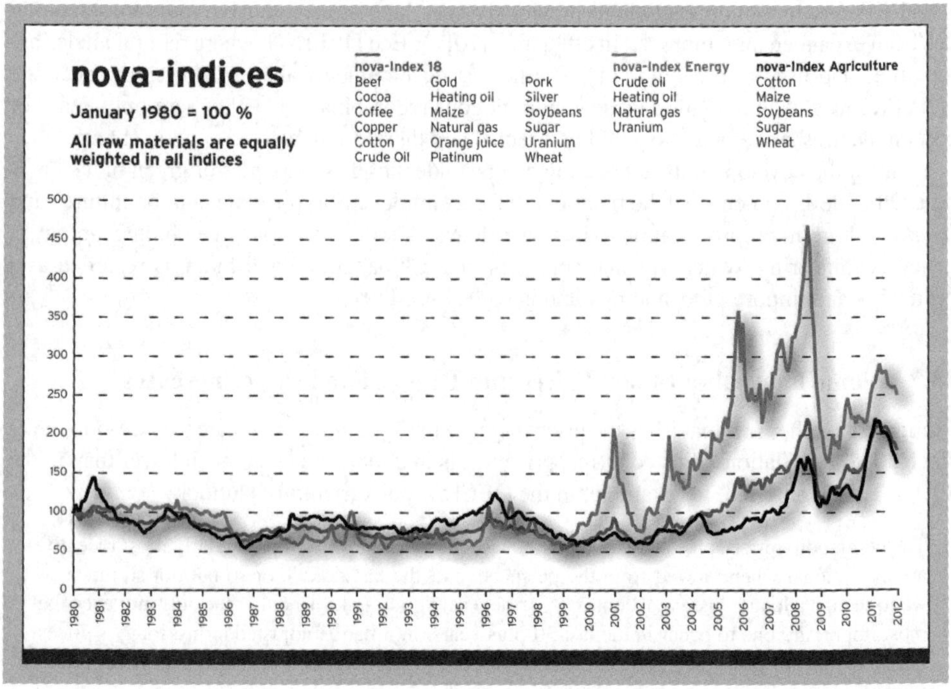

Figure 13.11 *Nova price-indices for agricultural and non-agricultural commodities and energy. (© nova-Institut GmbH, 2013.)*

prices; 75% of the world's hungry people are still living in rural areas and are dependent on agriculture for their livelihoods. Over time, high agriculture prices should benefit them. In poorer urban areas of the world, expenditure on food makes up, on average, about 50% of an individual's disposable income. As such, price increases in these regions have dramatic effects. This percentage climbs to 65% if the food prices rise by 30%. In wealthy countries, on the other hand, these effects will be limited to 1 to 2% of an individual's income. Apart from this, the hunger issue is, however, only partially attributable to the demand for biofuels and is much more attributable to bad policy and the poor performance of the markets [15].

To sum up, the target should be to cultivate crops that are using the land most efficiently for their intended purpose, regardless of whether these are food or nonfood crops. Even if an increasing share of arable land is used for energy and material use – there is still much scope for the expansion of agricultural areas and even more for productivity increases. However, biofuels have so far had a very small impact on food prices and the impact of bioplastics was, with a factor of 250 times less, clearly negligible. Even if they had a significant impact – a higher agricultural price level, together with the international commitment to support sustainable agricultural development, is necessary for more investments in the agriculture sector to increase production and secure the supply in the future. Although high food prices certainly have adverse effects for some, they will lead to the activation of agricultural land that is currently not in production and also to higher productivity on already cultivated land, which would increase the aggregate production of food, feed and renewable raw materials. Furthermore, high prices for agricultural products are necessary for poor farmers in developing countries to sustain their livelihoods.

13.8 Is there Enough Land for Food, Animal Feed, Bioenergy and Industrial Material Use, Including Bio-Based Plastics?

People have been using agricultural raw materials for energy and material purposes for as long as mankind has existed. It is quite common to use agricultural feedstock for biomaterials and this has been done on a large scale for decades. The additional impact of bioplastics is extremely small. Due to increasing demand for food and animal feed, and also for bioenergy and industrial material use, the crucial question is how to increase biomass production in a sustainable way. In fact, there are two strategies to reach this goal:

1. *Increasing the yields.* The tremendous potential for increasing yields in the developing countries is hindered by lack of technology and infrastructure, unfavourable agricultural policies such as no access to credits, an insufficient transmission of price incentives, and poorly enforced land rights.
2. *Expansion of arable land.* Some 0.6 to 1.6×10^9 hectares [2] could be added to the current 1.4×10^9 hectares of crop land (excluding forests, urban areas, protected areas). The figure shows that even in the year 2020 more than 200 million hectares of free arable land will be available.

Prerequisites to points 1 and 2 are political reforms and huge investments in agro-technologies. Compared to these potentials, the impact of GMO on the increase in biomass production will stay low.

On the other hand there is also a huge potential for using biomass more efficiently:

- On average, 10 g of vegetable protein are needed to generate 1 g of animal protein. For broiler production, the protein conversion efficiency is about 18%, for pork about 9%, and for beef about 6%. Therefore, compared with soybean production, biomass and land requirements are roughly a factor 6–17 larger for meat protein production [16]. To switch from meat to vegetarian food would set free a huge amount of arable land for other uses. Furthermore, the differences for water requirements for meat production are about a factor of 26 when there is enough rain for agricultural production or factor 4.4 when intensive irrigation is needed [16]. Although fossil fuel requirements for protein production depend on agricultural intensity and practices, they are about 2.5 to 50 times better for vegetable proteins, compared with animal husbandry. In Europe this difference will usually be a factor of 6–20 to the advantage of soybean-based protein food [16]. Getting proteins from cattle also demands 40 to 50 times the greenhouse gas emissions compared with proteins obtained from wheat or soya [17].
- Reducing food losses will also set free huge amounts of arable land. The results of a recent study showed that roughly one-third of food produced for human consumption is lost or wasted globally, which amounts to about 1.3×10^9 tonnes per year [18].
- Finally solar energy, which will be fully competitive in 10 to 15 years from now (2013), is 40 to 50 times more land efficient compared with bioenergy (and biofuels) and will mainly use nonarable land. This will also release huge areas of arable land, today used by bioenergy.

So the conclusion is, yes, there is enough feedstock – but there will be only enough feedstock for industrial material use, including bio-based plastics, if:

- we are able to activate strongly the potentially free areas ($0.6–1.6 \times 10^9$ hectares) for agriculture and to increase productivity in developing countries – this needs huge investments and political reforms;

- we switch to more vegetarian food and also reduce losses in the food chain;
- we switch from bioenergy to solar and wind energy and increase significantly the use of solar and wind energy; and
- we establish a new policy for equal support of bioenergy and industrial material use based on their efficiency, GHG reduction/hectare and employment/hectare.

Otherwise 'food and feed first', high subsidies for bioenergy and increasing population and meat consumption could mean no feedstock left for high-volume industrial material use, bio-based chemistry and bio-based plastics.

The reason for hunger is not a bottleneck in cropland for food or feed production. We have more than enough space to produce sufficient food and feed for everyone and we are doing this already. The main reasons for hunger are poor distribution, bad logistics and missing financial resources. Or in other words, mankind is producing enough food and feed and there are still huge areas free or unused. These areas can be used for energy and material use without any harm – without any impact on food and feed production. Using these areas for energy and material use will provide additional income to many farmers, who will be able to buy food for their families. After all, three out of four poor people in developing countries live in rural areas.

The type of crops to be cultivated for energy or material use in free agricultural areas should be only a question of efficiency, economy, ecology, sustainability, and so on – but not a question of whether the crop could be used as food or feed instead. Very often, food crops are the most efficient industrial crops too, because they have been optimized by breeding since the 1960s (Figure 13.12). Using a less efficient nonfood crop for energy or material use instead would mean misspending cropland.

In principle there are sufficient and sustainable biomass resources available for food, animal feed, bioenergy and industrial material use, including bio-based plastics, but we should change

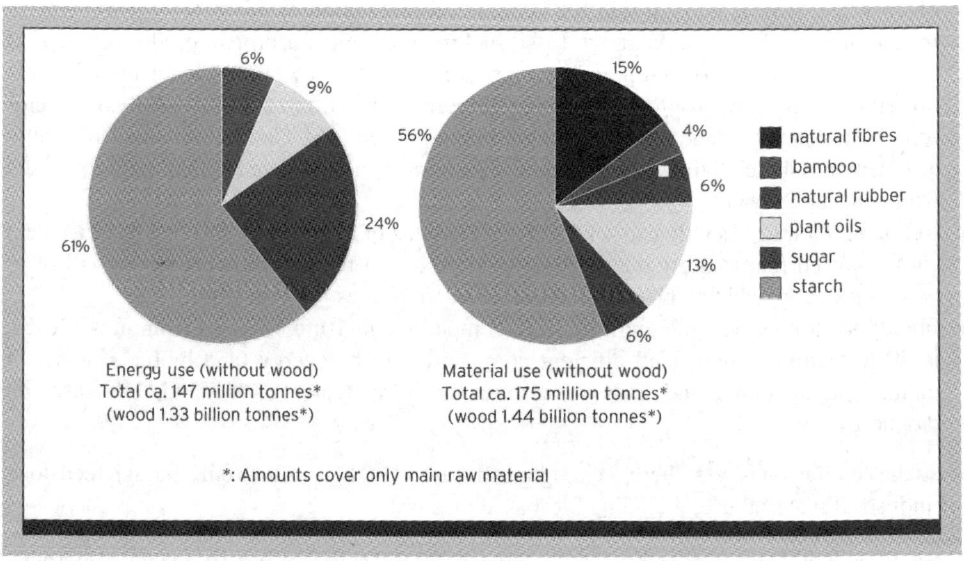

Figure 13.12 *Energetic and materials use of biomass worldwide 2008. (© nova-Institut GmbH, 2011.)*

and optimize biomass allocation and therefore the political framework. And we should invest in agriculture, induce political reforms in rural areas of the world, optimize our nutritional habits so that they are sustainable and switch from bioenergy to solar energy. These measures can secure an adequate and sustainable supply of biomass for at least the next 100 years.

References

1. Carus, M. and Raschka, A. (2011) Agricultural resources needed to grow feedstocks for bioplastics. *Bioplastics Quarterly*, **3** (11), 12–15.
2. Food and Agriculture Organisation of the United Nations (FAO) (2009) *How to Feed the World in 2050*, http://www.fao.org/fileadmin/templates/wsfs/docs/expert_paper/How_to_Feed_the_World_in_2050.pdf (accessed 22 June 2013).
3. Organisation for Economic Co-operation and Development (OECD), Food and Agriculture Organisation of the United Nations (FAO) (2008) Agriculture Outlook 2008–2017, OECD Publishing and FAO, Rome.
4. Food and Agriculture Organisation of the United Nations (FAO) (2009) *FAO Welcomes G8 Food Security Initiative*, at http://www.fao.org/news/story/en/item/24457/icode/ (accessed 22 June 2013).
5. Fischer, G., Van Velthuizen, H., Shah, M. and Nachtergaele, F.O. (2002) Global Agro-ecological Assessment for Agriculture in the 21st Century: Methodology and Results. Research Report, IIASA, Laxenburg.
6. Carrez, D., Albrecht, J., Cunningham, P. *et al.* (2010) The Knowledge Based Bio-Economy (KBBE) in Europe: Achievements and Challenges. Report for the Belgian Government, Clever Consult, Brussels 2010.
7. Bringezu, S. (2009) Biomass use for climate change mitigation and sustainable resource management. Presentation at the JRC Workshop 'Biomass resource assessment' Presentation Eberswalde, 8–9 December 2009.
8. Xu, K., Xu, X., Fukao, T. *et al.* (2006) Sub1A is an ethylene-response-factor-like gene that confers submergence tolerance to rice. *Nature*, **442**, 705–708.
9. Jacobsen, E. and Nataraja, Karaba N. (2008) Cisgenics – facilitating the second green revolution in India by improved traditional plant breeding. *Current Science*, **94**(11), 1365–1366.
10. Henikoff, S., Till, B.J. and Comai, L. (2004) TILLING. Traditional mutagenesis meets functional genomics. *Plant Physiology*, **135**, 630–636.
11. Comai, L., Young, K., Till, B.J. *et al.* (2004) Efficient discovery of DNA polymorphisms in natural populations by Ecotilling. *Plant Journal*, **37**, 778–86.
12. Shukla, V.K., Doyon, Y., Miller, J.C. *et al.* (2009) Precise genome modification in the crop species Zea mays using zinc-finger nucleases. *Nature*, **459**, 437–441.
13. Organisation for Economic Co-operation and Development (2008) Biofuel Support Policies – An Economic Assessment, OECD, Paris.
14. BiofuelsDigest (2009) Fat vs Fuel: BiofuelsDigest Special Report on Indirect Land Use Change and Biofuel Emissions. *BiofuelsDigest* (27 March), n.p.
15. Banse, M., Nowicki, P. and van Meijl, H. (2008) Why are current world food prices so high? A memo. Report 2008-040, http://edepot.wur.nl/28538/ (accessed 12 March 2012).
16. Reijnders, L. and Soret, S. (2003) Quantification of the environmental impact of different dietary protein choices. *American Journal of Clinical Nutrition*, **78**(3), 664S–668S.
17. Hirschfeld, J., Weiß, J., Preidl, M. and Korbun, T. (2008) The Impact of German Agriculture on the Climate, Report 189/08, Institute for Ecological Economy Research, Berlin.
18. von Otterdijk, R. and Meybeck, A. (2011) *Global Food Losses and Food Waste*, FAO, Rome, Italy.

14

The Promise of Bioplastics – Bio-Based and Biodegradable-Compostable Plastics

Ramani Narayan

Department of Chemical Engineering and Materials Science, Michigan State University, United States

Plastic materials are the dominant materials used in applications from agriculture, to electronics, to medical devices to packaging. Starting from a mere 1.65×10^6 tons in 1950 they reached 280×10^6 tons in 2011 and continue to show explosive growth as populous countries like India and China become more industrialized and consume more plastic materials.

Two major issues that arise from this extensive use of plastics, especially in relation to short-life disposable packaging, are its carbon footprint and end of life – what happens to the product after use when it enters the waste stream. This chapter describes how bioplastics can provide a value proposition to address the twin issues of carbon footprinting and end of life.

Bioplastics describe two separate but interlinked concepts:

- **Bio-based plastics** – plastics made from biomass/plant feedstocks as opposed to petro/fossil feedstocks – their 'beginning of life'. This concept refers to replacing petro/fossil carbon with bio-based carbon. Bio-based plastics derive their value proposition from having a zero material carbon footprint arising from the short (in balance) sustainable carbon cycle. This material carbon footprint differs from the process carbon footprint, which is the carbon and environmental footprint arising from converting the feedstock to products, using them and ultimately disposing of them [1, 2]. These materials are not necessarily biodegradable or

Bio-Based Plastics: Materials and Applications, First Edition. Edited by Stephan Kabasci.

compostable. The end-of-life option for these bio-based products needs to be identified and implemented.

- **Biodegradable-compostable plastics** – these are plastics designed to be **completely biodegradable** in the targeted disposal environment (composting, soil, marine, anaerobic digester) in a short defined time period. They are assimilated by micro-organisms present in the disposal environment as food to drive their life processes. They are not necessarily bio-based but can be petro / fossil based.

There are also additive-containing plastics that are claimed to be made 'biodegradable' using so-called oxo- and organic additives added at 1–2% levels to conventional polyethylenes (PE), polypropylene (PP), polystyrene (PS), polyethylene terephthalate (PET) or other plastics. However, biodegradability claims need to be substantiated by competent and reliable scientific evidence using international standards. It must be proven that micro-organisms present in the disposal environment are using the plastic carbon substrate in a defined and measurable time period [2–4]. Unfortunately, misleading and false biodegradability claims proliferate in the marketplace and several companies have been the subject of regulatory actions.

14.1 Value Proposition for Bio-Based Plastics

Switching the manufacturing base (the origins of the carbon) from petro / fossil to bio-based plant carbon feedstock offers an intrinsic zero material carbon footprint value proposition. This is readily apparent from reviewing nature's carbon cycle. Nature cycles carbon through various environmental compartments with specific rates and time scales, as shown in Figure 14.1.

Carbon is present in the atmosphere as inorganic carbon in the form of CO_2. The current level of CO_2 in the atmosphere is around 380 ppm (parts per million). The life-sustaining heat-trapping value of CO_2 in the atmosphere (which maintains the Earth's temperature) is becoming life threatening because of increasing man-made carbon dioxide and other heat-trapping gas emissions to the atmosphere. It is therefore prudent and necessary to try to

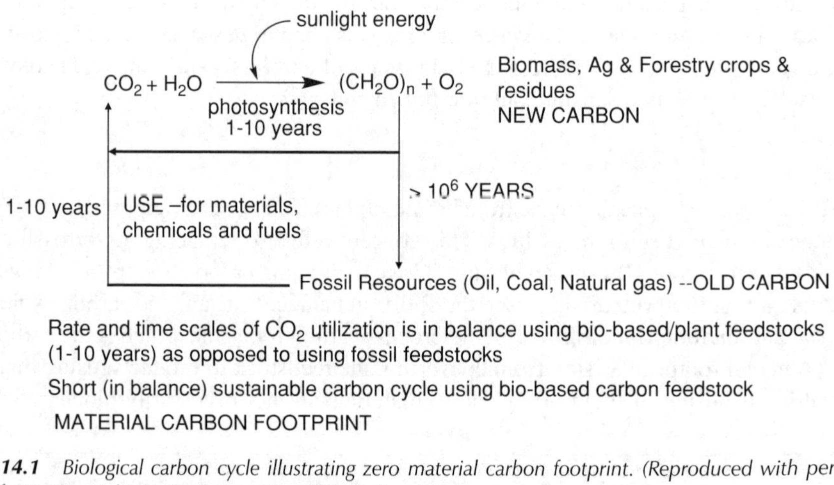

Figure 14.1 *Biological carbon cycle illustrating zero material carbon footprint. (Reproduced with permission from ref. [1]. Copyright © 2012, American Chemical society.)*

maintain current levels – the 'zero carbon' approach. This can best be done by using biomass / plant materials as feedstock to manufacture carbon based products, so that the CO_2 released at the end of life of the product is captured by planting new crops in the next season. Specifically the rate of CO_2 release to the environment at end of life equals the rate of photosynthetic CO_2 fixation by the next generation crops planted – a 'zero material carbon' footprint. In the case of fossil feedstock, the rate of carbon fixation is in millions of years while the end-of-life release rate into the environment is in 1–10 years – the mathematics is simple, this is not sustainable and results in more CO_2 release than fixation, resulting in an increased carbon footprint, and with it the attendant global warming and climate change problems [1,2].

14.2 Exemplars of Zero or Reduced Material Carbon Footprint – Bio-PE, Bio-PET and PLA

Basic stoichiometry teaches that for every 100 kg of polyolefin (polyethylene, PE; polypropylene, PP) manufactured, a net 314 kg of CO_2 is released into the environment at its end of life (100 kg of PE contains 85.7 kg carbon and upon combustion will yield $(44/12) \times 85.7 = 314$ kg of CO_2). Similarly, PET contains 62.5% carbon, which results in 229 kg of CO_2 released into the environment at end of life. However, if the carbon in the polyester or polyolefin comes from biomass feedstock, the net release of CO_2 into the environment is zero, because the CO_2 released is sequestered in a short time period by the next crop or biomass plantation (Figure 14.2). Thus, the fundamental value proposition for bio-based plastics arises from this intrinsic zero material carbon footprint and not necessarily from the process carbon footprint, which may be equal to or slightly better than current processes.

This approach is illustrated by Braskem who have built a 200 kton/year bio-PE plant using sugarcane as bio-based feedstock in Brazil. Sugar from sugar cane is fermented to

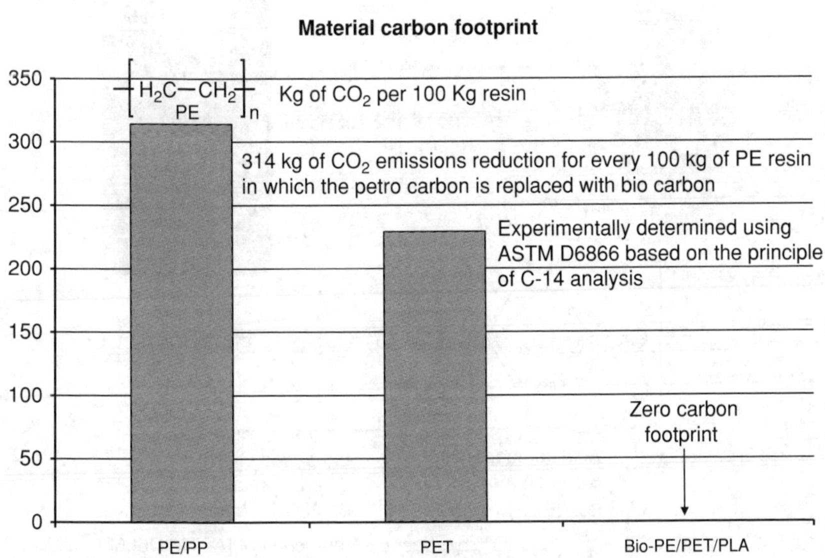

Figure 14.2 Material carbon footprint calculations for fossil and bio-based plastics. (Reproduced with permission from ref. [1]. Copyright © 2012, American Chemical society.)

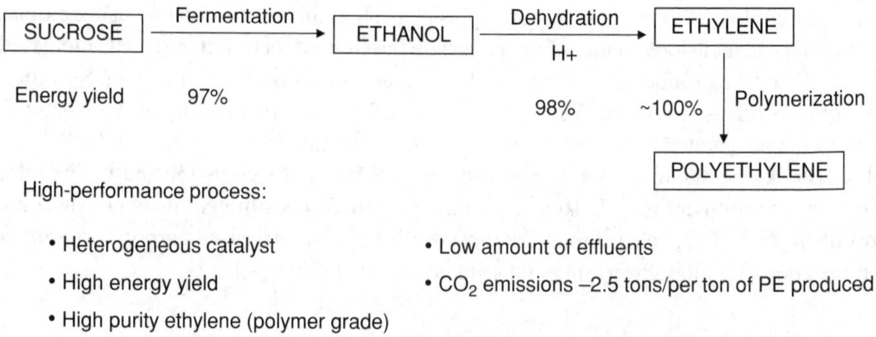

Figure 14.3 *Braskem 100% bio-PE (zero material carbon footprint) process. (Reproduced with permission from Braskem S.A.)*

ethanol which is dehydrated to ethylene and polymerized to bio-PE (Figure 14.3). In addition, the company plans to erect a plant manufacturing 30 ktons/year of bio-PP as well (http://www.braskem.com/site.aspx/green-products-USA, accessed 10 July 2013).

Another example is the switch by The Coca-Cola Company to bio-PET with is characterized by a 20% bio-based carbon content (http://www.coca-colacompany.com/our-company/introducing-plantbottle, accessed 10 July 2013). Polyethylene terephthalate bottles are extensively used for packaging beverages, water, and a number of other food and nonfood items (Figure 14.4). Polyethylene terephthalate is manufactured by condensation

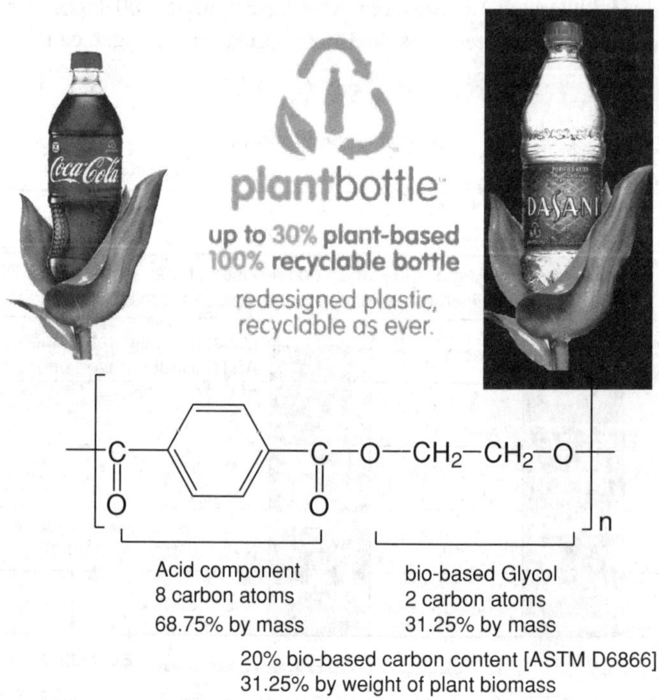

Figure 14.4 *Bio-PET Coca Cola Plant bottle. (© The Coca-Cola Company.)*

polymerization of terephthalic acid and ethylene glycol. In the bio-PET the glycol component is bio-based. There are also efforts under way to manufacture terephthalic acid from bio-based feedstocks, but currently it is still made from fossil feedstock.

As shown in Figure 14.4, there are two bio-based carbon atoms and eight fossil carbon atoms per PET molecule, which means a 20% bio-based carbon content. On a mass basis, there are 31.25% bio-based glycol components and 68.75% fossil terephthalic acid components in the polymer. From a material carbon footprint perspective the CO_2 arising from the glycol's two carbon atoms would have a zero carbon emissions impact, and only the eight fossil carbon atoms from the terephtahlic acid component would contribute to the carbon emissions impact.

About 37.5×10^6 tons of PET are used every year to make bottles. As discussed earlier, two of the ten carbons in PET coming from the glycol component would have zero carbon emissions impact. Therefore, an annual reduction of 17.2×10^6 tons of CO_2 emissions is realized by just introducing two bio-based carbon atoms in PET. This translates to saving 40×10^6 barrels of oil use per year. Recently, major brand owners The Coca-Cola Company, Ford Motor Company, H.J. Heinz Company, NIKE, Inc. and Procter & Gamble announced the formation of the Plant PET Technology Collaborative (PTC), a strategic working group focused on accelerating the development and use of 100% plant-based PET materials and fibres in their products. The end-of-life strategy for bio-PET is recycling. Bio-PET is structurally, chemically and physically identical to the current petro-fossil based PET molecule. Therefore, bio-PET can be readily recycled in the well established PET recycling infrastructures. PET bottles is the most recycled polymer material and continues to grow.

Poly(lactic acid) (PLA) offers 100% bio-based carbon content for a zero material carbon footprint and both compostability and recycling as viable end-of-life options. NatureWorks LLC (a Cargill subsidiary) manufactures 140 000 tons/year of PLA under the trade name Ingeo™ for packaging and industrial applications (see www.natureworksllc.com, accessed 23 June 2013).

14.3 Process Carbon Footprint and LCA

The fundamental intrinsic value proposition of a zero material carbon footprint arises from the origin of the carbon in the product as described in the earlier section – using bio-based in place of petro/fossil feedstock. However, this does not address the carbon emissions and other environmental impacts for the process of converting the feedstock to products, using them and ultimately disposing them – the process carbon footprint. LCA methodology and standards (ISO 14040 standards) [5] are the accepted tools to compute the process environmental footprint. Unfortunately, LCA focuses almost exclusively on the process (carbon and environmental) footprint. The impact of the carbon present in the product, the material carbon footprint, is treated as feedstock energy or embodied carbon energy for potential use in the next product cycle, or it is even not considered at all as would happen in a cradle-to-gate LCA.

Figure 14.5 shows the process and material carbon footprint of the above mentioned plastic products PLA, PE and PET. As can be seen from the diagram, the process carbon footprint for PE (values for PP are very close to the PE data) is around 200 kg CO_2 emitted per 100 kg plastic manufactured, whereas that for PLA is around 385 kg CO_2 emitted per 100 kg plastic manufactured – a much higher carbon footprint for the bio-based product (see www.plasticseurope.org/, accessed 23 June 2013) [6]. However, if one includes the material carbon footprint, then PLA has a lower total carbon footprint – 385 kg CO_2 emission per 100 kg plastic manufactured versus 519 kg CO_2 emission per 100 kg plastic manufactured. This is because the material

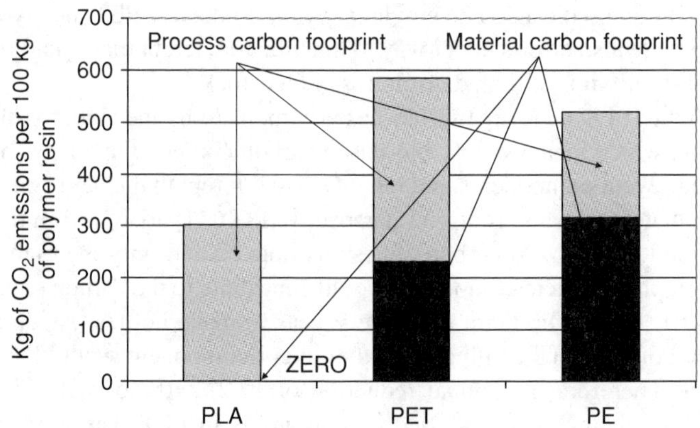

Figure 14.5 *Process and material carbon footprints for PLA, PET, and PE. (Reproduced with permission from ref. [2]. Copyright © Cambridge University Press.)*

carbon footprint for the PLA plastic is zero, whereas PE/PP has a material carbon footprint of 314 kg CO_2 emissions per 100 kg plastic manufactured [1, 2].

Of course it is important to calculate and report on the process carbon and environmental footprint using LCA tools and ensure that the process carbon and environmental footprint is equal or better than the process being replaced. However, the intrinsic fundamental value proposition for bio-based plastics arises from the zero material carbon footprint in harmony with time scales of the natural biological carbon cycle.

14.4 Determination of Bio-Based Carbon Content

A key requirement to document this material carbon footprint reduction is a transparent and accurate test method to determine unequivocally the amount of bio-based carbon present in a product. The bio-based carbon present in a material can be readily calculated as a percentage from the C-14 signature of the product. All bio-based carbon atoms have a radioactive signature that arises from the plant's CO_2 uptake out of the atmosphere in which a small but measurable amount of C-14 containing CO_2 is present due to reactions of nitrogen with cosmic radiation (see Figure 14.6). Fossil carbons have almost zero C-14 content as the half-life of this radioactive carbon isotope is 5730 years. The ASTM D6866 test method [7] can be used to calculate the percentage bio-based carbon content in materials. It involves combusting the test sample in the presence of oxygen to produce carbon dioxide (CO_2) gas. This gas is analysed to provide a measure of the product's $^{14}C/^{12}C$ content relative to the modern carbon-based oxalic acid radiocarbon Standard Reference Material (SRM) 4990c.

The US Congress passed the Farm Security and Rural Investment Act of 2002 (P.O. 107–171) requiring the purchase of bio-based products by the Federal Government. The U.S. Department of Agriculture (USDA) was charged with developing guidelines for designating bio-based products and publishing a list of designated bio-based product classes for mandated Federal purchase (www.biopreferred.gov, accessed 10 July 2013). In its rule-making, the USDA adopted the methodology described above for identifying and quantifying bio-based content prescribing the use of ASTM D6866 to establish bio-based content of products. More recently, a voluntary labelling programme, 'USDA Certified Biobased Product', has been

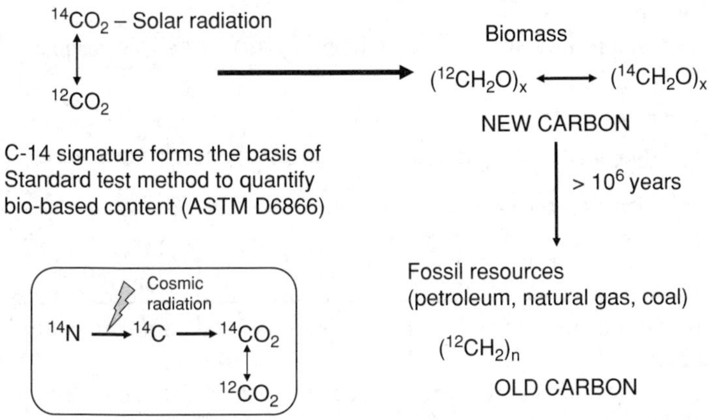

Figure 14.6 *Measurement of bio-based carbon content using radiocarbon analysis (ASTM D6866). (Reproduced with permission from ref. [1]. Copyright © 2012, American Chemical Society.)*

launched and managed by ASTM (Figure 14.7). The basis for the certification is to have the product tested for bio-based carbon content using ASTM D6866 and to meet several other requirements.

14.5 End-of-Life Options for Bioplastics – Biodegradability-Compostability

Bio-PE and bio-PET are bio-based plastics but not biodegradable-compostable. They can be advantageously integrated with the already existent PE or PET recycling streams. Bio-PE also has the advantage of high heat of combustion value which is advantageous for energy recovery. The end-of-life option for these bio-based plastics is recycling or energy recovery. On the other hand bio-based PLA and petro/fossil based polyesters are completely biodegradable under composting conditions.

Biodegradability is a very misused and abused term. False and misleading claims are proliferating in the market space. Using the terms 'biodegradable' or 'biodegradability' means nothing and is misleading, unless one defines the disposal system and the time required for complete biodegradation in that disposal system. Biodegradability means harnessing

Figure 14.7 *USDA biopreferred voluntary labelling program logo. (© USDA.)*

Aerobic process

$$\text{Glucose/C-bioplastic} + 6O_2 \longrightarrow 6CO_2\uparrow + 6H_2O;\ \Delta G^{0\prime} = -686\ \text{kcal/mol}$$

Anaerobic process

$$\text{Glucose/C-bioplastic} \longrightarrow 2\ \text{lactate} \longrightarrow 3CO_2\uparrow + 3CH_4\uparrow\ \Delta G^{0\prime} = -47\ \text{kcal/mol}$$

Scheme 1 *Basic biochemical processes for the microbial utilization of carbon substrates*

micro-organisms present in the disposal system to use the plastic carbon substrate as their food completely and thus to remove it from the environmental compartment in a safe, timely and efficacious manner.

Basic biology teaches how to measure the rate and extent of this microbial use and forms the basis for all the ASTM, ISO, and EN standards for measuring biodegradability. Under aerobic conditions, the carbon is biologically oxidized inside the cell to CO_2, releasing energy that is harnessed by the micro-organisms for their life processes. Under anaerobic conditions, CO_2 and CH_4 are produced. Thus, a measure of the rate and amount of CO_2 or CO_2 and CH_4 evolved as a function of total carbon input to the process is a direct measure of the amount of carbon substrate being used by the micro-organisms (percentage biodegradation) (See Scheme 1). So, one has to ensure complete biodegradability in a defined disposal system like composting or anaerobic digestion in a defined short time period.

It would seem obvious and logical from the above basic biology that, to make a claim of biodegradability, all one needs to do is the following: expose the test plastic substrate as the sole carbon source to micro-organisms present in the target disposal environment (like composting, or soil or anaerobic digestion or marine), and measure the CO_2 (aerobic) or CO_2 and CH_4 (anaerobic) evolved. A measure of the evolved gas provides a direct measure of the plastics substrate carbon being utilized by the micro-organisms present in the target disposal environment (percentage biodegradation). ASTM, EN, and ISO test methods teach how to measure the percentage biodegradability in different disposal environments based on the fundamental biochemistry described above.

The various biodegradability test method standards from ASTM and ISO are as follows:

- **ASTM D5338** – test method for measuring biodegradability under composting environment.
- **ASTM D5988** – test method for measuring biodegradability in soil environment.
- **ASTM D5511** – test method for measuring biodegradability in a high solids anaerobic digester.
- **ASTM D5526** – test method for measuring biodegradability in a landfill/bioreactor environment.
- **ISO 14852** – ultimate **aerobic biodegradability** of plastic materials in an aqueous medium – method by analysis of evolved carbon dioxide.
- **ISO 14853** – ultimate **anaerobic biodegradability** in an aqueous system – Method by measurement of biogas production.
- **ISO 14855**; Determination of the ultimate **aerobic biodegradability** of plastic materials under **controlled composting conditions** – Part 1: Method by analysis of evolved carbon dioxide and Part 2: Gravimetric measurement of carbon dioxide evolved in a laboratory-scale test.

However, these test methods have no pass / fail criteria, and so should not be used to claim biodegradability in any environment. Strict adherence to the test method's reporting requirements should be followed. The report, at a minimum must have the disposal environment, the percentage biodegradation achieved and the time frame to obtain the percentage biodegradation.

For the composting environment, specification standards with definite pass / fail criteria requirements are available as ASTM, ISO, and EN standards. All the standards are harmonized and have very minor differences but the basic principles and requirements are the same. EN 13432 and ISO 18606 are specification standards for packaging. The specification standards are as follows:

- **Biodegradability under composting conditions – Compostable Bioplastics**
  - **ASTM D6400** – Specification for compostable plastics
  - **ASTM D6868** – Specification for plastics coatings and modifiers on paper and other compostable substrates
  - **ISO 17088** – Specification for compostable plastics
  - **EN 13432** – Specification for compostable packaging – focus on packaging
  - **ISO 18606** – Packaging and the Environment – Organic recycling – focus on packaging
- **Biodegradability under marine environment**
  - **D7021** – Specification for non-floating biodegradable plastics in the marine environment
- **Biodegradability under soil environment**
  - **ASTM** – under development – 90% carbon assimilation by microorganism as measured by evolved CO_2 in 2 years or less.

Thus, one can measure the rate and extent of biodegradation or microbial utilization of the test plastic material by using it as the sole carbon source in a test system containing a microbially rich matrixlike compost in the presence of air and under optimal temperature conditions (preferably at 58 °C – representing the thermophilic phase). Figure 14.8 shows

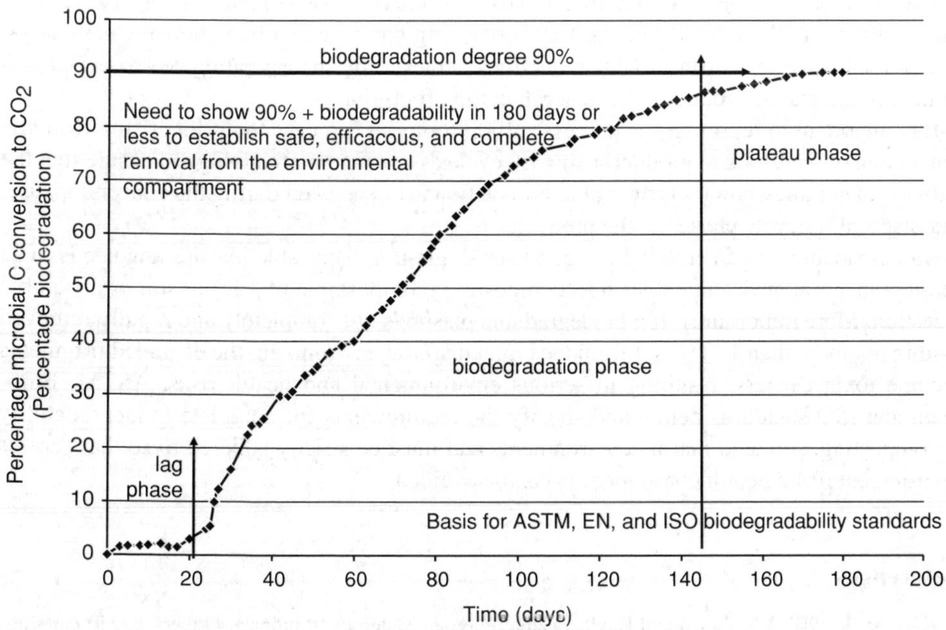

Figure 14.8 Experimental measurement of biodegradability using International Standards. (Reproduced with permission from [1]. Copyright © 2012, American Chemical society.)

a typical graphical output that would be obtained if one were to plot the percentage carbon converted to CO_2 as a function of time in days. First, a lag phase can be observed during which the microbial population adapts to the available test C-substrate. Then, the biodegradation phase follows during which the adapted microbial population begins to use the carbon substrate for its cellular life processes, as measured by the conversion of the carbon in the test material to CO_2. Finally, the output reaches a plateau when the substrate is almost completely utilized. Linear or any other form of data extrapolation from these complex biological systems is not acceptable and is very misleading because there is no credible scientific substantiation for an extrapolation model.

Claims of 'degradable', 'partially biodegradable', or 'eventually biodegradable' are not acceptable. For instance, in the case of the so-called oxo-degradable materials that are merely disintegrated to very small particles that one cannot see any more, it has been shown that these degraded fragments absorb toxins present in the environment, concentrating them and transporting them up the food chain. Therefore, complete removal from the disposal environment is a short time period of not more than 1–2 years is essential to eliminate potentially serious human health and environmental consequences [8]. So, verifiable scientifically valid evidence from approved third part laboratory is needed to document complete biodegradability in a defined disposal system like composting or anaerobic digestion in a short time period using the specified International standards.

14.6 Summary

Replacing petro / fossil carbon with bio-based carbon by using plant biomass feedstock in place of fossil feedstock for the manufacture of plastic materials offers a strong 'value proposition' for a zero material carbon footprint. It may also reduce the process carbon and environmental footprint. A methodology for quantification of 'bio-based carbon content' has been developed and codified into the ASTM Standard D6866. Using bio-based carbon content calculations, one can calculate the intrinsic CO_2 reductions achieved by incorporating bio-based carbon content into a plastic product – the material carbon footprint.

It is important to report on the process carbon footprint (process footprint arising from the conversion of feedstock to product) using lifecycle assessment methodology to ensure that the intrinsic material carbon footprint value proposition is not negated during the conversion, use, and disposal lifecycle phases of the product.

Biodegradability is an end-of-life option for single-use disposable plastics and needs to be tied to a disposal environment such as composting (compostable plastic) or soil or anaerobic digestion. More importantly, if a biodegradable plastic is not completely and rapidly removed (within not more than 1–2 years) from the target disposal environment the degraded fragments become toxin carriers, resulting in serious environmental and health risks. ASTM, European, and ISO standards define and specify the requirements for complete biodegradability in composting, soil, and marine environments and must be strictly adhered to so that serious environmental and health consequences can be avoided.

References

1. Narayan, R. (2012) Biobased and Biodegradable Polymer Materials: Rationale, Drivers, and Technology Exemplars. ACS (an American Chemical Society publication) Symposium Ser. 1114, Chapter 2, pp. 13–31

2. Narayan, R. (2011) Carbon footprint of bioplastics using biocarbon content analysis and life cycle assessment. MRS (Materials Research Society) Bulletin, **36**, pp. 716–721.
3. Narayan, R. (2009) Fundamental Principles and Concepts of Biodegradability – Sorting through the Facts, Hypes, and Claims of Biodegradable Plastics in the Marketplace. BioPlastics Magazine (Jan.).
4. Narayan, R. (2010) Misleading Claims and Misuse of Standards Continues to Proliferate in the Nascent BioPlastics Industry Space. BioPlastics Magazine (Jan.)
5. ISO (2009) *Environmental Management: The ISO 14000 Family of Standards,* ISO, Geneva, http://www.iso.org/iso/theiso14000family_2009.pdf (accessed 10 July 2013).
6. Vink, E.T.H., Glassner, D.A., Kolstad, J.J. *et al.* (2007) The eco-profiles for current and near future NatureWorks polylactide (PLA) production. *Industrial Biotechnology*, **3** (1), 58.
7. ASTM International (2010) *Annual Book of Standards*, ASTM International, Philadelphia, PA, vol. 8.03, Standards D 6866; D6400, D6868, D7021.
8. Thompson, R.C., Olsen, Y., Mitchell. R. P. *et al.* (2009) Plastics, the environment, and human health. *Philosophical Transactions of the Royal Society B*, **364**, 1973–1976.

Index

Bio-Based Plastics: Materials and Applications, First Edition. Edited by Stephan Kabasci.